GENETICALLY ENGINEERED TOXINS

EDITED BY
ARTHUR E. FRANKEL
Altamonte Springs, Florida

CRC Press
Taylor & Francis Group
Boca Raton London New York

CRC Press is an imprint of the
Taylor & Francis Group, an **informa** business

CRC Press
Taylor & Francis Group
6000 Broken Sound Parkway NW, Suite 300
Boca Raton, FL 33487-2742

First issued in paperback 2019

ISBN-13: 978-0-8247-8454-6 (hbk)
ISBN-13: 978-0-367-40277-8 (pbk)

Library of Congress Cataloging-in-Publication Data

Genetically engineered toxins / edited by Arthur E. Frankel. -- 1st
  ed.
      p.   cm. -- (Targeted diagnosis and therapy ; 7)
  Includes bibliographical references and index.
  ISBN 0-8247-8454-5 (alk. paper)
  1. Recombinant toxins.   I. Frankel, Arthur E.   II. Series.
  [DNLM: 1. Genetic Engineering--methods. 2. Molecular Biology-
-methods. 3. Toxins--genetics.   W1 TA579 v.7 / QW 630 G328]
  QP632.R43G46 1992
  615'.373--dc20
  DNLM/DLC   92-14608
  for Library of Congress   CIP

**Visit the Taylor & Francis Web site at
http://www.taylorandfrancis.com**

**and the CRC Press Web site at
http://www.crcpress.com**

This book is dedicated to my teachers, Edna Boon, Austin Riggs, Stanley G. Schultz, Jerome Lettvin, Peter Fischinger, George Todaro, Saul Rosenberg, Leonard Herzenberg, and Stephen Johnston. Each served to inspire a respect of life and truth.

Introduction to the Series

Targeted Diagnosis and Therapy is a series intended to collect new knowledge generated in the research and development of self-directed diagnostic and therapeutic agents. The powerful tools of recombinant DNA and monoclonal antibody technologies have contributed immensely to our understanding of the concept of molecular recognition as well as protein structure-function relationships. This has yielded a view of the future that includes the use of a variety of new pharmaceutical products. These products will have the property of localizing to a predetermined site, with a consequent diagnostic or therapeutic effect. The clinical use of these products will include the treatment in vivo of malignant organs or tissues as well as elimination of specific cell types in ex vivo bone marrow-purging procedures. Each volume will focus on one product or strategy, and contain the relevant preclinical and/or clinical experience. The list of near-term subjects includes a variety of antibody conjugates, the interferons, the interleukins, tissue plasminogen activator, and gene therapy. Other volumes will deal with the next generation of agents, such as genetically engineered toxins and fusion proteins. It is expected that the series will be useful for basic researchers and clinicians alike.

New frontiers lie ahead. The opportunities for research and development of important new pharmaceutical products are considerable. It is hoped that Targeted Diagnosis and Therapy will assist in the efforts to achieve this goal.

JOHN D. RODWELL
EDITOR

Preface

Malignant disease and autoimmune disorders are caused by uncontrolled proliferation or dysfunction of particular cell types in humans. The selective elimination of particular cell types in vivo has been a major effort of biomedical research in the last 2 decades. The development of hybridoma monoclonal antibodies and recombinant hormones in the last 15 years has provided pure ligands in adequate quantities to assist in targeted therapy. A number of laboratories have attached a variety of cell-killing moieties to these ligands, including human Fc receptor domains, bivalent antibodies with the ability to adhere human effector cells, radioisotopes, chemotherapeutic drugs, and finally peptide toxins. Toxins offer the most selective and deadly agents for targeted therapeutics, but a number of obstacles must be overcome so that the anticipated therapeutic index can be achieved. Many toxins possess normal tissue binding domains that must be chemically or genetically modified. Peptide toxins act intracellularly and must be properly internalized and translocated to the cytosol to be active. Receptors for new cell-selective ligands must be present on the vast majority of the target cell population. The size of the targeted toxin must be small enough to escape from the capillary bed to the tumor cells. Finally, the immunogenicity of the molecules must be low enough to permit the targeted toxin to circulate long enough to reach the tumor capillary bed. These complexities as well as the importance of the overall goal have led to a burst of scientific investigation into the structure-function relationships and protein engineering of peptide toxins. One goal of this monograph is to provide an update on the progress in this field.

The presence of lethal material in plants has been known for centuries. Soon after the discovery of microorganisms, toxins were found in these life forms as well. Details of toxin structure and physiology have only recently been defined. Plant and bacterial toxins must first bind to the cell surface. The plant toxins (ricin, abrin, viscumin, volkensin, modeccin) have

v

lectin domains that bind cell surface glycoproteins. The bacterial toxins (diphtheria toxin, *Pseudomonas* exotoxin, and *Shiga* toxin) bind unidentified cell surface receptors. Plant hemitoxins (pokeweed antiviral protein, momordin, and others) and the fungal toxins (α-sarcin, restrictocin) lack cell-binding domains and are only toxic to cells when they are linked to cell-binding peptides or permitted to leak into cells via complement-induced pores. Once bound, the toxins must internalize into intracellular compartments from which they can escape to the cytosol. Bacterial toxins cross to the cytosol from acidic endosomes. Plant toxins appear to reach the trans-Golgi before escape to the cytosol. Once they have reached the cytosol, many toxins catalytically inactivate protein synthesis. Some bacterial toxins (diphtheria toxin and *Pseudomonas* exotoxin) ADP-ribosylate elongation factor 2 (EF-2). Other bacterial toxins (*Shiga* toxin), plant holotoxins and hemitoxins, and the fungal toxins modify (nicking or depurinating) a conserved stem-loop rRNA structure on the 60S ribosomal subunit. In all cases, protein synthesis is inactivated by blocking the association of EF-2 with the ribosome. This prevents peptide elongation by impairing movement of the tRNAs from A to P sites. Recent work by Art Pardi and Harry Noller confirms the importance of the stem-loop structure for normal protein synthesis.

In the 1960s and 1970s, antitumor antibodies were conjugated to whole toxins. The normal tissue-binding domains of the toxins were not removed and there was high nonspecific toxicity. In the late 1970s and 1980s, peptide toxins were chemically modified to remove or block their binding domains. Unfortunately, in many cases, the translocation domains were inactivated as well. As a result, the antibody-toxin hybrids were often less cytotoxic than the unmodified toxins. In addition, the antibody-toxin conjugates were large and immunogenic and performed poorly in the clinic.

More recently, two developments have rekindled enthusiasm in the field. First, x-ray crystallographic three-dimensional molecular models are now (or soon to be) available for the major toxins—ricin, diphtheria toxin, and *Pseudomonas* exotoxin. This permits more rational drug design. Second, DNA cloning has been accomplished for a large number of the toxins. This permits specific modification of individual amino acid residues by genetic means.

The editor has spent the last 5 years attempting to use the new molecular biology tools to clone plant toxins, express toxins, and produce and purify chimeric toxins. While the new techniques are powerful, they are often much more difficult than the manuals imply. The second goal of this book is to provide a source of methods and experiences in toxin molecular biology. In particular, several problem areas are dealt with extensively. Cloning plant toxin genes is nontrivial even with extensive protein sequence information. Some hints and pitfalls are documented both in the Introduction and under the various toxins. Expression of peptide toxins can be impaired by self-intoxication (see Chapter 6), proteolysis in vivo (see Chapter 16), or weak promoters. Purification of chimeric toxins may be plagued with difficulties in proper folding. Both preclinical and clinical studies may be affected by selection of the proper biological model or disease entity (see Chapter 19).

The availability of a large number of mutant toxins and three-dimensional structure information has led to a number of hypotheses of structure-function relationships with individual amino acid residues and domains in the various toxins. We have attempted to assemble much of the data and the associated theories. We hope the text will stimulate further tests to define the function of particular protein toxin motifs. Further specialists outside the toxin field will hopefully be able to compare results. In particular, critical hydrogen bonds for sugar binding can be compared with other lectins. Electrostatic bonds for RNA binding can be compared with nontoxin RNA-binding proteins. Enzymatic mechanisms of ADP ribosylation and N-glycoside hydrolysis can be compared. Amphipathic helices or hydrophobic surfaces that may participate in membrane translocation can be defined. New toxins with similar physiological activity can be analyzed to reveal conserved amino acid residues with important functional properties. Thus, the third goal of the monograph is to encourage both broader and more in-depth protein structure-function studies.

Chimeric toxins represent the successful coalescence of protein engineering design based on structure-function work and the solution of molecular biology problems of expression, refolding, and purification of recombinant proteins. The generation of a novel protein with new binding specificities and intact membrane translocation and protein synthesis inhibitory activity is truly a remarkable achievement. However, extension of these successes to nonbacterial toxins and the successful application of these chimeras in the clinic are still unresolved. A final goal of the book is to speed the development of these drugs for patients by providing background information for pharmacologists and clinicians who are becoming introduced to these molecules at the late stages of testing. Although minimal in vivo and clinical information was available for most of the toxins, the data were included when available.

This volume includes contributions by most of the leading investigators in the field. While each laboratory has chosen a different toxin or an alternative engineering approach, assembling all these efforts into a single volume will hopefully provide a single source for those in the field. The struggles with these molecules are not unique to one group. The solutions often require skill, hard work, and patience. It is hoped that the field of genetically engineered toxins will continue to produce high-quality science and new pharmaceutics of use to humans. If this book can support this effort in some small way, then the commitment of the editor and contributors will not have been in vain.

I am grateful to the contributing scientists for their commitment and time. Finally, I wish to thank Ellen Vitetta for her careful review of the manuscript and Henry Boehm and Sandra Beberman at Marcel Dekker for their support throughout this project.

ARTHUR E. FRANKEL

Contents

VI *Pseudomonas* Exotoxins

VII Conclusions

Contributors

Paige Anderson   Department of Biomolecular Medicine, The University Hospital, Boston, Massachusetts

Yuji Aoyama   Tokushima University School of Medicine, Kuramoto-cho, Tokushima, Japan

Patricia Bacha   Seragen, Inc., Hopkinton, Massachusetts

Luca Benatti   Biotechnology Department, Farmitalia Carlo Erba, Milan, Italy

Bruce R. Blazar   Division of Bone Marrow Transplantation, Department of Pediatrics, University of Minnesota, Minneapolis, Minnesota

Aldo Ceriotti   Istituto Biosintesi Vegetali del CNR, Milan, Italy

Vijay K. Chaudhary   Laboratory of Molecular Biology, National Cancer Institute, National Institutes of Health, Bethesda, Maryland

G. Jiliani Chaudry   Molecular and Cell Biology Program, University of Texas at Dallas, Richardson, Texas

Royston C. Clowes†   Molecular and Cell Biology Program, University of Texas at Dallas, Richardson, Texas

John W. Crabb   Protein Chemistry Facility, W. Alton Jones Cell Science Center, Inc., Lake Placid, New York

Julian Davies   Institut Pasteur, Paris, France

Deborah Defeo-Jones   Department of Cancer Research, Merck Sharp & Dohme Research Laboratories, West Point, Pennsylvania

---

†Deceased.

Rockford K. Draper Molecular and Cell Biology Program, University of Texas at Dallas, Richardson, Texas

Gwynneth M. Edwards Department of Cancer Research, Merck Sharp & Dohme Research Laboratories, West Point, Pennsylvania

Yaeta Endo* Department of Biochemistry, Yamanashi Medical College, Tamaho, Nakakoma, Yamanashi, Japan

David FitzGerald Laboratory of Molecular Biology, National Cancer Institute, National Institutes of Health, Bethesda, Maryland

Arthur E. Frankel Altamonte Springs, Florida

Lawrence Greenfield Roche Molecular Systems, Alameda, California

Noriyuki Habuka Life Science Research Laboratory, Japan Tobacco, Inc., Yokohama, Kanagawa, Japan

David C. Heimbrook Department of Cancer Research, Merck Sharp & Dohme Research Laboratories, West Point, Pennsylvania

Walter K.-K. Ho Biochemistry Department, The Chinese University of Hong Kong, Shatin, Hong Kong

Priscilla L. Holmans† University of Texas at Dallas, Richardson, Texas

L. L. Houston Chiron Corporation, Emeryville, California

Richard Intres‡ Protein Chemistry Facility, W. Alton Jones Cell Science Center, Inc., Lake Placid, New York

Vicki Rubin Kelley Harvard Medical School, Boston, Massachusetts

Robert L. Kirkman Department of Surgery, Brigham and Women's Hospital and Harvard Medical School, Boston, Massachusetts

Robert J. Kreitman Laboratory of Molecular Biology, National Cancer Institute, National Institutes of Health, Bethesda, Maryland

Fadi G. Lakkis Department of Biomolecular Medicine, The University Hospital, Boston, Massachusetts

Bernard Lamy Microbial Engineering Unit, Institut Pasteur, Paris, France

Present affiliations:
*Department of Applied Chemistry, Faculty of Engineering, Ehime University, Bunkyo-cho, Matsuyama, Japan
†Department of Biochemistry, University of Texas Southwestern Medical Center at Dallas, Dallas, Texas
‡Molecular Biology Laboratory, Pathology Department, Berkshire Medical Center, Pittsfield, Massachusetts

Douglas A. Lappi Department of Molecular and Cellular Growth Biology, The Whittier Institute for Diabetes and Endocrinology, La Jolla, California

C. F. LeMaistre Department of Medicine, University of Texas Health Science Center at San Antonio, San Antonio, Texas

Bi-Yu Li Department of Pharmacology, University of Minnesota, Minneapolis, Minnesota

J. Michael Lord Department of Biological Sciences, University of Warwick, Coventry, England

Rolando Lorenzetti Department of Biotechnology, Marion Merrell Dow Research Institute—Lepetit Research Center, Gerenzano, Italy

Carole M. Meneghetti Department of Hematology, University of Texas Health Science Center at San Antonio, San Antonio, Texas

John R. Murphy Department of Medicine, The University Hospital, Boston, Massachusetts

Peter J. Nicholls Surgical Neurology Branch, National Institute of Neurological Disorders and Stroke, National Institutes of Health, Bethesda, Maryland

Jean C. Nichols Seragen, Inc, Hopkinton, Massachusetts

Gianpaolo Nitti Department of Biotechnology, Farmitalia Carlo Erba, Milan, Italy

Tatsuzo Oka Department of Nutritional Chemistry, Tokushima University School of Medicine, Kuramoto-cho, Tokushima, Japan

Allen Oliff Department of Cancer Research, Merck Sharp & Dohme Research Laboratories, West Point, Pennsylvania

Sjur Olsnes Biochemistry Department, Institute for Cancer Research at the Norwegian Radium Hospital, Oslo, Norway

Ira Pastan Laboratory of Molecular Biology, National Cancer Institute, National Institutes of Health, Bethesda, Maryland

Michael Piatak, Jr. Genelabs Incorporated, Redwood City, California

S. Ramakrishnan Department of Pharmacology, University of Minnesota, Minneapolis, Minnesota

Lynne M. Roberts Department of Biological Sciences, University of Warwick, Coventry, England

Jon D. Robertus   Department of Chemistry and Biochemistry, University of Texas, Austin, Texas

Daniel Schindler   Department of Chemical Immunology, Weizmann Institute of Science, Rehovoth, Israel

Paul Sehnke   Florida Hospital Cancer and Leukemia Research Center, Altamonte Springs, Florida

Michael E. Shapiro   Department of Surgery, Beth Israel Hospital and Harvard Medical School, Boston, Massachusetts

Pang-Chui Shaw   Biochemistry Department, The Chinese University of Hong Kong, Shatin, Hong Kong

Clay B. Siegall*   Laboratory of Molecular Biology, National Cancer Institute, National Institutes of Health, Bethesda, Maryland

Michela Solinas   Istituto Biosintesi Vegetali del CNR, Milan, Italy

Marco R. Soria   Director of Biotechnology Transfer, Department of Biotechnology, San Raffaele Research Institute, Milan, Italy

Steven M. Stirdivant   Department of Cancer Research, Merck Sharp & Dohme Research Laboratories, West Point, Pennsylvania

Terry B. Strom   Department of Medicine, Beth Israel Hospital and Harvard Medical School, Boston, Massachusetts

Philip E. Thorpe   Cancer Immunobiology Center, University of Texas Southwestern Medical Center, Dallas, Texas

Alexander Tonevitsky   All-Union Research Institute of Genetics and Selection of Industrial Microorganisms, Moscow, Russia

James W. Tregear   Department of Biological Sciences, University of Warwick, Coventry, England

Daniel A. Vallera   Department of Therapeutic Radiology-Radiation Oncology, University of Minnesota, Minneapolis, Minnesota

Johanna C. vanderSpek   Section of Biomolecular Medicine, Department of Medicine, The University Hospital, Boston, Massachusetts

Robert F. Weaver   Department of Biochemistry, The University of Kansas, Lawrence, Kansas

---

Present affiliation: Bristol-Myers Squibb Company, Evansville, Indiana

Marc Whitlow   Department of Protein Engineering, Enzon Incorporated, Gaithersburg, Maryland

Hin-Wing Yeung   Biochemistry Department and Chinese Material Research Center, The Chinese University of Hong Kong, Shatin, Hong Kong

Richard J. Youle   Biochemistry Section, Surgical Neurology Branch, National Institute of Neurological Disorders and Stroke, National Institutes of Health, Bethesda, Maryland

Mei-Hing Yung   Biochemistry Department, The Chinese University of Hong Kong, Shatin, Hong Kong

Rong-Huan Zhu*   Biochemistry Department, The Chinese University of Hong Kong, Shatin, Hong Kong

---

Present affiliation: Institute of Genetics, Academia Sinica, Beijing, People's Republic of China

I
Strategies and Techniques

1
General Methods

Robert F. Weaver *The University of Kansas, Lawrence, Kansas*

I. INTRODUCTION

This book is intended as a guide for scientists who are new to the toxin field. It presents aspects of molecular biology as applied to the investigation of toxins, especially the three best-studied examples: diphtheria toxin (DT), *Pseudomonas* exotoxin (PE), and ricin, or *Ricinus*, toxin (RT). We hope that what we have already learned about these three toxins will be useful to scientists tackling similar problems with other toxins, or to those who are extending our knowledge of the "big three."

The primary scientific literature is very helpful in showing how to do experiments and pointing the way to new avenues of investigation. However, it is the nature of this literature to describe successes, not failures. Thus, many fruitless lines of experimentation go unreported, and may be needlessly replicated in other laboratories. For this reason, we intend to include some of the problems one is likely to encounter in this field, as well as the methods that work.

The toxin field has an elegant history dating back to the nineteenth century. It encompasses three main areas: (1) protein chemistry—the purification and structure-function analysis of the toxins; (2) molecular biology—the cloning, manipulation, and expression of the toxin genes; and (3) clinical medicine—the attempt to direct the lethal power of these toxins against diseased cells. Thus, in this field, it is relatively easy to see a path from the laboratory to the clinic, which makes research more fun for many scientists. By bringing all these facets together in one volume, we hope to inspire new toxin researchers. By providing explicit instructions, we hope to make it easier for them to succeed.

Adapted from *Genetics*, 2nd edition by Robert F. Weaver and Philip W. Hedrick, Wm. C. Brown Publishers, Dubuque, IA, 1992.

II. MOLECULAR CLONING

Since molecular cloning techniques represent a common thread running through this book, we present here a short primer on the principles underlying these important procedures. Details of the methods themselves can be found in the individual reports, or in any of the excellent laboratory manuals now available.

A. Vectors

The primary purpose of gene cloning is to produce a particular gene in sufficient quantity to study. Frequently, this serves the secondary goal of producing that gene's product. The "classic" way to produce a large quantity of a gene, invented by Stanley Cohen and Herbert Boyer in 1973, is to place our gene of interest into a strain of bacteria and to allow these bacteria to replicate the gene for us. However, exogenous genes introduced into bacteria do not ordinarily survive for long. They are subject to immediate attack by degradative enzymes, and even if they are not destroyed, they do not have signals to tell the bacterium to replicate them. Thus, we need to couple our gene to a carrier DNA that is "at home" in a bacterial cell. It has an origin of replication that will direct the replication of the entire DNA—the gene along with the carrier. These carriers are called vectors, and they fall into two classes, plasmids and phage DNAs.

Plasmids as Vectors

In the early years of the cloning era, Boyer and his colleagues developed a set of very popular vectors known as the pBR plasmid series. One of these plasmids, pBR322 (Figure 1), contains genes that confer resistance to two antibiotics, ampicillin and tetracycline. Between these two genes lies the origin of replication. The plasmid has been engineered to contain only one cutting site for several of the common restriction enzymes, including *Eco*RI, *Bam*HI, *Pst*I, *Hind*III, and *Sal*I. This is convenient because it allows us to use each enzyme to create a site for inserting foreign DNA without losing any of the plasmid DNA.

For example, let us consider cloning a foreign DNA fragment into the *Pst*I site of pBR322 (Figure 2). First, we would cut the vector with *Pst*I to generate the sticky ends characteristic of that enzyme. In this example, we also cut the foreign DNA with *Pst*I, thus giving it *Pst*I sticky ends. Next, we combine the cut vector with the foreign DNA and incubate them with DNA ligase. As the sticky ends on the vector and on the foreign DNA base-pair momentarily, DNA ligase seals the nicks, attaching the two DNAs together covalently. Once this is done, they cannot come apart again unless they are recut with *Pst*I.

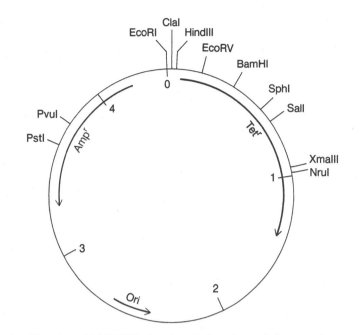

Figure 1 The plasmid pBR322, showing the locations of eleven unique restriction sites that can be used to insert foreign DNA. The locations of the two antibiotic-resistance genes (Amp^r = ampicillin resistance; Tet^r = tetracycline resistance) and the origin of replication (Ori) are also shown. Numbers refer to kilobase pairs (kb) from the *Eco*RI site.

In the following step, we transform *E. coli* with our DNA mixture. It would be nice if all the cut DNA had been ligated to plasmids to form recombinant DNAs, but that never happens. Instead, we get a mixture of re-ligated plasmids and re-ligated inserts, along with the recombinants. How do we sort these out? This is where the antibiotic resistance genes of the vector come into play. First, we grow cells in the presence of the tetracycline, which selects for cells that have taken up either the vector or the vector with inserted DNA. Cells that received no DNA, or that received insert DNA only, will not be tetracycline resistant and will fail to grow.

Next, we want to find the clones that have received recombinant DNAs. To do this, we screen for clones that are both tetracycline resistant and ampicillin sensitive. Figure 2 shows that the *Pst*I site, where we are inserting DNA in this experiment, lies within the ampicillin-resistance gene. There-fore, inserting foreign DNA into the *Pst*I site inactivates this gene and leaves the host cell vulnerable to ampicillin. How do we do the screening? One way is to transfer copies of the clones from the original tetracycline

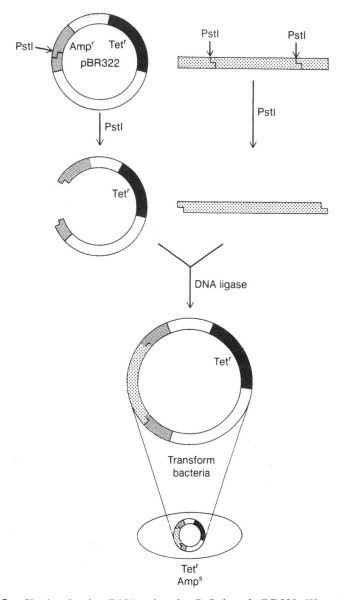

Figure 2 Cloning foreign DNA using the *Pst*I site of pBR322. We cut both the plasmid and the insert (▨) with *Pst*I, then join them through these sticky ends with DNA ligase. Next, we transform bacteria with the recombinant DNA and screen for tetracycline-resistant, ampicillin-sensitive cells. The plasmid no longer confers ampicillin resistance because the foreign DNA interrupts that resistance gene (▨).

plate to an ampicillin plate. This can be accomplished with a felt transfer tool as illustrated in Figure 3. We touch the tool lightly to the surface of the tetracycline plate to pick up cells from each clone, then touch the tool to a fresh ampicillin plate. This deposits cells from each original clone in the same relative positions as on the original clone in the same relative positions as on the original plate. We look for colonies that do *not* grow on the ampicillin plate and then find the corresponding growing clone on the tetracycline plate. Using a sterile toothpick or wire loop, we transfer cells from this positive clone to fresh medium for storage or immediate

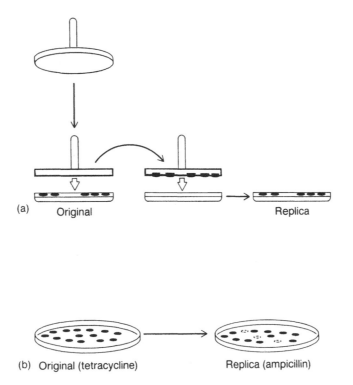

(a)   Original   Replica

(b) Original (tetracycline)   Replica (ampicillin)

Figure 3   Screening bacteria by replica plating. (a) The replica plating process. We touch a felt-covered circular tool to the surface of the first dish containing colonies of bacteria. Cells from each of these colonies stick to the felt and can be transferred to the replica plate in the same positions relative to one another. (b) Screening for inserts in the pBR322 ampicillin resistance gene by replica plating. The original plate contains tetracycline, so all colonies containing pBR322 will grow. The replica plate contains ampicillin, so colonies bearing pBR322 with inserts in the ampicillin resistance gene will not grow (these colonies are depicted by dotted circles). The corresponding colonies from the original plate can then be picked.

use. Notice that we did not call this procedure a selection because it does not remove unwanted clones automatically. Instead, we had to examine each clone individually. We call this more laborious process a screen.

Nowadays we can choose from a wide variety of plasmid cloning vectors with distinct advantages over the pBR plasmids. One useful class of plasmids is the pUC series (Figure 4). These plasmids are based on pBR322, from which almost half the DNA, including the tetracycline-resistance gene, has been deleted. Furthermore, the pUC vectors have their cloning sites clustered into one small area called a multiple cloning site (MCS). The pUC vectors contain the pBR322 ampicillin-resistance gene to allow selection for bacteria that received a copy of the vector. Moreover, to compensate for the loss of the other antibiotic-resistance gene, they have genetic elements that permit a very convenient screen for clones with inserts in their multiple cloning sites.

Figure 4(b) shows the multiple cloning sites of pUC18 and pUC19. Notice that they lie within a DNA sequence denoted *lacZ*, which codes for the amino terminal portion of β-galactosidase (the α-peptide). The host bacteria used with the pUC vectors carry a gene fragment that encodes the carboxyl portion of β-galactosidase. By themselves, the β-galactosidase fragments made by these partial genes have no activity. But they can complement each other by intracistronic complementation (called α-complementation in this case). In other words, the two partial gene products can cooperate to form an active enzyme. Thus, when pUC18 by itself transforms a bacterial cell carrying the partial β-galactosidase gene, active β-galactosidase is produced. If we plate these clones on medium containing a β-galactosidase indicator, colonies with the pUC plasmid will turn color. The indicator X-gal, for instance, is a synthetic, colorless galactoside; when β-galactosidase leaves X-gal, it releases galactose plus a dye that stains the bacterial colony blue.

On the other hand, if we have interrupted the plasmid's partial β-galactosidase gene by placing an insert into the multiple cloning site, the gene is usually inactivated. It can no longer make a product that complements the host cell's β-galactosidase fragment, so the X-gal remains colorless. Thus, in principle, it is a simple matter to pick the clones with inserts. They are the white ones; all the rest are blue. Notice that in contrast to screening with pBR322, this is a one-step process. We look simultaneously for a clone that (1) grows on ampicillin, and (2) is white in the presence of X-gal. The multiple cloning sites have been carefully constructed to preserve the reading frame of β-galactosidase. Thus, even though the gene is interrupted by 18 codons (placed in parentheses, with amino acid names in lower-case letters), a functional protein still results. Further

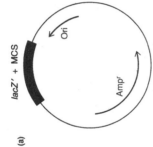

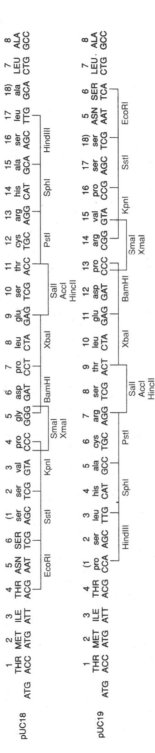

Figure 4 Architecture of a pUC plasmid. (a) The pUC plasmids retain the ampicillin-resistance gene and the origin of replication of pBR322. In addition, they include a multiple cloning site (MCS) inserted into a gene encoding the amino part of β-galactosidase (*lacZ'*). (b) The MCSs of pUC18 and pUC19. In pUC18, the MCS, containing 13 restriction sites, is inserted after the sixth codon of the *lacZ'* gene. In pUC19, it comes after the fourth codon and is reversed. In addition to containing the restriction sites, the MCSs preserve the reading frame of the *lacZ'* gene, so plasmids with no inserts still support active expression of this gene. Therefore, clones harboring the vector alone will turn blue in the presence of the synthetic β-galactosidase substrate X-gal, whereas clones harboring the vector plus an insert will remain white. (*Reprinted with permission of Life Technologies, Inc., Gaithersburg, MD.*)

interruption by large inserts, especially those that shift the reading frame, is usually enough to destroy the gene's function.

Even with the color screen, cloning into pUC can give false positives—white colonies without inserts. This can happen if the vector's ends are "nibbled" slightly by nucleases before ligation to the insert. Then, if these slightly degraded vectors simply close up during the ligation step, chances are two in three that the reading frame of the β-galactosidase has been changed, so white colonies will result. This underscores the importance of using clean DNA and enzymes that are free of exonuclease activity.

This phenomenon of vector re-ligating with itself can be a greater problem when we use vectors that do not have a color screen because it is impossible to distinguish colonies with inserts from those without, unless we laboriously analyze DNA from each colony. Even with pUC and related vectors, we would like to minimize vector re-ligation. A good way to do this is to treat the vector with alkaline phosphatase, which removes the 5′-phosphates that are necessary for ligation. Without these phosphates, the vector cannot ligate to itself, but can still ligate to the insert that retains its 5′-phosphates. Figure 5 illustrates this process. Notice that since only the insert has phosphates, two nicks (unformed phosphodiester bonds) remain in the ligated product. These are not a problem; they will be completed once the ligated DNA has made its way into a bacterial cell.

The multiple cloning site of the pUC vectors also allows us to cut them with two different restriction enzymes (say, *Eco*RI and *Bam*HI) and then to clone a piece of DNA with one *Eco*RI end and one *Bam*HI end. This is called directional cloning, or forced cloning, because we force the insert DNA into the vector in only one orientation. (Obviously, the *Eco*RI and *Bam*HI ends of the insert have to match their counterparts in the vector.) Knowing the orientation of an insert is important if we are trying to express a gene contained in that insert, as we will see later in this chapter. Forced cloning also has the advantage of preventing the vector's simply re-ligating by itself since its two restriction sites are incompatible.

On the other hand, suppose we clone a *Hind*III fragment into the *Hind*III site of a hypothetical plasmid vector, as illustrated in Figure 6. Since this is not directional, or forced, cloning, the fragment will insert into the vector in both possible orientations: with the *Bam*HI site on the right (left side of Figure 6); or with the *Bam*HI site on the left (right side of the Figure 6). How do we decide which orientation we have in a given clone? To answer this question, we locate a restriction site asymmetrically situated in the vector, relative to the *Hind*III cloning site. In this case, there is an *Eco*RI site only 0.3 kb from the *Hind*III site. This means that if we cut the cloned DNA pictured on the left with *Bam*HI and *Eco*RI, we will generate two fragments 3.6 and 0.7 kb long. On the other hand,

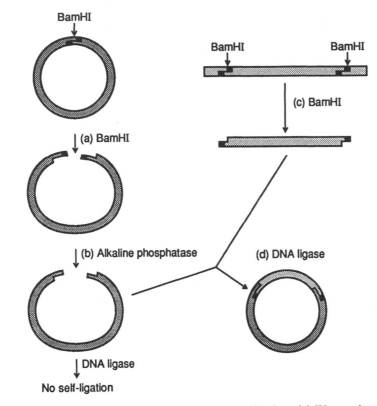

Figure 5   Alkaline phosphatase prevents vector re-ligation. (a) We cut the vector (▨ , top left) with *Bam*HI. This produces sticky ends with 5′-phosphate (◢). (b) We remove the phosphates with alkaline phosphatase, making it impossible for the vector to re-ligate with itself. (c) We also cut the insert (▨ , upper right) with *Bam*HI, producing sticky ends with phosphates that we do not remove. (d) Finally, we ligate the vector and insert together. The phosphates on the insert allow two phosphodiester bonds to form (▨), but leave two unformed bonds, or nicks (shown as gaps in the DNA here). These will be completed once the DNA is in the transformed bacterial cell.

if we cut the DNA pictured on the right with the same two enzymes, we will generate two fragments 2.8 and 1.5 kb in size. We can distinguish between these two possibilities easily by electrophoresing the fragments to measure their sizes, as shown at the bottom of Figure 6. Usually, we prepare DNA from several different clones, cut each of them with the two enzymes, and electrophorese the fragments side-by-side with one lane reserved for

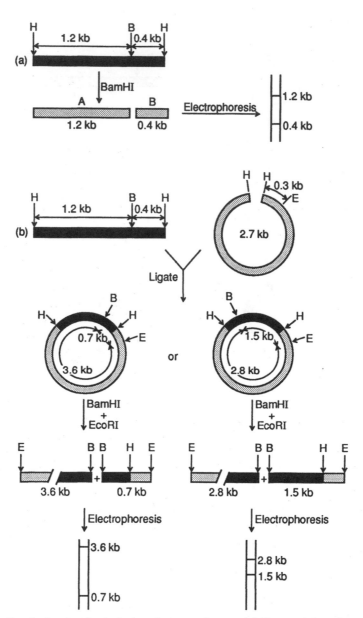

Figure 6 A simple physical mapping experiment. (a) Determining the position of a BamHI site. A 1.6-kb HindIII fragment is cut by BamHI to yield two subfragments. The sizes of these fragments are determined by electrophoresis to be 1.2 kb and 0.4 kb, demonstrating that BamHI cuts once, 1.2 kb from one end of the HindIII fragment, and 0.4 kb from the other end. (b) Determining the orientation of the HindIII fragment in a cloning vector. If we clone the 1.6-kb HindIII fragment

marker fragments of known sizes. On average, half of the clones will have one orientation, and the other half will have the opposite orientation.

Notice that the multiple cloning site of pUC18 is just the reverse of that of pUC19; that is, the restriction sites are in opposite order. This means we can clone our fragment in either orientation simply by shifting from one pUC plasmid to the other. Even more convenient vectors for the pUC series are now available. We will discuss some of them later in this chapter.

Phages as Vectors

Phages serve as natural vectors in transducing bacterial DNA from one cell to another. It was only natural, then, to engineer phages to do the same thing for *all* kinds of DNA.

Fred Blattner and his colleagues constructed the first phage vectors by modifying the well-known lambda phage. They took out the region in the middle of the phage DNA, which codes for proteins needed for lysogeny, but retained the genes needed for lytic infection. These phages are no longer capable of lysogenic infection, but their missing genes can be replaced with foreign DNA. Blattner named these vectors Charon phages after Charon, the boatman on the river Styx in classical mythology. Just as Charon carried souls to the underworld, the Charon phages carry foreign DNA into bacterial cells. Charon the boatman in pronounced "karen" but Charon the phage is usually pronounced "sharon."

One clear advantage of the lambda phages over plasmid vectors is that they can accommodate much more foreign DNA. For example, Charon 4 can accept up to about 20 kb of DNA, a limit imposed by the capacity of the phage head. When would we need such high capacity? One common use for labmda phage vectors is in constructing genomic libraries. Suppose we wanted to clone the entire *Ricinus cummunis* (castor plant) genome. This would obviously require a great many clones, but the larger the insert in each clone, the fewer total clones would be needed. In fact, such genomic libraries have been constructed for *Ricinus cummunis* and for a wide variety

---

into the HindIII site of a cloning vector, it can insert in either of two ways: (1) with the BamHI site near an EcoRI site in the vector, or (2) with the BamHI site remote from an EcoRI site in the vector. To determine which, we cleave the DNA with both BamHI and EcoRI and electrophorese the products to measure their sizes. A short fragment (0.7 kb) shows that the two sites are close together (left). On the other hand, a long fragment (1.5 kb) shows that the two sites are far apart (right).

of other organisms, and the Charon phages and more modern adaptations of the lambda have been popular vectors for this purpose.

Aside from their high capacity, some of the lambda vectors have the advantage of a minimum-size requirement for their inserts. Figure 7 illustrates the reason for this requirement: To get the Charon 4 vector ready to accept an insert, we cut it with *Eco*RI. This cuts at two sites near the middle of the phage DNA, yielding two "arms" and two "stuffer" fragments. Next, we purify the arms by ultracentrifugation and throw away the stuffers. The final step is to ligate the arms to our insert, which then takes the place of the discarded stuffers.

At first glance, it may appear that the two arms could simply ligate together without accepting an insert. Indeed, this may happen, but it will not produce a clone because the two arms constitute too little DNA and will not be packaged into a phage head. The packaging is done in vitro; we simply mix the ligated arms plus inserts with all the components needed to put together a phage particle. Nowadays one can buy the purified lambda arms, as well as the packaging extract. These cloning kits have trade names, such as lambda GEM. This extract has rather stringent requirements as to the size of DNA it will package. It must have at least 12 kb of DNA in addition to the arms, but no more than 20 kb or the phage head will overflow.

Since we can be sure that each clone in or genomic library has at least 12 kb, we know we are not wasting space with clones that contain insignificant amounts of DNA. This is an important consideration since, even at 12–20 kb per clone, we need about half a million clones to be sure of having each *Ricinus communis* gene represented at least once. It would be much more difficult to make a *Ricinus cummunis* genomic library in a pUC vector since plasmids tend to ligate selectively to small inserts, and bacteria selectively reproduce small plasmids. Therefore, most of the clones would contain inserts of a few thousand, or even just a few hundred base pairs. Such a library would have to contain many millions of clones to be complete.

Since *Eco*RI produce fragments whose average size is about 4 kb, and yet the vector will not accept any inserts smaller than 12 kb, it is obvious that we cannot cut of DNA completely with *Eco*RI, or most of the fragments will be too small to clone. Furthermore, *Eco*RI, or any other restriction enzyme, cuts in the middle of most eukaryotic genes one or more times, so a complete digest would contain only fragments of most genes. We can avoid these problems by preforming an incomplete digestion with *Eco*RI; if the enzyme cuts only about every fourth or fifth site, the average length of the resulting fragments will be about 16–20 kb, just the size the

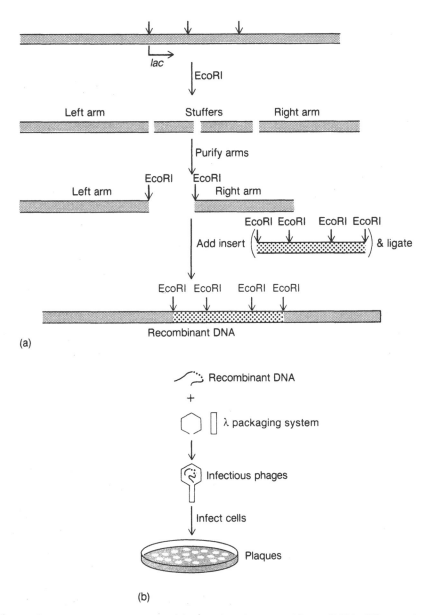

(a)

(b)

Figure 7   Cloning in Charon 4. (a) Forming the recombinant DNA. We cut the vector (▢) with *Eco*RI to remove the stuffer fragments, and save the arms. Next, we ligate partially digested insert DNA (▨) to the arms. (b) Packaging and cloning the recombinant DNA. We mix the recombinant DNA from (a) with an in vitro packaging extract that contains λ phage head and tail components and all other factors needed to package the recombinant DNA into functional phage particles. Finally, we plate these particles on *E. coli* and collect the plaques that form.

vector will accept and big enough to include the entirety of most eukaryotic genes, introns and all.

Another vector designed especially for cloning large DNA fragments is called a cosmid. Cosmids resist classification because they behave both as plasmids and as phages. They contain the cos sites, or cohesive ends, of lambda phage DNA, which allow the DNA to be packaged into lambda phage heads (hence the "cos" part of the name *cosmid*). They also contain a plasmid origin of replication, so they can replicate as plasmids in bacteria (hence the "mid" part of the name).

Because almost the entire lambda genome, except for the cos sites, has been removed from the cosmids, they have room for very large inserts (40–50 kb). Once these inserts are in place, the recombinant cosmids are packaged into phage particles. These particles cannot replicate as phages because they have almost no phage DNA, but they are infectious, so they carry their recombinant DNA into bacterial cells. Once inside, the DNA replicates as a plasmid, using its plasmid origin of replication.

A genomic library is very handy. Once it is established, we can search it for any gene we want. The only problem is that there is no card catalog for such a library, so we need some kind of probe to tell us which clone contains the gene of interest. An ideal probe would be labeled nucleic acid whose sequence matches that of the gene we are trying to find. If we have a phage library, we would then carry out a plaque hybridization procedure in which the DNA from each of the thousands of phages from our library is hybridized to the labeled probe. The DNA that forms a labeled hybrid is the right one.

Figure 8 shows how plaque hybridization works. We grow thousands of plaques on each of several Petri dishes (only a few plaques are shown here for simplicity). Next, we touch a filter made of DNA-binding material such as nitrocellulose to the surface of the Petri dish. This transfers phage DNA from each plaque to the filter. The DNA is then denatured with alkali and hybridized to the radioactive probe. When the probe encounters complementary DNA, which should be only the DNA from the clone of interest, it will hybridize, making that DNA spot radioactive. This radioactive spot is then detected by autoradiography. The black spot on the autoradiograph shows us where to look on the original Petri dish for the plaque containing our gene. In practice, the original plate may be so crowded with plaques that it is impossible to pick out the right one, so we pick several plaques from that area, replate at a much lower phage density, and rehybridize to find the positive clone.

M13 Phage Vectors.   Another phage frequently used as a cloning vector is the filamentous phage M13. Joachim Messing and his coworkers endowed the phage DNA with the same β-galactosidase gene fragment and multiple

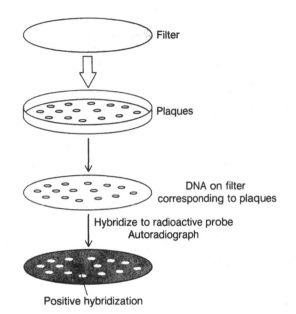

Figure 8   Selection of positive genomic clones by plaque hybridization. First, we touch a nitrocellulose or similar filter to the surface of the dish containing the Charon 4 plaques from figure 15.6. Phage DNA released naturally from each plaque will stick to the filter. Next, we denature the DNA with alkali and hybridize the filter to a radioactive probe for the gene we are studying, then autoradiograph to reveal the position of any radioactivity. Cloned DNA from one plaque near the center of the filter was hybridized, as shown by the dark spot on the autoradiograph.

cloning sites found in the pUC family of vectors. In fact, M13 vectors were engineered first; then the useful cloning sites were simply transferred to the pUC plasmids.

What is the advantage of the M13 vectors? The main factor is that the genome of this phage is a single-stranded DNA, so DNA fragments cloned into this vector can be recovered in single-stranded form. As we will see later in this chapter, single-stranded DNA is an invaluable aid to site-directed mutagenesis, by which we can introduce specific, premeditated alterations into a gene. It also makes it easier to determine the sequence of a piece of DNA.

Figure 9 illustrates how we can clone a double-stranded piece of DNA into M13 and harvest a single-stranded DNA product. The DNA in the phage particle itself is single-stranded, but after infecting an *E. coli* cell, it is converted to a double-stranded replicative form (RF). This double-stranded replicative form of the phage DNA is what we use for cloning.

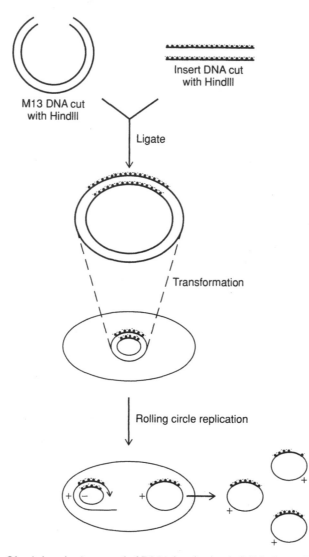

Figure 9   Obtaining single-stranded DNA by cloning in M13 phage. Foreign DNA
(····), cut with *Hind*III, is inserted into the *Hind*III site of the double-stranded
phage DNA. The resulting recombinant DNA is used to transform *E. coli* cells,
whereupon the DNA replicates by a rolling circle mechanism, producing many
single-stranded product DNAs. These product DNAs are called positive (+) strands,
by convention. The template DNA is therefore the negative (−) strand.

After it is cut by one or two restriction enzymes at its multiple cloning site, foreign DNA with compatible ends can be inserted. This recombinant DNA is then used to transform host cells, giving rise to progeny phages that bear single-stranded recombinant DNA. The phage DNA, along with phage particles, is secreted from the transformed cells and can be collected from the growth medium.

Phagemids.   Another class of vectors with single-stranded capability has now been developed. These are like the cosmids in that they have characteristics of both phages and plasmids; thus, they are sometimes called phagemids. One popular variety (Figure 10) goes by the trade name Bluescript. It has a multiple cloning site inserted into the *lacZ* gene, so clones with inserts can be distinguished by their white color from those with the unaltered vector, which are blue. This vector also has the origin of replication of the single-stranded phage f1, which is related to M13. This means that a cell harboring a recombinant phagemid, if infected by f1 helper phage, will produce and package single-stranded phagemid DNA. A final useful feature of this class of vectors is that the multiple cloning site is flanked by two different phage RNA polymerase promoters. For example, Bluescript has a T3 promoter on one side and a T7 promoter on ther other.

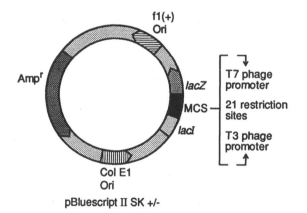

pBluescript II SK +/-

Figure 10   A Bluescript vector. This cloning vector has three important features. (1) Its multiple cloning site (MCS) interrupts the *lacZ* gene, allowing blue/white selection of recombinants. (2) It contains the f1 phage origin of replication, so it can produce single-stranded DNA in the presence of a helper phage. (3) Its MCS lies between the T7 and T3 promoters, so inserted genes can be transcribed in vitro from either direction, just by using either the T7 or T3 phage polymerase in the reaction. This gives sense or antisense transcripts, depending on the orientation of the inserted gene.

This allows us to isolate the double-stranded phagemid DNA and transcribe it in vitro with either of the phage polymerases to produce pure RNA transcripts corresponding to either strand.

B. Identifying a Specific Clone with a Specific Probe

We have already mentioned the need for a probe to identify the clone we want among the thousands we do not want. What sort of probe could be employ? Two different kinds are widely used: polynucleotides and antibodies. Both are molecules able to bind specifically to other molecules. We will discuss oligonucleotide and polynucleotide probes here and antibody probes later in this chapter.

Polynucleotide Probes

If the gene we are trying to locate is a very active one, appropriate cells may make its corresponding mRNA in large enough quantities to purify. In principle, we could use this mRNA as a probe, but it is more practical to make a cDNA copy by reverse transcription. Once we have made the cDNA, we can clone it or use it directly as a probe.

More often than not, the gene we want is not active enough to allow this direct approach. In that case, we might use the homologous gene from another organism if someone has already managed to clone it. For example, if we were after the human insulin gene, and another research group had already cloned the rat insulin gene, we could ask them for their clone to use as a probe. We would hope the two genes have enough similarity in sequence that the rat probe could hybridize to the human gene. This hope is usually fulfilled. However, we generally have to lower the stringency of the hybridization conditions so that the hybridization reaction can tolerate some mismatches in base sequence between the probe and the cloned gene.

Researchers use several means to control stringency. High temperature, high organic solvent concentration, and low salt concentration all tend to promote the separation of the two strands in a DNA double helix. We can, therefore, adjust these conditions until only perfectly matched DNA strands will form a duplex; this is high stringency. By relaxing these conditions (lowering the temperature, for example), we lower the stringency until DNA strands with a few mismatches can hybridize.

Without cDNA or homologous DNA from another organism, what could we use? There is still a way out if we know at least part of the sequence of the protein product of the gene. We faced a problem just like this in our lab when we cloned the preproricin gene. Fortunately, the entire amino acid sequences of both polypeptides of ricin were known. That meant we could examine the amino acid sequence and, using the genetic code, deduce the nucleotide sequence that would code for these amino acids. Then we could construct the nucleotide sequence chemically and use this synthetic probe to find the preproricin gene. This sounds easy, but there is a hitch. The genetic code is degenerate, so for most amino acids we would have to consider several different nucleotide sequences.

Fortunately, we were spared some inconvenience because one of the polypeptides of ricin includes this amino acid sequence. Trp-Met-Phe-Lys-Asn-Glu. The first two amino acids in this sequence have only one codon each, and the next three only two each. The sixth gives us two free bases because the degeneracy occurs only in the third base. Thus, we had to make only eight 17-base oligonucleotides (17-mers) to be sure of getting the exact coding sequence for this string of amino acids. This degenerate sequence can be expressed as follows:

```
              U    G    U
UGG  AUG  UUC  AAA  AAC  GA
Trp  Met  Phe  Lys  Asn  Glu
```

Using this mixture of eight 17-mers (UGGAUGUUCAAAAAACGA, UGGAUGUUUAAAAACGA, etc.), we quickly identified several ricin-specific clones.

C. Complementary cDNA Cloning

Molecular geneticists use a variety of techniques to clone cDNAs; here we will consider a fairly simple yet effective strategy, which is illustrated in Figure 11. We used this approach to make a *Ricinus communis* cDNA library. The central part of any cDNA cloning procedure is synthesis of the cDNA from a messenger DNA (mRNA) template using reverse transcriptase. This reverse transcriptase is like any other DNA-synthesizing enzyme in that it cannot initiate DNA synthesis without a primer; give it only mRNA and you get nothing in return. To get around this problem, we take advantage of the poly(A) tail at the end of most mRNAs and use oligo(dT) as the primer. Since oligo(dT) is complementary to poly(A), it binds at the 3' end of the mRNA and primes DNA synthesis, using the mRNA template.

After the mRNA has been copied, yielding a single-stranded DNA (the "first strand"), we remove the mRNA with alkali or ribonuclease H (RNase H). This enzyme degrades the RNA part of an RNA/DNA hybrid—just what we need to remove the RNA from our first-strand cDNA. Next, we must make a second DNA strand, using the first as a template. Again, we need a primer, and this time we do not have a convenient poly(A) to which to hybridize the primer. Instead, we build an oligo(dC)

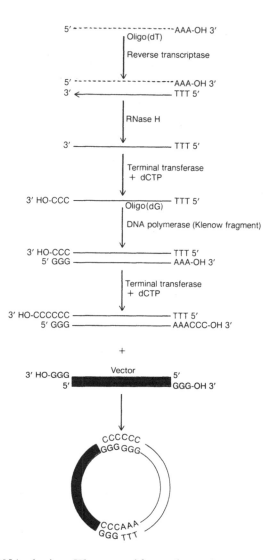

Figure 11 cDNA cloning. We start with a eukaryotic messenger RNA (----) having poly(A) at its 3′ end. Oligo(dT) hybridizes to the poly(A) and primes reverse transcription, forming the first cDNA strand. We remove the mRNA template, using RNase H, then add an oligo(dC) tail to the 3′ end of the cDNA, using terminal transferase. Oligo(dG) hybridizes to the oligo(dC) and primes second strand cDNA synthesis by reverse transcriptase or by the Klenow fragment of DNA polymerase I. In order to give the double-stranded cDNA sticky ends, we add oligo(dC) with terminal transferase, then anneal these ends to complementary oligo(dG) ends of a suitable vector (■). The recombinant DNA can then be used to transform bacterial cells.

tail at the 3' end of the first strand, using the enzyme terminal transferase and the substrate dCTP. The enzyme adds dCs, one at a time to the 3' end of the first strand. To this tail, we hybridize a short oligo(dG), which primes second-strand synthesis. We can use reverse transcriptase again to make the second strands, but DNA polymerase also works. Actually, the most successful enzyme is a fragment of DNA polymerase called the Klenow fragment. This piece of enzyme is generated by cleaving *E. coli* polymerase I with a proteolytic enzyme. The Klenow fragments contains the DNA polymerase activity and the 3'→5' exonuclease activity, but it lacks the 5'→3' exonuclease activity normally associated with DNA polymerase I. The latter activity is undesirable because it degrades DNA from the 5' end, which is damage the DNA polymerase cannot repair.

Once we have a double-stranded cDNA, we must ligate it to a vector. This was easy with our pieces of genomic DNA since they had sticky ends, but the cDNA has no sticky ends. This problem is easily solved. We simply tack sticky ends (oligo[dC]s) onto the cDNA, again using terminal transferase and dCTPs. In the same way, we attach oligo(dG)s ends to our vector and allow the oligo(dC)s to anneal to the oligo(dG)s. This brings the vector and cDNA together in recombinant DNA that can be used directly for transformation. The base-pairing between the oligonucleotide tails is strong enough that no ligation is required before transformation. The DNA ligase inside the transformed cells finally performs this task.

What kind of vector should we use? Several choices are available, depending on the way we wish to detect positive clones (those that bear the cDNA we want). We can use a simple vector such as one of the pUC plasmids; if we do, we usually identify positive clones by colony hybridization with a radioactive DNA probe. This procedure is analogous to the plaque hybridization described previously. Or we can use lambda phage, such as lambda g11, as a vector. This vector places the cloned cDNA under the control of the β-galactosidase promoter, so that transcription and translation of the cloned gene can occur. We can then use an antibody to screen directly for the protein product of the correct gene. We will describe this procedure in more detail later in this chapter.

D. Methods of Expressing Cloned Genes

Why would we want to clone a gene? An obvious reason is that cloning allows us to produce large quantities of pure eukaryotic (or prokaryotic) genes so we can study them in detail. Thus, the gene itself can be a valuable product of gene cloning. Another goal of gene cloning is to make a large quantity of the gene's product, either for investigative purposes or for profit.

Expression Vectors

The vectors we have examined so far are meant to be used primarily in the first stage of cloning—when we first put a foreign DNA into a bacterium and get it to replicate. By and large, they work well for that purpose, growing readily in *E. coli* and producing high yields of recombinant DNA. Some of them even work as expression vectors that can yield the protein products of cloned genes. For example, the pUC vectors place inserted DNA under the control of the *lac* promoter, which lies upstream from the multiple cloning site. If an inserted DNA happens to be in the same reading frame as the *lac* gene it interrupts, a fusion protein will result. It will have a partial β-galactosidase protein sequence at its amino end and another protein sequence, encoded in the inserted DNA, as its carboxyl end (Figure 12).

However, if we are interested in high expression of our cloned gene, specialized bacterial expression vectors usually work better. These typically have two elements that are required for active gene expression: The first of these is a strong promoter. The second element needed, since the cloned eukaryotic gene does not provide it, is a ribosome binding site that includes a Shine-Dalgarno sequence near an initiating ATG codon.

The main function of an expression vector is to yield the product of a gene—usually the more protein the better. Therefore, expression vectors are ordinarily equipped with very strong promoters; the rationale is that the more mRNA is produced, the more protein product will be made.

One vector that offers both potent expression and very tight control is pET-11a. This vector places the gene we want to express under the control of a late T7 phage promoter, which means that it will not be expressed until the T7 RNA polymerase is available (Chapter 7). We can do the initial cloning into pET711a in any bacterium we choose, then transfer the plasmid to a special bacterium harboring the T7 polymerase gene under the control of another promoter. One cell line, for example, has this gene next to the λ P_L promoter. This bacterium also carries a temperature-sensitive lambda repressor gene, so it cannot make T7 polymerase until we raise the temperature. However, when we do, the bacteria produce a burst of T7 polymerase, which then goes to work transcribing our gene. It is possible to get a tremendous amount of protein this way.

Inducible Expression Vectors.   It is usually advantageous to keep a cloned gene turned off until we are ready to express it. One reason is that eukaryotic proteins produced in large quantities in bacteria can be toxic. Even if these proteins are not actually toxic, they can build up to such

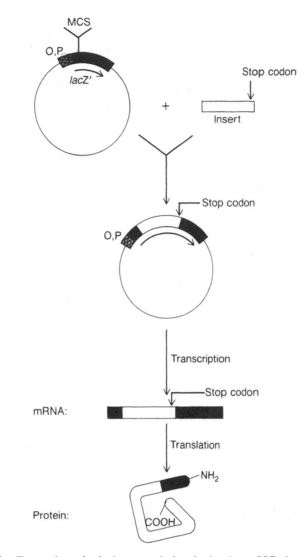

Figure 12 Formation of a fusion protein by cloning in a pUC plasmid. We insert foreign DNA (☐) into the multiple cloning site (MCS); transcription from the *lac* promoter (▨) gives a hybrid mRNA beginning with a few *lacZ* codons, changing to insert sequence, then back to *lacZ* (■). This mRNA will be translated to a fusion protein containing a few β-galactosidase amino acids at the beginning (amino end), followed by the insert amino acids for the remainder of the protein. Since the insert contains a translation stop codon, the remaining *lacZ* codons will not be translated.

great levels that they interfere with bacterial growth. In either case, if the cloned gene were allowed to remain turned on constantly, the bacteria bearing the gene would never grow to a great enough concentration to produce meaningful quantities of protein product. The solution is to keep the cloned gene turned off by placing it behind an inducible promoter.

The *lac* promoter is inducible to a certain extent, presumably remaining off until stimulated by the inducer allolactose or by its synthetic analog IPTG. However, the repression wrought by the *lac* repressor is incomplete, and some expression of the cloned gene will be observed even in the absence of inducer. One way around this problem is to express our gene in a host cell that has an overactive *lac* repressor gene, *lac*I. The excess repressor produced by such a cell keeps our cloned gene turned off until we are ready to induce it.

Another strategy is to use a very tightly controlled promoter such as the λ phage promoter P_L. Expression vectors with this promoter/operator system are cloned into host cells bearing a temperature-sensitive λ repressor gene (*c*I857). As long as the temperature of these cells is kept relatively low (32°C), the repressor functions, and no expression takes place. However, when we raise the temperature to the nonpermissive level (42°C), the temperature-sensitive repressor can no longer function and the cloned gene is induced. Figure 13 illustrates the expression vector pKC30, which uses this mechanism. One disadvantage of this approach is that raising the temperature to inactivate the repressor also induces the heat-shock genes of the host. Since some of these genes encode proteases, they can harm the cloned gene product.

Expression Vectors that Produce Fusion Proteins.   When most expression vectors operate, they produce fusion proteins. This might at first seem a disadvantage because the natural product of the inserted gene is not made. However, the extra amino acids at the amino terminals of the fusion protein can be useful. Consider the pUR series of expression vectors. These use the *lac* operator and promoter, and they have a multiple cloning site near the end of *lac*Z gene (Figure 14). This means the product will be a fusion protein with a large piece of β-galactosidase at its amino end.

How can fusion proteins be useful? One reason is that they are sometimes more stable in bacterial cells than normal eukaryotic proteins are. Furthermore, if we do not have a good way of purifying the protein product of our cloned gene, the long β-galactosidase addition can be a big help. For one thing, with a molecular weight of 116,000, the β-galactosidase monomer is already among the largest polypeptides in an *E. coli* cell and is correspondingly easy to purify by techniques that separate molecules by size. A fusion protein, with another polypeptide tacked onto the carboxyl-terminal, will be even larger, and it is usually a simple matter to identify

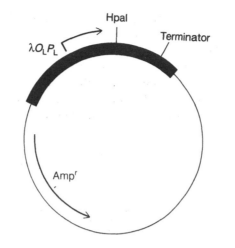

Figure 13 The inducible expression vector pKC30. A gene to be expressed is inserted into the unique *Hpa*I site, downstream from the $\lambda O_L P_L$ operator-promoter region. The host cell is a λ lysogen bearing a temperature-sensitive λ repressor gene (*cI*857). To induce expression of the cloned gene, the temperature is raised from 32 to 42°C, which inactivates the temperature-sensitive λ repressor, removing it from O_L and allowing transcription to occur.

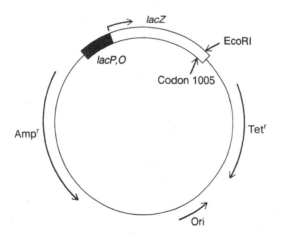

Figure 14 The pβ-ga113C expression vector. The unique *Eco*RI cloning site lies near the end of the *lacZ* coding region, just after codon number 1005. Therefore, products of genes cloned into this site will contain a 1005–amino acid tag of β-galactosidase at their amino ends.

and purify such a protein. Moreover, the β-galactosidase provides the fusion protein with a tag that allows purification with anti-β-galactosidase antibodies, as we will see.

The λ phages have also served as the basis for expression vectors. A λ phage that has been designed specifically as an expression vector is λ g11. This phage (Figure 15) contains the *lac* control region followed by the *lacZ* gene. The cloning sites are located within the *lacZ* gene, so products of a gene inserted into this vector will be fusion proteins with a leader of β-galactosidase.

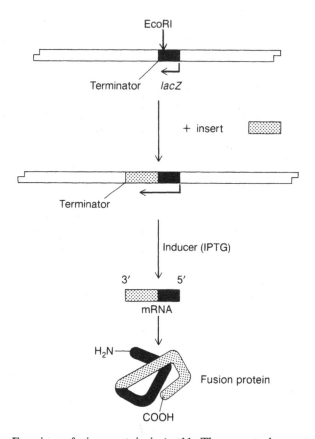

Figure 15 Forming a fusion protein in λgt11. The gene to be expressed (▒) is inserted into the *Eco*RI site near the end of the *lacZ* coding region (■) just before the transcription terminator. Thus, upon induction of the *lacZ* gene by IPTG, a fused mRNA results, containing the inserted coding region just downstream from that of β-galactosidase. This mRNA is translated by the host cell to a fusion protein.

The expression vector λ g11 has become a popular vehicle for making and screening cDNA libraries. In the examples of screening presented earlier, we looked for the proper DNA sequence by probing with a labeled oligonucleotide or polynucleotide. By contrast, λ g11 allows us to directly screen a group of clones for the expression of the right protein. The main ingredients required for this procedure are a cDNA library in g11 and an antiserum directed against the protein of interest.

Figure 16 shows how this works. We plate our λ phages with various cDNA inserts and blot the proteins released by each clone onto a support such as nitrocellulose. The nice thing about a clone in this regard is that

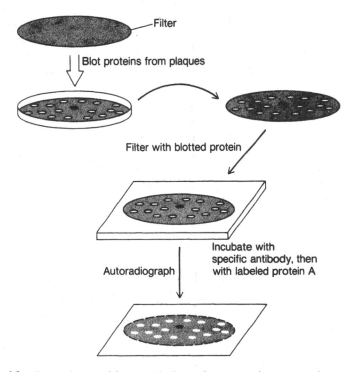

Figure 16   Detecting positive λgt11 clones by measuring expression. A filter is used to blot proteins from phage plaques on a Petri dish. One of the clones (◼) has produced a plaque containing a fusion protein including β-galactosidase and a part of the protein we are interested in. The filter with its blotted proteins is incubated with an antibody directed against our protein of interest, then with radioactive *Staphylococcus* protein A, which binds specifically to antibodies. It will, therefore, bind only to the antibody-antigen complexes at the spot corresponding to our positive clone. A dark spot on the autoradiograph of the filter reveals the location of our positive clone.

the host cells lyse, forming plaques. In so doing, they release their products, making it easy to transfer proteins from thousands of clones simultaneously simply by touching a nitrocellulose filter to the surface of a Petri dish containing the plaques.

Once we have transferred the proteins from each plaque to nitrocellulose, we probe with our antiserum. Next, we probe for antibody bound to protein from a particular plaque, using, for example, radiolabeled protein A from *Staphylococcus aureus*. This protein binds tightly to antibody and makes the corresponding spot on the nitrocellulose radioactive. We detect this radioactivity by autoradiography, then go back to our master plate and pick the corresponding plaque. Note that we are detecting a fusion protein, not the cloned protein by itself. Furthermore, it does not matter if we have cloned a whole cDNA or not. Our antiserum is a mixture of antibodies that will react with several different parts of our protein, so even a partial gene will do, as long as its coding region is cloned in the same orientation and reading frame as the leading β-galactosidase coding region.

One disadvantage of the method outlined here for purifying a fusion protein is its reliance on an antibody to bind to the protein in the affinity chromatography step. The problem is that the antibody and fusion protein bind together so tightly that it usually takes harsh conditions to separate them. Typically, this means reducing the pH to about 2.5, which does indeed tear the antibody-antigen complex apart, but it is also very hard on some proteins. Thus, in the process of purifying your protein, you risk destroying it.

To circumvent this problem, molecular biologists have designed some other very useful expression vectors. One of these is called pGEX-2T, whose map is presented in Figure 17. This vector produces fusion proteins that have the glutathione-S-transferase (GST) at their amino-terminals. The beauty of this is that GST has a high affinity for its substrate, glutathione, so the fusion protein will bind to a glutathione-affinity column. Then, instead of having to use stringent conditions to remove the fusion protein from the column, we simply strip it off with glutathione—a very mild procedure.

Another disadvantage of producing a fusion protein is the frequently unwelcome additon of extra amino acids to the protein we are really interested in. This is indeed a potential problem with vectors like pUR278, but not with more modern vectors like pGEX-2T. In the latter case, the tag GST at the beginning of the fusion protein can be easily removed because the designers of this vector put the coding region for a thrombin-sensitive site right after the GST coding region. This means that the protease thrombin will cut the fusion protein immediately after the GST part,

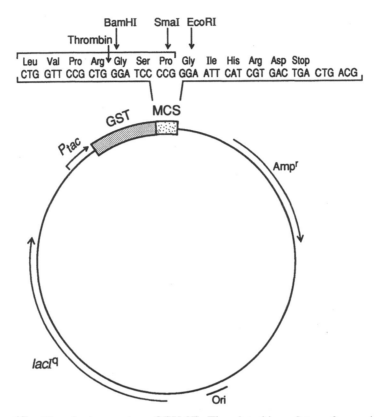

Figure 17 The cloning vector pGEX-2T. The glutathione-S-transferase (GST) coding region is shown at 11:00 o'clock on the circular plasmid, under the control of the P_{tac} promoter, which is represented by the short arrow. The multiple cloning site is shown at top in brackets. This sequence also includes the coding region for the thrombin-sensitive site, which lies between the Arg and Gly in the sequence Leu-Val-Pro-Arg-Gly-Ser. Three stop codons, in all three reading frames, follow the multiple cloning site. This ensures that translation termination will occur at the end of the fusion protein, no matter which cloning site is used. Other unique restriction sites are given around the periphery of the circle.

releasing the protein we really want. This protein can then be purified from GST by one more pass through the glutathione column. This time it will not stick, but GST will. Thrombin is an excellent choice for the enzyme to do the clipping because it cuts only after a specific sequence of six amino acids. The chance of finding this sequence in any given protein is extremely remote. Another pGEX vector, pGEX-2X, uses a site for another protease, blood clotting factor Xa.

If you do not like the idea of a lot of extra baggage on the amino end of your fusion protein, even temporarily, and you would like a larger selection of cloning sites than the three pGEX vectors offer, you may want to consider the vector pFLAG (Figure 18). Like pGEX, this vector uses the *tac* promoter, a combination of two promoters (*trp* and *lac*) that is quite powerful, and inducible. Instead of GST, this vector places a short, eight–amino acid peptide called the flag peptide on the amino end of the fusion protein. You can then purify the fusion protein using the affinity chromatography with an antibody directed against the flag peptide. However, this antibody binds to the flag peptide only in the presence of calcium,

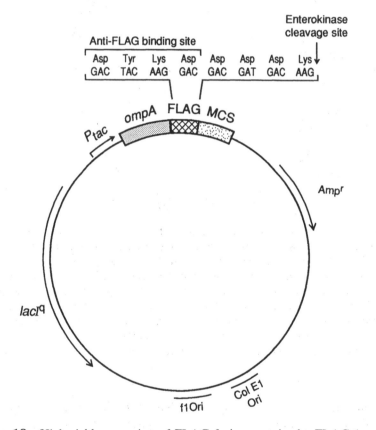

Figure 18 High-yield expression of FLAG fusion proteins btpFLAG-1 vector. The pFLAG-1 expression vector is used for the fusion of the FLAG peptide to the N-terminal of cloned proteins that are expressed from the strong *tac* promoter and secreted from the cytoplasm via the ompA signal peptide.

so the fusion protein can be removed from the antibody relatively easily, just by removing calcium. Finally, you can release your protein of interest from the flag peptide with the protease enterokinase. This enzyme recognizes a five–amino acid sequence that has been added to the end of the flag peptide. As with thrombin, the chances of this sequence appearing in any given protein are small, so your protein of interest would probably be safe.

Eukaryotic Expression Systems. Eukaryotic genes are not really "at home" in prokaryotic cells, even when they are expressed under the control of their prokaryotic vectors. One reason is that *E. coli* cells frequently recognize the protein products of cloned eukaryotic genes as outsiders and destroy them. Another is that prokaryotes do not carry out the same kinds of postranslational modifications as do eukaryotes. For example, a protein that would ordinarily be coupled to sugars in a eukaryotic cell will be expressed as a bare protein when cloned in bacteria. This can affect a protein's activity or stability, or at least its response to antibodies. A more serious problem is that the interior of a bacterial cell is not as conducive to proper protein folding as the interior of a eukaryotic cell. Frequently, the result is improperly folded, inactive products of cloned genes. Perhaps related to this phenomenon is the fact that we can sometimes express a cloned gene at a stupendously high level in bacteria, but the product forms highly insoluble, inactive granules that are of no use unless we can somehow get the protein to dissolve and regain its activity. Finally, it is obviously hopeless to try to express eukaryotic genes with introns in prokaryotes since only eukaryotes have the machinery to splice these introns out.

In order to avoid the incompatibility between a cloned gene and its host, we can express our gene in a eukaryotic cell. In such cases, we usually do the initial cloning in *E. coli.* using a shuttle vector that can replicate in both bacterial and eukaryotic cells. We then transfer the recombinant DNA to the eukaryote of choice by transformation. The traditional eukaryote for this purpose is yeast. It shares the advantages of rapid growth and ease of culture with bacteria, yet it is a eukaryote, and thus it carries out the protein folding and glycosylation (adding sugars) expected of a eukaryote. In addition, by splicing our cloned gene to the coding region for a yeast export signal peptide, we can usually ensure that the gene product will be secreted to the growth medium. This is a great advantage in purifying the protein. We simply remove yeast cells in a centrifuge, leaving relatively pure secreted gene product behind in the medium. The yeast vectors are based on a plasmid, called the 2-micron plasmid, that normally inhabits yeast cells. It provides the origin of replication needed by any vector that must replicate in yeast. Yeast–bacterial shuttle vectors also contain the pBR322 origin of replication, so they can replicate in E. coli. In addition, of course, a yeast expression vector must contain a strong yeast promoter.

Another eukaryotic vector that has been remarkably successful is derived from a baculovirus, the nuclear polyhedrosis virus (NPV) that infects the caterpillar known as the alfalfa looper. Viruses in this class have a rather large circular DNA genome, approximately 130 kb in length. The major viral structural protein, polyhedrin, is made in copious quantities in infected cells. In fact, it has been estimated that when a caterpillar dies of NPV infection, up to 10% of the dry mass of the dead insect is this one protein. This indicates that the polyhedrin gene must be very active, and indeed it is, in part owing to its powerful promoter.

Max Summers and Lois Miller and their colleagues, working separately, first developed successful vectors using the polyhedrin promoter in 1983. Since then, many other NPV vectors have been constructed using this and other viral promoters. At their best, these vectors can produce up to 0.5 g/L of protein from a cloned gene—a large amount indeed. Figure 19 shows how a typical baculovirus expression system works.

First, we clone the gene we want to express in one of the vectors. In this example, let us assume that we are using a vector with the polyhedrin promoter. Most such vectors have a unique *Bam*HI site directly after the promoter, so we can cut with *Bam*HI and use DNA ligase to insert a fragment with *Bam*HI-compatible ends into the vector and thus under control of this promoter. Next, we mix the recombinant plasmid (vector plus insert) with wild-type viral DNA and transfect insect cells with this mixture. Because the vector has extensive homology with the regions flanking the polyhedrin gene, recombination can occur within the transfected cells. This transfers our gene into the viral DNA, still under the control of the polyhedrin promoter.

Since the recombination frequency is low, often less than 1%, most progeny viruses from this cotransfection will be wild type. This means that we must perform a plaque assay and screen a number of plaques to find a few recombinants, but the recombinants are easy to distinguish because they lack polyhedrin, and therefore cannot make the polyhedral inclusion bodies that are characteristic of the wild-type virus. We can readily detect these refractile polyhedral bodies in the nuclei of infected cells with a light microscope. Some experienced investigators can even see the difference between an recombinant and a wild-type plaque with the naked eye. We go through at least three cycles of plaque purification to be sure our virus is uncontaminated with wild-type virus. Now we can use this recombinant virus to infect cells and then harvest the protein we want after these cells enter the very late phase of infection, during which the polyhedrin promoter is most active.

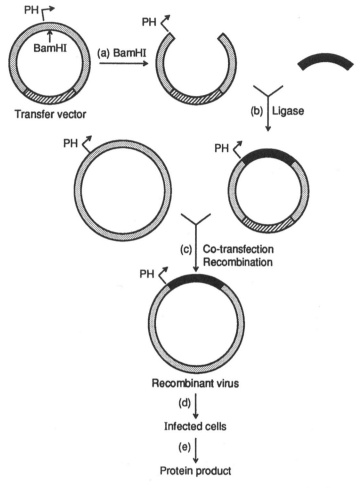

Figure 19 Forming a recombinant virus for expression of a foreign gene in a baculovirus. (a) First, we digest the transfer vector with BamHI, which cuts just after the polyhedrin promoter. (b) Next, we ligate an insert (black), containing the gene we want to express, into this BamHI site. (c) Next, we co-transfect insect cells with the recombinant vector and wild-type baculovirus DNA. These two DNAs are not drawn to scale. The viral DNA is actually almost 15 times the size of the vector. The regions of homology (gray, stippled) between the vector and the viral DNA promote homologous recombination within the co-transfected cells. This places the gene we want to express into the viral DNA, and under the control of the powerful polyhedrin promoter. After we plaque-purify the recombinant virus from the great majority of wild-type viruses resulting from the co-transfection, we can use it to infect insect cells (d) and harvest the protein product from these infected cells (e).

E. Manipulating Cloned Genes

Besides simply obtaining the products of cloned genes, we can put them to many uses. For one thing, we do not have to be satisfied with the natural product of a cloned gene; once the gene is cloned, we can change it any way we want and collect the correspondingly changed gene product.

Protein Engineering with Cloned Genes

Traditionally, protein biochemists have relied on chemical methods to alter certain amino acids in the proteins they study; they can then observe the effects of these changes on protein activities. But chemicals are rather crude tools for manipulating proteins; it is difficult to be sure that only one amino acid, or even one kind of amino acid, has been altered. Cloned genes make this sort of investigation much more precise, allowing us to perform microsurgery on a protein. By replacing specific bases in a gene, we also replace amino acids at selected spots in the protein product and observe the effects of those changes on the protein's function.

How do we perform such site-directed mutagenesis? First, we need a cloned gene whose base sequence is known. Then, we need to obtain our gene in single-stranded form. This can be done by cloning it into M13 phage and collecting the single-stranded progeny phage DNA (plus strands) as described above, or by an analogous procedure in a phagemid.

Our next task is to change a single codon in this gene. Let us suppose the gene contains the sequence of bases given in Figure 20, which codes for a sequence of amino acids that includes a tyrosine. The amino acid tyrosine contains a phenolic group. To investigate the importance of this phenolic group, we can change the tyrosine codon to a phenylalanine codon. If the tyrosine phenolic group is important to a protein's activity, replacing it with phenylalanine's phenyl group should diminish that activity.

Fritz Eckstein and colleagues developed a convenient method for site-directed mutagenesis, which is described in Figure 22. (Other methods are shown in Figures 21 and 23.) We want to change the DNA codon TAC (Tyr) to TTC (Phe) in the protein described above. The simplest way is to use an automated DNA synthesizer to make an oligonucleotide (a "21-mer") with the following sequence:

3'-AGTCTGCCAAAGCATGTATAG-5'

This has the same sequence as a piece of the original minus strand except that the central triplet has been changed from ATG to AAG. Thus, this oligonucleotide will hybridize to the plus strand we harvested from the M13 phage, except for the one base we changed, which will cause A-A mismatch. We have used a 21-mer here for convenience. It would probably

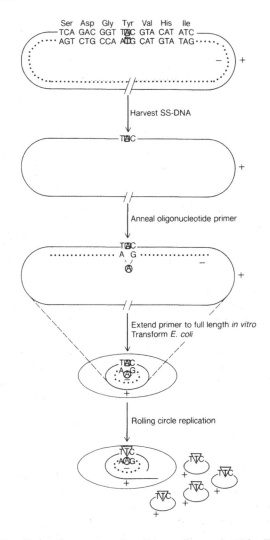

Figure 20 Site-directed mutagenesis using an oligonucleotide. We begin with the gene we want to mutagenize, cloned into double-stranded phage M13 DNA. Our goal is to change the tyrosine codon TAC (□) to the phenylalanine codon TTC. We transform *E. coli* cells with the recombinant phage DNA and harvest single-stranded (plus strand) progeny phage DNA. To this plus strand, we anneal a synthetic primer that is complementary to the region of interest, except for a single base change, an A (○) for a T. This creates an A–A mismatch ($\frac{\square}{\bigcirc}$). We extend the primer to form a full-length minus strand in vitro and use this double-stranded DNA to transform *E. coli* cells. The minus strand with the single base change then serves as the template for making many copies of the plus strand, each bearing a TTC phenylalanine codon (▽) in place of the original TAC tyrosine codon.

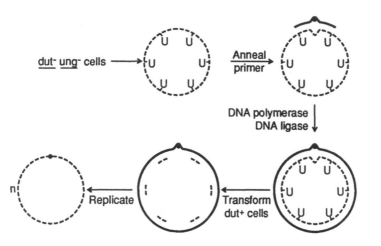

Figure 21 Site-directed mutagenesis using uridine-substituted DNA. The template single-stranded DNA comes from *dut⁻ ung⁻* cells that are deficient in both dUTPase and uracil *N*-glycosidase. Such cells do not degrade dUTP, so uridine (U) is incorporated into the DNA-positive strand (dotted) in place of thymidine. Once incorporated, there is no glycosidase to remove the uridine residues, so they persist. After primer annealing and extension, the double-stranded product DNA with the mismatch caused by the mutant primer is introduced into *dut⁺* cells by transformation. These cells contain the uracil *N*-glycosidase, so the wild-type parental strand is destroyed before it can be replicated. This leaves only the mutant negative strand (solid) to serve as template for the production of many copies of the mutant positive strand.

work, but is a bit short for good hybridization. To play safe, we could use a 31-mer. This allows 15 base-pairs on either side of the mismatch.

The oligonucleotide in this hybrid can then serve as a primer for DNA polymerase to complete the negative strand in vitro. We include a sulfur-substituted nucleotide such as dCTP α-S, as shown in Figure 21*a*. After producing the double-stranded DNA, we filter the DNA through nitrocellulose to remove any remaining uncopied single-stranded wild-type DNA. The double-stranded DNA passes through the filter. DNA that contains phosphorothionates is not cleaved by certain restriction enzymes. Thus, we can use one of these ezymes (NciI, for example) to cleave our double-stranded DNA. It will selectively attack the wild-type, parental strand, which has no thionucleotides, but leave the mutated strand alone. This produces a double-stranded DNA with nicks in the wild-type strand and an intact mutant strand. Next, we treat the DNA duplex with exonuctease III, which starts at the nicks and degrades the wild-type strand. Since the

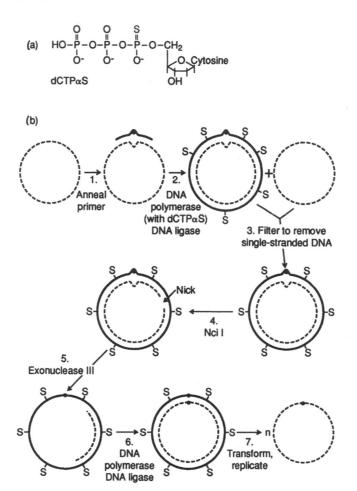

Figure 22 Site-directed mutagenesis using sulfur-substituted DNA. (a) The structure of dCTPαS. Note the substitution of sulfur for one of the oxygen atoms in the α-phosphate. (b) Outline of the method. In step 1, we anneal the mutant primer to wild-type single-stranded DNA. In step 2, we complete the negative strand (solid) in the presence of dCTPαS, which introduces sulfur-containing cytosine nucleotides into this DNA. In step 3, we filter the product through nitrocellulose to remove any uncopied wild-type positive strands (dotted). In step 4, we use the restriction enzyme NciI to nick the positive strand. The negative strand, because of the sulfur-substituted residues, is not nicked. In step 5, we use exonuclease III to degrade the nicked strand past the mismatch. In step 6, we re-synthesize the positive strand with DNA polymerase and DNA ligase, this time, the mutant strand is copied, so both strands now have the mutation. Finally, in step 7, we transform cells with this mutated double-stranded DNA. The DNA replicates to produce many copies of the mutant positive strand.

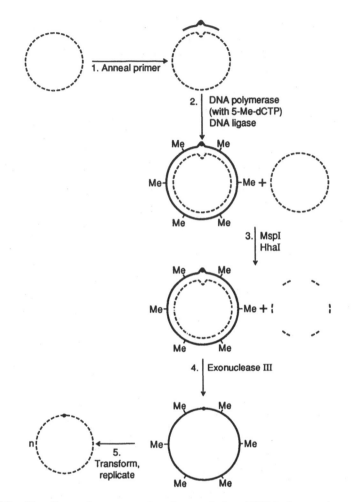

Figure 23 Site-directed mutagenesis using methylated DNA. In step 1, we anneal the mutant primer to form a mismatch. In step 2, we complete the negative strand (solid) in the presence of 5-Me-dCTP, which introduces methyl-C residues into this strand. In step 3, we cleave the DNA with MspI and HhaI, which destroy any uncopied positive strands (dotted), and nick the unmethylated positive strand (dotted) in the duplex DNA. The mutant negative strand, because it is methylated, escapes nicking by the restriction enzymes. In step 4, we degrade the nicked positive strand with exonuclease III, leaving only the mutant negative strand. In step 5, we transform cells with the mutant negative strand, which replicates to yield many copies of mutant positive strand.

mutant strand has no nicks, it escapes degradation. Next, we use this mutant single strand as a template for the synthesis of a mutant complementary strand in vitro.

In the final step of the process, we can collect the mutant RF molecules produced by these cells, cut out the mutant gene with restriction enzymes, and clone it into an appropriate expression vector. The protein product will be identical to the wild-type product except for the single change of tyrosine to phenylalanine.

Note the precision of this technique compared to traditional chemical mutagenesis. With chemical mutagenesis, we assault organisms with an often fatal dose of mutagen, then examine progeny organisms for the desired mutant characteristics. Before DNA sequencing was possible, we could not be certain of the molecular nature of such a mutation without laborious protein sequencing. Even with modern DNA sequencing techniques, we would have to determine the sequence of an entire mutant gene before we could be sure what specific damage a chemical had caused.

By contrast, site-directed mutagenesis enables us to decide in advance what parts of a protein we want to change and to tailor our mutants accordingly. Nevertheless, in spite of its lack of precision, traditional mutagenesis still plays an important role; one such experiment can quickly and easily create a rich variety of mutants, some of which involve changes we might never think of—or never get around to making by site-directed mutagenesis. The result is a double-stranded DNA with mutations in *both* strands, which we can use to transform cells. The percentage of mutant single-stranded DNA product should be quite high.

These then are some of the fundamental tools needed for genetic engineering of toxins. Additional techniques are described in later chapters, including the polymerase chain reaction in the cloning strategy chapter (Chapter 3) and recombinant protein purification in the chimera chapters (Chapters 9 and 24). Detailed methodology for electrophoresis, plasmid DNA purification, and DNA fragment purification blotting and hybridization can be found in the many molecular biology laboratory manuals available or in the original literature. We hope this brief introduction will provide some insight into the rationale behind the toxin gene cloning, expression, and modification strategies used in the rest of the book.

2
Cloning Strategies

Richard Intres* and John W. Crabb *Protein Chemistry Facility, W. Alton Jones Cell Science Center, Inc., Lake Placid, New York*

I. INTRODUCTION

The synergism of protein chemistry and molecular biology is indisputable. Investigators with little or no training in protein chemistry can now use gel electrophoresis followed by electroblotting onto solid supports such as polyvinylidene difluoride (PVDF) membranes to prepare a protein sample for sequence analysis in a biomolecular resource facility. Based upon the resulting sequence analysis, a corresponding peptide and/or oligonucleotide can be synthesized. The synthetic structure can be used for structure-function studies and as an antigen to elicit specific antibodies. The oligonucleotide probe can be used to screen cDNA libraries or to confirm clones isolated with antibody probes. For those genes characterized without first having isolated the corresponding protein, recombinant expression or synthesis of peptides deduced from the sequence allows antibodies to be made for affinity purification of the natural gene product, providing another route to evaluation of biological function as well as posttranslational modifications and native three-dimensional structure.

Molecular cloning strategies depend largely upon what is already known about the protein or gene of interest, the size and abundance of the mRNA, and the availability of specific screening tools. Strategies beyond the scope of this brief overview include differential and subtractive cloning methods, functional and ligand screening assays, and polymerase chain reaction (PCR) techniques for target genes where no sequence information is known. Most of the toxins described in this volume have been cloned using classic strategies based upon the availability of partial amino acid sequence for oligonucleotide probe design, specific antibodies to the protein, or both. These classic strategies utilizing both protein chemistry and molecular bi-

*Present affiliation: Berkshire Medical Center, Pittsfield, Massachusetts.

ology are still the most efficient and certainly the best long-term approach for understanding the structure and function of the proteins. This overview emphasizes the complementarity of these biotechnologies using as examples toxin studies and some of our work. In addition, straightforward PCR strategies are highlighted that can enhance progress in cooperative structural studies.

II. COMPLEMENTARITY OF PROTEIN/DNA BIOTECHNOLOGIES

We have been involved in a variety of collaborative cloning projects, including, for example, the *E. coli* xylose isomerase gene (1), phage T4 glucosyltransferase (2), rabbit cDNAs encoding the α- (3) and β- (4) subunits of skeletal muscle phosphorylase kinase, the rat seminal vesicle secretion II gene (5), and rabbit and human cDNAs encoding serum paraoxonase (6). Results of the SVS II study (5) clearly illustrate the complementarity of protein chemistry and molecular biology. Isolation and direct characterization of the SVS II protein allowed: (1) the identity of the unproven cDNA probe and the isolated genomic clone to be confirmed; (2) the structure of the mature N^α-pyroglutamyl–blocked protein to be largely determined; (3) definition of intron-exon borders in the gene; and (4) corroboration of the DNA-deduced sequence. Characterization of the SVS II gene (1) revealed a leader peptide associated with the nascent polypeptide; (2) allowed deduction of the difficult to obtain overlapping sequences in repeating regions of protein structure; (3) corroborated the COOH–terminal protein characterization; and (5) revealed the location of conserved elements potentially involved in gene regulation.

Our work with the visual tissue-specific protein, cellular retinaldehyde-binding protein (CRALBP), has included determining the complete protein sequence directly (7), cloning the bovine and human cDNAs (8), and cloning the human gene encoding CRALBP (9). Bovine CRALBP cDNA was cloned from a retinal cDNA expression library using polyclonal and monoclonal anti-CRALBP antibodies raised against purified protein. The monoclonal antibody was directed in part against an NH_2-terminal assembly epitope (10) and was key to isolating a full-length bovine cDNA, which was used in turn to clone the human CRALBP cDNA. Using multiple cDNA probes, the human CRALBP gene was cloned in part from a human leukocyte genomic DNA library and a human chromosome 15-specific library (11,12). The polymerase chain reaction was then used to amplify two missing regions and connect the genomic fragments. Definition of the eight exons within the CRALBP gene has complemented current efforts to identify functional domains within the tertiary structure of the protein (10).

The toxin studies presented in this volume have benefited from comparatively advanced knowledge of protein primary structure for purposes of both probe design and for definitive identification of the clones isolated. For example, the fungal ribotoxin gene from *Aspergillus restrictus* encoding putative alleles for restrictocin and mitogillin were isolated by Davies and coworkers (Chapter 13) using degenerate oligonucleotides designed from amino acid sequence. On the other hand, α-sarcin was cloned from an *Aspergillus giganteus* expression library using specific antibodies. The deduced α-sarcin sequence confirmed the previously determined protein sequence and identified a 27-residue leader peptide on the nascent polypeptide (13).

As outlined by Lord and coauthors (Chapter 5) ricin cDNA and genomic clones have been isolated from the castor bean by a variety of strategies that relied largely upon oligonucleotide probes designed from ricin A and B chain protein sequence. Direct analyses of this prototype plant toxin have revealed the mature protein structure and provided sequence information critical for cloning, whereas DNA studies have demonstrated that the A and B chains are synthesized together as a single polypeptide containing both leader and linker sequences. Continued joint protein/ molecular biological analyses are needed for identifying the functional domains, critical residues, and posttranslational modifications within the ricin multigene family.

Cloning and expression studies in progress concerning *Luffa* ribosome inhibitory proteins are described by Li and Ramakrishnan in Chapter 12. A genomic expression library was constructed and screened with antibodies to the purfied protein based upon the assumption that the *Luffa* toxin genes would lack introns like those encoding many of the other plant ribosome inhibitor proteins.

Another plant toxin, saporin-6, was cloned, as described by Soria et al. (Chapter 10), using a combination of long and short synthetic oligonucleotides derived from direct analysis of purified protein. Analysis of clones demonstrated a signal peptide not present in the mature protein. Direct analyses of saporin-6 peptide fragments as well as carboxypeptidase data and pulse-labeled immunoprecipitates from developing seeds suggest posttranslational COOH-terminal proteolytic processing.

Plant toxin α-momorcharin cDNA was isolated by Shaw and colleagues (Chapter 14) by screening an expression library with antibodies raised against the succinylated protein. A portion of the α-momorcharin cDNA was in turn used to clone the related plant toxin trichosanthin. Identification of the trichosanthin clones and confirmation of the deduced sequence was supported with the directly determined complete protein sequence. In the case of trichosanthin, posttranslational COOH–terminal proteolytic processing has been demonstrated (14a). *Pseudomonas* exotoxin A was first cloned with protein-based oligonucleotides and the original DNA sequence used, as described by Draper and coauthors (Chapter 21), to clone the gene from several other strains of *Pseudomonas* and extend structural knowledge of the toxin. A classic genetic analysis of the identification and cloning of the diphtheria toxin gene is presented by Greenfield (Chapter 15). Diphtheria toxin cloning efforts were also supported by earlier direct protein sequence reports that have now been corrected and significantly extended by characterization of the gene isolated from multiple strains.

III. PCR STRATEGIES

The polymerase chain reaction (PCR) technique, developed at Cetus Corporation (14b,c), is a simple method for amplifying nucleic acids for cloning or sequencing (Figure 1). A double-stranded DNA sample (the template) is *denatured* by incubation at high temperature. The two strands, now disassociated, remain free in solution until the temperature is lowered sufficiently to allow annealing. Synthetic oligonucleotide primers are *annealed* to specific sites flanking the region to be amplified. Because the primers are in large excess over the DNA templates, formation of a primer template complex will be favored over reassociation of the DNA strands on lowering the temperature. Thermostable *Taq* DNA polymerase is then used to extend the DNA $5' \rightarrow 3'$ from the primer. After a number of cycles, the vast majority of amplified product consist of the region between and including the two primers.

Detailed and specialized PCR protocols have been published and will not be reviewed here (15,16). Rather we would like to emphasize the advantages of PCR technology to both cloning and protein microstructural studies, including the applicability of the methodology to the earliest stages of the project and the resulting time and labor reductions it can provide. This PCR strategy depends upon first obtaining some amino acid sequence, including internal sequence. Methods for protein microcharacterization (17) and for obtaining internal sequence in particular from electroblotted samples have evolved into reliable procedures (18–21). A recent study demonstrated that it is reasonable to expect to productively sequence by Edman degradation three or more tryptic peptides having an average length of 12 residues following fragmentation and peptide purification from 70 pmol or greater amounts of PVDR-immobilized protein (21). Having NH_2-

terminal sequence is advantageous for PCR but often inaccessible by Edman degradation since 80–90% of eukaryotic proteins may be NH_2-terminally blocked (22,23). Mass spectrometric analyses can also provide rapid, high-sensitivity protein sequence determination and the methodology is ideal for identifying posttranslational modifications (24–26). It is feasible to obtain 30–70% of a protein sequence in less than a week by mass spectrometry at the picomole level (27). This powerful technology is rapidly becoming more accessible to investigators (28,29).

The PCR strategy outlined in Figure 1 is neither unique nor complicated. Double- or single-stranded DNA can be amplified by PCR. RNA can also serve as a PCR template following reverse transcription into cDNA. A new DNA polymerase from *Thermus thermophilus* (*Tth*) can reverse transcribe RNA at high temperatures, allowing cDNA synthesis and PCR amplification in a single tube reaction (30). As the first segments of peptide sequence become available, oligonucleotide primers are designed for PCR. Guidelines and recommendations for using degenerate primers in the amplification of cDNA may be found in Lee and Caskey (31). Fewer oligonucleotides need to be synthesized when both NH_2-terminal and internal sequence information is available. However, with no indication of peptide orientation, both upstream and downstream PCR should be performed with all combinations of available sequence and larger PCR products selected for subcloning. This comparative approach will also orient the peptide fragments without the need to isolate overlapping peptides and consequently enhances progress in the direct protein analysis. With several partial sequences in hand, hints regarding peptide orientation and productive PCR primer combinations may be obtained by comparing the peptide sequences with known potentially related structures. This approach has been helpful in the cloning and direct structural analysis of mistletoe lectin A chain because of homology with ribosome-inactivating proteins (J. W. Crabb and A. E. Frankel, personal communication). Once a central segment of the cDNA has been amplified and characterized, the source cDNA can be cleaved outside this region with restriction enzymes, circularized, and the remainder of the coding region and flanking sequences determined by "inverse PCR" (32). Of course, the PCR products may also be used as specific screening probes, as recently demonstrated in the cloning of α-trichosanthin by Chow et al. (14). If only one peptide sequence has been obtained and it is reasonably large (~15 residues), PCR primers can be designed from the NH_2- and COOH-terminal parts of the peptide and the resulting PCR product used as a cDNA screening probe. With a complete protein sequence available, PCR can also be used to clone homologs or isoenzymes as recently demonstrated in the isolation of multiple abrin A-chain genes directly from genomic DNA (40).

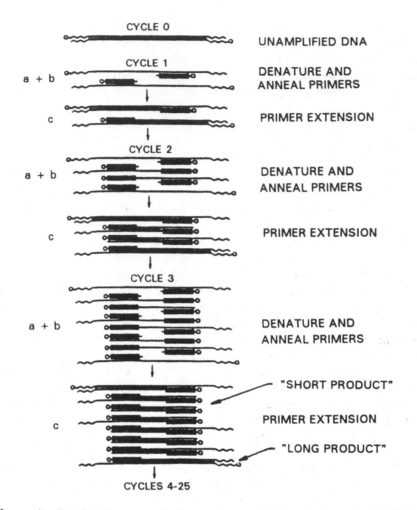

Figure 1 (Left) Polymerase chain reaction: principles of the method. (Right) PCR cloning strategies. (A) Oligonucleotide primers (→) derived from internal peptides (▭) of unknown orientation are used in multiple PCR combinations and the longest product (▭) selected, subcloned, and sequenced. (B) The cDNA template is then cleaved at convenient or synthetic restriction enzyme sites (R) outside the region determined (▭) and the fragments circularized by ligation. Initial flanking (—) and NH_2- and COOH-terminal coding sequences (▬) may now be amplified with new primers (P_i and P_{ii}) by inverse PCR. (Left panel from Ref. 41.)

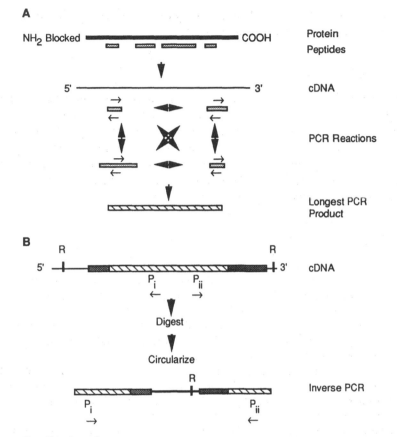

Figure 1 Continued

Finally, a note of caution is warranted in the use of individual PCR clones as the sole source of cDNA for characterization because of the possibility of introducing errors during the amplification reactions. The *Taq* DNA polymerase used in many PCR protocols lacks 3'–5' exonuclease activity or "proofreading" capability (33). However, with careful attention to reaction conditions, errors with *Taq* polymerase can be minimized to 10^{-4}–10^{-5} per DNA doubling (34,35). Notably, if the lowest reported error rate of 10^{-5} per DNA doubling is not achieved, significant misincorporation may result (36–38). For example, a 30-cycle PCR with 10^{-4} misincorporations per cycle would contain a 0.3% error, or 3 misincorporations per 1000 nucleotides. We have observed 0.3% misincorporation in a 998-bp CRALBP cDNA prepared by PCR that resulted in three amino acid changes and a misengineered protein.

IV. CONCLUSIONS

The future of biotechnology in the field of toxins and immunotoxins appears bright (39). Advances in protein miocrocharacterization methods and mass spectrometry coupled with ever-improving PCR technology provide a strong foundation for continued molecular cloning and protein engineering studies. These complementary strategies will continue to advance our understanding of the structure and function of toxins and help to design targeted toxins for therapeutic purposes.

REFERENCES

1. Schellenberg, G. D., Sarthy, A., Larson, A. E., Backer, M. P., Crabb, J. W., Lidstrom, M., Hall, B. D., Furlong, C. E. Xylose Isomerase from *Escherichia coli*: Characterization of the protein and the structural gene. J. Biol. Chem., *259*: 6826–6832, 1983.
2. Tomaschewski, T., Gramm, H., Crabb, J. W., and Ruger, W. T4-Induced α- and β-Glucosyltransferase: Cloning of the genes and a comparison of their products based on sequencing data. Nucleic Acids Res., *13*: 7551–7568, 1985.
3. Zander, N. F., Meyer, H. E., Hoffmann-Posorske, E., Crabb, J. W., Heilmeyer, L. M. G. Jr., and Kilmann, M. W. cDNA Cloning and complete primary structure of skeletal muscle phosphorylase kinase (α-subunit). Proc. Nat. Acad. Sci. U.S.A., *85*: 2929–2933, 1988.
4. Kilimann, M. W., Zander, N. F., Kuhn, C. C., Crabb, J. W., Meyer, H. E., and Heilmeyer, L. M. G. Jr. The α and β subunits of phosphorylase kinase are homologous: cDNA cloning and primary structure of the β subunit. Proc. Nat. Acad. Sci. U.S.A., *85*: 9381–9385, 1988.
5. Harris, S. E., Harris, M. A., Johnson, C. M., Bean, M. F., Dodd, J. G., Matusik, R. J., Carr, S. A., and Crabb, J. W. Structural characterization of the rat seminal vesicle secretion II protein and gene. J. Biol. Chem. *265*: 9896–9903, 1990.
6. Hassett, C., Richter, R. J., Humbert, R., Chapline, C., Crabb, J. W., Omiecinski, C. J., and Furlong, C. E. Characterization of cDNA clones encoding rabbit and human serum paraoxonase: The mature protein retains its signal sequence. *Biochemistry*, *30*: 10141–10149, 1991.
7. Crabb, J. W., Johnson, C. M., Carr, S. A., Armes, L. G. and Saari, J. C. The complete primary structure of the cellular retinaldehyde-binding protein from bovine retina. J. Biol. Chem., *23*: 18678–18687, 1988.
8. Crabb, J. W., Goldflam, S., Harris, S. E. and Saari, J. C. Cloning of the cDNAs encoding the cellular retinaldehyde-binding protein from bovine and

human retina and comparison of the protein structures. J. Biol. Chem., *263*: 18688–18692, 1988.

9. Intres, R., Goldflam, S., Cook, J. R., and Crabb, J. W. Cloning and sequence analysis of the human gene encoding cellular retinaldehyde-binding protein. Invest. Ophthal. Vis. Sci., *32*(Suppl.): 1011, 1991.

10. Crabb, J. W., Gaur, V. P., Garwin, G. G., Mark, S. V., Chapline, C., Johnson, C. M., and Saari, J. C. Topological and epitope mapping of the cellular retinaldehyde-binding protein from retina. J. Biol. Chem., *266*: 16674–16683, 1991.

11. Crabb, J. W., Heinzmann, C., Mohandas, T., Goldflam, S., Saari, J. C., and Sparkes, R. S. Assignment of the gene for cellular retinaldehyde-binding protein to human chromosome 15. Invest. Ophthalmol. Vis. Sci., *30*(Suppl.): 43, 1989.

12. Sparkes, R. S., Heinzmann, C., Goldflam, S., Kojis, T., Saari, J. C., Mohandes, T., Klisak, I., Bateman, J. B., and Crabb, J. W. Assignment of the gene for cellular retinaldehyde-binding protein (CRALBP) to human chromosome 15q26 and mouse chromosome 7. Genomics, *12*: 58–62, 1992.

13. Oka, T., Natori, Y., Tanaka, S., Tsurugi, K., and Endo, Y. Complete nucleotide sequence of cDNA for the cytotoxin alpha sarcin. Nucleic Acid Res., *18*: 1897, 1990.

14a. Chow, T. P., Feldman, R. A., Lovett, M., and Piatak, M. Isolation and DNA sequence of a gene encoding alpha-trichosanthin, a type I ribosome-inactivating protein. J. Biol. Chem., *265*: 8670–8674, 1990.

14b. Saiki, R. K., et al. Science, *230*: 1350–1354, 1990.

14c. Mullis, K. B. and F. A. Falloona. Methods Enzymology, *155*: 335–350, 1987.

15. Erlich, H. A. (ed.). PCR Technology, Applications for DNA Amplifications. New York: Stockton Press, 1989.

16. Innis, M. A., Gelfland, D. H., Sninsky, J. J., and White, T. J. (eds.) PCR Protocols, A Guide to Methods and Applications. San Diego: Academic Press, 1990.

17. Shively, J. E., Paxton, R. J., and Lee, T. D. Highlights of protein structural analysis. Trends Biochem. Sci., *14*: 246–252, 1989.

18. Abersold, R. H., Leavitt, J., Saaverdra, R. A., Hood, E. E., and Kent, S. B. Internal amino acid sequence analysis of proteins separated by one- or two-dimensional gel electrophoresis after in situ protease digestion on nitrocelluse. Proc. Natl. Acad. Sci. U.S.A., *84*: 6970–6974, 1987.

19. Yuen, S. W., Chiu, A. H., Wilson, K. J., and Yuan, P. M. Microanalysis of SDS-PAGE electroblotted proteins. BioTechniques, *7*: 74–82, 1989.

20. Matsudaira, P. T. (ed.). A Practical Guide to Protein and Peptide Purification for Microsequencing. San Diego: Academic Press, 1989.

21. Stone, K. L., McNulty, D. E., LoPresti, M. L., Crawford, J. M., DeAngelis, R., and Williams, K. R. Elution and internal amino acid sequencing of PVDF-blotted proteins. *In*: R. H. Angeletti, ed., Techniques in Protein Chemistry III. San Diego: Academic Press, pp. 23–34, 1992.

22. Brown, J. L., and Roberts, W. K. Evidence that approximately eighty per

cent of the soluble proteins from Ehrlich ascites cells are N$^\alpha$-acetylated. J. Biol. Chem., *251*: 1009–1019, 1976.

23. Driessen, H. P., deJong, W. W., Tesser, G. I., and Bloemendal, H. The mechanism of N-terminal acetylation of proteins. *In*: (G. O., Fasman, ed.), Critical Reviews in Biochemistry, Vol. 18, pp. 281–325. Boca Raton, FL: CRC Press, 1985.

24. Fenn, J. B., Mann, M., Meng, C. K., Wong, S. F., and Whitehouse, L. M. Electrospray ionization for mass spectrometry of large biomolecules. Science, *246*: 64–71, 1989.

25. Burlingame, A. L., and McClosky, J. A. (eds.). Biological Mass Spectrometry. New York: Elsevier Press, 1990.

26. Carr, S. A., Hemling, M. E., Bean, M. F., and Roberts, G. D. Integration of mass spectrometry in biopharmaceutical research and development. Anal. Chem., *63*: 2802–2824, 1991.

27. Hunt, D. F., Alexander, J. E., McCormick, A. L., Martino, P. A., Michel, H., Shabanowitz, J., and Sherman, N. Mass spectrometry methods for protein and peptide sequence analysis. *In*: (J. J. Villafranca, ed.), Techniques in Protein Chemistry II, San Diego: Academic Press, pp. 441–454, 1991.

28. Villafranca, J. J. (ed.). Mass spectrometry workshop. *In*: Techniques in Protein Chemistry II, pp. 419–572, San Diego: Academic Press, 1991.

29. Angeletti, R. H. (ed.). Techniques in Protein Chemistry III, San Diego: Academic Press, 1992.

30. Meyers, T. W., and Gelfand, D. H. Reverse transcription and DNA amplification by a *Thermus thermophilus* DNA polymerase. Biochemistry, *30*: 7661–7666, 1991.

31. Lee, C. C., and Caskey, C. T. cDNA cloning using degenerate primers. *In*: M. A. Innis, D. H. Gelfand, J. J. Sinsky, and T. J. White, eds., PCR Protocols, A Guide to Methods and Applications, pp. 46–53, San Diego: Academic Press, 1990.

32. Ochman, H., Medhora, M. M., Garza, D., and Hartl, D. L. Amplification of flanking sequences by inverse PCR. *In*: M. A. Innis, D. H. Gelfand, J. J. Sinsky and T. J. White, eds., PCR Protocols, A Guide to Methods and Applications, pp. 219–227, San Diego: Academic Press, 1990.

33. Tindall, K. R., and Kunkel, T. A. Fidelity of DNA synthesis by *Thermus aquaticus* DNA polymerase. Biochemistry, *27*: 6008–6013, 1988.

34. Erlich, H. A., Gelfand, D., and Sinsky, J. J. Recent advances in the polymerase chain reaction. Science, *252*: 1643–1651, 1991.

35. Gelfand, D. H., and White, T. J. Thermostable DNA polymerases. *In*: M. A. Innis, D. H. Gelfand, J. J. Sinsky and T. J. White, eds., PCR Protocols, A Guide to Methods and Applications, pp. 129–141, San Diego: Academic Press, San Diego, 1990.

36. Dunning, A. M., Talmud, P., and Humphries, S., Scharf, S. J., Higuchi, R., Horn, G. T., Errors in the polymerase chain reaction. Nucleic Acids Res., *16*: 10393, 1988.

37. Saiki, R. K., Gelfand, D. H., Stoffel, S., Mullis, K. B., and Erlic, H. A. *Science*, *239*: 487–491, 1988.
38. Karlovsky, P. Misuse of PCR. Trends in Biochem. Sci., *15*: 419, 1990.
39. Oetlmann, T. N., and Frankel, A. E. Advances in immunotoxins. FASEB J., *5*: 2334–2337, 1991.
40. Evensen, G., Mathiesen, A., and Sudan, A. J. Biol. Chem., *266*: 6848–6852, 1991.
41. Oste, C. BioTechniques, *6*: 162–167, 1988.

3
Protein Engineering Strategies

Marc Whitlow *Enzon Incorporated, Gaithersburg, Maryland*

I. INTRODUCTION

The principal objective in engineering a chimeric immunotoxin is to replace the native toxin binding specificity without effecting the membrane translocation and cytosolic release or the toxin's ability to inactivate protein synthesis. This has been accomplished for two bacterial toxins, diphtheria toxin and *Pseudomonas* exotoxin, because for each of these toxins the intoxication functions have been identified as separate elements in the native toxin gene. In *Pseudomonas* exotoxin, for which the crystal structure has been determined (1), the cell-binding domain is followed by the membrane translocation domain, which in turn is followed by the adenosine diphosphate (ADP)–ribosyltransferase domain (2). In the middle of the membrane translocation domain there is a trypsinlike proteolytic clip site bridged by a disulfide bond, which is important for the release of the cytotoxic domain. Because of the independence of the N-terminal–binding domain from the rest of the toxin, one can remove the binding domain and replace it with a new binding specificity. The domains of diphtheria toxin are in the opposite order from those of *Pseudomonas* exotoxin (see Figure 1 in Chapter 24). Thus, one replaces the C-terminal domain with a new binding specificity. Although the crystal structure of the plant toxin ricin is known, ricin present a more difficult protein-engineering problem because (1) the portions of the molecule involved in translocation are unknown; and (2) mammalian proteases cannot remove the peptide between the A and B chains of preproricin, which is required for cytosolic release.

II. BINDING AFFINITY AND SPECIFICITY

Enhancing the cell-binding specificity or the membrane-translocation activity of an immunotoxin is likely to result in an enhanced toxicity to the target cells. Since a single diphtheria toxin molecule is capable of killing a cell (3), there is no need to enhance the inhibition of protein synthesis. The higher the affinity and specificity of an immunotoxin for its target cell type, the more likely it is to be therapeutically useful. The success of the hormone immunotoxins is in part due to the high affinity of hormones for their natural receptors. Interleukin-2 (IL-2) is 50% displaced from the high-affinity IL-2 receptor (p55, p75) by 8.1×10^{-9} M chimeric immunotoxin DAB_{486}–IL-2 (see Table 2 in Chapter 18). In addition, the binding of a hormone may activate the endocytotic mechanism, an important first step in the translocation and release of a cytotoxic domain into the cytosol. A disadvantage of the hormone immunotoxins is that they often activate biological processes that one is trying to stop, such as cell division and proliferation.

III. TRANSLOCATION

The efficiency of translocation of an immunotoxin across the appropriate membrane and release into the cytosol is an important step; thus, it should be considered in the design and testing of a chimeric immunotoxin. The translocation of a number of toxins occurs via acidified endosomes. For example, diphtheria toxin has two hydrophobic helices that become exposed when the endosomes becomes acidified. There is some evidence suggesting that two or more molecules of diphtheria toxin associate, probably through the exposed helicies, and form a pore through which the toxin is translocated into the cytosol. Prior to translocation, an endosome protease clips the arginine-rich disulfide-bonded loop. After translocation, the cytotoxic domain is released by reduction of the disulfide bond. *Pseudomonas* exotoxin has a very similar mechanism of translocation to that of diphtheria toxin. A number of successful chimeric immunotoxins have been produced in which the cell-binding portion of *Pseudomonas* exotoxin or diphtheria toxin has been replaced, leaving the cytotoxic and translocation domains intact.

Plant toxins are produced in plants as preproteins. During biogenesis a peptide between the disulfide-linked A and B chains is removed. Thus, the first obstacle in the design of a plant toxin–derived chimeric immunotoxin expressed in bacteria, is the need for a cleavable linker between the cytotoxic (A chain) and binding domains. As described in Chapter 5, Lord et al. have designed two strategies to address this problem for the plant toxin ricin. First they inserted the arginine-rich disulfide loop of diphtheria toxin between the ricin A chain and a protein A–binding domain. Without the arginine-rich disulfide loop the IC_{50} for the immunotoxin was 10^{-7} M or greater, whereas with the loop it was 8×10^{-11} M, demonstrating the importance of a cleavable linker between the binding and cytotoxic domains of ricin-derived immunotoxin. In the second strategy for a chimeric ricin immunotoxin, the proteolytic clip site in the disulfide loop of preproricin has been replaced by a factor X cleavage site, and most of the B chain has been replaced with an alternative cell-binding region, such as a single-chain Fv. The factor X site can thus be cleaved by factor X prior to use.

In the design of a ricin-derived chimeric immunotoxin, one would like to include portions of the hydrophobic interface between the A and B chains of ricin to facilitate membrane translocation while eliminating the galactose-binding activity. The hydrophobic interface between the A and B chain of ricin is made up of residues Tyr-183, Leu-207, Phe-240, Ile-247, Pro-250, and Ile-251 of the A chain and residues Phe-140, Phe-218, Pro-260, and Phe-262 of the B chain (4,5). The simplest strategy would be to keep the B chain of ricin but destroy its galactose-binding activity by site-directed mutagensis. Vitetta and Yen have shown this can be accomplished by changing Asn-255 to Ala (6). Other key residues involved in galactose binding such as Asp-22, Asn-46, and Asp-234 could also be changed. Other strategies could be envisioned in which only the hydrophobic portion of the B chain was kept. This would be a difficult protein-engineering problem because the B chain residues involved in the interface between the A and B chains are in the second galactose-binding domain, a good distance from the C-terminus of the A chain.

IV. SINGLE-CHAIN Fvs

Immunotoxins have traditionally been antibodies that have been chemically crosslinked to a toxin. With the advent of single-chain Fvs (sFv), chimeric antibody-based immunotoxins can be constructed (7,8). The first chimeric sFv immunotoxin reported contains an anti-*tac* single-chain Fv, which binds to the p55 subunit of the IL-2 receptor, fused at its C-terminus to the translocation and cytotoxic domains of *Pseudomonas* exotoxin (PE40) (9). A single-chain Fv domain has also been demonstrated to be active when fused at its N-terminal to protein A (10). Thus, one should be able to construct an active single-chain diphtheria toxin–Fv immunotoxin. We have recently reviewed the design, construction, and production of single-chain Fv proteins and their fusion proteins (11), which is briefly summarized below.

Single-chain Fv proteins can and have been constructed in either one of two ways. Either V_L is the N-terminal domain followed by the linker and V_H (a V_L–linker–V_H construction) or V_H is the N-terminal domain followed by the linker and V_L (V_H-linker-V_L construction).* Both types of sFv proteins have been successfully constructed and purified, and both have shown binding affinities and specificities similar to the antibodies from which they were derived. All of the V_H-linker-V_L sFv proteins reported to date have used a single linker design by Huston et al. (8). The (Gly-Gly-Gly-Gly-Ser)$_3$ linker was designed to bridge the 3.5-nm gap between the C-terminus of V_H and the N-terminus of V_L, without exhibiting any propensity for ordered secondary structure.

We have designed a number of sFv linkers using the Fab crystal structures of MCPC603 and 4-4-20 (13,14). These sFv linkers were constructed and tested in antifluorescein 4-4-20 sFv proteins. A series of "linear" linkers, having an extended or linear structure, has been designed. The linear linkers were designed to span the 3.7-nm distance between the C-terminus of the V_L domain and the N-terminus of the V_H domain with a minimum length of 12 amino acid residues. Single-chain Fv proteins have been designed and constructed with linear linker lengths of 12, 14, 16, and 18 residues; are designated 202', 212, 214, and 216, respectively (Table 1). These linkers have been designed to be flexible, having an underlying sequence of alternating Gly and Ser residues. To enhance the solubility of these linkers and their associated sFv proteins, three or four charged residues have been introduced into the linker—two positively charged residues (Lys) and one or two negatively charged residues (Glu). One of the Lys residues has been placed close to the N-terminus of the V_H domain to replace the positive charge lost upon forming the peptide bond between the linker and the V_H domain. The 205 linker design was based on four repeats of the helical peptide sequence Asp-Asp-Ala-Lys-Lys found in protein G, with flexible ends to facilitate joining of the linker to the variable domains (see Table 1). Marqusee and Baldwin have shown that sequences which contain a negatively charged residue, such as Glu, followed three or four residues later by a positively charged residue, such as Lys, are helix stabilizing (15).

We have tested the helical linker (205) and three of the linear linkers (202', 212, and 216) in 4-4-20 sFv proteins. The binding affinities at room temperature of the antifluorescein sFv proteins are 0.5×10^9 M^{-1}, 1.1×10^9 M^{-1}, 1.3×10^9 M^{-1}, and 1.2×10^9 M^{-1} for the three linear linkers 4-4-20/202', 4-4-20/212, and 4-4-20/216, and for the helical linker 4-4-20/205, respectively (16). Under similar conditions, the 4-4-20 Fab-binding

---

*The variable light chain domain (V_L) extends from residue 1 to residue 107 for the lambda light chain and to residue 108 for kappa light chains, and the variable domain of the heavy chain (V_H) extends from residue 1 to residue 113 (12).

Table 1 Linker Designs

V_L	Linker sequence	V_H	Name	Reference
–KLEIK[a]		EVQLV–[a]		
–KLEI.	EGKSSGSGSESKSTQ	. .KLD–	202'	6
–KLEIK	SSADDAKKDDAKKDDAKKDDAKKDG	DVKLD–	205	12
–KLEIK	GSTSGSGKSSEGKG	EVKLD–	212	12
–KLEIK	GSTSGSGDSSEGKG	EVKLD–	213	
–KLEIK	GSTSGSGKSSEGSGSTKG	EVKLD–	216	

[a]Consensus sequences, residues L103–L107 and H1–H5.

affinity is $1.7 \times 10^9 \, M^{-1}$. The stability of these sFv proteins was determined in the denaturants urea and guanidine HCl. The helical linker, 205, was shown to be more stable than any of the linear linkers, and the longer linear linker, 216, was shown to be more stable than the shorter linear linkers, 202' and 212. Estimates of the free energy of unfolding in urea for the 4-4-20/202', 4-4-20/212, and 4-4-20/205 sFv proteins are 3.8, 5.1, and 5.4 kcal/mol, respectively (16).

A third property that is important in engineering an sFv protein is proteolytic stability. We began to examine this issue when we found that our 212 linker had been clipped in a crystallization experiment which we were unable to reproduce with intact 4-4-20/212 sFv. We were able to reproduce the crystallization by proteolytically treating the sFv with subtilisin BPN'. The proteolytic clip in the 212 linker occurred between Lys-8 and Ser-9 of the linker (see Table 1). In our first attempt to correct this problem, we replaced Lys-8 with an Asp, and designated this linker as 213. After treatment with subtilisin BPN', in which the 4-4-20/212 sFv was completely converted to Fv, the 213 linker showed about a 50% conversion to Fv. Moreover, a different Fv was formed, indicating that the proteolytic site after residue 8 in the 213 linker no longer remained. The replacement of Lys-8 in the 212 linker with Asp resulted in a twofold drop in the binding affinity of the 4-4-20/213 sFv, but did not effect its stability in denaturants. This demonstrates the long-distance effects that electrostatics can have on the binding of a charged ligand, such as fluorescein.

Only one comparison of the various sFv designs and their performance as immunotoxins has been published. Batra et al. compared four different sFv immunotoxin constructions (17). All four had the truncated *Pseudomonas* exotoxin (PE40) as the C-terminal domain and an N-terminal anti-Tac sFv domain. They compared three V_L–linker–V_H–PE40 immunotoxins and a V_H–linker–V_L–PE40 immunotoxin for their ability to inhibit [3H]leucine incorporation in a variety of cell lines. All of the sFv immunotoxins had about the same activity. These experiments would suggest that the choice of linker has little effect on the in vivo performance of an sFv immunotoxin.

The consideration of solubility, stability, and proteolytic susceptibility which we have examined in designing linkers between the variable domains of a single-chain Fv proteins, may also be applied to the design of linkers between other domains in an immunotoxin. In the engineering of the DAB_{389}–IL-2 immunotoxin, amino acids 2–8 of IL-2 were duplicated, effectively increasing the linker length between the diphtheria toxin and IL-2 domains (see Chapter 18). The resulting DAB_{389}–$(1–10)_2$–IL-2 immunotoxin has a 10-fold lower IC_{50} than that of DAB_{389}–IL-2, 5×10^{-12} M compared to $2–5 \times 10^{-11}$ M, respectively. This result and the results of our studies on linker length suggest that it is important to have a sufficiently long linker between functional domains. The longer linker lengths probably allow a domain to function with less interference from the other domains in a fusion protein.

V. PRODUCTION OF CHIMERIC IMMUNOTOXINS

The principal difficulty in the production of chimeric immunotoxins is obtaining properly folded proteins. All of the chimeric immunotoxins discussed in the subsequent chapters are produced in *Escherichia coli* bacterial expression systems. One would expect an increasing level of difficulty in producing chimeric immunotoxins in *E. coli* as one moves from bacterial proteins, such as diphtheria toxin and *Pseudomonas* exotoxin, to plant proteins, such as ricin and abrin, and mammalian proteins such as growth factors, cytokines, and antibodies. The probability of producing a fully functional protein will decrease as the number of domains in a fusion protein increases because if one domain in a given molecule is incorrectly folded, and thus inactive, the entire molecule is of little use. Factors such as individual domain stability and the length of the polypeptide linker between two functional domains may also effect production yield.

REFERENCES

1. Allured, V. S., Collier, R. J., Carroll, S. F., and McKay, D. B. Structure of exotoxin A of *Pseudomonas aeruginosa* at 3.0-Ångstrom resolution. Proc. Natl. Acad. Sci. U.S.A., *83*: 1320–1324, 1986.
2. Hwang, J., Fitzgerald, D. J., Adhya, S., and Pastan, I. Functional domains of Pseudomonas exotoxin identified by deletion analysis of the gene expressed in *E. coli*. Cell, *48*: 129–136, 1987.
3. Yamaizumi, M., Mekada, M., Uchida, T., and Okada, Y. One molecule of diphtheria toxin fragment A introduced into a cell can kill the cell. Cell, *15*: 245–250, 1978.

4. Rutenber, E., Katzin, B. J., Ernst, S., Collins, E. J., Mlsna, D., Ready, M. P., and Robertus, J. D. Crystallographic refinement of ricin to 2.5Å. Proteins, *10*: 240–250, 1991.

5. Rutenber, E., and Robertus, J. D. Structure of ricin B-chain at 2.5Å resolution. Proteins, *10*: 260–269, 1991.

6. Vitetta, E. S., and Yen, N. Expression and functional properties of genetically engineered ricin B chain lacking galactose binding activity. Biochim. Biophys. Acta, *1049*: 151–157, 1990.

7. Bird, R. E., Hardman, K. D., Jacobson, J. W., Johnson, S., Kaufman, B. M., Lee, S.-M., Lee, T., Pope, S. H., Riordan, G. S., and Whitlow, M. Single-chain antigen-binding proteins. Science, *242*: 423–426, 1988.

8. Huston, J. S., Levinson, D., Mudgett-Hunter, M., Tai, M.-S., Novotny, J., Margolies, M. N., Ridge, R. J., Bruccoleri, R. E., Haber, E., Crea, R., and Oppermann, H. Protein engineering of antibody binding sites: Recovery of specific activity in an anti-digoxin single-chain Fv analogue produced in *Escherichia coli*. Proc. Natl. Acad. Sci. U.S.A., *85*: 5879–5883, 1988.

9. Chaudhary, V. K., Queen, C., Junghans, R. P., Waldmann, T. A., FitzGerald, D. J., and Pastan, I. A recombinant immunotoxin consisting of two antibody variable domains fused to Pseudomonas exotoxin. Nature, *339*: 394–397, 1989.

10. Tai, M.-S., Mudgett-Hunter, M., Levinson, D., Wu, G.-M., Haber, E., Oppermann, H., and Huston, J. S. A bifunctional fusion protein containing Fc-binding fragment B of Staphylococcal protein A amino terminal to antidigoxin single-chain Fv. Biochemistry, *29*: 8024–8030, 1990.

11. Whitlow, M., and Filpula, D. Single-chain Fvs and their fusion proteins. Methods, *2*: 97–105, 1991.

12. Kabat, E. A., Wu, T. T., Reid-Miller, M., Perry, H. M., and Gottesman, K. S. Sequences of Proteins of Immunological Interest, Ed. 4, Washington, D.C., U.S. Department of Health and Human Services, 1987.

13. Satow, Y., Cohen, G. H., Padlan, E. A., and Davies, D. R. Phosphocloline binding immunoglobulin Fab McPC603 an X-ray Diffraction study at 2.7Å. J. Mol. Biol., *190*: 593–604, 1986.

14. Herron, J. N., He, X.-M., Mason, M. L., Voss, E. W., Jr., and Edmundson, A. B. Three-dimensional structure of a fluorescein-Fab complex crystallized in 2-methyl-2,4-pentanediol. Proteins, *5*: 271–280, 1989.

15. Marqusee, S., and Baldwin, R. L. Helix stabilization by Glu- . . . Lys+ salt bridges in short peptides of de novo design. Proc. Natl. Acad. Sci., U.S.A., *84*: 8898–8902, 1987.

16. Pantoliano, M. W., Bird, R. E., Johnson, L. S., Asel, E. D., Dodd, S. W., Wood, J. F., and Hardman, K. D. Thermodynamic stability, protein folding, and ligand binding affinity of single chain Fv immunoglobulin fragments expressed in *E. coli*. Biochemistry, *30*: 10117–10125, 1990.

17. Batra, J. K., FitzGerald, D., Gately, M., Chaudhary, V. K., and Pastan, I. Anti-Tac(Fv)-PE40, a single chain antibody pseudomonas fusion protein directed at interleukin 2 receptor bearing cells. J. Biol. Chem., *265*: 15198–15202, 1990.

4
General Strategies in In Vivo Animal Modeling

Daniel A. Vallera and Bruce R. Blazar *University of Minnesota, Minneapolis, Minnesota*

I. A NEW GENERATION OF TOXIN CONJUGATES

The bioengineering of immunotoxins (ITs) and cytokine fusion toxins described in this book clearly represents the epitomy of technology in new drug development. The technology involving the cloning of plant and bacterial toxins and human antibody genes from hybridomas, site-specific alteration of these genes to improve the quality of the biological agent, complex expression systems, and even the synthesis of toxin conjugates by polymerase chain reaction (PCR) is staggeringly sophisticated in comparison to the technology involved in the synthesis of the first chemically linked polyclonal antibody-toxin conjugates. Despite the dazzling display of genetic engineering described in this text, it would seem that the clinical advancement of toxin conjugates has not exactly kept pace with the rapid technological advancement. Many of the clinical studies that have been undertaken have been halted because of unexpected toxicities. In the studies that have not been halted or are currently underway, the clinical administration of ITs has been limited by unacceptably low maximal tolerated doses (MTDs). It would appear that the future of the toxin conjugate field as a whole will likely depend on a clearer understanding of the mechanism of conjugate toxicity and on our ability to somehow increase the MTDs of these agents.

One possible means of understanding conjugate toxicity is through clinical investigation. Unfortunately, waiting for the emergence of clinical toxicities and then looking for potential solutions to these clinical problems is not particularly practical. Some of the more recent antibody-toxin conjugate clinical trials have encountered unexpected neurological toxicities that have resulted in the termination of the study after only a few patients.

63

Also, the minimal doses of conjugate administered did not approach levels necessary to demonstrate antitumor effects. Unless the means can be discovered to increase MTDs and to deal with some of the observed clinical toxicities, the results may have profound negative effects in the field.

II. A ROLE FOR TOXIN CONJUGATES IN EXPERIMENTAL CANCER ANIMAL MODELS

Historically, animal models have played an important role in the clinical development of immunoconjugates. Some of the earlier animal work provided direct evidence that ITs could be efficacious in cancer treatment. For example, in one early study (1) murine tumor BCL1, a murine equivalent of a prolymphocytic variant of human chronic leukemia, was treated with an anti–B-cell monoclonal antibody (mAb) linked to RTA (A chain of ricin toxin) and found to be highly effective in protecting mice against lethal doses of tumor. The study served as a basis for a recent phase I clinical study in which an anti-CD22–ricin A chain IT was used to treat patients with refractory B cell leukemia (2). In a study of 15 patients with B-cell lymphoma, 50% of patients with >50% CD22+ tumor cells achieved partial remissions.

For T-cell malignancies investigators developed a murine model of T-cell malignancy in which the T-cell leukemia WEHI-7 was treated with a pan–T-cell mAb, anti-Thy1.2 linked to ricin toxin A chain (3). This group used this information several years later in human clinical studies in which leukemia patients with acute lymphoblastic leukemia (T-ALL) were treated with a pan–T-cell anti-CD5 mAb (T101) linked to RTA (4). In addition to the syngeneic mouse studies, the heterotropic xenogeneic nude mouse model was useful in determining the efficacy of T101 prior to clinical use. Investigators produced 11 complete regressions and 18 partial regressions by injecting 45 μg of T101-RTA (5). In the clinical studies (6), one of two patients treated showed a 40% reduction in lymphocyte count, but the effect was transient. Similar results were observed when the same IT was given to five chronic lymphocytic leukemia patients. Despite the lack of sustained clinical benefit, the study showed that the conjugate bound and saturated leukemia cells in vivo.

The nude mouse model has become one of the most popular models for evaluating potential anti-cancer reagents prior to clinical trial. This could in part be due to the requirement of the U.S. Food and Drug Administration (FDA) that prior to investigational new drug (IND) approval, efficacy must be demonstrated in some type of animal model. Thus, studies utilizing nude mice to study RTA conjugates have been common. For example, in addition to T-cell malignancy, IT against B leukemia cell lines

(7), melanoma cell lines (8), and colon cancer (9) have been tested in the nude mouse, often preceding clinical testing. However, to date, no antitumor efficacy studies with nude mouse models have been reported with genetically engineered toxins.

III. A ROLE FOR TOXIN CONJUGATES IN EXPERIMENTAL MODELS OF BONE MARROW TRANSPLANTATION

Bone marrow transplantation is regarded as a serious and aggressive therapy for leukemia and other hematological disorders. The field of marrow transplantation has provided an excellent arena for toxin conjugate testing because it has been necessary to remove immunocompetent T cells from the donor graft that are capable of reacting against HLA- and non−HLA-expressing cells in the immunosuppressed host. This graft-versus-host disease (GVHD) is a life-threatening disorder, with the skin, gut, and liver as the predominant target organs (reviewed in Ref. 10). Patients at risk for GVHD were among the first to receive immunotoxin-treated bone marrow cells. Animal models helped determine the efficacy of this approach. For example, pretreatment of donor bone marrow with a brief (2–3 hr) exposure to anti-Thy1.2−ricin was found to protect mice from GVHD generated across an aggressive major histocompatibility barrier difference ($H-2^d$ into $H-2^b$) (11,12). Several aspects of the animal studies helped to precipitate a clinical trial. (1) The animal studies showed that a single course of treatment resulted in efficacy. (2) A dose range of 0.1–1.0 µg/ml was established in mice and later used clinically. (3) The animals demonstrated hematological and immunological recovery of the donor graft. (4) Recovering mice demonstrated classic immunological tolerance to donor skin grafts, suggesting that the less mature donor T cells had undergone reeducation in the recipient's thymus. Clinical studies using this approach have demonstrated protection against GVHD onset. However, the approach was studied only in high-risk patients (13–15).

It is important that this animal model has been useful in identifying problems in the prophylaxis setting. For example, the clinical elimination of donor T cells has been associated with unexpected problems in engraftment likely associated with the loss of T cells as a source of engraftment-promoting cytokines. This problem was observed in animal models (16,17) and later clinically (18,19). On the one hand, elimination of T cells in mice minimized GVHD. On the other hand, these same cells which secrete many different cytokines play a major role in hematopoietic and immunological reconstitution, and their elimination facilitates engraftment problems. We have created a murine model of donor T cell depletion to understand the complex events post–bone marrow transplantation (17). Already, the model is providing useful information concerning the biology of engraftment (20–24). Studies reveal that administration of cytokines that have been removed by eliminating donor T cells for GVHD prevention may facilitate engraftment.

In more recent studies, the in vivo administration of an immunoconjugate was made by linking RTA to the murine homolog of antihuman CD5 which had previously demonstrated efficacy in clinical trials treating steroid refractory GVHD (25). In this case, the clinical studies preceded animal studies. However, recent animal studies (26) have revealed exciting parallels to clinical studies. Most notably, efficacy, transiency of effect, elevated neutrophil levels, decreased lymphocyte levels, and similar pathological effects have been noted in both mouse and human studies. Since no application of fusion toxins has been reported in models of GVHD, it is probable that the animal model will be useful in improving efficacy in future clinical trials. Some of the most recent data in murine bone marrow transplant models suggest that immunoconjugates may also be useful in promoting alloengraftment by eliminating the resistant host immunological cells that contribute to the rejection of the donor graft (27).

Since animal studies have had an impact on our clinical studies, it would indeed be logical for us to turn to them again in solving some of our recent toxicity problems. In the future, it will be important to obtain a greater understanding of the toxicities of biochemically linked antibody-toxin conjugates, as well as genetically engineered fusion toxins. This will be particularly important since some fusion proteins are targeted to hormone receptors in vivo. The following is a brief discussion of some of the major factors contributing to the dose-limiting toxicities of clinical toxin conjugates and some ideas of how animal systems may be of value in finding potential solutions to the problems.

IV. VASCULAR LEAK SYNDROME

Several factors might contribute to conjugate toxicity in vivo. In the majority of clinical trials (2,25,28), the vascular leak syndrome (VLS) has occurred. In VLS, the breakdown of the integrity of the vascular endothelium is accompanied by weight gain, peritoneal effusion, and a drop in plasma protein levels (hypoalbuminemia). It will be extremely important to determine in future studies whether damage to the vascular endothelium is directly or indirectly mediated by the toxin conjugate. How might this be determined? It is possible to enrich for endothelial cells (ECs) and then to test the binding of immunoconjugates directly with the cell population by immunoperoxidase techniques or flow cytometry. It is also possible to label endothelial cell populations with ^{51}Cr and measure endothelial damage by isotope release (29). Such methods have been useful in determining the effect of IL-2 and lymphokine-activated killer (LAK) cell damage in endothelial cells, and one could directly assess immunoconjugate damage. Although these procedures could be performed in vitro on human ECs, they are limited by the necessity for collagenase treatment to isolate endothelial populations. Thus, animal models may be extremely useful in assessing immunoconjugate-induced vascular damage in vivo.

Animal models have been used to assess vascular permeability in vivo. Some investigators have assessed VLS in murine tissues by measuring the extravasation of [^{125}I]bovine serum albumin (30). Other investigators have used the implantation of bilateral chambers to assess vascular changes in tumor tissue implanted in the ears of rabbits (31). Recently, we have reported a model (26) in the bone marrow transplant setting in which irradiation and syngeneic bone marrow were given to mice. Mice given anti–Thy1–RTA (a pan-mouse T-cell immunoconjugate) demonstrated VLS quantifiable by weight gain, significantly reduced plasma protein levels, and peritoneal effusions. Two other immunoconjugates, anti–Ly1–RTA (pan-mouse T-cell, anti-CD5 homolog) and H65-RTA (antihuman CD5) did not demonstrate VLS at the doses tested. We do not yet know whether anti-Thy1.2–RTA is contributing to VLS in a direct or indirect manner. The expression of Thy1 on epidermal cells is somewhat controversial (32–35). Preliminary experiments have shown that anti-Thy1.2–RTA did not efficiently induce VLS in irradiated congeneic mice which express a different allelic form of Thy (Thy1.1) that is not recognized by the anti-Thy1.2 immunoconjugate. These data argue that the occurrence of VLS may be in part related to the specificity of the antibody chosen for study. This model displays all of the classic symptoms of clinical VLS and may be a valuble tool in studying new approaches of limiting VLS-related toxicity, such as the use of anticytokine antibodies or soluble cytokine receptors discussed below.

V. CYTOKINES AND TOXICITY

It has been shown from clinical and animal data that elevated cytokine levels are associated with toxicity. Toxic shock and vascular damage have been associated with elevations of IL-1, IL-2, and tumor necrosis factor alpha (TNFα) (36–40). In the IL-2 trials, IL-2 immunotherapy induces vigorous VLS characterized by fever, nausea, vomiting, hypoxemia, and the development of increased vascular endothelial permeability. These effects are accompanied by decreases in circulating plasma volume (36,37). The increased endothelial permeability results in pulmonary edema, decreased arterial O_2 tension, and dyspnea. These effects are reversible with discontinuation of IL-2 administration and are attenuated by administration of phospholipase A_2 inhibitors (i.e., corticosteroids) (41). Both TNFα and IL-1, other important cytokines, can cause endothelial injury and are involved in the pathogenesis of shock (38–40). Perhaps toxin conjugate administration induces certain cell populations to secrete cytokines which act in the amplification of toxic responses. The identification of such cytokines would help us select anticytokine antibodies to diminish toxicity and increase dose. For example, inhibition of TNFα action by the infusion of anti-TNF–protected mice from lethal toxic shock (42,43). Thus, a role of anticytokine antibodies in protecting animals from vascular toxicities associated with toxic shock has been established in animals, and clinical trials with either antibodies or chemicals that inhibit cytokine secretion (such as pentoxifylline (reviewed in Ref. 44)) are currently under clinical investigation.

VI. ACTIVATION AND TOXICITY

Activation of cells triggered by the binding of mAb or cytokine, especially to cells of the immune system, can have toxic consequences. For example, OKT3, a mAb binding to the human T-cell receptor, has been useful for the promotion of renal grafts, but at the same time has displayed toxic side effects (45,46). Investigators have recently tested a murine model measuring the promotion of bone marrow engraftment (27). A mAb called 145-2C11 (47–49), which recognizes the murine homolog of human CD3, was tested in the model. Although highly efficacious in its graft-promoting capabilities, the mAb displayed the same toxicities as have been observed in human clinical trials. An anti-CD3–RTA was even more toxic than the unconjugated antibody. Both agents induced elevations in TNFα levels. It is believed that TNFα levels contribute at least in part to toxicity in human trials. From the murine studies, it was also clear that the anti-CD3 agents activated cells of the immune system. The investigators noted significant elevations in neutrophils and lymphocytes and accompanying increases in cytokines such as IL-3 and GM-CSF. The studies indicated that triggering can still occur despite the delivery of a toxic signal.

In the anti-CD3–RTA studies (27), removal of the Fc region of the IT resulted in an anti-CD3F(ab′)2 IT that did not trigger IT activation, did not release cytokines, but still promoted engraftment. Most important, toxicity normally associated with the administration of this IT was not present. Together, these studies illustrate the importance of animal models in exploring mechanisms of immunoconjugate toxicity. They prove that toxicity may not always be related to the presence of the toxin moiety.

Triggering activation may be a problem with other immunoconjugates, particularly those that bind determinants on cells that are associated with activation of T cells or B cells.

Activation may also be an issue when testing cytokine fusion toxins in which the cytokine still may have the capability of triggering activation prior to delivery of a toxic signal. The knowledge derived from testing these reagents in animal models may be extremely important when attempting to limit toxicity through molecular design.

A. The Existence of "Unexpected Specificities"

The existence of unique tumor antigen has been contested for many years. Groups that have spent considerable time in the screening and careful testing of mAb which they believe to be largely tumor restricted have been thwarted by the unexpected reactivity of their immunotoxins with normal tissues, predominantly neurological. For example, investigators produced an anti–breast cancer IT using the mAb 260F9, specific for a 55-kd antigen expressed on approximately 50% breast cancer cells (50). Extensive preclinical testing showed that an immunotoxin made by linking 260F9 to recombinant ricin A chain was selective, potent, and displayed acceptable toxicity profiles. Unexpected severe peripheral neurological toxicity emerged during a trial with 260F9-RTA in breast cancer patients (51), prompting discontinuation of the trial. Immunohistochemical techniques revealed reactivity of 260F9 with peripheral nerve Schwann cells and/or myelin sheath. In a different phase trial involving *Pseudomonas aeruginosa* exotoxin A linked to OVB3, a mAb against human ovarian cancer, dose-limiting CNS toxicity occurred after repeated doses of 5 and 10 μg/kg (52). Based on antibody localization studies using frozen sections of human tissue, pancreatic and thyroid toxicity was anticipated. Toxicity to these organs was not observed. Instead, neurocortical toxicity occurred. When OVB3 was tested against fresh samples from various portions of normal brain, it was weakly reactive with cells in the molecular layer of the cerebellum. Presumably, IT, even in small amounts, entered the cerebrospinal fluid (CSF) and damaged critical cells in the brain. In a recently completed phase I intraperitoneal clinical trial (52) in which ovarian cancer patients were treated with an antitransferrin receptor IT, 454A12-rRTA, central neurological toxicity associated with capillary endothelial damage in the basal ganglia resulted in the termination of the study. It has been suggested that transferrin receptor may be expressed on these cells.

Since several clinical trials have now shown neurological side effects, we will need to understand more about the mechanism of neurological toxicities. Animal models may not prove as useful in situations in which toxicity is determined by unexpected specific reactivity of an antitumor immunoconjugate with neurological tissue. Instead, there may be no substitute for rigorous immunopathological analysis of panels of human neurological tissues, including brain, spinal cord, and CSF. In other instances, animal models may be more useful. For example, toxicity could be attributed to the instability of the immunoconjugate in vivo and the reactivity of toxin with nervous tissue. Certain toxins, such as diphtheria toxin (DT) and ricin, have well-defined neurological effects, and these can be directly assessed in an animal model. For example, investigators have shown that diphtheria toxin can cause sensory-motor neuropathy (53) resulting from a classic segmental demyelination of nerve fibers (54). Schwann cells and oligodendrocytes are also sensitive to DT in vitro (55,56). Investigators have recently shown that diphtheria toxin and diphtheria toxin mutants specifically kill cerebellar Purkinje cells with no detectable toxicity to other neurons when administered in the CSF (57). Purkinje cells transmit the only efferent impulses from the cerebellar cortex.

Perhaps animal models may be of greatest use in understanding how the neurological toxicity of toxin, and toxin conjugates, can be used to our advantage. For example, several inherited diseases, such as ataxia-telangectasia and certain lysosomal storage disorders, result in extensive loss of Purkinje cells. Investigators have used the intrathecal injection of diphtheria toxin mutant as a model to study the biology and physiology of Purkinje cell loss (57). Also, since an interaction of IT with the vascular endothelium might cause a break in the blood-brain barrier and promote more indiscriminate neurological toxicity, such an effect might be more amenable to study in an animal system. However, such a system has not yet been described. It should also be noted that animal models will be most useful in studying the utility of antineurological immunoconjugates for purposes of treating brain tumors and other neurological disorders.

Fc-Related Toxicity

In certain studies, investigators have suspected that toxicity was related at least in part to the presence of the Fc fragment on the mAb portion of their immunoconjugate. For example, in a recent phase I trial evaluating anti–breast cancer IT (mAb linked to recombinant RTA), investigators observed VLS (28). To investigate a possible Fc-related mechanism, they incubated human monocytes with IT and found that there was nonspecific binding. Binding was inhibited by preincubation of the monocytes with pooled human immunoglobulin containing Fc, suggesting that binding was Fc receptor mediated. Although informative, these in vitro studies could provide only limited information. As discussed above (27), investigators have shown that elimination of the Fc region from certain immunotoxins results in dramatic reductions in toxicity when immunoconjugates are administered in vivo.

Fusion Toxins

Fusion toxins are constructed by fusing cytokine genes or antibody genes to toxin genes. Their synthesis and expression are discussed in great detail in the body of this text. This new fusion proteins can be directed against a variety of target cells expressing cytokine receptors. Since the cytokine–cell receptor interaction is usually high affinity and regulates the proliferation of target cell populations, the approach should be highly useful for destruction of the rapidly proliferating cells of the immune system (such as T cells in graft rejection and autoimmune disease) or even cytokine-dependent tumor cells.

Animal models have already proven useful in evaluating the efficacy of fusion toxins in regulating the immune system in the transplant situation. Shapiro et al. have reviewed the current experience of IL-2 fusion toxins as immunosupressive agents in Chapter 19. They have proven effective in in vivo rodent models, suppressing delayed hypersensitivity responses (58), and have also successfully prolonged the survival of murine heterotropic heart allografts (59,60) and pancreatic islet grafts (61) transplanted across major histocompatibility differences. In addition to their use in the field of transplantation, fusion toxins may be very useful for regulating the immune system. Animal models have demonstrated the potential of fusion toxins in the therapy of immunological disorders, such as the autoimmune diseases allergic encephalomyelitis and adjuvant arthritis (62).

Although the promise is great, these agents will likely provide a different subset of problems that also can be examined in the context of animal models. For example, a major problem of toxin conjugates is maintaining in vivo levels adequate to saturate target cells. Although theoretically a single toxin molecule in the cell is cytotoxic, it may take hundreds to thousands selectively bound to the target cell in order for one toxin molecule to penetrate the cell's complex microsomal network. In the case of cytokine fusion toxins, it could be feasible for the cytokine to trigger its receptor without enough conjugate being present to trigger toxicity. In the case of a lymphoproliferative disease such as leukemia, which might express low levels of the cytokine receptor, the result could be disasterous; i.e., stimulation of the neoplasia rather than the delivery of a lethal hit. In the case of fusion proteins that are directed against immunological targets such as IL-4, the result could be lymphoid activation, with the same limiting toxicities that were discussed above. Animal models are available in which a variety of different tumors can be tested, whether expressing or not expressing a given cytokine receptor.

In Chapter 24, FitzGerald et al. mention another potential problem worth considering. The attachment of a toxin to a cytokine may be somewhat analogous to the attachment of a hapten molecule to a carrier. The union may create new specificities that elicit vigorous immunological responses in patients. Already, the ability of immunoconjugates to induce immunological responses in patients is established. Although chimerization of antibodies may reduce the immunological response to these immunoconjugates, gene fusion to create hybrid proteins may create an even broader base of specificities to which the immune system might respond. Again, animal systems defining the immunogenicity of immunoconjugates are readily available.

VII. CONCLUSIONS

Molecular biology has given new life to the immunoconjugate field. On the one hand, toxin conjugates have major limitations in achieving doses necessary for saturation of clinical target sites and curing a given disease. On the other hand, we now have the means to design our own molecules which can be tailored to fit a given clinical situation. However, we cannot afford mistakes in bringing these new agents to clinical trial. Many of the toxin conjugates that are now under clinical investigation are currently brought to trial under the corporate umbrella. Several years and tens of millions of dollars must be invested in order to obtain FDA licensure, and failure to obtain that approval would be devastating. No toxin conjugate has yet achieved that approval, and failure for the first applicant will likely discourage others. One might speculate that even if the large corporations abandon the approach, we could rely on independent investigators and National Institutes of Health (NIH) funds. But in today's fiscal environment, that would be highly unlikely.

Clearly, it will be important that we use all of our means and particularly animal models to better understand in vivo administration of immunoconjugates and their toxicities before we undertake their evaluation in the clinic. If immunotoxins are to be effective drugs, then we must increase the astoundingly low maximum tolerated doses that are currently being reported in the literature. To do so, we must more thoroughly understand all of the complex factors contributing to their toxicity. In the mouse, monoclonal antibodies have been produced recognizing structures to which homologous antibodies exist in humans. Also, many homologous cytokines have been cloned in the mouse and human. But even if toxin conjugates which are not homologous in the mouse and human are used for in vivo studies, the information will likely be important. Once we understand conjugate toxicities more thoroughly, it should be possible to use the knowledge of molecular biology, described in part in this text, to construct new agents with reduced clinical toxicities. It may also be possible to combine immunoconjugate therapy with other approaches that have been researched in animal systems and are designed to enhance the tolerated dose of toxin conjugates.

REFERENCES

1. Krolick K., Uhr, J. W., Shuin, S., and Vitetta, E. S. In vivo therapy of murine B cell tumor (BCL 1) using antibody-ricin A chain immunotoxins. J. Exp. Med., *155*: 1979–1809, 1982.
2. Vitetta, E. S., Stone, M., Amlot, P., Fay, J., May, R., Till, M., Newman, J., Clark, P., Collins, R., Cummingham, D., Ghetie, V., Uhr, J. W., and Thorpe, P. E. Phase I immunotoxin trials in patients with B-cell lymphoma. Cancer Res., *51*: 4052–4058, 1991.
3. Blythman, H. E., Casellas, P., Gros, O., Gros, P., Jansen, F., Paolucci, F., Pau, B., and Vidal, H. Immunotoxins: Hybrid molecules of monoclonal antibodies and A toxin subunit specifically kill tumour cells. Nature, *290*: 145–146, 1981.
4. Laurent, G., Pris, J., Farcet, J., Carayon, P., Blythman, H., Casellas, P., Poncelet, P., and Jansen, P. K. Effects of therapy with T101 ricin A-chain immunotoxin in two leukemia patients. Blood, *67*: 1680–1687, 1986.
5. Leonard, J. E., Johnson, D. E., Shawler, D. L., and Dillman, R. O. Inhibition of human T-cell tumor growth by T101-ricin A-chain in an athymic mouse model. Cancer Res., *48*: 4862–4867, 1988.
6. Frankel, L. G., Hertler, A. A., Schlossman, D. M., Casellas, P., and Jansen, F. K. Treatment of leukemia patients with T101 ricin A chain immunotoxins. Cancer Treat Res., *37*: 483–491, 1988.
7. Hara, H., Luo, Y., Haruta, Y., and Seon, B. K. Efficient transplantation of human non-T-leukemia cells into nude mice and induction of complete regression of the transplanted distinct tumors by ricin A-chain conjugates of monoclonal antibodies SN5 and SN6. Cancer Res., *48*: 4673–4680, 1988.
8. Trowbridge, I. S., and Domingo, D. L. Anti-transferrin receptor monoclonal antibody and toxin-antibody conjugates affects growth of human tumour cells. Nature, *294*: 171–173, 1981.
9. Byers, V. S., Pimm, M. V., Scannon, P. J., Pawluczyk, I., and Baldwin, R. W. Inhibition of growth of human tumor xenografts in athymic mice treated with ricin toxin A chain-monoclonal antibody 791T/36 conjugates. Cancer Res., *47*: 5042–5046, 1987.
10. Grebe, S. C., and Streilein, W. J. Graft-versus-host reactions: a review. Adv. Immunol., *22*: 119, 1976.
11. Vallera, D. A., Youle, R. J., Neville, D. M., Jr., and Kersey, J. H. Bone marrow transplantation across major histocompatibility barriers. V. Protection

of mice from lethal GVHD by pretreatment of donor cells with monoclonal anti-Thy-1.2 coupled to the toxin ricin. J. Exp. Med., *155*: 949–954, 1982.

12. Vallera, D. A., Youle, R. J., Neville, D. M., Jr., Soderling, C. C. B., and Kersey, J. H. Monoclonal antibody-toxin conjugates for experimental GVHD prophylaxis. Reagents selectively reactive with T cells and not murine stem cells. Transplantation, *36*: 73–80, 1982.

13. Filipovich, A. H., Vallera, D. A., Youle, R. J., Quinones, R. R., Neville, D. M., Jr., and Kersey, J. H. Ex vivo treatment of donor bone marrow with anti-T cell immunotoxins for the prevention of graft-versus-host disease. Lancet, 8375: 469–472, 1984.

14. Filipovich, A. H., Vallera, D. A., Youle, R. J., Haake, R., Blazar, B. R., Arthur, D., Neville, D. M., Jr., Ramsay, N. K. C., McGlave, P., and Kersey, J. H. Graft-versus-host disease prevention in allogeneic bone marrow transplantation from histocompatible siblings: A pilot study using immunotoxins for T cell depletion of donor bone marrow. Transplantation, *44*: 62–69, 1987.

15. Filipovich, A. H., Vallera, D., McGlave, P., Polich, D., Gajl-Peczalska, K., Haake, R., Lasky, L., Blazar, B., Ramsay, N. K. C., Kersey, J., and Weisdorf, D. T cell depletion with anti CD5 immunotoxin in histocompatible bone marrow transplantation: The correlation between residual CD5 negative T cells and subsequent GVHD. Transplantation, *50*: 410–415, 1990.

16. Vallera, D. A., Soderling, C. C. B., Carlson, G. J., and Kersey, J. H. Bone marrow transplantation across major histocompatibility barriers in mice. II. T cell requirements for engraftment in TLI-conditioned recipients. Transplantation, *33*: 243–248, 1982.

17. Soderling, C. C. B., Song, C. W., Blazar, B. R., and Vallera, D. A. A correlation between conditioning and engraftment in recipients of MHC mismatched T cell depleted murine bone marrow transplants. J. Immunol., *135*: 941–946, 1985.

18. Butturini, A., and Gale, R. P. The role of T-cells in preventing relapse in chronic myelogenous leukemia. Bone Marrow Transplant., *2*: 351, 1988.

19. Poynton, C. H. T Cell depletion in bone marrow transplantation. Bone Marrow Transplant., *3*: 265, 1988.

20. Blazar, B. R., Widmer, M. B., Soderling, C. C. B., Urdal, D. L., Gillis, S., Robison, L. L., and Vallera, D. A. Augmentation of donor bone marrow engraftment in histoincompatible murine recipients by granulocyte macrophage colony-stimulating factor. Blood, *71*: 320–328.

21. Vallera, D. A., and Blazar, B. R. Depressed leukocyte reconstitution and engraftment in murine recipients of T-cell-depleted histoincompatible marrow pretreated with interleukin 3. Transplantation, *46*: 616–620, 1988.

22. Blazar, B. R., Kersey, J. H., McGlave, P. B., Vallera, D. A., Lasky, L. C., Haake, R., Bostrom, B., Weisdorf, D. R., Epstein, C., and Ramsay, N. K. C. In vivo administration of recombinant human granulocyte/macrophage colony-stimulating factor in acute lymphoblastic leukemia patients receiving purged autografts. Blood, *73*: 849–857, 1989.

23. Blazar, B. R., Widmer, M. B., Cosman, D., Sassenfeld, H. M., and Vallera, D. A. Improved survival and leukocyte reconstitution without detrimental

effects on engraftment in murine recipients of human recombinant granulocyte colony-stimulating factor following transplantation of T-cell depleted histoincompatible bone marrow. Blood, *74*: 2264–2269, 1989.

24. Blazar, B. R., Thiele, D. L., and Vallera, D. A. Pretreatment of murine donor grafts with L-leucyl-L-leucine methyl ester: Elimination of graft-versus-host disease without detrimental effects on engraftment. Blood, *75*: 798–805, 1990.

25. Byers, V. S., Henslee, P. J., Kernan, N. A., Blazar, B. R., Gingrich, R., Philips, G. L., LeMaistre, C. F., Gililand, G., Antin, J. H., Martin, P., Tutscha, P. J., Trown, P., Ackerman, S. K., O'Reilly, R. J., and Scannon, P. J. Use of an anti-pan T-lymphocyte ricin A chain immunotoxin in steroid-resistant acute graft-versus-host disease. Blood, *75*: 1426–1432, 1990.

26. Vallera, D. A., Carroll, S. F., Snover, D., Carlson, G. J., and Blazar, B. R. Toxicity and efficacy of anti-T cell ricin toxin A chain immunotoxins in a murine model of established graft-versus-host disease induced across the major histocompatibility barrier. Blood, *77*: 182–194, 1991.

27. Blazar, B. R., Hirch, R., Gress, R., Carroll, S. F., and Vallera, D. A. In vivo administration of anti-CD3 monoclonal antibodies for immunotoxins in murine recipients of allogeneic T-cell depleted marrow for the promotion of engraftment. J. Immunol., *147*: 1492–1503, 1991.

28. Weiner, L. M., O'Dwyer, J., Kitson, J., Comis, R. L., Frankel, A. E., Bauer, R. J., Konrad, M. S., and Groves, E. S. Phase I evaluation of an anti-breast carcinoma monoclonal antibody 260F9-recombinant ricin A chain immunoconjugate. Cancer Res., *49*: 4062-4067, 1989.

29. Kotasek, D., Vercellotti, G. M., Ochoa, A. C., Bach, F. H., White, J. G., and Jacob, H. S. Mechanism of cultured endothelial injury induced by lymphokine-activated killer cells. Cancer Res., *48*: 5528–5532, 1988.

30. Ettinghausen, S. E., Puri, R. K., and Rosenberg, S. A. Increased vascular permeability in organs mediated by the systemic administration of lymphokine-activated killer cells and recombinant interleukin-2 in mice. J. Natl., Cancer Inst., *80*: 177–188, 1988.

31. Nugent, L. J., and Jain, R. K. Monitoring transport in the rabbit ear chamber. Microvasc. Res., *24*: 204–209, 1982.

32. McKenzie, I., and Potter, T. Murine lymphocyte surface antigens. Adv. Immunol., *27*: 217, 1979.

33. Scheild, M., Boyse, E. A., Carswell, E. A., and Old, L. J. Serologically demonstrable alloantigens of mouse epidermal cells. J. Exp. Med., *135*: 938–955, 1972.

34. Lesley, J. F., and Lennon, V. A. Transitory expression of Thy-1 antigen in skeletal muscle development. Nature, *268*: 163–165, 1977.

35. Berman, J. W., and Basch, R. S. Thy-1 antigen expression by murine hematopoietic precursor cells. Exp. Hematol., *13*: 1152–1156, 1985.

36. Lotze, M. T. New approaches to the immunotherapy of cancer using interleukin-2. Ann. Intern. Med., *108*: 853–864, 1988.

37. Rosenberg, S. A., Lotze, M. T., Muul, L. M., Chang, A. E., Avis, F. P., Leitman, S., Linehan, W. M., Robertson, C. N., Lee, R. E., and Rubin, J. T. A progress report on the treatment of 157 patients with advanced cancer

using lymphokine-activated killer cells and interleukin-2 or high-dose interleukin-2 alone. N. Engl. J. Med., *316*: 889, 1987.

38. Braquet, P., Touqui, L., Shen, T. Y., and Vargaftig, B. B. Perspectives in platelet-activating factor research. Pharmacol. Rev., *39*: 97–145, 1987.

39. Braquet, P., Paubert-Braquet, M., Koltai, M., Bourgain, R. H., Bussolino, F., and Hosford, D., Is there a case for PAF antagonists in the treatment of ischemic states? Trends Pharmacol. Sci., *10*: 23–30, 1989.

40. Dubois, C., Bissonnette, E., and Rola-Pleszczynski, M. Platelet-activating factor (FAF) enhances tumor necrosis factor production by alveolar macrophages. J. Immunol., *143*: 964–970, 1989.

41. Butler, L. D., Mohler, N. K., Layman, N. K., Cain, R. L., Riedl, P. E., Puckett, L. D., and Bendele, A. M. Interleukin-2 induced systemic toxicity: Induction of mediators and immunopharmacologic intervention. Immunopharmacol. Immunotoxicol., *11*: 445–487, 1989.

42. Beutler, B., Milsark, I. W., and Cerami, A. C. Passive immunization against cachecctin/tumor necrosis factor protects mice from lethal effect of endotoxin. Science, *229*: 869–871, 1985.

43. Tracey, K. J., Fong, Y., Hesse, D. G., Manogue, K. R., Lee, A. T., Kuo, G. C., Lowry, S. F., and Lerami, A. Anti-cachectin/TNF monoclonal antibodies prevent septic shock during lethal bacteraemia. Nature, *330*: 662–664, 1987.

44. Bianco, J. A., Appelbaum, F. R., Nemunaitis, J., Almgran, J., Andrews, F., Kettner, P., Shields, A., and Singer, J. W. Phase I–II trial of pentoxifylline for the prevention of transplant-related toxicities following bone marrow transplantation. Blood, *78*: 1205–1211, 1991.

45. Chatenoud, L., Ferran, C., Legendre, C., Thouard, I., Merite, S., Reuter, A., Gevaert, Y., Kreis, H., Franchimont, P., and Bach, J. In vivo cell activation following OKT3 administration. Transplantation, *49*: 697–702, 1990.

46. Abramowicz, D., Schandene, L., Goldman, M., Crusiaux, A., Vereerstraeten, P., De Pauw, L., Wybran, J., Kinnaert, P., Dupont, E., and Toussaint, C. Release of tumor necrosis factor, interleukin-2 and gamma-interferon in serum after injection of OKT_3 monoclonal antibody in kidney transplant recipients. Transplantation, *47*: 606–608, 1989.

47. Quintans, J., Yokoyama, A., Evavold, B., Hirsch, R., and Mayforth, R. D. Direct activation of murine resting T cells by Con A or anti-CD3 Ig. J. Mol. Cell. Immunol., *4*: 225.

48. Hirsch, R., Gress, R. E., Plutnik, D. H., Eckhaus, M., and Bluestone, J. A. Effects of in vivo administration of anti-CD3 monoclonal antibody on T cell function in mice. J. Immunol., *142*: 737–743, 1989.

49. Flamand, V., Abramowicz, M., Goldman, M., Biernaux, C., Huez, G., Urbain, J., Moser, M., and Leo, O. Anti-CD3 antibodies induce T-cells from unprimed animals to secrete IL-4 both in vitro and in vivo. J. Immunol., *144*: 2875–2882, 1990.

50. Frankel, A. E., Ring, D. B., Tringale, F., and Hsieh-MA, S. T. Tissue distribution of breast cancer-associated antigens defined by monoclonal antibodies. J. Biol. Response Mod., *4*: 273–286, 1985.

51. Gould, B. J., Borowitz, M. J., and Groves, E. S. Phase I study of an anti-breast cancer immunotoxins by continuous infusion: Report of a targeted toxic effect not predicted by animal studies. J. Natl. Cancer Inst., *81*: 775–781, 1989.
52. Bookman, M. A., Godfrey, S., and Padavic, K. Anti-transferrin receptor immunotoxin (IT) therapy: A phase-I intraperitoneal (i.p.) trial. Proc. Am. Soc. Clin. Oncol., *9*: 187, 1990.
53. Fisher, C. M., and Adams, R. D. Diphtheritic polyneuritis—a pathological study. J. Neuropathol. Exp. Neurol., *15*: 243–268, 1956.
54. Webster, H. deF., Spiro, D., Waksman, B., and Adams, R. D. Phase and electron microscopic studies of experimental demyelination. J. Neuropathol. Exp. Neurol, *20*: 5–34, 1961.
55. Pappenheimer, A. M., Jr., Harper, A. A., Moynihan, M., and Brockes, J. P. Diphtheria toxin and related proteins: Effect of route of injection on toxicity and the determination of cytotoxicity for various cultured cells. J. Infect. Dis., *145*: 94–102, 1981.
56. Murray, K., and Nobel, M. In vitro studies on the comparative sensitivities of cells of the central nervous system to diphtheria toxin. J. Neurol. Sic., *70*: 283–293, 1985.
57. Riedel, C. J., Muraszko, K. M., and Youle, R. J. Diphtheria toxin mutant selectively kills cerebellar purkinje neurons. Proc. Natl. Acad. Sci. USA, *87*: 5051–5055, 1990.
58. Kelley, V. E., Bacha, P., Pankewycz, O., Nichols, J. C., Murphy, J. R., and Strom, T. B. Interleukin 2-diphtheria toxin fusion protein can abolish cell-mediated immunity in vivo. Proc. Natl. Acad. Sci. U.S.A., *85*: 3980–3984, 1988.
59. Kirkman, R. L., Bacha, P., Barrett, L. V., Forte, S., Murphy, J. R., and Strom, T. B. Prolongation of cardiac allograft survival in murine recipients treated with a diphtheria toxin-related interleukin-2 fusion protein. Transplantation, *47*: 327–330, 1989.
60. Lorberboum-Galski, H., Barrett, L. V., Kirkman, R. L., Ogata, M., Willingham, M. C., Fitzgerald, D. J., and Pastan, I. Cardiac allograft survival in mice treated with IL-2-PE40. Proc. Natl. Acad. Sci. U.S.A., *86*: 1008–1012, 1989.
61. Pankewycz, O., Mackie, J., Hassarjian, R., Murphy, J. R., Strom, T. B., and Kelley, V. E. Interleukin-2-diphtheria toxin fusion protein prolongs murine islet cell engraftment. Transplantation, *47*: 318–322, 1989.
62. Case, J. P., Lorberbaum-Galski, H., Lafyatis, R., Fitzgerald, D., Wilder, R. L., and Pastan, I. Chimeric cytotoxin IL2-PE40 delays and mitigates adjuvant-induced arthritis in rats. Proc. Natl. Acad. Sci. U.S.A., *86*: 287–291, 1989.

II
Ricin

5
Molecular Cloning of Ricin

Lynne M. Roberts, James W. Tregear, and J. Michael Lord
University of Warwick, Coventry, England

I. INTRODUCTION

Ricin is a type II ribosome-inactivating protein which occurs in the seeds of the castor oil plant (*Ricinus communis*). It is a heterodimeric glycoprotein comprising subunits with apparent molecular weights around 30 kd. The ribosome-inactivating polypeptide of ricin toxin (the A chain, or RTA) is an RNA-specific N-glycosidase which depurinates RNA within a highly conserved stem-loop present in 26S or 28S ribosomal ribonucleic acid (rRNA) (1). The A chain is covalently linked through a disulfide bond to a galactose-binding lectin subunit of ricin toxin (the B chain, or RTB). The B chain is responsible for binding the heterodimeric cytotoxin to the surface of cells; the important first step in the delivery of RTA into the cytosol where the rRNA substrate is located.

In castor oil seeds, ricin occurs together with *Ricinus communis* agglutinin (RCA), a tetrameric galactose-specific lectin consisting of two ricinlike heterodimers. Several forms of ricin and RCA are known to exist among seed varieties and even within a single variety. From molecular weight and isoelectric point analyses, at least three forms of ricin and two of RCA have been tentatively identified (2). The first form of ricin to be directly sequenced by Edman degradations is described as ricin D (3,4). More recently, the B chain of a variant form of ricin, termed ricin E, has been directly sequenced (5). The sequence reveals clear differences between the B chains of ricin E and ricin D; that of the former appearing to be a hybrid ricin/RCA sequence.

Research into the biosynthesis of the members of this lectin family have established that ricin and RCA isoforms are:

1. Synthesized in the same tissue- and development-specific manner within endosperm cells during the later stages of seed maturation when the seed coat (testa) has formed (6).
2. The A and B chains of each lectin are initially synthesized together as part of a single proprotein precursor (7).
3. Synthesis is accompanied by segregation into the lumen of the rough endoplasmic reticulum from where core glycosylated prolectin molecules are transported, via the Golgi complex, to protein body organelles. Deposition in protein bodies is accompanied by endoproteolytic processing to generate the mature A and B subunits (8).

Ricin or purified RTA has frequently been used as the toxic component of chimeric conjugates such as immunotoxins (ITs). Immunotoxins offer an alternative biotherapeutic approach to the treatment of various disease states, including certain cancers and autoimmune diseases. There have been several recent reviews describing the current and potential applications of ITs (e.g., Refs. 9, 10), from which it is clear that despite promising in vitro selectivity and potency, many ITs have limitations, that, as yet, preclude their successful application in vivo. While many of these limitations concern the specificity and properties of the antibody moiety, some wholly or partly concern the toxin component. The inability of ITs to permeate solid tumors, or to extravasate effectively, and the induction of an immune response are problems which can be partly addressed by reducing the size of IT components. Rapid clearance can be avoided if nonglycosylated toxin is utilized, and it is speculated that increased specificity and potency would result from the use of holotoxin in which the RTB subunit is defective in sugar binding. Finally, to produce a toxin of maximum potency it is crucial that the structural features related to all the different toxin functions are precisely determined and any interplay or synergy between them defined at the molecular level. The only satisfactory route toward the streamlining of toxins and the obvious approach to address structure-function relationships is to produce and characterize wild-type and genetically engineered recombinant toxins. An essential prerequisite is, thus, the isolation of complementary DNA (cDNA) (11–13) or genomic (14–16) preproricin clones as well as cDNA clones for RTA (17) and RTB (18) alone. Cloning is almost invariably followed by attempts to express and subsequently manipulate the encoding DNA. This has also been accomplished in recent years using a variety of prokaryotic and eukaryotic expression systems (17–24).

II. STRATEGIES FOR THE SYNTHESIS AND CLONING OF RICIN-ENCODING cDNAs

Utilizing size-fractionated polyadenylated RNA extracted from developing posttesta castor beans, double-stranded (DS) cDNAs have been synthesized using standard procedures involving oligo(dT)-primed reverse transcription, DNA-dependent second-strand synthesis, and S1 nuclease treatment (11). Oligo(dC) homopolymer–tailed cDNAs were subsequently fractionated on a linear sucrose density gradient and those with lengths greater than 1 kbp were annealed with *Pst*I-digested, oligo(dG)-tailed pBR322. After transformation of *Escherichia coli*, ricin-related sequences were localized in situ using radiolabeled 20-mer oligonucleotide probes representing all 16 possible DNA sequences predicted from a seven–amino acid stretch (residues 214–220) in the published primary sequence of mature RTB chain (4). Eighty of the 1600 clones screened gave strong hybridization signals, and subsequently several of these were sequenced (11) and manipulated for expression purposes (19,21–24). The mRNA sequence encoding the ricin precursor was deduced from two overlapping cDNA sequences which revealed, sequentially, a putative signal sequence of 24 codons, an A chain of 267 codons, a linker of 12 codons, and a B chain of 262 codons. Although the subunit coding sequences revealed some differences and were slightly longer (see below) than those determined previously by direct amino acid sequencing (3,4), they showed a sufficiently high degree of homology (approximately 94.4% absolute amino acid identity) for the cDNAs to be assigned as ricin encoding (Figure 1). The two sequences described (11) were subsequently joined at a common site in the overlapping region and subcloned into pUC8 to generate pRCL617, a proricin-encoding plasmid. RCA clones were also characterized from the same library. The clones were confirmed as RCA encoding since the deduced N-terminal sequence corresponded exactly with that determined for purified RCA protein across a region where several residue differences occur in the RTB protein (25). It is important to be able to distinguish the agglutinin form since, in addition to a different quaternary structure, the agglutinin protein has a more restricted sugar-binding property (being specific only for galactose, whereas ricin is specific for galactose and N-acetylgalactosamine) and possesses an A chain which is less active toward ribosomes in vitro (36). In the cloning of other toxin genes with agglutinin counterparts (e.g., abrin), it will likewise be important to distinguish the two types of sequences.

Applying a similar overall cloning strategy DS cDNAs were synthesized from affinity-purified polyadenylated castor bean mRNA (12). To enrich for cDNAs encoding full-length RTA sequences, an internal first-strand oligonucleotide primer was utilized which was complementary to a 5′ stretch of the RTB coding sequence (bases 876–892 according to the numbering in Figure 1). The single-stranded (SS) cDNAs produced using reverse transcriptase were then tailed to give oligo(dC) cDNAs. This allowed oligo(dG) to be used as a primer for second-strand cDNA synthesis using reverse transcriptase (12). Tailed DS cDNAs were then annealed with *Pst*I-

```
1                                                          50
ATA TTC CCC AAA CAA TAC CCA ATT ATA AAC TTT ACC ACA GCG GGT GCC ACT GTG CAA AGC TAC ACA AAC TTT ATC AGA GCT
Ile Phe Pro Lys Gln Tyr Pro Ile Ile Asn Phe Thr Thr Ala Gly Ala Thr Val Gln Ser Tyr Thr Asn Phe Ile Arg Ala
1                             10                        20

              100                                          150
GTT CGC GGT CGT TTA ACA ACT GGA GCT GAT GTG AGA CAT GAT ATA CCA GTG TTG CCA AAC AGA GTT GGT TTG CCT ATA AAC CAA CGG TTT
Val Arg Gly Arg Leu Thr Thr Gly Ala Asp Val Arg His Asp Ile Pro Val Leu Pro Asn Arg Val Gly Leu Pro Ile Asn Gln Arg Phe
         30                        40        Glu                     50

              200                                          250
ATT TTA GTT GAA CTC TCA AAT CAT GCA GAG CTT TCT GTT ACA TTA GCC CTG GAT GTC ACC AAT GCA TAT GTG GTC GGC TAC CGT GCT GGA
Ile Leu Val Glu Leu Ser Asn His Ala Glu Leu Ser Val Thr Leu Ala Leu Asp Val Thr Asn Ala Tyr Val Val Gly Tyr Arg Ala Gly
         60            Gln                   70    Ser                80

              300                                          350
AAT AGC GCA TAT TTC TTT CAT CCT GAC AAT CAG GAA GAT GCA GAA GCA ATC ACT CAT CTT TTC ACT GAT GTT CAA AAT CGA TAT ACA TTC
Asn Ser Ala Tyr Phe Phe His Pro Asp Asn Gln Glu Asp Ala Glu Ala Ile Thr His Leu Phe Thr Asp Val Gln Asn Arg Tyr Thr Phe
         90                            100                        110

              400
GCC TTT GGT GGT AAT TAT GAT AGA CTT GAA CAA CTT GCT GGT AAT CTG AGA GAA AAT ATC GAG TTG GGA AAT GGT CCA CTA GAG GAG GCT
Ala Phe Gly Gly Asn Tyr Asp Arg Leu Glu Gln Leu Ala Gly Asn Leu Arg Glu Asn Ile Glu Leu Gly Asn Gly Pro Leu Glu Glu Ala
         120                       130                        140

   450                      500
ATC TCA GCG CTT TAT TAT TAC AGT ACT GGT GGC ACT CAG CTT CCA ACT CTG GCT CGT TCC TTT ATA ATT TGC ATC CAA ATG ATT TCA GAA
Ile Ser Ala Leu Tyr Tyr Tyr Ser Thr Gly Gly Thr Gln Leu Pro Thr Leu Ala Arg Ser Phe Ile Ile Cys Ile Gln Met Ile Ser Glu
   150                        160                        170

              550                                          600
GCA GCA AGA TTC CAA TAT ATT GAG GGA GAA ATG CGC ACG AGA ATT AGG TAC AAC CGA AGA TCT GCA CGG GAT CCT AGC GTA ATT ACA CTT
Ala Ala Arg Phe Gln Tyr Ile Glu Gly Glu Met Arg Thr Arg Ile Arg Tyr Asn Arg Arg Ser Ala Arg Asp Pro Ser Val Ile Thr Leu
         180                       190                        200

              650                                          700
GAG AAT AGT TGG GGG AGA CTT TCC ACT GCA ATT CAA GAG TCT AAC CAA GGA GCC TTT GCT AGT CCA ATT CAA CTG CAA AGA CGT AAT GGT
Glu Asn Ser Trp Gly Arg Leu Ser Thr Ala Ile Gln Glu Ser Asn Gln Gly Ala Phe Ala Ser Pro Ile Gln Leu Gln Arg Arg Asn Gly
         210                       220                        230    --- Asp

              750                                          800
TCC AAA TTC AGT GTG TAC GAT GTG AGT ATA TTA ATC CCT ATC ATA GCT CTC ATG GTG TAT AGA TGC GCA CCT CCA CCA TCG TCA CAG TTT
Ser Lys Phe Ser Val Tyr Asp Val Ser Ile Leu Ile Pro Ile Ile Ala Leu Met Val Tyr Arg Cys Ala Pro Pro Pro Ser Ser Gln Phe
         240                Leu    250                        260    ---

                                 850
TCT TTG CTT ATA AGG CCA GTG GTA CCA AAT TTT AAT GCT GAT GTT TGT ATG GAT CCT GAG CCC ATA GTG CGT ATC GTA GGT CGA AAT GGT
Ser Leu Leu Ile Arg Pro Val Val Pro Asn Phe Asn Ala Asp Val Cys Met Asp Pro Glu Pro Ile Val Arg Ile Val Gly Arg Asn Gly
(--- --- --- --- --- --- --- --- --- --- --- --- ---)        280                        290

   900                      950
CTA TGT GTT GAT GTT AGG GAT GGA AGA TTC CAC AAC GGA AAC GCA ATA CAG TTG TGG CCA TGC AAG TCT AAT ACA GAT GCA AAT CAG CTC
Leu Cys Val Asp Val Arg Asp Gly Arg Phe His Asn Gly Asn Ala Ile Gln Leu Trp Pro Cys Lys Ser Asn Thr Asp Ala Asn Gln Leu
   300            Asn            Asn His    310                        320

              1000                                         1050
TGG ACT TTG AAA AGA GAC AAT ACT ATT CGA TCT AAT GGA AAG TGT TTA ACT ACT TAC GGG TAC AGT CCG GGA GTC TAT GTG ATG ATC TAT
Trp Thr Leu Lys Arg Asp Asn Thr Ile Arg Ser Asn Gly Lys Cys Leu Thr Thr Tyr Gly Tyr Ser Pro Gly Val Tyr Val Met Ile Tyr
---       330                        340                Pro Ser

              1100                                         1150
GAT TGC AAT ACT GCT GCA ACT GAT GCC ACC CGC TGG CAA ATA TGG GAT AAT GGA ACC ATC ATA AAT CCC AGA TCT AGT CTA GTT TTA GCA
Asp Cys Asn Thr Ala Ala Thr Asp Ala Thr Arg Trp Gln Ile Trp Asp Asn Gly Thr Ile Ile Asn Pro Arg Ser Ser Leu Val Leu Ala
         360            Thr    Asp        --- Glu        Asn        380

                                 1200
GCG ACA TCA GGG AAC AGT GGT ACC ACA CTT ACG GTG CAA ACC AAC ATT TAT GCC GTT AGT CAA GGT TGG CTT CCT ACT AAT AAT ACA CAA
Ala Thr Ser Gly Asn Ser Gly Thr Thr Leu Thr Val Gln Thr Asn Ile Tyr Ala Val Ser Gln Gly Trp Leu Pro Thr Asn Asn Thr Gln
         390                        400                        Pro        Phe

                        1300
CCT TTT GTT ACA ACC ATT GTT GGG CTA TAT GGT CTG TGC TTG CAA GCA AAT AGT GGA CAA GTA TGG ATA GAG GAC TGT AGC AGT GAA AAG
Pro Phe Val Thr Thr Ile Val Gly Leu Tyr Gly Leu Cys Leu Gln Ala Asn Ser Gly Gln Val Trp Ile Glu Asp Cys Ser Ser Glu Lys
Trp      420                        430                Val 440            Ser Cys

   1350                                 1400
GCT GAA CAA CAG TGG GCT CTT TAT GCA GAT GGT TCA ATA CGT CCT CAG CAA AAC CGA GAT AAT TGC CTT ACA AGT GAT TCT AAT ATA CGG
Ala Glu Gln Gln Trp Ala Leu Tyr Ala Asp Gly Ser Ile Arg Pro Gln Gln Asn Arg Asp Asn Cys Leu Thr Ser Asp Ser Asn Ile Arg
   450                            Ser        Asn    Asn        Arg        470

   1450                                 1500
GAA ACA GTT GTT AAG ATC CTC TCT TGT GGC CCT GCA TCC TCT GGC CAA CGA TGG ATG TTC AAG AAT GAT GGA ACC ATT TTA AAT TTG TAT
Glu Thr Val Val Lys Ile Leu Ser Cys Gly Pro Ala Ser Ser Gly Gln Arg Trp Met Phe Lys Asn Asp Gly Thr Ile Leu Asn Leu Tyr
   480                        490        Glu                500

              1550                                         1600
AGT GGA TTG GTG TTA GAT GTG AGG GGA GAT CCG AGC CTT AAA CAA ATC ATT CTT TAC CCT CTC CAT GGT GAC CCA AAC CAA ATA TGG
Ser Gly Leu Val Leu Asp Val Arg Arg Ser Asp Pro Ser Leu Lys Gln Ile Ile Leu Tyr Pro Leu His Gly Asp Pro Asn Gln Ile Trp
   510                Ala    520                        Trp    *            --- ---

                  1650                                                   1700
TTA CCA TTA TTT TGA TAGACAGATT ACTCTCTTGC AGTGTGTGTG TCCTGCCATG AAAATAGATG GCTTAAATAA AAAGGACATT GTAAATTTTG TAACTGAAAG
Leu Pro Leu Phe ***
*Leu Pro
              1750                              1780
GACAGCAAGT TATTGCAGTC CAGTATCTAA TAAGAGCACA ACTATTGTCT TGTGCATTCT AAATTT-Poly(A)
```

restricted homopolymer-tailed pBR322 followed by transformation of *E. coli*. A library of approximately 5000 transformants was screened with a mixture of radiolabeled oligonucleotide 35 mers based on the published RTA amino acid sequence (3), and which are almost exactly complementary to bases 268–302 in Figure 1. Sequencing of a full-length RTA insert (from clone pRA123) revealed a putative signal sequence of 35 codons and an A chain of 267 codons.

In a different approach (26), *Sal*I linkers were ligated to the free ends of the cDNA "hairpins" created during cDNA synthesis. After S1 nuclease treatment of the hairpin regions, *Eco*RI linkers were ligated. Subsequent digestion of the DS cDNAs with both restriction enzymes produced fragments suitable for forced ligation into *Eco*RI- and *Sal*I-restricted pUC13. After transformation of *E. coli*, transformants were screened using a mixture of the 16 RTB-specific oligonucleotides complementary to bases 1483–1502 (see Fig. 1). Approximately 50 clones from the 5000 screened gave hybridization signals. A variety of cDNA clones containing various regions of the RTB coding sequence were thereby obtained (26). In vitro manipulation of RTB sequences generated full-length clones as convenient cassettes suitable for subcloning and expression purposes. Full-length ricin cDNA clones were also obtained using the method of Okayama and Berg (27) and resulting transformants probed with the same 35-mer RTA-specific oligonucleotides described earlier. In this way, a clone encoding an almost full-length preproricin cDNA was obtained (12).

RTB-encoding cDNAs have also been synthesized from polyadenylated castor oil seed mRNA using the earlier method of Okayama and Berg (28). *E. coli* was transformed and RTB-containing clones identified in a two-step screening procedure which first used the RTB-specific 20-mer probes already described (complementary to bases 1483–1502, Figure 1), and subsequently utilized a unique probe complementary to a sequence at the 5' end of RTB (18). One clone, subsequently used for manipulations

---

Figure 1 Primary structure of the mRNA encoding proricin. Nucleotide and derived amino acid sequences of the cDNA clone pRCL617. The sequence shown commences at the start of the mature A chain. Differences with the earlier published protein sequences (3,4) are given beneath. Residues absent from the earlier sequences are indicated by the dashed line and the position of unique additional residues in the earlier sequence are indicated by asterisks. The dashed line beneath the 12–amino acid linker is bracketed. Amino acids and bases are numbered. Potential N-glycosylation sites are boxed and poly(A) signals underlined. The leader sequence is omitted from the figure. (Adapted from Ref. 11.)

leading to transient RTB expression in COS cells, contained a full-length RTB cDNA sequence.

Ricin E cDNA clones have been obtained using a method involving homopolymer tailing of SS cDNAs and subsequent priming of second-strand synthesis with a complementary homopolymer (13). The DS cDNAs were then tailed in the presence of dCTP and annealed with *Pst*I-digested, oligo(dG)-tailed pBR322. The resulting library of clones was screened with an RTB-specific probe (bases 1483–1499, Figure 1). One of the resulting clones was found to contain a ricinlike A chain with a B chain sequence which was ricinlike 5' to codon 150 and RCA-like to the 3' side of this site. The predicted B chain sequence (13) is almost identical with the reported sequence of a ricin E protein (5).

III. ANALYSIS OF CLONED GENES

The DNA sequence coding for a protein includes the promoters, ribosome-binding sites (for prokaryotic genes), start codon (AUG), signal sequences (for secreted proteins), coding sequences, introns (for eukaryotic genes), and 3' untranslated transcribed DNA. An analysis of those sequences permits rational adaption of the cloned gene for expression in *E. coli* and eukaryotic cells as well as insights into gene function and regulation.

Bacterial promoters consists of two hexonucleotide consensus sequences separated by 17 nucleotides upstream of the transcription start. Eukaryotic promoters include an upstream promoter element which extends for about 100 nucleotides, including the TATA box and further upstream enhancer sequences. Bacterial ribosome binding is mediated by the Shine-Dalgorno sequence located 3–11 nucleotides upstream of the initiation codon. In contrast, eukaryotic mRNA is capped with a 7-methylguanidine residue at the 5' end and ribosome binding occurs at the first AUG triplet in the messenger RNA in the eukaryotic initiation consensus sequence:

A/GNNAUGG

Nucleic acids coding for secreted proteins have a signal sequence after the AUG start codon which codes for 15–60 residues. The signal peptide has an amino-terminal basic region with 1–3 positively charged amino acid residues in 4–7 total residues, a hydrophobic core region of 14–20 amino acid residues punctuated by 1–2 prolines or glycines and often ends in serine or threonine and a cleavage region following the von Hiejne rules. The residue at −1 must be small (A,S,G,C,T, or Q); the residue at −3 must not be aromatic (F,H,Y,W), charged (D,E,K,R), or large and polar (N, Q); proline must be absent from −3 to +1. The signal peptide is usually cleaved (in the ER in eukaryotes). The resulting N-terminal amino acid may be unstable in the cytoplasm (amino acid residues other than M,S,T,A,V,C,G, or P) but are stable in the protease-deficient endoplasmic reticulum (ER) or periplasm. A number of proteins require secretion across membranes to facilitate folding and posttranslational processing.

The coding sequence may contain signals for N-linked glycosylation (N-X-S/T) or retention in the ER (C-terminal KDEL). The coding sequence ends with a TAA, TAG, or TGA stop codon. The 3' untranslated region may include a consensus sequence for RNA cleavage and polyadenylation: AAUAAA located 10–30 bases upstream from the cleavage site for animal genes and 25–44 bases downstream from the stop codon for plant genes.

Many eukaryotic genes (but so far no toxin genes except those encoding ribotoxins) have introns which are spliced from the heterogeneous nuclear RNA in the cell nucleus. Many genes, including toxin genes, are part of the multigene families which code for several related proteins.

IV. GENOMIC CLONING OF RICIN

A DNA sequence for a genomic ricin clone has been published (14). In the construction of the library, genomic DNA from castor oil seeds was partially digested with *Eco*RI, size fractionated, ligated in lambda Charon 4 arms, packaged in vitro, and *E. coli* strain K802 was transformed. Plaque hybridization using a ricin cDNA probe revealed a positive clone containing the complete preproricin gene. The gene was shown to contain no introns, but it did possess consensus promoter sequences and polyadenylation sites in the 5' and 3' flanking regions, respectively. There are just 10 nucleotide differences between this genomic sequence and the cDNA sequence originally published (11), only two of which result in a codon change, equivalent to residues 41 and 516 (see Figure 1).

A second ricin genomic clone with a coding sequence apparently differing in only 1 base position from that above has also been described (15). Once again, partial *Eco*RI fragments were prepared and these were ligated into λgt10 arms, packaged in vitro, and *E. coli* strain BNN102 (*hfl*A) was infected. Plaques were screened with an oligonucleotide probe complementary to the 5' end of RTB. The one positive clone identified contained the complete ricin-coding sequence. However, as the sequence obtained was not given and S1 nuclease or expression experiments were not performed, there is no direct evidence that this particular gene is expressed within the plant.

A third genomic clone has been fully characterized from partial *Sau*3A genomic DNA fragments ligated into Charon 35 arms, packaged, and introduced into *E. coli* strain K803 (16). Screening of the library with a ricin cDNA clone produced 17 lectin-positive clones which could be classified into five groups on the basis of Southern blot analyses. Five of these clones were selected for further analysis. A combination of detailed restriction mapping, sequencing, and RNase protection experiments revealed that only one of these clones contained a functional ricin gene. This differed by only 2 base changes in the coding region when compared with the sequence published by Halling et al. (14); these being a T to A change at base 360 and an A to C change at base 645 (see Figure 1). In the 5' flanking region, there are nine base differences over the 310-bp sequence compared, and nine differences in 321 bp of 3' flanking sequence. In the promoter region, none of the differences lie within the putative TATA or CAAT/ AGGA consensus sequences. A sequence that might function in ricin gene regulation was present at −172 of this ricin gene (pCBG3H1 [16]), where the numbering is relative to the start of the ricin coding sequence.

$$\begin{array}{ccc} & -172 & -159 \\ \text{pCBG3H1} \quad 5' & \dot{\text{C}}\text{ A T G C A T C T T C C G }\dot{\text{T}} & 3' \\ & * & \end{array}$$

This sequence is highly homologous with an upstream sequence of the castor bean 2S albumin gene promoter (S. Irwin, unpublished), where the only base difference is indicated by an asterisk. This element, in both these genes, resembles the CATGCATG RY repeat found in the promoter region of a large number of seed protein genes (29). The numbers and positions of the elements has been shown to vary considerably. The RY repeat seen here is not absolutely conserved with the equivalent sequence in the pAKG (14) which reads 5' CATTGCATTCTT 3'. However, a more upstream element, TATGCATA, is absolutely conserved. Whether or not the RY sequence divergence causes the two ricin genes to be expressed at different levels or with differing developmental patterns remains to be determined.

Evidence from the earlier work by Cawley et al. (2) suggested that the castor oil seed lectins are encoded by several genes. Southern analysis of *Hin*dIII and *Eco*RI-digested genomic DNA revealed several hybridizing bands estimated to represent approximately six lectin genes (14). Our own estimates are similar. In contrast to the distribution of *Eco*RI sites within the various lectin gene members, only one of the five genes analyzed in our laboratory possessed an internal *Hin*dIII site. Consequently, the number of hybridizing *Hin*d fragments seen in the blot shown in Figure 2 most likely represents the size of the ricin/RCA gene family. This appears to contain approximately eight members. Differential hybridization of the cDNA probe or the occurrence of multiple hybridizing fragments of similar size could possibly account for the varying intensities of the visualized

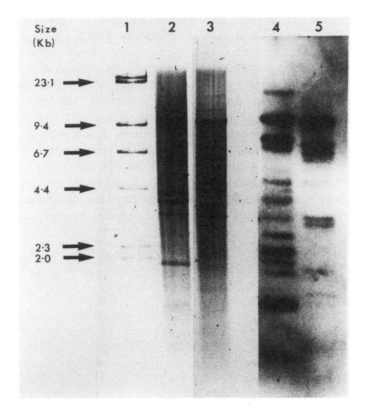

Figure 2 Southern blot analysis of genomic fragments containing lectin-hybridizing sequences. Lane 1, λ DNA *Hind*III size markers; lanes 2 and 3, 10-μg aliquots of castor oil seed genomic DNA restricted with *Eco*RI and *Hind*III, respectively; lanes 4 and 5, autoradiograph of blotted DNA fragments from lanes 2 and 3, respectively, which had been probed with a ricin cDNA and washed at low stringency (1 × SSC at 55°C for 1 hr). (From Ref. 16; used with permission.)

bands. Several of the gene members of this family are variously scrambled and nonfunctional (16).

V. THE RICIN LEADER SEQUENCE

The genomic sequences (14,16) reveal that an in-frame ATG at −105 (relative to the start of the mature RTA sequence) was the most 5′ ATG downstream from the putative transcription start point(s). Upon transcription, the resulting AUG lies in a favorable translation initiation context (30) and would generate a leader peptide of 35 amino acid residues. This was also predicted from a cDNA sequence analysis (12). However, a signal peptide with a molecular weight of approximately 4.5 kd would not be predicted from size comparisons of in vitro–synthesized preproricin and deglycosylated proricin (Figure 3). From these data, it was estimated that the size of the cleaved signal peptide was approximately 1.5 kd (equivalent to approximately 13 amino acid residues). The cDNA originally described was missing the most 5′ ATG (11). This was later introduced into the cDNA by site-directed mutagenesis and has been shown to function as the true start codon by the ability of the translated protein to translocate across ER-derived microsomes and to become N-glycosylated (Figure 4, lanes 4 and 5). In contrast, translation commencing at the in-frame AUG at −72 does not generate a protein with a functional signal peptide (Figure 4, lanes 2 and 3). Protease treatment reveals that the glycosylated proricin is pro-

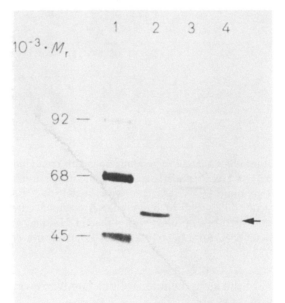

Figure 3 Endo-N-acetylglucosaminidase H treatment of nascent *R. communis* lectin polypeptides. Castor bean mRNA was translated in reticulocyte lysates in the presence or absence of pancreatic microsomes. Lane 1, molecular weight markers; lane 2, lectin polypeptides immunoprecipitated after synthesis in the absence of microsomes; lane 3, immunoprecipitated lectins after synthesis in the presence of microsomes (glycosylated); lane 4, lectins (arrows) made as in lane 3 and subsequently treated with Endo H. Analysis was by SDS-PAGE and fluorography. (From Ref. 37.)

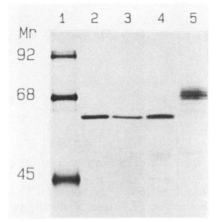

Figure 4 In vitro translocation of preproricin across microsomal membranes. Preproricin transcripts were synthesized in vitro (as in Ref. 24) and translated in rabbit reticulocyte lysates in the presence or absence of pancreatic microsomal membranes. Lane 1, molecular weight–size markers; lanes 2 and 3, proricin transcripts possessing an AUG at -72 translated in the absence and presence of membranes, respectively; lanes 4 and 5, preproricin transcripts possessing an AUG at -105 translated in the absence and presence of membranes, respectively. Analysis was by SDS-PAGE and fluorography.

tected within the vesicles, whereas the nonglycosylated preproricin made in the absence of membranes is unprotected and totally degraded (data not shown). Since it is clear from the data above and subsequent in vivo expression experiments (21,24) that translation does commence with amino acid -35, how can this be reconciled with the conclusions drawn from Figure 3 and the fact that the apparent signal peptide cleavage site at residue -1 does not obey the -3 to -1 rule proposed by von Heijne (31)? Applying the general rules of von Heijne (32) to the ricin leader peptide (Figure 5), it would appear that the alanine residue at -20 or the glycine at -14 are the most likely cleavage points. Asparagine, such as that at -1, has never been observed at a signal peptide cleavage point but is a common residue at the posttranslational cleavage sites of many plant protein body proteins (33). Taken together, this seems to indicate that the signal peptidase cleavage site lies within the 35-residue leader sequence and that the remaining N-terminal extension may subsequently be removed by protein body–located, asparagine-specific endoproteases. The hypothesis that the N-terminal extension of proricin might function as a vacuolar targeting signal is currently being tested. The significance of this relates to

```
-105
 .
ATG AAA CCG GGA GGA AAT ACT ATT GTA ATA TGG ATG
Met Lys Pro Gly Gly Asn Thr Ile Val Ile Trp Met
 -35
                  ↓                       ↓
TAT GCA GTG GCA ACA TGG CTT TGT TTT GGA TCC ACC
Tyr Ala Val Ala Thr Trp Leu Cys Phe Gly Ser Thr
             -20                      -14
                                          -1 +1
                                           .  .
TCA GGG TGG TCT TTC ACA TTA GAG GAT AAC AAC ATA
Ser Gly Trp Ser Phe Thr Leu Glu Asp Asn Asn Ile
                                          -1     +1
```

Figure 5 N-terminal leader sequence of preproricin. The amino-terminal residue of mature ricin A chain is numbered +1 beneath the sequence and those comprising the leader peptide are denoted by negative numbers. The base numbers are similarly denoted above the nucleotide sequence. Arrows indicate possible signal peptidase cleavage sites (based on data from Ref. 30).

the use of the ricin leader sequence in heterologous expression systems where it is important to define the protein product made. It is suspected, though not yet proven, that constructs containing the 35-residue leader sequence to direct import into the secretory pathway of eukaryotic cells might be processed to leave a mature protein with an extension comprising the C-terminal part of the leader sequence.

VI. COMPARISON OF CLONED SEQUENCES

From the sequences of cloned cDNAs and genes (11–18), it can be concluded that the ricin precursor is a preproprotein of 576 amino acid residues, of which 35 constitute an N-terminal leader, 267 comprise the A chain, 12 the linker, and 262 the B chain. In all the cloned sequences (exemplified

Figure 6 Primary structure of the mRNA encoding proRCA I and its comparison with proricin. The upper nucleotide sequence is that determined for RCA I cDNAs and the lower sequence is that determined previously for proricin (11). The deduced amino acid sequence of proRCA I is shown above the upper nucleotide sequence, whereas amino acid differences between this and proricin are indicated beneath the lower nucleotide sequence as the corresponding proricin residues. The linker sequence is underlined, N-glycosylation sites are boxed, and polyadenylation signals are underlined with a broken line. The N-terminal leader sequences are omitted from the figure. (From Ref. 25; used with permission.)

```
                    A→
    ⁺¹
    Ile Phe Pro Lys Gln Tyr Pro Ile Ile Asn Phe Thr Thr Ala Asp Ala Thr Val Glu Ser Tyr Thr Asn Phe Ile Arg Ala
    ATA TTC CCC AAA CAA TAC CCA ATT ATA AAC TTT ACC ACA GCA GAT GCC ACT GTG GAA AGC TAC ACA AAC TTT ATC AGA GCT 81
    ATA TTC CCC AAA CAA TAC CCA ATT ATA AAC TTT ACC ACA GCG GGT GCC ACT GTG GAA AGC TAC ACA AAC TTT ATC AGA GCT
    ⁺¹                                                Gly              Gln

    28
    Val Arg Ser His Leu Thr Thr Gly Ala Asp Val Arg His Glu Ile Pro Val Leu Pro Asn Arg Val Gly Leu Pro Ile Ser Gln Arg Phe
    GTG CGC AGT CAT TTA ACA ACT GGA GCT GAT GTG AGA CAT GAA ATA CCA GTG TTG CCA AAC AGA GTT GGT TTG CCT ATA AGC CAA CGG TTT 171
    GTT CGC GCT CCT TTA ACA ACT GGA GCT GAT GTG AGA CAT GAA ATA CCA GTG TTG CCA AAC AGA GTT GGT TTG CCT ATA AAC CAA CGG TTT
            Gly Arg                                      Asp                                              Asn

    58
    Ile Leu Val Glu Leu Ser Asn His Ala Glu Leu Ser Val Thr Leu Ala Leu Asp Val Thr Asn Ala Tyr Val Val Gly Cys Arg Ala Gly
    ATT TTA GTT GAA CTC TCA AAT CAT GCA GAG CTT TCT GTT ACA TTA GCC CTG GAT GTC ACC AAT GCA TAT GTG GTC GGC TAC CGT GCT GGA 261
    ATT TTA GTT GAA CTC TCA AAT CAT GCA GAG CTT TCT GTT ACA TTA GCC CTG GAT GTC ACC AAT GCA TAT GTG GTC GGC TAC CGT GCT GGA
                                                                                                          Tyr

    88
    Asn Ser Ala Tyr Phe Phe His Pro Asp Asn Gln Glu Asp Ala Glu Ala Ile Thr His Leu Phe Thr Asp Val Gln Asn Ser Phe Thr Phe
    AAT AGC GCC TAT TTC TTT CAT CCT GAC AAT CAA GAA GAT GCA GAA GCA ATC ACT CAT CTT TTC ACG GAT GTT CAA AAT TCA TTT ACA TTC 351
    AAT AGC GCA TAT TTC TTT CAT CCT GAC AAT CAG GAA GAT GCA GAA GCA ATC ACT CAT CTT TTC ACT GAT GTT CAA AAT CGA TAT ACA TTC
                                                                                                          Arg Tyr

    118
    Ala Phe Gly Gly Asn Tyr Asp Arg Leu Glu Gln Leu  -  Gly Gly Leu Arg Glu Asn Ile Glu Leu Gly Thr Gly Pro Leu Glu Glu Asp Ala
    GCC TTT GGT GGT AAT TAT GAT AGA CTT GAA CAA CTT     GGA GGT CTG AGA GAA AAT ATT GAG TTG GGA ACT GGT CCA TTA GAG GAC GCT 438
    GCC TTT GGT GGT AAT TAT GAT AGA CTT GAA CAA CTT GCT GGT GGT AAT CTG AGA GAA AAT ATC GAG TTG GGA ACT GGT CCA CTA GAG GAG GCT
                                                    Ala Asn                                        Asn              Glu

    147
    Ile Ser Ala Leu Tyr Tyr Tyr Ser Thr Cys Gly Thr Gln Ile Pro Thr Leu Ala Arg Ser Phe Met Val Cys Ile Gln Met Ile Ser Glu
    ATC TCA GCG CTT TAT TAT TAT AGT ACT TGT GGC ACT CAG ATT CCA ACT CTG GCT CGT TCC TTT ATG GTT TGC ATC CAA ATG ATT TCA GAA 528
    ATC TCA GCG CTT TAT TAT TAC AGT ACT TGT GGC ACT CAG ATT CCA ACT CTG GCT CGT TCC TTT ATA ATT TGC ATC CAA ATG ATT TCA GAA
                                Gly              Leu                              Ile Ile

    177
    Ala Ala Arg Phe Gln Tyr Ile Glu Gly Glu Met Arg Thr Arg Ile Arg Tyr Asn Arg Arg Ser Ala Pro Asp Pro Ser Val Ile Thr Leu
    GCA GCA AGA TTC CAG TAC ATT GAG GGA GAA ATG CGG ACG AGA ATT AGG TAC AAC CGG AGA TCT GCA CCA GAT CCT AGC GTA ATT ACA CTT 618
    GCA GCA AGA TTC CAA TAT ATT GAG GGA GAA ATG CGC ACG AGA ATT AGG TAC AAC CGG AGA TCT GCA CCA GAT CCT AGC GTA ATT ACA CTT

    207
    Glu Asn Ser Trp Gly Arg Leu Ser Thr Ala Ile Gln Glu Ser Asn Gln Gly Ala Phe Ala Ser Pro Ile Gln Leu Gln Arg Asn Gly
    GAG AAT AGT TGG GGG AGA CTT TCC ACT GCA ATT CAA GAG TCT AAC CAA GGA GCC TTT GCT AGT CCA ATT CAA CTG CAA AGA CGT AAC GGT 708
    GAG AAT AGT TGG GGG AGA CTT TCC ACT GCA ATT CAA GAG TCT AAC CAA GGA GCC TTT GCT AGT CCA ATT CAA CTG CAA AGA CGT AAT GGT
                                                                                                          ←————— A

    237
    Ser Lys Phe Asn Val Tyr Asp Val Ser Ile Leu Ile Pro Ile Ile Ala Leu Met Val Tyr Arg Cys Ala Pro Pro Pro Ser Ser Gln Phe
    TCC AAA TTC AAT GTG TAC GAT GTG AGT ATA TTA ATC CCT ATC ATA GCT CTC ATG GTG TAT AGA TGC GCA CCT CCA CCG TCG TCA CAG TTT 798
    TCC AAA TTC AGT GTG TAC GAT GTG AGT ATA TTA ATC CCT ATC ATA GCT CTC ATG GTG TAT AGA TGC GCA CCT CCA CCA TCG TCA CAG TTT
    └——┘       Ser
    
    267                               ←———— L      B →
    Ser Leu Leu Ile Arg Pro Val Val Pro Asn Phe Asn Ala Asp Val Cys Met Asp Pro Glu Pro Ile Val Arg Ile Val Gly Arg Asn Gly
    TCT TTG CTT ATA AGG CCA GTG GTG CCA AAT TTT AAT GCT GAT GTT TGT ATG GAT CCT GAG CCC ATA GTG CGT ATC GTA GGT CGA AAT GGT 888
    TCT TTG CTT ATA AGG CCA GTG GTA CCA AAT TTT AAT GCT GAT GTT TGT ATG GAT CCT GAG CCC ATA GTG CGT ATC GTA GGT CGA AAT GGT
                                 ——————————————————————————————

    297
    Leu Cys Val Asp Val Thr Gly Glu Glu Phe Phe Asp Gly Asn Pro Ile Gln Leu Trp Pro Cys Lys Ser Asn Thr Asp Trp Asn Gln Leu
    CTA TGT GTT GAT GTT ACA GGT GAA GAA TTC TTC GAT GGA AAC GCA ATA CAA TTG TGG CCA TGC AAA TCT AAT ACA GAT TGG AAT CAG TTA 978
    CTA TGT GTT GAT GTT AGG GAT GGA AGA TTC CAC GAC GGA AAC GCA ATA CAG TTG TGG CCA TGC AAG TCT AAT ACA GAT TGG AAT CAG TTA
                             Arg Asp Gly Arg     His Asn         Ala                                        Ala

    327
    Trp Thr Leu Arg Lys Asp Ser Thr Ile Arg Ser Asn Gly Lys Cys Leu Thr Ile Ser Lys Ser Ser Pro Arg Gln Gln Val Val Ile Tyr
    TGG ACT TTG AAA AGA GAC AAT ACT ATT CGA TCT AAT GGA AAG TGT TTA ACT ACT TAC CGG TAC AGT CCG ACA CAG GTC TAT GTG ATC TAT 1068
    TGG ACT TTG AAA AGA GAC AAT ACT ATT CGA TCT AAT GGA AAG TGT TTA ACT ACT TAC GGG TAC AGT CCG ACA CAG GTC TAT GTG ATG ATC TAT
                 Lys Arg     Asn                              Thr Tyr Gly Tyr            Gly Val Tyr     Met

    357
    Asn Cys Ser Thr Ala Thr Val Gly Ala Thr Arg Trp Gln Ile Trp Asp Asn Arg Thr Ile Ile Asn Pro Arg Ser Gly Leu Val Leu Ala
    AAT TGC AGT ACC GCT ACA GTT GGT GCC ACC CGT TGG CAA ATA TGG GAC AAT AGA ACC ATC ATA AAT CCC CGA TCT GGT CTA GTT TTG GCA 1158
    GAT TGC AAT ACT GCT GCA ACT GAT GCC ACC CGC TGG CAA ATA TGG GAT AAT GGA ACC ATC ATA AAT CCC AGA TCT AGT CTA GTT TTA GCA
    Asp     Asn         Ala Thr Asp                     Asp     Gly                          Ser

    387
    Ala Thr Ser Gly Asn Ser Gly Thr Lys Leu Thr Val Gln Thr Asn Ile Tyr Ala Val Ser Gln Gly Trp Leu Pro Thr Asn Asn Thr Gln
    GCC ACA TCA GGG AAC AGT GGT ACC AAA CTT ACA GTG CAA ACC AAC ATT TAT GCC GTT AGT CAA GGT TGG CTT CCT ACT AAT AAT ACA CAA 1248
    GCG ACA TCA GGG AAC AGT GGT ACC ACA CTT ACG GTG CAA ACC AAC ATT TAT GCC GTT AGT CAA GGT TGG CTT CCT ACT AAT AAT ACA CAA
                                     Thr

    417
    Pro Phe Val Thr Thr Ile Val Gly Leu Tyr Gly Met Cys Leu Gln Ala Asn Ser Gly Lys Val Trp Leu Glu Asp Cys Thr Ser Glu Lys
    CCT TTT GTG ACA ACC ATT GTT GGG CTA TAT GGC ATG TGC TTG CAA GCA AAT AGT GGA AAA GTA TGG TTA GAG GAC TGT ACC AGT GAA AAG 1338
    CCT TTT GTT ACA ACC ATT GTT GGG CTA TAT GGT ATG TGC TTG CAA GCA AAT AGT GGA CAA GTA TGG ATA GAG GAC TGT AGC AGT GAA AAG
                                             Leu                                 Gln             Ile             Ser

    447
    Ala Glu Gln Gln Trp Ala Leu Tyr Ala Asp Gly Ser Ile Arg Pro Gln Gln Asn Arg Asp Asn Cys Leu Thr Thr Asp Ala Asn Ile Lys
    GCT GAA CAA CAG TGG GCT CTT TAT GCA GAT GGT TCA ATA CGT CCT CAG CAA AAC CGA GAT AAT TGC CTT ACA ACT GAT GCA AAT ATA AAA 1428
    GCT GAA CAA CAG TGG GCT CTT TAT GCA GAT GGT TCA ATA CGT CCT CAG CAA AAC CGA GAT AAT TGC CTT ACA ACT GAT TCT AAT ATA CGG
                                                                                         Ser         Ser         Arg

    477
    Gly Thr Val Val Lys Ile Leu Ser Cys Gly Pro Ala Ser Ser Gly Gln Arg Trp Met Phe Lys Asn Asp Gly Thr Ile Leu Asn Leu Tyr
    GGA ACA GTT GTC AAG ATC CTC TCT TGT GGC CCT GCA TCC TCT GGC CAA CGA TGG ATG TTC AAG AAT GAT GGA ACC ATT TTA AAT TTG TAT 1518
    GAA ACA GTT GTT AAG ATC CTC TCT TGT GGC CCT GCA TCC TCT GGC CAA CGA TGG ATG TTC AAG AAT GAT GGA ACC ATT TTA AAT TTG TAT
    Glu

    507
    Asn Gly Leu Val Leu Asp Val Arg Arg Ser Asp Pro Ser Leu Lys Gln Ile Ile Val His Pro Phe His Gly Asn Leu Asn Gln Ile Trp
    AAT GGA TTG GTG TTA GAT GTG AGG CGA TCG GAT CCG AGC CTT AAA CAA ATC ATT GTT CAC CCT TTC CAT GGA AAC CTA AAC CAA ATA TGG 1608
    AGT GGA TTG GTG TTA GAT GTG AGG CGA TCG GAT CCG AGC CTT AAA CAA ATC ATT TAC CCT CTC CAT GGT GAC CCA AAC CAA ATA TGG
    Ser                                                             Leu Tyr     Leu     Asp Pro

    537    B →
    Leu Pro Leu Phe ***                                                                                            ⁺⁶⁸⁹
    TTA CCA TTA TTT TGA TAGACAGATT ACTCTCTTGC AGTGTGTATG TCCTGCCACT AAAATAGATG GCTTAAATAA AAAGGA - C-tail
    TTA CCA TTA TTT TGA TAGACAGATT ACTCTCTTGC AGTGTGTATG TCCTGCCATG AAAATAGATG GCTTAAATAA AAAGGACATT GTAAATTTTG TAACTGAAAG

    GACAGCAAGT TATTGCAGTC CAGTATCTAA TAAGAGCACA ACTATTGTCT TGTGCATTCT AAATTT - Poly(A) - C-tail
```

by that shown in Figure 1), there are four potential asparagine-linked glycosylation sites at residues 10 and 236 in RTA and 374 and 414 in RTB. One of these sites was not evident from the original published protein sequence of ricin A chain (3) where an aspartate was apparently detected at position 236. This could have represented an amino acid identification error or a genuine varietal or isoform difference. It is clear from analysis of ricin preparations that a variant or "heavy" form of A chain occurs which contains two oligosaccharide side chains. From Figure 1, it can be seen that there are a number of other differences between the derived sequence and the protein sequenced directly (3,4). The accumulation of nucleotide sequence data has tended to confirm the derived cDNA sequence shown. However a number of base changes between different cloned DNAs are evident. These are not extensive and rarely cause amino acid substitutions. They probably reflect heterogeneity between different seed varieties or isoforms within a given seed.

As expected, there are rather more differences between proricin and proRCA (Figure 6). In this comparison, 59 amino acids are different, of which 22 might not be expected to radically alter the properties of the proteins. It is perhaps noteworthy that the RCA A chain has two cysteine residues (84 and 156), which are absent in ricin A, and that RCA B chain has an additional N-glycosylation site (residues 357–359). It has been shown that a variant form of ricin, ricin E, possesses a B chain that appears to be a hybrid of ricin and RCA (5,13). The derived sequence (13) exhibits only three differences with the published protein sequence of the ricin E B chain (5). One of these is an amino acid transposition (Ser-Pro in [13], Pro-Ser in [5]), most likely a protein sequencing error; whereas the others are substitutions (Val-210 in [13], Ala-210 in [5]; and Phe-250 in [13], Val-250 in [5]). The switch from a ricinlike B chain sequence to an RCA-like sequence occurs at residue 150, with possibly a second switch in the 3′ untranslated region (13). It seems probable that the ricin E gene has resulted from a gene conversion event.

VII. SUMMARY

A variety of strategies have been used to obtain cDNA and genomic clones encoding ricin. Since their isolation these sequences have been manipulated to allow expression of A chain (19) and A chain mutants (15,20,34), B chain (14,21–23) and proricin (24). Utilizing structural information (35), precise changes have been introduced into both A and B chains with the aim of probing catalytic and sugar-binding residues, respectively. In the longer term, such manipulations, coupled with successful expression and purification schemes, will allow the delineation of functional residues and domains, ensuring that ricin remains the prototype plant toxin with which to study cellular intoxication and ribosome inactivation and to utilize in pharmaceutical product development.

REFERENCES

1. Endo, Y., and Tsurugi, K. RNA N-glycosidase activity of ricin A chain J. Biol. Chem., *262*: 8128–8130, 1987.
2. Cawley, D. B., Hedblom, M. L., and Houston, L. L. Homology between ricin and *Ricinus communis* agglutinin: Amino terminal sequence analysis and protein synthesis inhibition studies. Arch. Biochem. Biophys., *190*: 744–755, 1978.
3. Yoshitake, S., Funatsu, G., and Funatsu, M. Isolation and sequences of peptic peptides and the complete sequence of Ile chain of ricin D. Agric. Biol. Chem., *42*: 1267–1274, 1978.
4. Funatsu, G., Kimura, M., and Funatsu, M. Primary structure of Ala chain of ricin D. Agric. Biol. Chem., *43*: 2221–2224, 1979.
5. Araki, T., and Funatsu, G. The complete amino acid sequence of the B chain of ricin E isolated from small-grain castor bean seeds. Ricin E is a gene recombination product of ricin D and *Ricinus communis* agglutinin. Biochim. Biophys Acta, *911*: 191–200, 1987.
6. Roberts, L. M., and Lord, J. M. Protein biosynthetic capacity in the endosperm tissue of ripening castor bean seeds. Planta, *152*: 420–427, 1981.
7. Butterworth, A. G., and Lord, J. M. Ricin and *Ricinus communis* agglutinin subunits are all derived from a single size polypeptide precursor. Eur. J. Biochem., *137*: 57–65, 1983.
8. Lord, J. M., Precursors of ricin and *Ricinus communis* agglutinin: Glycosylation and processing during synthesis and intracellular transport. Eur. J. Biochem., *146*: 411–416, 1985.
9. Lord, J. M., Spooner, R. A., Hussain, K., and Roberts, L. M. Immunotoxins: Properties, applications and current limitations. Advanced Drug Delivery Reviews, *2*: 297–318, 1988.
10. Blakey, D. C., and Thorpe, P. E. An overview of therapy with immunotoxins containing ricin or its A chain. Antibody, Immunoconjugates and Radiopharmaceuticals, *1*: 1–16, 1988.
11. Lamb, F. I., Roberts, L. M., and Lord, J. M. Nucleotide sequence of cloned cDNA coding for preproricin. Eur. J. Biochem., *148*: 265–270, 1985.
12. Houston, L. L., Lane, J. A., Piatak, M., and Clark, R. Production of ricin toxins in a baculovirus-insect cell express system. International patent application WO89/01037.
13. Ladin, B. F., Murray, E. A., Halling, A. C., Halling, K. C., Tilakaratne, N., Long, G. L., Houston, L. L., and Weaver, R. F. Characterization of a cDNA encoding ricin E, a hybrid ricin-RCA gene from the castor plant, *Ricinus communis*. Plant Mol. Biol., *9*: 287–295, 1987.

14. Halling, K. C., Halling, A. C., Murray, E. M., Ladin, B. F., Houston, L. L., and Weaver, R. Genomic cloning and characterization of a ricin gene from *Ricinus communis*. Nucl. Acids Res., *13*: 8019–8033, 1985.

15. Sundan, A., Evensen, G., Hornes, E., and Mathiesen, A. Isolation and *in vitro* expression of the ricin A chain gene: Effect of deletions on biological activity. Nucl. Acids Res., *17*: 1717–1732, 1989.

16. Tregear, J. W. The Lectin Gene Family of *Ricinus Communis*. Ph.D. thesis, University of Warwick, England, 1989.

17. Piatak, M., Lane, J. A., Laird, W., Bjorn, M. J., Wang, A., and Williams, M. Expression of soluble and fully functional RTA in *E. coli* is temperature sensitive. J. Biol. Chem., *263*: 4837–4843, 1988.

18. Chang, M-S., Russell, D. W., Uhr, J. W., and Vitetta, E. S. Cloning and expression of recombinant functional ricin B chain. Proc. Natl. Acad. Sci. U.S.A., *84*: 5640–5644, 1987.

19. O'Hare, M., Roberts, L. M., Thorpe, P. E., Watson, G. J., Prior, B., and Lord, J. M. Expression of ricin A chain in *E. coli*. FEBS Lett., *216*: 73–78, 1987.

20. Frankel, A., Schlossman, D., Welsh, P., Hertler, A., Withers, D., and Johnston, S. Selection and characterization of ricin toxin A chain mutations in *Saccharomyces cerevisiae*. Mol. Cell Biol., *9*: 415–420, 1989.

21. Richardson, P. T., Gilmartin, P., Colman, A., Roberts, L. M., and Lord, J. M. Expression of functional ricin B chain in *Xenopus* oocytes. Biotechnology, *6*: 565–570, 1988.

22. Richardson, P. T., Roberts, L. M., Gould, J. H., and Lord, J. M. The expression of functional ricin B chain in *Saccharomyces cerevisiae*. Biochim. Biophys. Acta, *950*: 385–394, 1988.

23. Hussain, K., Bowler, C., Roberts, L. M., and Lord, J. M. Expression of ricin B chain in *E. coli*. FEBS Lett., *244*: 383–387, 1989.

24. Richardson, P. T., Westby, M., Roberts, L. M., Gould, J. H., Colman, A., and Lord, J. M. Recombinant proricin binds galactose but does not depurinate 28S rRNA. FEBS Lett., *255*: 15–20, 1989.

25. Roberts, L. M., Lamb, F. I., Pappin, D. J. C., and Lord, J. M. The primary sequence of *Ricinus communis* agglutinin. J. Biol. Chem., *260*: 15682–15686, 1985.

26. Piatak, M., Emerick, A. W., and Houston, L. L. Recombinant ricin toxin fragments. International patent application WO88/07081.

27. Okayama, H., and Berg, P. A cDNA cloning vector that permits expression of cDNA inserts in mammalian cells. Mol. Cell. Biol., *3*: 280–289, 1983.

28. Okayama, H., and Berg, P. High efficiency cloning of full length cDNA. Mol. Cell. Biol., *2*: 161–170, 1992.

29. Dickinson, C. D., Evans, R. P., and Nielsen, N. C. RY repeats are conserved in the 5' flanking regions of legume seed protein genes. Nucl. Acids Res., *16*: 371–376, 1988.

30. Kozak, M. Compilation and analysis of sequences upstream from the translational start site in eukaryotic mRNAs. Nucl. Acids Res., *12*: 857–872, 1984.

31. von Heijne, G. A new method for predicting signal sequence cleavage sites. Nucl. Acids Res., *14*: 4683–4690, 1986.

32. von Heijne, G. Signal sequences. The limits of variation. J. Mol. Biol., *184*: 99–105, 1985.

33. Lord, J. M., and Robinson, C. Role of proteolytic enzymes in the posttranslational modification of proteins. *In*: M. J. Dalling (ed.), Plant Proteolytic Enzymes, Vol. II, pp. 74–80. CRC Press, Boca Raton, Florida: 1986.

34. May, M., Hartley, M. R., Roberts, L. M., Krieg, P. A., Osborn, R. W., and Lord, J. M. Ribosome inactivation by ricin A chain: a sensitive method to assess the activity of wild-type and mutant polypeptides. EMBO J., *8*: 301–308, 1989.

35. Montfort, W., Villafranca, J. E., Monzingo, A. F., Ernst, S. R., Katzin, B., Rutenber, E., Xuong, N. H., Hamlin, R., and Robertus, J. D. The three-dimensional structure of ricin at 2.8 A. J. Biol. Chem., *262*: 5398–5403, 1987.

36. O'Hare, M., Roberts, L. M., and Lord, J. M. Biological Activity of recombinant *Ricinus communis* agglutinin A chain produced in *Escherichia coli*. FEBS Lett. (in press).

37. Roberts, L. M., and Lord, J. M. The synthesis of *Ricinus communis* agglutinin: Co-translational and post-translational modification of agglutinin peptides. Eur. J. Biochem., *119*: 31–40, 1981.

6

Expression of Plant-Derived Ribosome-Inactivating Proteins in Heterologous Systems

Michael Piatak, Jr.   *Genelabs Incorporated, Redwood City, California*

Noriyuki Habuka   *Japan Tobacco, Inc., Yokohama, Kanagawa, Japan*

I. INTRODUCTION

The purpose of this chapter is to review the expression of plant-derived ribosome-inactivating proteins (RIPs) in heterologous systems as it applies to the production of these proteins and to their study. Several reviews are available that discuss the isolation, characteristics, activities, and distribution of these proteins (1–4), and the reader is referred to these and to other chapters in this book for those details. In consideration of expressing these proteins in heterologous systems, however, several features are worth noting.

Foremost, these proteins are potent inhibitors of eukaryotic protein synthesis. They function as N-glycosidases to remove a specific adenine residue, A^{4324} in rat liver rRNA, from 28S rRNA (5,6), leading to an impaired ability of the ribosome to bind elongation factors (7,8). Different RIPs have selective activities on different ribosomes, but in general all eukaryotic ribosomes are affected to some degree (9,10,80).

These proteins have been generally grouped into two types (4). Type I RIPs consist of a single peptide chain that may or may not be glycosylated and have an M_r of 23–32K. They exhibit a typical alkaline pI of 8.0–10.0. Type II RIPs consist of an A chain, which is essentially equivalent to a type I RIP, disulfide linked to a lectinlike B chain that binds the RIP to cell surfaces and facilitates entry of the A chain into the cytosol. Thus, the type II RIPs are potent toxins and are typified by the proteins ricin and abrin, the most familiar and best described of the group. Overall, the type

II proteins are generally glycosylated and exhibit a relatively neutral pI. The A chain, or type I analog, in this case may be acidic or basic. The type II proteins are typically about M_r 60K.

There is a third class of related proteins, the agglutinins, that consist of two A chains and two B chains and resemble dimers of type II proteins. Because of their structural arrangement, they are bifunctional lectins and agglutinate cells very efficiently. They are also relatively nontoxic to cells, although the A chain components are effective in inactivating ribosomes in vitro (78; J. M. Lord, personal communication).

Both the type I and the type II RIPs have received considerable interest because of their use in the development of chimeric toxins as therapeutic agents (discussed in other Chapters 9 and 24) and the topic of the first book in this series (11) (also see Refs. 12,13). The type I RIPs and the A chains of the type II RIPs alone also have been shown to have antiviral properties (14–19), which have not been exploited until recently. The type I RIP, α-trichosanthin (α-TCS), has been shown to be effective in blocking replication of human immunodeficiency virus 1 (HIV-1) in T cells and macrophages in vitro (19), and is currently being tested in patients with acquired immune deficiency syndrome (AIDS). These applications in human therapeutics and the potential applications for chimeric toxins and antiviral activity in other areas, for example, in veterinary medicine and agriculture, should further stimulate interest and research in this large and diverse group of proteins.

The plant-derived ribosome-inactivating proteins are widely distributed throughout the plant kingdom. They may be found in virtually all parts of a plant and may accumulate to significant amounts, from 0.1% to greater than 1% of total tissue weight (see reviews in Refs. 1–4, 20; B. Almassian, personal communication). These numbers may be dramatically higher if taken as a percent of total protein, particularly when the organ source is low in overall protein content, for example, tubers. Given such abundant sources of native protein, why clone and express these proteins in heterologous systems? The obvious reason is to be able to manipulate a gene sequence and thereby functionally dissect the activities and properties of a protein. This is particularly applicable in the case of ricin since a crystallographic model has been determined (21) and structure/function assignments might be made. Energy-minimized molecular models for α-TCS and abrin A chain (abrA) also have been generated based on the ricin model (22), making these proteins candidates for similar analyses. There are several other reasons, however, for expression of RIPs in heterologous systems that apply not only to facilitating their study, but to their utility as therapeutic and other agents.

One reason is to have a homogeneous, sequence-defined protein for study. Likely without exception, RIPs exist in multigene families in plant genomes (23,24). Expression may be temporally regulated and tissue specific, but it is common to identify multiple, different RIPs in one plant. Given the similarity in their physical and chemical characteristics, separation of closely related but otherwise distinct proteins may not be achieved. In addition, posttranslational modifications, such as glycosylation, may occur and these may not be uniform, further contributing to the heterogeneity. For example, ricin A chain contains two potential N-linked glycosylation sites, but both singly and doubly glycosylated species have been isolated from the same preparation of toxin (25). Variants of the same plant may also yield RIPs of slightly different primary sequence and extent of glycosylation, requiring that plant source material be rigorously controlled or cultivated from one seed source (see Chapter 5).

In studying the separate chains of the type II RIPs and, in particular, in using them for therapeutics, protein purity is an important issue that may be resolved by expression of cloned genes. It is difficult to purify A chains from B chains, and it is estimated that one molecule of contaminating B chain per 1000 molecules of A chain can result in unacceptable levels of toxicity in vivo.

Another reason for expressing cloned genes is to be able to control the type and extent of glycosylation. If used as a therapeutic, RIPs that are naturally glycosylated may be cleared from circulation through recognition and binding of the sugar moiety; this may also lead to selective organ toxicity. As an example, immunotoxins prepared with native ricin A chain (RTA) were selectively cleared through mannose-dependent recognition in the liver and showed liver toxicity independent of the specificity of the conjugated antibody (26). Immunotoxins prepared with chemically deglycosylated RTA, on the other hand, had longer serum half-lives and did not show the liver uptake of the former (27). The RIPs expressed in *Escherichia coli* would not be glycosylated, but it would also be possible to engineer the RIP gene to eliminate any glycosylation sites for expression in other systems. Since there are RIPs that are naturally not glycosylated, it seems likely that the sugar additions are dispensable. As noted above, RTA has been shown to be functional in both states; an aglycosylated recombinant form has also been used in the preparation of active immunotoxins (28,29).

II. EXPRESSION OF RIPs IN *E. COLI*

Although it would be the obvious first choice for most any protein, *E. coli* holds particular advantages for the expression of RIPs. With only one known exception, *Mirabilis* antiviral protein (MAP), *E. coli* ribosomes are essentially insensitive to the action of RIPs, and thus high levels of expressed product might be accumulated intracellularly. The other advantage to *E. coli* is that its proteins are not glycosylated.

A. Intracellular expression

Ricin A chain (RTA) was first expressed from a modified cDNA clone by O'Hare et al. (30). Ricin is naturally produced as a preproprotein containing the sequences for both the A and the B chain (see Chapter 5) and undergoes processing at the amino end and at a linker peptide joining the A chain to the B chain sequence. The cDNA was first modified, then, to allow for expression of a maturelike A chain. The modified sequence was placed under control of the strong coliphage T5 promoter P_{N25} and the *E. coli lac* operator. The A chain was expressed as a short fusion protein, containing a Met from the vector followed by five residues from a polylinker and then the last four residues of the ricin amino-terminal leader. Nonetheless, the expressed product was enzymatically active and functionally analogous to native RTA. The estimated level of expression was about 10% of cell protein.

Piatak et al. (28) also demonstrated expression of RTA in *E. coli*, and further showed that the production of soluble, fully functional protein was temperature sensitive. Three promoter systems were studied:

1. The major leftward promoter of λ phage, P_L, regulated by the temperature-sensitive repressor, cI_{857}, which is inactivated at 42°C
2. The *trpE* promoter, typically induced by starvation for tryptophan
3. The *phoA* promoter, induced by depletion of phosphate.

Depending on the conditions, expression levels up to 10% of cell protein could be achieved with each system. Amino-terminal sequencing of purified product indicated that the starting Met residue placed ahead of the mature sequence was removed in 60% of the case. Ben-Bassat et al. (31) showed that this could be improved to 91% if *E. coli* methionine aminopeptidase is coexpressed in the cells. Hence, preparations of RTA that are 91% identical to the native primary sequence are achievable by intracellular expression.

The most significant observation that Piatak and his colleagues made, however, was that production of soluble, presumably correctly folded RTA, is temperature sensitive. Particular features of the P_L system employed allowed them to study the expression of RTA at 37°C as well as at 42°C, and also to perform appropriate shift-up/shift-down experiments. The gene for RTA was placed in a plasmid under control of the P_L promoter and

N gene ribosome-binding site. The plasmid also contained a temperature-sensitive replicon which resulted in approximately a 20-fold increase in copy number when cultures were shifted from 30 to 42°C. Incubation at 37°C yielded an intermediate copy number. The host strain employed contained one copy of the λ prophage, $\lambda N_7 N_{53} cI_{857} SusP80$, which also carried a functional *cro* gene. This last feature is important in that the product of *cro*, the expression of which is also repressed by cI_{857}, functions to repress further synthesis of cI_{857}. Hence, a brief shift to 42°C to inactivate existing cI_{857} is sufficient to allow for full induction of P_L since *cro* will also be expressed and will prevent further production of the cI_{857} repressor. Additionally, continued expression at 37°C is also ensured because of the increase in template and operator copy number from the temperature-sensitive replicon which serves to titrate any available repressor. Once induced by a brief exposure to 42°C, or by a longer incubation at 37°C, the system is "autoinduced" and protein expression may be studied at different temperatures.

The λ bacteriophage is a temperate phage that can either undergo lytic growth (reproducing itself manyfold while it kills cells) or lysogenic growth (its DNA combines into the bacterial chromosome and persists passively). After phage infection of a bacterium, the phage DNA circularizes and RNA synthesis starts from P_L and P_R and makes mRNA for N and *cro* genes. N acts as an antiterminator so RNA polymerase extends to cII, O, and P. cII protein stimulates synthesis of λ repressor at P_{RE} and synthesis of integrase from P_i. If enough λ repressor (cI) is made, it binds O_R and O_L and blocks RNA synthesis and stimulates more repressor protein from P_{RM}. Lysogeny is established and maintained. On the hand, if *cro* protein dominates early, then *cro* occupies O_R and O_L and blocks repressor synthesis from P_{RM} but allow Q protein to accumulate. Q stimulates transcription of phage late genes and lytic infection occurs. The opposite effects of cI and *cro* on the promoters P_L and P_R is mediated by their differential binding to the suboperators O_{R1}, O_{R2}, and O_{R3} and the analagous O_{L1}, O_{L2}, and O_{L3}. cI binds the O_{R1} and O_{R2} sites, repressing the rightward promoter and activating P_{RM} for cI RNA synthesis. At high cI levels, O_{R3} is bound and P_{RM} is blocked. Conversely, *cro* binds O_{R3} most, which blocks P_{RM} and only binds O_{R1} and O_{R2} at higher *cro* levels. So the rightward promoter stays active long enough for lytic infection. The ricA gene placed under the control of the P_L promoter in a host bacterial strain carrying a λ prophage with cI_{857} with a functional *cro* gene will also have a functional switch for expression. cI_{857} constitutively makes λ repressor, which represses *cro* synthesis. At high temperature, the cI is inactivated, *cro* synthesis is permitted, and further cI synthesis is blocked. The P_L is released to permit continuous RTA mRNA synthesis.

By manipulating this system, Piatak et al. (28) showed that RTA expressed at 42°C was insoluble and active against ribosomes only when assayed in the presence of SDS; in contrast, RTA expressed at 37°C was soluble and fully functional, presumably because it was folded properly. This effect was demonstrated in vivo only, however. Soluble material in lysates was not converted to insoluble material in vitro, although the incubations were carried out for only 30 min. It seems likely, however, that RTA and perhaps type I RIPs can take on different conformations at different temperatures. A conformational analysis has been performed on the type I RIP trichosanthin. The results suggested that structural changes begin at about 35°C, and that an alternative conformation is achieved at 42°C (Wang Jiahuai, personal communication). These observations strongly suggest that high culture temperatures be avoided and that expression at lower temperatures, even less than 37°C, be tried. O'Hare et al. (30) observed that the RTA molecule that they produced was aggregated and less active when expressed at 37°C, but was fully active when expressed at 30°C. Piatak et al. (28) also reported that the highest levels of expression of RTA in their systems were achieved at 30°C.

Piatak et al. (28) also showed expression of RTA in a cistronlike arrangement with the *phoA* signal peptide (see Figure 5, B[5]). The potential advantage of such a system is that translation initiation and elongation over a short stretch of a naturally expressed *E. coli* sequence would presumably be optimal. The cistron arrangement avoids the generation of a fusion protein, however, and the need to proteolytically cleave it before final purification. With some manipulation, very high levels of expressed protein are possible (32).

Expression of a protein in fusion with another does have advantages, however, in both ease of purification and stability. Schlossman et al. (33) exploited a system developed by Germino and Bastia (34) and expressed RTA as a tripartite fusion protein. It consisted of MetAspProAsn added to the amino-terminus of RTA and Leu instead of Phe at the carboxyl-end of RTA followed by 60 amino acids of chicken proα-2 collagen and β-galactosidase. Expression was controlled by the λ phage P_R promoter regulated as noted above by the cI_{857} repressor, which was also carried on the plasmid. Induction was accomplished by brief incubation at 42°C followed by continued incubation at 37°C. The fusion protein was readily purified and monitored by assaying for β-galactosidase activity. RTA was then released by digestion with collagenase and isolated by size exclusion chromatography. The weakest aspect of this approach lies with the collagenase digestion. Highly purified preparations of enzyme are essential and the conditions for digestions need to be carefully controlled to assure efficient specific cleavage but minimal protein degradation. The overall expression level of the fusion protein was comparable to that noted above for RTA alone and, despite the somewhat involved process, the yield of purified RTA was about 2–3 mg/L culture. Functionally, the purified recombinant RTA was comparable to native protein, although reproducible differences were noted. In a translation inhibition assay, the recombinant protein was more than twofold less active, showing an ID_{50} concentration of 4.5 × 10^{-11} M compared to 2 × 10^{-11} M for the native protein. In reconstitution experiments with native ricin B chain, five times more recombinant than native protein was required to form an equivalent amount of holotoxin. The reasons for these differences are unclear but may reflect subtle variations in protein folding or be due to the additional amino-terminal residues and/or to residual collagen-derived sequences at the carboxyl-end of the recombinant protein. Nevertheless, the expressed product was analogous in activity to the native protein and provided a base for comparison of activities of sequence-altered proteins also produced in the system.

The first expression of a type I RIP, *Mirabilis* antiviral protein (MAP), was doubly significant as MAP was later found to have activity against *E. coli* ribosomes. As shown in Figure 1, MAP shows an ID_{50} concentration of about 200 nM in an *E. coli*–derived in vitro translation system. RTA in the same system did not show any effect on translation at up to 2 μM (data not shown). Habuka et al. (35) first determined the primary amino acid sequence of MAP, isolated from *Mirabilis jalapa*, and then designed a synthetic gene for expression in *E. coli*, an approach that holds particular advantages and is worth describing here.

The synthetic gene was designed with optimal codon usage for *E. coli* and to incorporate multiple, unique restriction sites at fairly even spacing to facilitate later gene manipulations. The primary amino acid sequence and synthetic gene sequence for MAP are shown in Figure 2. MAP was found to consist of 250 amino acids and to have a calculated molecular weight of 27,800 and a calculated pI of 9.43, consistent with the general characteristics of a type I RIP. There is one potential N-linked glycosylation site at position 16, but the native protein is not reported to contain any sugar (35). There are two cysteine residues, positions 36 and 220, that would appear to be disulfide linked in the native molecule based on analysis of cleavage peptides. Formation of this disulfide linkage within *E. coli* will not occur, however, as the intracellular environment is a reducing one and may form correctly only with low efficiency as the protein is exposed to more oxidizing environments during isolation. The formation of intermolecular disulfide linkages are also possible. This emphasizes a major limitation of intracellular expression in *E. coli* and should be seriously considered if expression of RIPs containing multiple cysteine residues, such

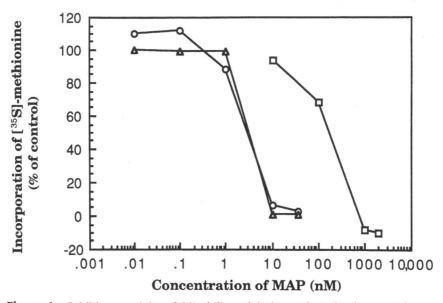

Figure 1 Inhibitory activity of *Mirabilis* antiviral protein to in vitro protein synthesis. The inhibitory activity of MAP was determined in rabbit reticulocyte (circles) and wheat germ (triangles) systems (commercially available) using tobacco mosaic virus RNA as mRNA. Inhibitory activity in an *E. coli* (boxes) system (74) was determined using MS2 phage RNA prepared after Shimura et al. (75). The results are presented as percent of control incorporation of [³⁵S]methionine in samples without added MAP.

as the pokeweed antiviral proteins (73), are of interest. In the case of MAP, a recombinant molecule was produced, but the formation of the disulfide linkage and the comparative activity of the recombinant to the native protein was not assessed because of limitations in the amount of material produced, as noted below.

The MAP gene was designed to have 759 bp, including a methionine codon at the 5′ end for initiation of translation and two termination codons (TAATAA) at the 3′ end. The gene originally reported had 13 unique restriction sites; the improved version shown in Figure 2 has 21 unique sites spaced, on the average, 36 bp apart. The gene was composed of 30 synthetic oligonucleotides. It was constructed with nine blocks of double-stranded DNA flanked by unique restriction sites to allow for sequential ligation of each block in the correct position and alignment in the final gene sequence. Each block comprised two or four complementary oligonucleotides (in the case of four oligonucleotides, there was a small overlapping sequence between the two major double-stranded regions) and

```
                      XbaI                    30                                          60
              ATG GCG CCT ACT CTA GAA ACC ATC GCT TCT CTG GAC CTG AAC AAC CCG ACC ACC TAC CTG
              Met Ala Pro Thr Leu Glu Thr Ile Ala Ser Leu Asp Leu Asn Asn Pro Thr Thr Tyr Leu

                      SspI    SplI            90                                         120
              TCT TTC ATA ACG AAT ATT CGT ACG AAA GTC GCA GAC AAA ACC GAA CAG TGT ACC ATC CAG
              Ser Phe Ile Thr Asn Ile Arg Thr Lys Val Ala Asp Lys Thr Glu Gln Cys Thr Ile Gln

                                             150              BglII        SacI     180
              AAA ATC TCT AAA ACC TTC ACC CAG CGT TAC TCT TAC ATA GAT CTG ATC GTG AGC TCG ACG
              Lys Ile Ser Lys Thr Phe Thr Gln Arg Tyr Ser Tyr Ile Asp Leu Ile Val Ser Ser Thr

                      NheI                   210                                         240
              CAG AAA ATC ACG CTA GCT ATC GAC ATG GCT GAC CTG TAC GTT CTG GGT TAC TCT GAC ATC
              Gln Lys Ile Thr Leu Ala Ile Asp Met Ala Asp Leu Tyr Val Leu Gly Tyr Ser Asp Ile

              NruI                           270         AatII                           300
              GCG AAT AAC AAG GGT CGT GCT TTC TTC TTC AAA GAC GTC ACT GAG GCT GTT GCG AAC AAT
              Ala Asn Asn Lys Gly Arg Ala Phe Phe Phe Lys Asp Val Thr Glu Ala Val Ala Asn Asn

                      XmaI                   330                                         360
              TTC TTC CCG GGA GCT ACA GGT ACT AAT CGT ATC AAA TTA ACC TTT ACA GGT TCT TAT GGC
              Phe Phe Pro Gly Ala Thr Gly Thr Asn Arg Ile Lys Leu Thr Phe Thr Gly Ser Tyr Gly

                      XhoI                   390              AvrII                       420
              GAT CTC GAG AAA AAC GGC GGA CTA CGT AAG GAC AAT CCC CTA GGT ATC TTC CGT CTG GAA
              Asp Leu Glu Lys Asn Gly Gly Leu Arg Lys Asp Asn Pro Leu Gly Ile Phe Arg Leu Glu

                      HpaI                450NaeI                                        480
              AAC TCG ATA GTT AAC ATT TAT GGC AAA GCC GGC GAC GTT AAA AAA CAG GCT AAA TTC TTC
              Asn Ser Ile Val Asn Ile Tyr Gly Lys Ala Gly Asp Val Lys Lys Gln Ala Lys Phe Phe

                                    510 BssHII                                           540
              TTA CTG GCT ATC CAG ATG GTT TCG GAG GCT GCG CGC TTT AAG TAT ATC AGT GAC AAA ATC
              Leu Leu Ala Ile Gln Met Val Ser Glu Ala Ala Arg Phe Lys Tyr Ile Ser Asp Lys Ile

                                    570SalI                    Eco47III           600
              CCG TCT GAA AAA TAC GAA GAA GTT ACC GTC GAC GAA TAC ATG ACA GCG CTG GAA AAC AAC
              Pro Ser Glu Lys Tyr Glu Glu Val Thr Val Asp Glu Tyr Met Thr Ala Leu Glu Asn Asn

                      Eco52I                 630                                         660
              TGG GCT AAA CTG TCT ACG GCC GTA TAC AAC TCT AAG CCT TCT ACC ACC ACC GCT ACC AAA
              Trp Ala Lys Leu Ser Thr Ala Val Tyr Asn Ser Lys Pro Ser Thr Thr Thr Ala Thr Lys

              PvuII                  BstEII  690                                         720
              TGT CAG CTG GCT ACC TCT CCG GTT ACC ATC TCT CCG TGG ATA TTC AAA ACC GTC GAG GAA
              Cys Gln Leu Ala Thr Ser Pro Val Thr Ile Ser Pro Trp Ile Phe Lys Thr Val Glu Glu

                          AflII 750                     BanIII
              ATC AAA CTG GTT ATG GGT CTG CTT AAG TCT TCT TAA TAA ATC GAT
              Ile Lys Leu Val Met Gly Leu Leu Lys Ser Ser *** ***
```

Figure 2 Synthetic gene for MAP. The designed DNA sequence and the determined protein sequence for MAP are shown. The DNA sequence is numbered for reference. Incorporated restriction sites are noted above the sequence.

contained *Eco*RI and *Hind*III cohesive ends for ligation into pUC19. Adjacent to the *Eco*RI and *Hind*III sites were dinucleotide spacers followed by the unique restriction sites used in condensing the gene (Figure 3). Each block was cloned and sequenced in pUC19. It was then relatively straightforward to sequentially clone each block using the unique restriction sites available into a cohesive gene. To complete the MAP construct for expression, the final *Eco*RI–*Hind*III fragment was appropriately inserted together with a linker containing the methionine codon for initiation of translation into pKK223-3 (36), which has a *tac* promoter and a ribosome-binding site.

Transformants of *E. coli* JM109 were cultured and appropriately induced by addition of isopropyl-β-D-thiogalactopyranoside (IPTG). An enzyme-linked immunosorbent assay (ELISA) on cell lysates showed that 3.6 μg of recombinant MAP was expressed per liter of culture. The expressed MAP was identical in size and immunologically similar to native MAP by Western blot analysis and was active in an anti–plant virus assay (Habuka et al., unpublished results).

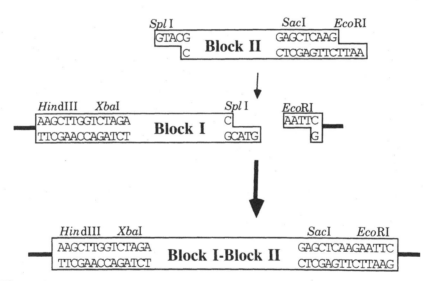

Figure 3 Design and ligation of synthetic DNA blocks of the constructed MAP gene. An example of two synthetic DNA blocks and their ligation are shown. The blocks were subcloned into plasmids as *Hind*III-*Eco*RI cassettes. The plasmid containing Block I is digested with *Spl*I and *Eco*RI and ligated with *Spl*I-*Eco*RI digested Block II. The resultant Block I–Block II construct is then ready for further ligation with a *Sac*I-*Eco*RI digested Block III, and so on.

The extremely low level of expression and an observed lower growth rate for transformants carrying the expression construct suggested that the MAP protein might be toxic to the cells. Consistent with this, only transformants of *E. coli* strains carrying the strong *lacI*q repressor, such as JM109, JM105, and MV1184, and not strains carrying only *lacI*+ could be obtained. As shown in Figure 1, MAP was subsequently found to inhibit *E. coli* protein synthesis.

In one approach to circumvent this problem, the MAP gene was expressed as a fusion protein with β-galactosidase, although in an alternative arrangement than employed by Schlossman et al. (33) for ricA. The MAP gene was linked to the carboxyl-terminal coding sequence of β-galactosidase through a linker coding for the cleavage site of the blood coagulation factor Xa (37). The chimeric gene was inserted into the plasmid pTTQ9 (38) downstream of the *tac* promoter and was appropriately induced. In this instance, the plasmid also carries a copy of *lacI*q for tighter regulation of the promoter. An ELISA assay for MAP showed that about 150 μg/L was produced, a significant increase over that noted above. This recombinant MAP was not characterized further, however, as an alternative expression system yielded better results as discussed below.

α-Trichosanthin (α-TCS) is the second type I RIP, and only the third overall, that has now been expressed in *E. coli*. Piatak and his colleagues (manuscripts in preparation) took two separate approaches to obtain expression clones for α-TCS. The first approach was to construct a synthetic gene as was done for MAP. For this, the complete primary amino acid sequence for α-TCS was first determined (22). The sequence obtained differed significantly from that reported by Wang et al. (39), most notably in the inclusion of an additional 21-residue tryptic peptide in the carboxyl region of the sequence. A gene was then designed and constructed from synthetic oligonucleotides. Differences from the approach of Habuka and his colleagues were that the coding sequence for the first 40 amino acids was made as A + T rich as possible to reduce the possibility of forming stable RNA secondary structures which could reduce translation efficiency and a two-plasmid system was employed to facilitate the condensation of the gene. The basic approach in the condensation is outlined in Figure 4. The plasmids depicted are pBluescript SK (40) and a mini-plasmid derived from pACYC184 (41; Piatak et al., manuscript in preparation). These plasmids are compatible and may be stably carried in the same cell. In each plasmid, a unique polylinker containing the sites placed at the end of each synthetic block was first introduced. Alternate, synthetic fragments having unique restriction site overlaps are then cloned into the respective plasmids and sequenced. As shown in Figure 4, to condense the gene, two

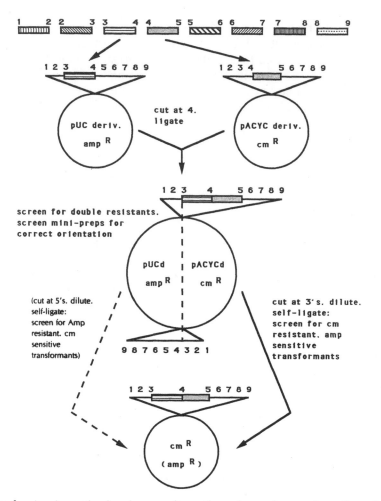

Figure 4 A schematic showing an alternative scheme for condensation of two cloned, synthetic blocks of DNA as part of a total synthetic gene.

compatible plasmids carrying adjacent fragments are opened at one site and ligated. Transformants carrying the fused plasmids are readily selected as $amp^R cm^R$. Restriction analysis of mini-prep DNA indicates those in which the fragments are fused correctly and the fused fragment is then subcloned in either original plasmid as appropriate by cutting at one of the duplicated, flanking sites to release the other plasmid sequence. The correct plasmid is selected as either $amp^S cm^R$ or $amp^R cm^S$ and is then ready for the second round of condensation. In the scheme shown in Figure 4, eight individual subclones may be condensed to four, which are condensed to two and then to 1. A synthetic gene for α-TCS was obtained as an *Nco*I–*Sac*I fragment and was placed in pKK233-2 (42) under control of a synthetic *trp-lac* promoter. This vector also contains the *rrn*B transcription termination sequence downstream of the cloned inserts to prevent run-on transcripts and to improve mRNA stability.

The second, concurrent approach was to modify a genomic clone for α-TCS (24). The gene sequence for α-TCS showed that, like the *Ricinus* proteins, this protein was also produced as a preproprotein that undergoes amino- and carboxyl-terminal processing to release the mature protein sequence (see Chapter 10). For expression, then, an ATG start codon was placed at the beginning of the mature sequence and a double termination sequence, TAATAA, was placed at the end of the mature sequence. This sequence, as a final *Nco*I–*Hin*dIII cassette, was also placed in pKK233-2.

Both constructs were induced and the expression of recombinant α-TCS was detected by Western blot analysis. The overall level of expression in each case, however, was no greater than about 1% of cell protein. Although the synthetic gene was designed with considerations for optimizing translational efficiency, the level of expression obtained was not significantly higher than the genomic. The reason for the low level of expressed α-TCS relative to that obtained for ricA is not obvious.

Purified, recombinant α-TCS was obtained at a yield of about 1 mg/L culture and was found to be identical to the nature protein in translation inhibition and anti–HIV-1 activities.

III. EXPRESSION OF RIPs AS SECRETED PROTEINS

Analysis of genomic and cDNA clones for several RIPs showed that they are produced as secreted proteins. It is likely that this is true of all RIPs, given their activity against ribosomes. Although ribosomes from the plant producing a given RIP are very resistant to its action, they are not absolutely resistant (10,80), and one mechanism to prevent self-intoxication would be to place active, mature protein in a compartment apart from the cytosol. Considerations of proper folding and of alignment and formation of disulfide bridges supply strong arguments for expressing a naturally secreted protein as a secreted protein in a heterologous system. The argument for secreting MAP, which has activity against *E. coli* ribosomes, is obvious.

Ricin B chain (RTB) and MAP have been successfully expressed as secreted proteins in *E. coli* (43,79). In both instances, the signal sequence from *ompA*, which encodes a major outer membrane protein, was employed (Figure 5, A). This is noteworthy as RTA designed for secretion

```
A.
(1) MetLysLysThrAlaIleAlaIleAlaValAlaLeuAlaGlyPheAlaThrValAlaGlnAla GlyIleProSerLeuAspProGluPro
    -------------------------------------  ------------------------------  -^ linker -- -^ rlcB -->
                    ompA

(2) ------------------------  ompA  ---------------------------  AlaProThrLeuGlu
                                                                ---- MAP --->
B.
(3) MetLysGlnSerThrIleAlaLeuAlaLeuLeuProLeuLeuPheThrProValThrLysAla IlePheProLysGln
    -------------------------------------  ------------------------------  ---- rlcA --->
                    phoA

(4) ------------------------  phoA  ---------------------------  AlaPheProLysGln
                                                                - rlcA --->
(5)
           -------------------  phoA peptide  ------------------------------------
       MetLysGlnSerThrIleAlaLeuAlaLaLeuLeuProLeuLeuPheThrProValThrLysAlaIleSerLeuTER
                                                               ---TTATGATAT
                                                                  MetIlePheProLys
                                                                  - rlcA --->
```

Figure 5 Expression constructs utilizing signal peptide sequences from *E. coli*. Shown are the amino acid sequences covering the fusion regions in those constructs thus far tested. Dashed lines indicate the extent of a particular peptide feature. ∧ indicates the position at which signal peptidase is expected to cleave. A. Fusions of RIP sequences to the *omp*A signal peptide. (1) The fusion used to successfully secrete a RTB sequence. (2) The fusion used to successfully *excrete* MAP. B. Fusions of RIP sequences to the *phoA* signal peptide. (3) A fusion to RTA which did not result in secretion of RTA. (4) Same as (3) except for conversion of the N-terminal residue of RTA from Ile to Ala. (5) A fusion to RTA designed in a cistronlike arrangement and resulting in successful expression of RTA.

using the signal sequence from *phoA* (Figure 5, B[3,4]), which codes for a periplasmic protein, was not secreted at all (28). All prokaryotic signal sequences are structurally similar (see review in Ref. 44), and it seems unlikely that switching this sequence would have a qualitative effect on secretion, but it would be of interest to test an *ompA*–RTA fusion. In most secreted proteins, including *ompA*, sequences within the mature protein are important to secretion. In heterologous fusions, it cannot be predicted if similar cooperativity between the signal peptide and the mature proteins would exist. Hence, one is left with the option of testing several fusions to find one that may work. The success with using the *ompA* signal to secrete two different RIP sequences, however, warrants further trials with this particular signal sequence. Recently, Frankel and colleagues have expressed RTA and RTB in the pFLAG system described in Chapter 1. Proper secretion and processing of the *ompA*–Flag–protein fusion leads to a Flag-protein which will react with a monoclonal antibody specific for the N-terminal amino acids. Improperly processed and secreted proteins are unreactive. The *ompA*–Flag–RTA fusion protein was secreted and processed properly (unpublished data). In contrast, the *ompA*–Flag–RTB fusion yielded insoluble inclusion bodies unreactive with the N-terminal FLAG-specific monoclonal antibody. However, only JM109 cells were tested with both constructs.

RTB contains nine Cys residues, eight of which are found in intramolecular disulfide linkages. Intracellularly expressed RTB is aggregated and presumably improperly folded and is devoid of any biological activity (J.M. Lord and colleagues, unpublished results). If secreted, however, it was thought that the molecule might fold properly and that the correct disulfide linkages would form. Hussain et al. (43) constructed an *ompA*–RTB fusion in which the first five residues of RTB were replaced with five residues, GlyIleProSerLeu, from a polylinker sequence. Under control of an *lpp-lac* promoter, the sequence was constitutively expressed in transformed cells grown in medium containing IPTG at either 30 or 37°C. Periplasmic protein was released by osmotic shock in the presence of 1 mM lactose. Western blot analyses generally showed a doublet of immunoreactive proteins. The faster migrating species corresponded in size to deglycosylated, native B chain; the size of the slower migrating form was consistent with it being unprocessed precursor. Expression in three different *E. coli* strains and under various culture conditions, including ± 1 mM lactose and growth at 18, 20, 25, 27, or 30°C in nutrient broth versus defined medium was later studied. The results indicated a substantial effect on the efficiency of precursor processing depending on the host strain employed. Expression in the *E. coli* strain JA221 (*lpp*, *hsdM*$^+$, *trpT5*, *leuB6*, *lacY*, *recA1* [F′, *lacI*^{q}, *lac*, *pro*]) gave essentially completely processed, mature-like RTB product. Using this host strain, Hussain et al. (43) then showed that the optimal yield of recombinant product was obtained by growing the transformed cells in defined medium containing lactose at 18°C. The differences in observed yield under different conditions were dramatic, from 1 μg/L in nutrient broth at 30°C to 1000 μg/L at optimal conditions. The overall yield in the latter case is actually a minimal estimate as the radioimmunoassay used for detection only quantifies biologically active product. This was defined as the ability of recombinant product to bind to immobilized asialofetuin.

The suggestion noted above that RTA and type I RIPs be expressed at low temperatures also seems to apply, then, to the production of functional RTB. In addition, when attempting to secrete RIP sequences in *E. coli*, the results above strongly suggest that a variety of host strains and culture conditions be screened.

The type I RIP, MAP, was found to inhibit protein synthesis in *E. coli* and this was suspected to be the main reason for its low expression as an intracellular protein. To reduce this influence, Habuka and his colleagues designed a construct in which MAP would be directed for secretion using, again, the *ompA* signal sequence (Figure 5, A[2]) (79). They placed the fusion sequence under control of the λ P_L promoter and, to provide for tighter control of its induction, also placed the cI$_{857}$ repressor on the same plasmid. Cultures were induced for expression by shifting them from 30 to 42°C and MAP expression was then monitored in various cell fractions by ELISA. The overall level of MAP expression obtained was still low, but significantly higher than that observed for intracellular expression, 140 μg/L compared to about 3 μg/L. Surprisingly, however, the secreted MAP was detected in the culture medium and not in the periplasm. The kinetics of MAP production and its localization was followed after induction. MAP was found in the cytosol 1 hr after induction and gradually secreted into the culture medium for up to 3 hr after induction. At no time was MAP detected in the periplasm.

After induction, the transformed cells were arrested in growth within 1 hr, suggesting that active MAP might have been expressed intracellularly or that some secreted MAP, perhaps that transiently present in the periplasm, is able to reenter the cell. This might be tested for directly by isolating RNA from induced versus uninduced cells and treating it with aniline as described by Endo et al. (5,6) to test for release of a 3′ terminal fragment from 26S rRNA, analogous to that seen with eukaryotic 28S rRNA.

The secreted MAP was purified in a two-step procedure using CM-Sepharose and Blue-Sepharose with a final yield of 50–70%. The secreted MAP was identical in size to the native protein by SDS-PAGE analysis, had the same amino-terminal sequence, and showed the same inhibitory activity in reticulocyte, wheat germ, and *E. coli* in vitro translation systems.

The reason for the excretion of MAP from cells is not clear. It would be useful to test similar constructs for secretion of another type I RIP and an A chain from a type II RIP. As noted above, a *phoA*–RTA fusion yielded only precursor protein, although the culture medium was not assayed (28). The synthetic MAP sequence used in the construction will also facilitate mutational analyses to study the sequence and structural requirements of this phenomenon.

The above examples demonstrate the potential applications of secreting RIP proteins in *E. coli*. In general, the maximum level of production will not be as high as that possible in intracellular expression, but it holds distinct advantages for the proper folding of naturally secreted proteins and the formation of correct disulfide linkages. Purification of research quantities of protein is also facilitated in that the protein is placed in a compartment (or in the medium) that does not contain many other proteins.

IV. EXPRESSION OF RIP SEQUENCES IN EUKARYOTIC CELLS

Although this group of proteins function to inactivate eukaryotic ribosomes and some of them are potent toxins, eukaryotic expression systems do hold particular utilities, as specifically described below, for production of toxin subunits, selection of mutant RIPs, potential selection of cellular mutations in ribosomes, and the study of RIP precursors and processing. If cells with resistant ribosomes can be obtained or, as discussed in Section V, RIP sequences can be introduced into non–RIP-containing plant systems, high-level production might be feasible as well.

A. Yeasts

Yeasts, in particular *Saccharomyces cerevisiae*, are the eukaryotic counterpart to *E. coli* for the cloning and manipulation of heterologous sequences. Their genetics is well documented and understood, they are easily grown and handled, and there is an extensive list of available vectors (45), all of which can also be propagated in *E. coli*. Yeast systems have seen limited but important use in the expression of RIP sequences. Because of the inherent sensitivity of their ribosomes to the action of A chains and type I RIPs, yeasts are not a reasonable choice for the production of these proteins. Rather, they afford an ideal system for the functional dissection of RIPs as they can be used to select for protein mutants deficient in ribosome action.

Frankel and his colleagues (46) took just such an approach to isolate several mutants of RTA deficient in their ability to inhibit protein translation. They placed an appropriately modified sequence encoding RTA into a low copy number CEN plasmid and a high copy number 2-μm plasmid, pBM150 (47) and pRY131 (48), respectively. Both plasmids carried the *URA*3 gene for selection and the *GAL*1–10 promoter to direct expression of the RTA sequences. In the latter vector, a RTA fusion protein containing 40 amino acids of *GAL*1 followed by several amino acids coded by a polylinker sequence added to the amino-terminal of RTA was specified. This was later thought to be significant as this protein, and not mature RTA, was able to be expressed in the yeast cells.

Frankel et al. (46) first examined the growth rate of transformed cells on three different carbon sources affecting promoter activity. Both types of transformants grew on glucose which strongly represses *GAL*1. When galactose was added to the cultures, growth was promptly arrested, consistent with the expression of active RTA. Only transformants containing the construct which would specify the RTA fusion protein, however, were able to grow on glycerol. This is a noninducing, nonrepressive carbon source, and these data suggested that the *GAL*1–10 promoter was leaky and that the RTA fusion protein did not bear full RTA activity. This also correlated with the levels of expressed protein noted in galactose-induced cultures. Functional RTA protein could be detected only in transformants carrying the fusion protein construct, approximately 250 ng/mg of yeast protein, although this is a minimal estimate as it was determined by activity in a translational assay. Immunoassay measurements based on reactivity with antisera to native protein gave much lower estimates, perhaps due to incomplete cross-reactivity with the fusion protein.

Transformants expressing mutant forms of RTA much reduced in antiribosome activity were selected by growth on galactose. The frequency of colonies growing on galactose medium was initially found to be 10^{-7} for either type of expression construct. Of 22 independent isolates from each type of transformant, all were found to have mutations in the RTA sequence. Subsequent passage of the expression plasmids through a *mut*D *E. coli* strain before transformation into yeast increased the frequency of colonies able to grow on galactose to 10^{-3}. Of 72 isolates, 12 were found to produce RTA immunoreactive material by Western blot analysis and immunoassay. Four mutants were from pBM150-RTA–transformed cells. Subsequent analyses showed that these did indeed contain mutations in the RTA sequence (see Chapters 7 and 8 for the utility of this approach and of expression in yeast to facilitate studies on RTA). This approach should also be directly applicable to the study of other RIPs for which molecular clones are available, such as MAP, SAP, and α-TCS. For MAP, which inhibits *E. coli* ribosomes, mutant selection can be done in that organism as well. In this instance, it would be particularly interesting to compare mutants isolated from *E. coli* to those isolated from yeast to see if similar residues are modified.

A yeast expression system was also used in an attempt to produce functional RTB. Richardson et al. (49) expressed a portion of preproricin containing RTB as an intracellular protein. The expressed product consisted of an initiator methionine followed by the last 12 residues of ricA, the linker region, and RTB. An immunoreactive product of appropriate size for an aglycosylated protein was detected but it was not reported to be active.

There are several vectors for yeast that can specify expression of secretable proteins and that might be exploited for the expression of RIPs and their subunits. The arguments noted above for secretion in *E. coli* are doubly appropriate in this system because of the ribosome sensitivity. It is also possible that the plant processing signals might be recognized properly enabling coexpression of type II subunits, for example. Expression of the precursor for phaseolin, the major storage protein in *Phaseolus vulgaris*, in *S. cerevisiae* resulted in the production of mature, glycosylated phaseolin at levels up to 3% of the total soluble protein (50). In this case, partial or complete substitution of the signal peptide sequence from yeast acid phosphatase (PHO5) for the phaseolin sequence resulted in inefficient or no processing of the preprotein and no secretion of mature phaseolin.

This demonstrated that plant-specific processing sequences are not only capable of being recognized in yeast, but may actually be essential for expression of those mature proteins. As proricin is not active in inhibiting ribosomes, this could afford a way of expressing active holotoxins in these ribosome-sensitive cells. Galactose could be added to the culture medium to prevent intoxication of the cells through B chain receptor binding.

B. Mammalian Cells

There has been one example of using mammalian cells to express a RIP subunit. Chang et al. (51) used a derivative of the pCDX cDNA cloning vector of Okayama and Berg (52), containing the SV40 early promoter, in monkey kidney COS-M6 cells to express RTB. They fused a cDNA sequence encoding RTB to an amino-terminal sequence coding for the 21 amino acid secretory signal sequence for the low-density lipoprotein receptor (LDLR) (53). The resulting vector construct was transfected into COS-M6 cells. Galactose was included in the medium during expression to preserve the lectin function and conformational integrity of the B chain and to prevent it from binding to cell receptors. Their results showed that RTB was secreted to the medium to a concentration of 0.03–0.5 nM. The expressed RTB was slightly larger than B chain from plants, probably reflecting the addition of different glycosyl structures. Significantly, this recombinant RTB appeared to be fully functional. It formed active ricin when mixed with RTA and bound asialofetuin as effectively as native B chain.

C. Insect Cells

The baculovirus/insect cell expression systems are capable of producing very high yields of recombinant proteins (see reviews in Refs. 54 and 55; 56). In these virus vectors, sequences for the nonessential polyhedrin gene are replaced with the sequence of interest; expression is directed by the strong polyhedrin promoter. As the polyhedrin protein is normally produced at very high levels, greater than 1 mg/ml, the potential is there for production quantities of recombinant material.

The expression of ricin and its subunits has been studied in this system (57). The cells used for infection, *Spodoptera frugiperda*, were found to be completely resistant to ricin tested at 10 μg/ml. This is approximately 10,000-fold higher than the ID_{50} concentration of ricin on susceptible cell lines and raised the possibility that this system could be useful for the expression of active A chains and holotoxins. Ricin was shown to bind to the cells by immunofluorescence. However, the binding was only partially blocked by lactose and not blocked at all by polyclonal sera to RTA or

RTB. This suggested that the binding did not occur through a normal receptor interaction and that there might be a defect in toxin internalization that could account for the cell resistance. Insect cell ribosomes were later found to be sensitive to the action of RTA (81).

Piatak et al. (57) used a derivative of pAcC3 (58) to test for expression of preproricin, intracellular RTA and RTB, and secreted RTA and RTB. In all constructs, the initiating methionine codon was placed at exactly the same position as the polyhedrin Met codon in the mRNA. Recombinant viruses were obtained for preproricin and RTB but not for RTA in any form. This is consistent with the sensitivity of the insect ribosomes.

The results from expression of preproricin were of particular interest. Proteins corresponding to proricin, RTA1, RTA2, and RTB were detected separately by Western blot analysis of culture supernatants. The overall level of expression was rather low, although functional ricin activity at about 100 ng/ml was detected. The A chains migrated slightly faster than the native proteins in SDS-PAGE, probably due to the presence of the shorter oligosaccharides typical of insect cell proteins. Despite this difference in glycosylation, though, the secreted B chain migrated *slower* than the native protein. Subsequent analysis with antibody prepared against the linker peptide suggested that this was due to incomplete processing and that the linker peptide remained associated with the B chain. The appearance of proricinlike material in the supernatant also emphasizes the inefficiency with which the plant signals were recognized. Nevertheless, this expression construct did afford the opportunity to study active holotoxin, at least measured by activity.

RTB was also produced as an intracellular protein and as a secreted protein fused to the signal peptide sequence from macrophage colony-stimulating factor (M-CSF); expression of M-CSF had previously been demonstrated using a similar vector (59). The intracellularly expressed RTB was not soluble in crude extracts and did not show any activity. Expression from the secreted construct, however, did yield active RTB, but at very low levels. Approximately 10 ng/ml of RTB was detected in the culture supernatant, and all of this appeared to be active when assayed in the presence of excess RTA against susceptible cells. This activity was completely blocked by the addition of lactose. However, about 2 μg/ml of RTB was detected associated with the cells, and of this only about 10 ng/ml was estimated to be active in the presence of added RTA. Western blot analysis showed that this bulk of the expressed protein was larger than native protein, with an apparent M_r of 36K. The secreted material and a small quantity of the cell associated material had an apparent M_r of 32K. Studies with inhibitors of glycosylation led to the conclusions that almost all of the expressed RTAB had been translocated into the endoplasmic reticulum where it was core-glycosylated and partially processed. Only a tiny amount of this material, however, that with an apparent M_r of 32K, was completely processed and transported through the Golgi. This latter material accounted for all the active RTB detected.

It would appear, then, that insect cells, as well as the mammalian system noted above, have severe limitations in utility for expressing RIP sequences. Active product has been obtained from both systems, but at extremely low levels. Active RTA (and presumably type I proteins) can be obtained only as a processed subunit of a presumably inactive precursor and is prohibited from expression alone. Also, given the long generation times for these cells, the efforts and time required to isolate recombinant virus or transformed cells, and the high cost of cell culture and maintenance, these systems would not appear to be viable options for the study of expressed RIPs.

D. *Xenopus* Oocytes

Although subject to the same limitations of other eukaryotic systems, *Xenopus* oocytes offer certain advantages for expression. They are more easily manipulated once the operation is established and give a more rapid turnaround time for assaying expressed products. Metabolic labeling studies are also facilitated since one can work with the oocytes in a limited volume of defined bathing medium. Becoming involved with this system, however, is a serious consideration for those laboratories not already equipped. The initial set-up costs are high, involving the establishment of a frog facility which adheres to strict containment requirements, the employment and training of at least part-time personnel to maintain it and to become adept at micromanipulation, and purchase of a micromanipulater. Obviously not a production system, oocytes nevertheless can produce enough material for study, in particular, for examining structure-function relationships in RIPs. Expression is achieved by injecting mRNA directly into individual oocytes, a tedious task but one that can be mastered and greatly accelerated with practice to accomplish injection of close to 100 oocytes in a hour.

RTB and proricin have been expressed in oocytes. Richardson et al. (60) eliminated the RTA sequence from preproricin, thereby fusing the signal peptide sequence directly to the RTB sequence. This construct was placed into pGEM-3^R (61) downstream of the T7 promoter, and transcripts for injection were generated by adding T7 RNA polymerase. Injected oocytes accumulated RTB primarily in the particulate vesicle fraction, although about 10% of the total material was recovered from the bathing medium. Translation of the pre-RTB transcripts in the presence of tunicamycin confirmed that the RTB produced was translocated and glycosylated. Comparison of this material to translation products obtained from a wheat germ in vitro system also confirmed that the signal sequence had been processed. Active RTB was detected and measured by binding to asialofetuin and by reconstitution with RTA to form active toxin. Approximately 7 ng of fully functional product per oocyte was produced.

Similarly, Richardson et al. (62), in separate experiments, produced proricin from a preproricin transcript. This was the predominant protein species detected, although minor amounts of active RTA were released, presumably owing to nonspecific endoproteinase activity. Expression of the plant storage globulin, legumin, from *Vicia fabia* had earlier been demonstrated in oocytes and, like the results for proricin, processing to release the α and β subunits was not detected (63). Although a limitation in one sense, the lack of specific processing of the proricin molecule offered Richardson et al. (62) the opportunity to assess the activity of this molecule. This was significant in that assays demonstrated that this molecule did not function to inactivate ribosomes but RTA released from it did. The proricin molecule was glycosylated and contained intramolecular disulfide bonds which were apparently properly aligned as the molecule showed complete activity in binding lactose and asialofetuin. The results from this work, then, furthered the study of how ricin might be posttranslationally processed and how the plant protects its own ribosomes from inactivation. Earlier results from the same laboratory showed that proricin is first secreted out of the cytosol and sequestered from ribosomes and is then processed to yield fully active holotoxin (64,65).

The above examples demonstrate some of the utility of *Xenopus* oocytes to generate different RIP sequences for study. Although the overall levels of production are quite low, sufficient amounts of active material may be obtained for assays. The results above are similar to those obtained in the insect cell system, but the time and effort taken to achieve them was significantly less. In addition, the sequences to be introduced for expression are easily manipulated through site mutagenesis. Thus, altered proteins may be just as easily expressed for study.

V. CONCLUSIONS AND NEW DIRECTIONS

Table 1 summarizes the current experiences with expression of RIP sequences in heterologous systems.

The results described above indicate that, with certain limitations, RIPs and their subunits may be expressed in a variety of systems. *E. coli* shows the most utility and should be the first choice for expression of any newly isolated RIP sequence. There are a multitude of different expression vectors and a multitude of ways in which to express the sequence as either an intracellular or a secreted product. RTA, RTB, and α-TCS have all been expressed in active form and at reasonable levels for research studies. In fact, RTA has been produced at levels high enough, about 10% of cell protein, to make the system an economically viable alternative for production. In addition, despite the sensitivity of *E. coli* ribosomes to MAP, assayable levels of this protein were also obtained by directing the protein for secretion. The major difference in proteins expressed in *E. coli* is that they are not glycosylated. This would be an advantage for therapeutic applications, as noted in Section I, unless the added sugars are important for product solubility or distribution. In that instance, however, one would want human or mammalian glycosylation patterns and structures.

The eukaryotic expression systems are inherently limited in not being applicable to the expression of active A chains or type I RIPs and, with the exception of yeasts and *Xenopus* oocytes, have not shown any significant advantages over *E. coli*. Yeasts have a short doubling time (about 90 min on rich medium), are easily manipulated, and much advantage may be taken of the extensive genetics available. There is also well-established fermentation technology, and it is possible that they could be used for production of B chains. For this purpose, secretable constructs might be tried, perhaps regulated by the *GAL*1 promoter. Using galactose for induction would serve the dual purposes of facilitating B chain folding and of preventing the expressed B chain from binding to receptors in the cells. Yeasts have also shown a particular utility for the selection of mutant A chains, as demonstrated for RTA, or type I RIPs with impaired activities, thus facilitating the study of functional relationships in these protein sequences. Induction of active A chains in yeast cells might also allow for selection of chromosomal mutations leading to RIP-resistant ribosomes which would be valuable for studying ribosome structure and function. Although the concept and approach are sound, several laboratories that have attempted just this have been unsuccessful thus far (46; J. Bodley, J. Gould, personal communications).

Although these proteins are naturally produced in plants, this is one potential expression system that has not yet been exploited. Introduction of RIP genes into plants or plant cells not known to contain them would have several obvious advantages and applications. Production is clearly one as these proteins are normally found at rather high levels in various plant tissues. This might be accomplished either by cultivating transgenic plants or by fermentation of transformed plant cells. Ikeda et al. (20) established callus cultures of *M. jalapa* and demonstrated that the type I RIP, MAP, was produced at levels comparable to that in the intact plant. Tobacco cells, for which there is a considerable amount of experience, might be transformed with a RIP gene and cultured for expression. Establishing expression of a RIP in a heterologous plant system would also

Table 1 Summary of RIP Expression in Heterologous Systems as of 4/90

System	RIP sequence	Approximate[a] yield	Cellular localization	Regulation	Biological activity	Comments	References
E. coli							
	RTA	15–25 mg/L	Intracellular	T5 P_{N25}-*lac*	Yes	10 additional amino acids on mature N-terminal; aggregates formed at 37°C, active product at 30°C	30
	RTA	15–25 mg/L	Intracellular	λ P_L; *trpE*; *phoA*	Yes	Inactive aggregates formed at 42°C, active product at 37°C, highest yield at 30°C; cistron arrangement with *phoA*	28
	RTA	2–3[b] mg/L	Intracellular	λ P_R	Yes	Tripartite fusion, RTA–collagen–β-*gal*; RTA released with collagenase; additional residues on N- and C-terminals	33
	MAP	3–4 μg/L	Intracellular	P_{tac}	Yes[c]	Synthetic gene; intramolecular disulfide bond in native protein; *E. coli* ribosomes sensitive to MAP	35
	MAP	150 μg/L	Intracellular	P_{tac}	N.D.	Tripartite fusion, β-gal–Factor Xa–MAP	*
	MAP	140 μm/L[d]	Secreted to medium	lambda P_L	Yes	*ompA* signal peptide; excreted to medium	79
	α-TCS	1 mg/L	Intracellular	P_{trc}	Yes	Genomic and synthetic sequences expressed	†

	Yield	Localization	Promoter	Active	Comments	Ref.
RTB	1 mg/L	Periplasm	$P_{lpp-lac}$	Yes	Yield dependent on host cell and culture conditions, optimal in JA221 in defined medium at 18°C; first five residues of RTB replaced by polylinker coded sequence	43
Yeasts						
RTA	25 µg/L	Intracellular	$P_{GAL1-10}$	Yes	Yield limited by ribosome sensitivity; system applied to selection of impaired activity RTA proteins	46
RTB	N.D.	Intracellular	P_{PGK}	No	Portion of proricin expressed, last 12 AA of RTA, linker, RTB	49
Mammalian						
RTB	15 µg/L	Secreted to medium	SV40 early promoter	Yes	COS-M6 cells; signal sequence from low density lipoprotein receptor; protein glycoslated	51
Insect cells/baculovirus						
RTB	2 mg/L	Secreted; ER/Golgi, medium	Polyhedrin promoter	Yes[c]	M-CSF signal sequence; only 10 µg/L active product in medium, 10 µg/L cell–associated; glycosylated	57
proricin RTA RTB	total N.D.	Secreted	Polyhedrin promoter	Yes[b]	100 µg/L active ricin; partial processing of proricin; A chain differentially glycosylated	57

Table 1 Continued

System	RIP sequence	Approximate[a] yield	Cellular localization	Regulation	Biological activity	Comments	References
Xenopus oocytes							
	RTB	7 ng/oocyte	Secreted, 10% in medium		Yes	Ricin signal sequence; protein glycosylated	60
	proricin	N.D.	Secreted		N.A.	Unprocessed proricin; some protease cleavage release active A chain; protein glycosylated and contained intramolecular disulfide bonds	62

[a]Estimated yield based on reported results; for standard laboratory culture conditions.
[b]Purification method not optimized; overall expression likely to be higher.
[c]Anti viral activity only.
[d]May be low; ELISA used may not recognize inactive product.
[e]Assessed solely as cytotoxicity to a susceptible cell line in the presence of excess A-chain.
[f]Assessed as cytotoxicity. N.D., not determined.
N.A., not applicable. *, personal communication, N. Habuka; †, Piatak and colleagues, manuscripts in preparation.

facilitate study of the synthesis and metabolism of RIPs and perhaps offer some clues as to their functional significance in plants. It is generally thought that these proteins play some role in pest control. It is clear that they exhibit antiviral activity in vitro and when tested on plant tissue, but it has not yet been established if that is their role in the intact plant. One also has to wonder as to the significance of typically finding multiple, different RIPs selectively distributed among various tissues in any one plant. In *T. kirilowii*, for example, α-TCS predominates in tubers and trichokirin predominates in seeds (66). A transgenic plant expressing RIPs in various tissues could be tested against control plants for resistance to viruses and to other pests, for example, nematodes. This could also lead to the development of agronomically valuable plants, potentially the most significant application for these proteins.

There has been considerable progress in recent years in developing gene transfer systems and vectors for higher plants. These have been ably review by experts in this field and the reader is referred to these reviews and the literature cited in them for details (67–69). The systems available include *Agrobacterium*-mediated transformation and free DNA delivery methods, such as microinjection, electroporation, and particle gun techniques. The list of plants and plant cells that have been successfully transformed is ever growing and includes both monocotyledonous and dicotyledonous varieties (69). Since these are plant genes to begin with, it would be reasonable to assume that stable transformation of RIP sequences and their expression in plant cells is achievable. Gene sequences should be designed for secretion, however, and should also include any sequences coding for posttranslationally processed peptides. Plant ribosomes are more resistant to the enzymatic activity of these proteins, but the resistance is not absolute (10). Although it was originally thought that ribosomes in *Phytolacca americana* were completely resistant to the action of the pokeweed antiviral protein (70), recent data indicate that they are indeed sensitive (80). Type I and type II RIPs are produced as preproproteins. Proricin, consisting of A chain linked to B chain via a 12 amino acid peptide sequence is not enzymatically active, and it is this protein form that is translocated to the endoplasmic reticulum. Active A chain is subsequently released during processing in the endoplasmic reticulum and the Golgi and is not accessible to ribosomes. Proprotein forms of the type I RIPs α-TCS and SO-6 (24,71) contain a carboxy extension which might function in targeting the mature protein to a specific organelle as well as to maintain the RIP sequence in an inactive form. This hypothesis has yet to be tested, however.

There has also been much progress toward understanding and obtaining regulated gene expression in plants (see review in Ref. 72). The RIP sequences might then be placed under different promoters to test for their distribution and activity in different tissues. Conversely, RIP promoter regions might be tested with a reporter gene, such as chloramphenicol acetyltransferase, to evaluate their regulation and specificity. Ultimately, RIP promoters might prove valuable for the regulated expression of other genes in transgenic plants.

As a final comment, there are a considerable number of vectors available for expression of genes in all of the systems noted above and variations and new vectors are continually being generated. It would be difficult to list all of these or to make any recommendations for a particular vector system as the selection of an expression vector is dependent upon the particular objectives and needs of the researcher and on the nature of the sequence being expressed. For those unfamiliar with the different systems and vectors, a reasonable compilation of published vectors is given in *Cloning Vectors*, edited by Pouwels et al. (45) and updated yearly. It will at least provide the researcher with an information base on most of the standardly applied and available vectors. Any of these vectors can be redesigned to meet the particular requirements for gene expression through standard recombinant methods. In particular, synthetic oligonucleotides in double-strand blocks of up to about 100 bp are readily produced and provide an easy and direct way of fusing sequences into proper alignment and/or modifying gene sequences for expression.

ACKNOWLEDGMENTS

We thank Dr. Arthur E. Frankel for the opportunity to write this chapter and for his support. We also thank Dr. Jeffery Lifson, Genelabs Inc.; Drs. Toshiaki Kudo and Koki Horikoshi, Riken Institute, for their support and helpful discussions; Drs. Takashi Matsumoto, Masashi Miyano, and Masana Noma, Life Science Research Institute, and Terry P. Chow, Genelabs Inc., for their critical reading of this manuscript; and our colleagues noted throughout the manuscript for gratefully allowing us to cite their unpublished results.

REFERENCES

1. Olsnes, S. Abrin and ricin: Two toxic lectins inactivating eukaryotic ribosomes. *In*: A. W. Bernheimer (ed.), Perspectives in Toxicology, pp. 121–147. New York: John Wiley & Sons, 1977.
2. Olsnes, S., and Pihl, A. Toxic lectins and related proteins. *In*: P. Cohen and S. Van Heyningen (eds.), Molecular Action of Toxins and Viruses, pp. 51–105. New York: Elsevier, 1982.
3. Jimenez, A., and Vasquez, D. Plant and fungal protein and glycoprotein toxins inhibiting eukaryote protein synthesis. Ann. Rev. Microbiol., *39*: 649–672, 1985.

4. Stirpe, F., and Barbieri, L. Ribosome-inactivating proteins up to date. FEBS Lett., *195*: 1–8, 1986.

5. Endo, Y., Mitsui, K., Motizuki, M., and Tsurugi, K. The mechanism of action of ricin and related toxic lectins on eukaryotic ribosomes. J. Biol. Chem., *262*: 5908–5912, 1987.

6. Endo, Y., and Tsurugi, K. RNA *N*-glycosidase activity of ricin A-chain. J. Biol. Chem., *262*: 8128–8130, 1987.

7. Nilsson, L., and Nygard, O. The mechanism of the protein-synthesis elongation cycle in eukaryotics. Effect of ricin on the ribosomal interaction with elongation factors. Eur. J. Biochem., *161*: 111–117, 1986.

8. Brigotti, M., Rambelli, F., Zamboni, M., Montanaro, L., and Sperti, S. Effect of α-sarcin and ribosome-inactivating proteins on the interaction of elongation factors with ribosomes. Biochem. J., *257*: 723–727, 1989.

9. Stirpe, F., and Hughes, C. Specificity of ribosome-inactivating proteins with RNA N-glycosidase activity. Biochem. J., *262*: 1001–1002, 1989.

10. Harley, S. M., and Beevers, H. Ricin inhibition of in vitro protein synthesis by plant ribosomes. Proc. Natl. Acad. Sci. U.S.A., *79*: 5935–5938, 1982.

11. Frankel, A. E. (ed.) Immunotoxins. Boston: Kluwer Academic Publishers, 1988.

12. Olsnes, S., and Pihl, A. Chimeric toxins. Pharmacol. Ther., *15*: 355–381, 1982.

13. Vitetta, E. S., and Uhr, J. W. Immunotoxins: Redirecting nature's poisons. Cell, *41*: 653–654, 1985.

14. Tomlinson, J. A., Walker, V. M., Flewett, T. H., and Barclay, G. R. The inhibition of infection by cucumber mosaic virus and influenza virus by extracts from *Phytolacca americana*. J. Gen. Virol., *22*: 225–232, 1974.

15. Ussery, M. A., Irvin, J. D., and Hardesty, B. Inhibition of poliovirus replication by a plant antiviral peptide. Ann. N.Y. Acad. Sci., *284*: 431–440, 1977.

16. Aron, G. M., and Irvin, J. D. Inhibition of Herpes simplex virus multiplication by the pokeweed antiviral protein. Antimicrob. Agents Chemother., *17*: 1032–1033, 1980.

17. Fernandez-Puentes, C., and Carrasco, L. Viral infection permeabilizes mammalian cells to protein toxins. Cell, *20*: 769–775, 1980.

18. Foa-Tomasi, L., Campadelli-Fiume, G., Barbieri, L., and Stirpe, F. Effect of ribosome-inactivating proteins on virus-infected cells. Inhibition of virus multiplication and of protein synthesis. Arch. Virol., *71*: 322–332, 1982.

19. McGrath, M. S., Hwang, K. M., Caldwell, S. E., Gaston, I., Luk, K.-C., Wu, P., Ng, V. L., Crowe, S., Daniels, J., Marsh, J., Deinhart, T., Lekas, P. V., Vennari, J. C., Yeung, H.-W., and Lifson, J. D. Proc. Natl. Acad. Sci. U.S.A., *86*: 2844–2848, 1989.

20. Ikeda, T., Takanami, Y., Imaizumi, S., Matsumoto, T., Mikami, Y., and Kubo, S. Formation of anti-plant viral protein by *Mirabilis jalapa* L. cells in suspension culture. Plant Cell Reps, *6*: 216–218, 1987.

21. Montfort, W., Villafranca, J. E., Monzingo, A. F., Ernst, S. E., Katzin, B., Reutenber, E., Xuong, N. H., Hamlin, R., and Robertus, J. D. J. Biol. Chem., *262*: 5398–5403, 1987.

22. Collins, E. J., Robertus, J. D., LoPresti, M., Stone, K. L., Williams, K. R.,

Wu. P., Hwang, K., and Piatak, M. Primary amino acid sequence of α-trichosanthin and molecular models for abrin A-chain and α-trichosanthin. J. Biol. Chem., *265*: 8665–8669, 1990.

23. Halling, K. C., Halling, A. C., Murray, E. E., Ladin, B. F., Houston, L. L., and Weaver, R. F. Genomic cloning and characterization of a ricin gene from *Ricinus communis*. Nucleic Acids Res., *13*: 8019–8033, 1985.

24. Chow, T. P., Feldman, R. A., Lovett, M., and Piatak, M. Isolation and DNA sequence of a gene encoding α-trichosanthin, a Type I ribosome-inactivating protein. J. Biol. Chem., *265*: 8670–8674, 1990.

25. Foxwell, B. M. J., Donovan, T. A., Thorpe, P. E., and Wilson, G. The removal of carbohydrates from ricin with endoglycosidases H, F and D and alpha-mannosidase. Biochim. Biophys. Acta, *840*: 193–203, 1985.

26. Blakey, D. C., Skilleter, D. N., Price, R. J., and Thorpe, P. E. Uptake of native and deglycosylated ricin A-chain immunotoxins by mouse liver parenchymal and non-parenchymal cells in vitro and in vivo. Biochim. Biophys. Acta, *968*: 172–178, 1988.

27. Blakey, D. C., and Thorpe, P. E. Effect of chemical deglycosylation on the in vivo fate of ricin A-chain. Cancer Drug Delivery, *3*: 189–196, 1986.

28. Piatak, M., Lane, J. A., Laird, W., Bjorn, M. J., Wang, A., and Williams, M. Expression of soluble and fully functional ricin A chain in *Escherichia coli* is temperature-sensitive. J. Biol. Chem., *263*: 4837–4843, 1988.

29. Ramakrishnan, S., Bjorn, M. J., and Houston, L. L. Recombinant ricin A chain conjugated to monoclonal antibodies: Improved tumor cell inhibition in the presence of lysosomotropic compounds. Cancer Res., *49*: 613–617, 1989.

30. O'Hare, M., Roberts, L. M., Thorpe, P. E., Watson, G. J., Prior, B., and Lord, J. M. Expression of ricin A chain in *Escherichia coli*. FEBS Lett., *216*: 73–78, 1987.

31. Ben-Bassat, A., Bauer, K., Chang, S.-Y., Myambo, K., Boosman, A., and Chang, S. Processing of the initiation methionine from proteins: Properties of the *Escherichia coli* methionine aminopeptidase and its gene structure. J. Bacteriol., *169*: 751–757, 1987.

32. Schoner, B. E., Belagaje, R. M., and Schoner, R. G. Translation of a synthetic two-cistron mRNA in *Escherichia coli*. Proc. Natl. Acad. Sci. U.S.A., *83*: 8506–8510, 1986.

33. Schlossman, D., Withers, D., Welsh, P., Alexander, A., Robertus, J., and Frankel, A. Role of glutamic acid 177 of the ricin toxin A chain in enzymatic inactivation of ribosomes. Mol. Cell. Biol., *9*: 5012–5021, 1989.

34. Germino, J., and Bastia, D. Rapid purification of a cloned gene product by genetic fusion and site-specific proteolysis. Proc. Natl. Acad. Sci. U.S.A., *81*: 4692–4696, 1984.

35. Habuka, N., Murakami, Y., Noma, M., Kudo, T., and Horikoshi, K. Amino acid sequence of *Mirabilis* antiviral protein, total synthesis of its gene and expression in *Escherichia coli*. J. Biol. Chem., *264*: 6629–6637, 1989.

36. Brosius, J., and Holy, A. Regulation of ribosomal RNA promoters with a synthetic *lac* operator. Proc. Natl. Acad. Sci. U.S.A., *81*: 6929–6933, 1984.

37. Fujikawa, K., Legaz, M. E., and Davie, E. W. Bovine factor X_1 (Stuart factor). Mechanism of activation by protein fro Russell's viper venom. Biochemistry, *11*, 4892–4899, 1972.

38. Stark, M. J. R. Multicopy expression vectors carrying the *lac* repressor gene for regulated high-level expression of genes in *Escherichia coli*. Gene, *51*: 255–267, 1987.

39. Wang, Y., Qian, R.-Q., Gu, Z.-W., Jin, S.-W., Zhang, L.-Q., Xia, Z.-X., Tian, G.-Y., and Ni, C.-Z. Scientific evaluation of Tian Hua Fen (THF)—history, chemistry and application. Pure and Appl. Chem., *58*: 789–798, 1986.

40. Stratagene Cloning Systems, La Jolla, California.

41. Chang, A. C. Y., and Cohen, S. N. Construction and characterization of amplifiable multicopy DNA cloning vehicles derived from the p15A cryptic miniplasmid. J. Bacteriol., *134*: 1141–1156, 1978.

42. Amann, E., and Brosius, J. 'ATG vectors' for regulated high-level expression of cloned genes in *Escherichia coli*. Gene, *40*: 183–190, 1985.

43. Hussain, K., Bowler, C., Roberts, L. M., and Lord, J. M. Expression of ricin B chain in *Escherichia coli*. FEBS Lett., *244*: 383–387, 1989.

44. Oliver, D. Protein secretion in *Escherichia coli*. Ann. Rev. Microbiol., *39*: 615–648, 1985.

45. Pouwels, P. H., Enger-Valk, B. E., and Brammar, W. J. Cloning Vectors. A Laboratory Manual. Amsterdam: Elsevier, 1985.

46. Frankel, A., Schlossman, D., Welsh, P., Hertler, A., Withers, D., and Johnston, S. Selection and characterization of ricin toxin A-chain mutations in *Saccharomyces cerevisiae*. Mol. Cell. Biol., *9*: 415–420, 1989.

47. Johnston, M., and Davis, R. Sequences that regulate the divergent gall-10 promoter in *Saccharomyces cerevisiae*. Mol. Cell. Biol., *4*: 1440–1448, 1984.

48. Yocum, R., Hanley, S., West, R., and Ptashne, M. Use of *lacZ* fusions to delimit regulatory elements of the inducible divergent gall-10 promoter in *Saccharomyces cerevisiae*. Mol. Cell. Biol., *4*: 1985–1998, 1984.

49. Richardson, P. T., Roberts, L. M., Gould, J. H., Smith, A. B., and Lord, M. The expression of ricin B-chain in *Saccharomyces cerevisiae*. Biochem. Soc. Trans., *15*: 903–904 1987.

50. Cramer, J. H., Lea, K., Schaber, M. D., and Kramer, R. A. Signal peptide specificity in posttranslational processing of the plant protein phaseolin in *Saccharomyces cerevisiae*. Mol. Cell. Biol., *7*: 121–128, 1987.

51. Chang, M.-S., Russell, D. W., Uhr, J. W., and Vitetta, E. S. Cloning and expression of recombinant, functional ricin B chain. Proc. Natl. Acad. Sci. U.S.A., *84*: 5640–5644, 1987.

52. Okayama, H., and Berg, P. A cDNA cloning vector that permits expression of cDNA inserts in mammalian cells. Mol. Cell. Biol., *3*: 280–289, 1983.

53. Yamamoto, T., Davis, C. G., Brown, M. S., Schneider, W. J., Casey, M. L., Goldstein, J. L., and Russell, D. W. The human LDL receptor: a cysteine-rich protein with multiple Alu sequences in its mRNA. Cell, *39*: 27–38, 1984.

54. Luckow, V. A., and Summers, M. D. Trends in the development of Baculovirus expression vectors. Bio/Technology, *6*: 47–55, 1988.

55. Miller, L. K. Baculoviruses as gene expression vectors. Ann. Rev. Microbiol., *42*: 177–199, 1988.

56. Luckow, V. A., and Summers, M. D. High level expression of non-fused foreign genes with *Autographa californica* nuclear polyhedrosis virus expression vectors. Virology, *170*: 31–39, 1989.

57. Piatak, M., Lane, J. A., O'Rourke, E., Clark, R., Houston, L. L., and Apell, G. Expression of ricin and ricin B-chain in insect cells. *In*: K. Brew, F. Ahmad, H. Bialy, S. Black, R. E. Fenna, D. Puett, W. A. Scott, J. Van Brunt, R. W. Voellmy, W. J. Whelan, and J. F. Woessner (eds.), Advances in gene technology: Protein engineering and production. ICSU Short Reports, Vol. 8, p. 62. Oxford: IRL Press, 1988.

58. Devlin, J. J., Devlin, P. E., Clark, R., O'Rourke, E. C., Levinson, C., and Mark, D. F. Novel expression of chimeric plasminogen activators in insect cells. Bio/Technology, *7*: 286–292, 1989.

59. Maiorella, B., Inlow, D., Shauger, A. and Harano, D. Large-scale insect cell-culture for recombinant protein production. Bio/Technology, *6*: 1406–1410, 1988.

60. Richardson, P. T., Gilmartin, P., Colman, A., Roberts, L. M., and Lord, J. M. Expression of functional ricin B chain in Xenopus oocytes. Bio/Technology, *6*: 565–570, 1988.

61. Promega Corporation, Madison, Wisconsin.

62. Richardson, P. T., Westby, M., Roberts, L. M., Gould, J. H., Colman, A., and Lord, J. M. Recombinant proricin binds galactose but does not depurinate 28 S ribosomal RNA. FEBS Lett., *255*: 15–20, 1989.

63. Bassuner, R., Huth, A., Manteuffel, R., and Rapoport, T. A. Secretion of plant storage globulin polypeptides by Xenopus laevis oocytes. Eur. J. Biochem., *133*: 321–326, 1983.

64. Butterworth, A. G., and Lord, J. M. Ricin and *Ricinus communis* agglutinin subunits are all derived from a single-sized polypeptide precursor. Eur. J. Biochem., *137*: 57–65, 1983.

65. Lord, J. M. Precursors of ricin and *Ricinus communis* agglutinin. Glycosylation and processing during synthesis and intracellular transport. Eur. J. Biochem., *146*: 411–416, 1985.

66. Casellas, P., Dussossoy, D., Falasca, A. I., Barbieri, L., Guillemot, J. C., Ferrara, P., Bolognesi, A., Cenini, P., and Stirpe, F. Trichokirin, a ribosome-inactivating protein from the seeds of *Trichosanthes kirilowii* Maximowicz. Eur. J. Biochem., *176*: 581–588, 1988.

67. Cocking, E. C., and Davey, M. R. Gene transfer in cereals. Science, *236*: 1259–1262, 1987.

68. Schell, J. S. Transgenic plants as tools to study the molecular organization of plant genes. Science, *237*: 1176–1183, 1987.

69. Gasser, C. S., and Fraley, R. T. Genetically engineering plants for crop improvement. Science, *244*: 1293–1299, 1989.

70. Owens, R. A., Bruening, G., and Shepherd, R. J. A possible mechanism for the inhibition of plant viruses by a peptide from *Phytolacca americana*. Virology, *56*: 390–393, 1973.

71. Benatti, L., Saccardo, M. B., Dani, M., Nitti, G., Sassano, M., Lorenzetti, R., Lappi, D. A., and Soria, M. Nucleotide sequence of cDNA coding for saporin-6, a type-1 ribosome-inactivating protein from *Saponaria officinalis*. Eur. J. Biochem., *183*: 465–470, 1989.

72. Benfey, P. N., and Chua, N.-H. Regulated genes in transgenic plants. Science, *244*: 174–181, 1989.

73. Irvin, J. D. Pokeweed antiviral protein. Pharmacol. Ther., *21*: 371–387, 1983.

76. Nirenberg, M. W., and Matthaei, J. H. Characteristics and stabilization of DNAase-sensitive protein synthesis in *E. coli* extracts. Proc. Natl. Acad. Sci. U.S.A., *47*: 1588–1602, 1961.

77. Shimura, Y., Moses, R. E., and Nathans, D. Coliphage MS2 containing 5-fluorouracil. J. Mol. Biol., *12*: 266–279, 1965.

78. Cawley, D. B., Hedbloom, M. L., and Houston, L. L. Homology between ricin and *Ricinus communis* agglutinin: Amino terminal sequence analysis and protein synthesis inhibition studies. Arch. Bioch. Biophys., *190*: 744–755, 1978.

79. Habuka, N., Akiyama, K., Tsuge, H., Miyano, M., Matsumoto, T., and Noma, M. Expression and secretion of *Mirabilis* antiviral protein by *Escherichia coli* and its inhibition to in vitro eukaryotic and prokaryotic protein synthesis. J. Biol. Chem., *265*: 10988–10992, 1990.

80. Taylor, B. E., and Irvin, J. D. Depurination of plant ribosomes by pokeweed antiviral protein. FEBS Letters, *273*: 144–146, 1990.

81. Maruniak, J. E., Fresler, S., and McGuire, P. M. Susceptibility of insect cells ribosomes to ricin. Comp. Biochem. Physiol., *96B*: 543–548. 1990.

7

The Structure of Plant Toxins as a Guide to Rational Design

Jon D. Robertus *University of Texas, Austin, Texas*

I. INTRODUCTION

A variety of higher plants contain heterodimeric cytotoxins or single-chain ribosome–inactivating proteins (RIPs). These probably serve a defensive role to the plant. For example, seed-eating mammals would be discouraged from devouring the seeds of *Ricinus communis*, which contains the cytotoxin ricin. It has been proposed that pokeweed antiviral protein (PAP) acts as an antiviral agent in *Phytolacca americana* (1). The RIP is stored in the cell wall space and enters the cell when the wall is breached, inhibiting its host ribosomes and retarding viral replication in the cell.

Biochemists and medical researchers have made extensive use of both forms of protein, which we shall refer to interchangeably as RIPs or plant toxins, in the search for therapeutic agents. The toxins have been conjugated to antibodies to create a class of agents called immunotoxins. These hybrids are meant to act as "magic bullets" in which the antibody binds the complex to a target cell. After endocytosis, the toxin attacks the target ribosomes and kills the cell. This aspect of toxin use has been extensively reviewed (2), and the reference describes a number of immunotoxin systems aimed at cancer and other target cells. More recently, it has been observed that the RIP trichosanthin appears to preferentially attack HIV-infected cells (3). Because of the evolutionary relationship (4) and common mode of action of the toxins (5), it is likely that this anti-AIDS activity will be observed for the plant toxins in general. Largely as a result of its use in disease treatment, the scientific interest in the entire class of proteins will continue to increase. Efforts are already underway, as works in this volume show, to use modern genetic engineering methods to redesign these proteins in the hope of improving their performance in specific therapeutic tasks. Knowledge of the three-dimensional structure of a protein is essential

133

if we are to understand the mechanism of its action at the atomic level. The structure is also vital to guide the rational design of specific genetic mutations to be created by molecular biology techniques. In this chapter, we provide a review of the structure of the heterodimeric toxin ricin, aimed at serving as a guide to the rational engineering of the protein. This is the first plant toxin to have its structure determined to high resolution. Because of its anticipated similarity to other toxins, the ricin model has also been used to create models of other toxins and can be used to aid engineering efforts on those proteins (6). Site-directed mutations fall into two broad classes, analytical mutations and custom product mutations. Analytical mutations are those that are made to explore the mechanism by which the protein acts or maintains its structural integrity. These are the main types of mutations made early in the systematic study of a protein. The second class of mutations, custom product mutations, are aimed at producing quantities of a specialty protein which has novel and desirable properties. These might include enhanced thermal stability, additional surface thiols for linking to other groups, or altered enzyme activity.

II. BACKGROUND

A vast literature is accumulating on the biochemistry of plant toxins, and it has been reviewed extensively in several chapters of References 2 and 7. For the sake of orientation in the structural discussions which follow, some key points of ricin biochemistry will be briefly described. Ricin is isolated as a heterodimer, consisting of a 267-residue, 32,000-d A chain glycoprotein (RTA) linked by a disulfide bond to a 262-residue, 32,000-d B chain glycoprotein (RTB). RTB is a lectin with a binding preference for galactosides, although it binds much more strongly to complex galactosides from cell surface carbohydrates than to simple sugars. Binding of the B chain to cell surface receptors appears to trigger endocytotic uptake of the protein. RTA and RTB have a reasonably strong affinity for one another (Kass $\approx 10^6$), mediated by hydrophobic forces, and association of the chains is necessary for toxicity. The disulfide link between the chains does not appear to be critical for toxicity, except in maintaining protein-protein interactions at very low toxin concentration. RTB is required for rapid cellular uptake, and also appears to assist RTA in its escape from the endosome. The mechanism of assistance is unknown, it may be simply a lectin-mediated routing of RTA to a porous region of the Golgi (8), or it may be more active, say aiding in pore formation.

Once the A chain of ricin reaches the cytoplasm of a typical eukaryotic cell, it enzymatically attacks the 60S ribosomal subunit and disrupts protein synthesis. Ricin has been shown to have a $K_m = 0.1$ μM for ribosomes and shows a $k_{cat} = 1500$ min^{-1} (9). It has recently been shown that the enzyme is an N-glycosidase, removing a specific adenine base from a very conservative region of the 28S rRNA (10,11). Bacterial ribosomes are not susceptible to ricin intoxication.

III. OUTLINE OF CRYSTALLOGRAPHIC REFINEMENT

The 2.8-Å x-ray structure of ricin was reported several years ago (12). Although correct in its overall description, that model was not refined in a crystallographic sense. Refinement refers to a process of model improvement by adjustment of the atom positions. The process is monitored in several ways.

1. The raw data of x-ray diffraction are a series of observed reflections, Fo, which are often displayed as ordered "spots" on photographic film. There are over 20,000 such reflections to 2.5-Å resolution for ricin, and they contain the truth about the correct protein structure. Given a model of the protein it is possible to calculate mathematically what the value of its reflections, Fc, would be. These can be compared to the observed values as a residual, or R factor, of the form,

$$R = \frac{\Sigma \mid |Fo| - |Fc| \mid}{\Sigma |Fo|}$$

where the summation is over all reflections.

The initial ricin model had an R of 48%, typical of structures solved by the isomorphous heavy atom method. Over the past several years of rebuilding and automated energy refinement, the R has dropped to 21%. This is a very good value and indicates that the model is nearly correct. It is very difficult to bring the R factor much lower for protein crystals, largely because water, which occupies about half the volume, has no regular structure, and because portions of the protein itself are not highly ordered and may take up several positions.

2. The standard bond lengths and bond angles between atoms in peptides are known to great precision from very high–resolution x-ray crystallography of small molecules, from spectroscopy, and from energy calculations. The bond lengths and angles of any given protein model can be compared to these standard values. A correct model should have all such parameters within several standard deviations of these values and indeed, a list of deviant parameters serves as a guide to troublesome areas of a model.

3. A correct protein model should be stereochemically reasonable in a more qualitative sense as well. This means that in general nonpolar atoms will be sheltered from solvent and that polar atoms inside the protein,

particularly charged atoms, are hydrogen bonded or salt bonded to other polar atoms.

4. The overall energy of the protein model has been calculated as part of the refinement process. An incorrect model will exhibit energetically unfavorable contacts, such as the clashing of atoms, and these are systematically dealt with.

Refinement of ricin to 2.5 Å is now complete and a description of the process (13) and of both protein chains is being prepared (14,15). The refinement has altered the initial model. The folding topology is unchanged, and for many purposes the alterations may seem unimportant. However, for a subtle manipulation of the molecular architecture, the refined structure will be necessary. In several areas, the polypeptide chain has been adjusted so that entire residues have been slid along the electron density by one place. This problem arises frequently in initial protein models. The unrefined map contains electron density which is not always distinct. For example, even in good regions of the map, density for a Val or a Ser cannot really be distinguished. On the other hand, density for a Phe or a Trp will usually serve as a marker. The density is strong because the residue is electron rich, it does not "smear" because it is usually internal and immobile, and its planarity gives it a distinctive disk shape. The density between landmarks is occasionally ambiguous, particularly if it corresponds to an outside loop, and positioning atoms in these regions is often difficult. Such problems occurred in at least three places in the initial ricin model. In one case, an adjustment of residues 226–242 of ricin A chain had a dramatic effect on the interpretation of which residues interacted with B chain residues to hold the two proteins together.

IV. THE REFINED MODEL OF RICIN

A description of the refined ricin model will now be presented, with an eye to understanding features of the molecule which are important to maintaining the folding or the catalytic properties. Comparison of the amino acid sequences between related proteins is a potentially powerful tool in understanding which residues are important for structure or function. Figure 1 shows the sequence alignment of six plant toxins and two bacterial toxins. Nine residues are absolutely invariant and others are conserved to varying extent. In our discussion we will attempt to rationalize these observations, and to provide other structural insights as well.

In native ricin, RTA and RTB are connected by a disulfide bridge between Cys-259 of RTA and Cys-4 of RTB. The thermodynamic parameters of the association between ricin chains has been evaluated using analytical ultracentrifuge data (16). The association constant was found to

```
                  10          20          30          40          50          60
RTA      IFPKQYPIINFTTA    GATVQSYTNFIRAVRGRLTTGADVRHEIPVLPNRVGLPINQRFILV
ATA      EDRPI KFSTE       GATSQSYKQFIEALRERLRGGL I HDIPVLPDPTTLQERNRYITV
TCS      DVSFRLS           GATSSSYGVFISNLRKALPNERKL YDIPLL  RSSLPGSQRYALI
MAP      APTLETIASLD LNNPT TYLSFITNIRTKVADKTEQ CTIQKI  SKTF  TQRYSYI
SO6      VTSITLD LVNPTAGQYSSFVDKIRNNVKDPNLK YGGTDI  AVIGPPSKEKFLR
BPSI     AAKMAKNVDKPLFTATFNVQASSADYATFIAGIRNKLRNPAHFSHNEPVLPPVEPNVPPSRWFHV
SLTA1    KEFTLDFS          TAKTYVDSLNVIRSAIGTPLQTISSGGTSLLMIDSGSGDNLFAV
SLTA2    KEFTIDFS          TQQSYVSSLNSIRTEISTPLEHISQGTTSVSVINHTHGS YFAV

                  70          80          90          100         110
RTA      ELSNHAELS VTLALD   VTNAYVVGYRAGNS    AYFFHPDNQEDAEAITHLF TDVQ
ATA      ELSNSDTES IEVGID   VTNAYVVAYRAGTQ    SYFLRDAPSS  ASDYLF TGTD
TCS      HLTNYADET ISVAID   VTNVYIMGYRAGDT    SYFFNEASATE AAKYVF KDAM
MAP      DLIVSSTQK ITLAID   MADLYVLGYSDIANNKGR AFFFKDVTEAV ANNF FPGATG
SO6      INFQSSRGT VSLGLK   RDNLYVVAYLAMDNTNVNRAYYF RSEITSAESTALFPEATT
BPSI     VLKASPTSAGLTLAIR   ADNIYLEGFKSSDG    TWWELT      PGLI
SLTA1    DVRGIDPEEGRFNNLRLIVERNNLYVTGFVNRTNNVFYRFADF   SHVTFPGTT
SLTA2    DIRGLDVYQARFDHLRLIIEQNNLYVAGFVNTATNTFYRFSDF   THISVPGVT

                  120         130         140         150         160         170
RTA      NRYTFAFGGNYDRLEQLAGNLRENIELGNGPLEEAISALYYY  STGGTQLPT LARS FII
ATA      QH SLPFYGTYGDLERWAHQSRQQIPLGLQALTHGIS  FF  SGGNDNEE KART LIV
TCS      RKVTLPYSGNYERLQTAAGKIRENIPLGLPALDSAITTLFYY   NANSA ASA LMV
MAP      TNRIKLTFTGSYGDLEKNGG LRKDNPLGIFRLENSIVNIYGK   AGDVKKQ ALF FLL
SO6      ANQKALEYTEDYQSIEKNAQITQGDQSRKELGLGIDLLSTSMEAVNKKARVVKD EARF LLI
BPSI     PGATYVGFGGTYRDLLGDTDKL TNVALGRQQLEDAVTALHGRTKADKASGPKQQQAREAVTT
SLTA1    A  VTLSGDSSYTTLQRVAGISRTGMQINRHSL  TTSYLDLMSHSGTSLTQSVAR AMLR
SLTA2    T  VSMTTDSSYTTLQRVAALERSGMQISRHSL  VSSYLALMEFSGNIMTRDASR AVLR

                  180         190         200         210         220
RTA      CIQMISEAARFQ  YIEGEMRTRIRY NRRSAPDPS VITLENSWGRLSTAIQE  SNQGAF
ATA      IIQMVAEAARFR  YISNRVRVSIQTG TAFQPDAA MISLENNWDNLRG VQE  SVQDTF
TCS      LIQSTSEAARYK  FIEQQIGKRVDKT  FLPSLA IISLENSWSALSKQIQIASTNNGQF
MAP      AIQMVSEAARFK  YISDKIPSEKYEE  VTVDEY MTALENNWAKLSTAVYNSKPS TTT
SO6      AIQMTAEAARFR  YIQNLVIKNFPNK  FNSENK VIQFEVNWKKISTAIYG DAKNGVF
BPSI     LLLMVNEATRFQTVSGFVAGLL HPKAVEKKSGKIGNE MKAQVNGWQDLSAALLKTDVKPPPG
SLTA1    FVTVTAEALRFR  QIQRGFRTTLDDLSGRSYVMTAEDVDLTLNWGRLSSVLPD  YHGQDS
SLTA2    FVTVTAEALRFR  QIQREFRQALSE TAPVYTMTPGDVDLTLNWGRLDAALGE  YRGEDG

                  230         240         250         260
RTA      ASPIQLQRRNGSKFSVYDVS  ILIPIIALMVYRCAPPPSSQF
ATA      PNQVTLTNIRNEPVIVDSLSH PTVAVLALMLFVCNPPN
TCS      ETPVVLINAQNQRVMITNVDAGVVTSNIALLLNRNNMA
MAP      ATKCQLATSPVTISPWIFKTVEEIKLVMGLLKSS
SO6      NKDYDFGFGKVRQVKDLQMGLLMY    LGKPKSSNEAN
BPSI     KSPAKFTEKM    GVRTAEQAAATLGILLFVEVPGGLTVAKALELFHASGGKPI
SLTA1    VRVGRISF  GSINAILGSVALILNCHHHASRVARMA  SDEFRSMCPADGRVRGITHNKILW
SLTA2    VRVGRISF  NNISAILGTVAVILNCHHQGARSVRAV  NEESQPECQITGDRPVIKINNTLW


SLTA1    DSSTLGAILMRRTISS
SLTA2    ESNTAAAFLNRKSQFLYTTGK
```

Figure 1 Sequence comparison of ribosome-inhibiting proteins. The amino acid sequences of five proteins from dicotyledonous plants are aligned using the one-letter code. RTA is the ricin A chain (30,31), ATA is the abrin A chain (32), TCS is trichosanthin (6), MAP is *Mirabilis* antiviral protein (33), and SO6 is saporin-6 (34), BPSI is the barley inhibitor (35), SLA1 is one form of *Shiga*-like inhibitor (36), whereas SLA2 is an antigenically distinct form (37), also from *E. coli*. The numbers along the top are RTA sequence numbers. Residues in the shaded columns are invariant.

measure 1.72×10^6 M^{-1} and the values for entropy and enthalpy were determined to be positive, suggesting that hydrophobic interactions are responsible for chain association. As predicted by the data above, the interface between RTA and RTB contains aromatic rings and aliphatic side chains which interact in a rather disordered manner. Somewhat surprising is the amount of trapped solvent in the interface, so that the hydrophobic contacts between the chain resembles a bridge across a small stream. This is illustrated in Figure 2. Within the hydrophobic bridge, the side chain of Tyr-262 from RTB makes strong contacts with such RTA residues as Phe-140 and Phe-240, and site-directed alteration of these groups could be expected to effect the stability of the heterodimer. Table 1 gives a list of contacts between the chains.

RTA and other members of the family of plant RIPs are N-glycosidases (5,10). They hydrolytically remove a specific adenine base from a highly conserved loop region of 28S rRNA (A$_{4324}$ in rat). No detailed mechanism for this family of N-glycosidases has been proposed, however Schramm and coworkers have used kinetic isotope effects to investigate the mech-

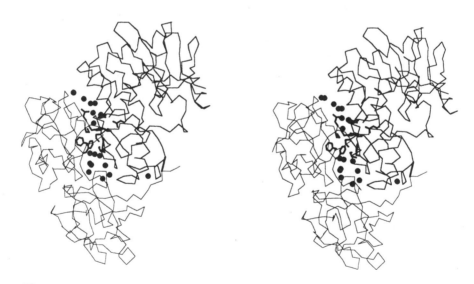

Figure 2   The A-B interface of ricin. The alpha carbon backbone of the protein is displayed as line segments. RTA is shown in the upper right as the heavier line, whereas RTB is the lighter structure. The dark spheres between the chains mark the positions of solvent waters trapped in the interface. The side chains of hydrophobic residues which interact across the interface are shown as the heaviest bonds, near the center of the interface. To the right center, the disulfide linking the chains is also shown as a heavy bond.

Table 1   Interactions Across the A/B Interface

Polar interactions		Distance
B chain	A chain	(Å)
Ala-1 N	Arg-258 O	3.10
Asp-2 O	Ala-260 N	3.17
Asp-94 OD2	His-40 ND1	2.93
Val-141 O	Arg-234 NH1	2.79
Lys-219 NZ	Glu-41 OE2	2.77
Asn-220 O	Gln-182 NE2	2.87
Asn-220 ND2	Ile-249 O	2.78
Asn-220 ND2	Ile-252 O	3.01
Phe-262 O	Arg-235 N	2.88

Hydrophobic residues[a]	
A-chain	B-chain
Tyr-183	Phe-262
	Pro-260
Leu-207	Phe-262
Phe-240	Phe-140
	Phe-262
Ile-247	Phe-140
Pro-250	Phe-218
	Pro-260
Ile-251	Pro-260
	Phe-262

[a]Residues are listed when carbon atoms from an RTA side chain make van der Waals contact with side chain carbons from RTB.

anism of a nucleosidase which hydrolyzes the adenine base from adenosine monophosphate (AMP) (17). Their results are best explained by a transition state in which there is considerable oxycarbonium character in the ribose ring, protonation of the leaving adenine, partial bonding between the ribose and the base, and participation of a water nucleophile attacking at C1′ of ribose. This is probably a reasonable working model for the participants in the ricin mechanism as well.

Figure 3 shows a close-up view of the active site region (14). It contains Tyro-80, Tyr-123, Glu-177, Arg-180, and Trp-211, residues which are in-

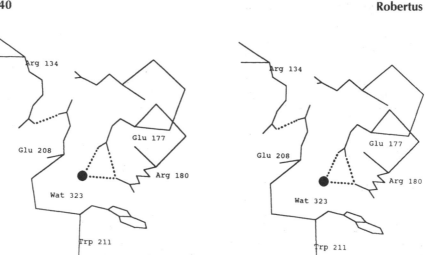

Figure 3 The active site of ricin A chain. A convergent stereogram shows invariant and conserved groups in the active site cleft.

variant in all the toxins, including those from bacteria. The hydroxyl of Tyr-80 makes a strong (2.5 Å) hydrogen bond to O_{121} and helps secure the chain in this area. The other invariant residues participate in a network structure. The side chain of Arg-180, including the resonant stabilized guanidinium group, lies parallel to the plane of the indole ring of Trp-211. The guanidinium protrudes beyond the Trp ring where N_{H1} forms a strong hydrogen bond to O_{78}. N_{H2} appears to donate two bonds, one to an active site water molecule (#323), and a second to O_{E2} of Glu-177. Glu-177 is involved in two additional bonds besides the link to Arg-180. It bonds to water #323 and also to the hydroxyl of Tyr-183.

It seems likely that some of these residues are involved in either binding the rRNA loop which is the immediate substrate, or act in catalyzing base excision. It is tempting to speculate that Glu-177 may be the anion which stabilizes the putative oxycarbonium intermediate and that water #323 is the hydrolytic attacking group. It may be activated by its position between the strong poles of Glu-177 and Arg-180, two charged and invariant active site residues.

In addition to the invariant residues discussed above, Asn-78, Arg-134, Gln-173, Glu-208, and Asn-209 are highly conserved active site residues and may play less specific roles in maintaining the active site structure or in the catalytic mechanism. Glu-208 forms an ion pair with Arg-134 and also hydrogen bonds to the side chain of Gln-173. In this position, the methylene carbons of Glu-208 contact those of the side chain of Glu-177, and appear to form part of one wall of the active site cleft. Gln-173 is very conservative and is a polar group inside the protein. Presumably it plays a key role in anchoring Glu-208, and it also bonds to a trapped water molecule (#331). Asn-209 appears to sit on the molecular surface at the edge of the active site cleft, as does Asn-78. They may be involved with RNA base recognition or related function.

These invariant and conservative residues are the target of site directed mutagenesis efforts in our laboratory and in several others. Table 2 shows a list of mutations which have been made to date. Conversion of Arg-180 to Lys has little effect on catalysis, but conversion to His reduces enzyme activity nearly 1000-fold, attesting to the importance of this residue in the mechanism of action (18). Conversion of Glu-177 to Gln decreases activity 170-fold, again suggesting an important role in the mechanism of action (19). Glu-177 has also been converted to Ala (E177A) and to Asp with only 20-fold and 80-fold reduction in activity, respectively (20). The small, and seemingly anomalous kinetic effect observed for E177A may be due to a partial substitution for Glu-177 by the nearby Glu-208, which can only occur when the carboxylate of 177 has been removed (18). Conversion of Tyr-80 to Phe decreases activity roughly 14-fold, suggesting that it is a fairly important residue in the enzyme. Conversion of Asn-209 to Ser or conversion of Tyr-123 to Phe reduces activity three and sevenfold, respectively (19), whereas conversion of Trp-211 to Phe drops activity tenfold (39). This suggests the residues play a relatively nonspecific or only a minor role in ribosome inactivation. It is important to remember that enzymes typically increase the rate of biological reactions by 10^8–10^{12}-fold. On this

Table 2 Site Directed Mutants of RTA

Protein	Relative activity	Reference
Wild type	100.00	
E208D	56.	18
R180K	38.	18
R180H	0.1	18
E177A	5.0	20
E177D	1.3	20
E177Q	0.6	19
E177A/E208D	>0.1	18
N209S	33.	19
Y80F	7.	19
Y123F	15.	19
W211F	11.	39

mechanistic scale, a residue contributing 10-fold to the activity is quite minor; under biological selection the residue may still be conserved because a protein with only 10% activity is at a disadvantage.

Ricin A chain can excise a single base from eukaryotic ribosomes, from naked 28S rRNA or from a 35-base synthetic oligonucleotide which contains the conserved stem and loop sequence of the target rRNA (11). However, the enzyme is on the order of 10,000 times more efficient at attacking ribosomes than either naked rRNA or the synthetic substrate. It has been suggested that the toxin may recognize higher-order RNA structure and not simply a sequence of bases. The active site cleft appears to large enough to accommodate a loop of single-stranded RNA but probably cannot bind a double helical stem. Ribosomal recognition may be accomplished at some other site. At this time we have no idea which portions of RTA might be involved in this putative extended site. It is interesting to consider however, that the binding to RNA might involve arginine residues which can form ion pairs with a polynucleotide backbone. On the back side of the molecule is an apparent second cleft which, like the enzymatic active site, is bordered by Arg residues and might serve this recognition function (14).

Although the active site is a primary target for analytical mutations, there are many other properties of the protein which should also be studied. An interesting observation is the invariance of Tyr-21 and Arg-29 in the sequences shown in Figure 1. A careful analysis of the RTA structure shows that a central alpha helix is bent near its C-terminus, and this bending allows Glu-177 and Arg-180 to reach the solvent of the active site cleft (14). This important helix bending disrupts the normal α-helical bonding pattern. However, the resulting structure is stabilized by new hydrogen bonds to the side chains of Tyr-21 and Arg-29. Alteration of these residues could, therefore, affect the folding rate or thermodynamic stability of the enzyme.

A. The Ricin B Chain

As described previously, RTB is made of two globular domains, each with one lactose binding site (12,21). Each domain is composed of four peptides designated λ, α, β, and γ. The subdomains of domain 1 (residues 1–135) are labeled 1λ, 1α, 1β, and 1γ; domain 2 is labeled in a corresponding manner. The subdomains are each approximately 40 residues in length and are homologs (21).

Several research groups have investigated the sugar-binding properties of ricin using a variety of methods. Equilibrium dialysis revealed two lactose-binding sites with association constants 2800 M^{-1} and 35,000 M^{-1} at 4°C and 3000 M^{-1} and 19,000 M^{-1} at 25°C (22,23). However, a study using fluorescent galactose analogs showed nearly equal affinities for the two sites (24). A difference Fourier analysis of ricin with and without bound lactose also suggested that the sites are occupied unequally (25). Ultraviolet spectroscopy suggested that a Tyr residues is involved in the stronger binding site (26), whereas chemical modifications suggested a Trp residue was important to sugar binding in the lower affinity site (27). It has also been shown (25–27) that RTB binds only one molecule of N-acetylgalactosamine (NAGal).

X-ray analysis shows that each sugar-binding site may be described as a shallow pocket (15). The top of the pocket is simply the side chain of an aromatic residue. The bottom of the pocket is formed by a three-residue kink in the polypeptide chain. The galactosyl moiety of lactose is oriented and secured by hydrogen bonds to the side chains of several polar residues which lie at the back of the pocket, whereas the glucosyl moiety extends freely into solvent and makes no specific interaction with the protein. The aromatic residues in sites 1 and 2, respectively, are Trp-37 and Tyr-248, whereas the three-residue kink is seen in residues 24–26 and 236–238.

In the shallow cleft, the bound galactose makes several specific protein interactions. A conserved aspartic acid residue, Asp-22 in site 1 and Asp-234 in site 2, accepts hydrogen bonds from both the C_3 and C_4 hydroxyls of the bound sugar. These carboxylates are anchored into their correct positions by a hydrogen bond from N_{E2} of a conserved amide, Gln-47 in site 1 and Gln-256 in site 2. Each C_3 hydroxyl also forms a strong hydrogen bond with N_{D2} of a conserved amide, Asn-46 in site 1 and Asn-255 in site 2 (15). Indeed, site-directed mutagenesis of these residues has confirmed that they are important in galactose binding, and loss of the side group may reduce binding by several orders of magnitude (41).

The hydrogen bond network of site 1 is more complex than that of site 2 in that N_{E2} of Gln-35 makes a bifurcated hydrogen bond to the C_4 and C_6 hydroxyls. In site 2, the analog of Gln-35 is Ile-246 and the hydrogen bonding interaction seen in site 1 is replaced by a hydrophobic contact between this residue and C_6 of the sugar. In contrast to the galactose in site 1, the C_6 hydroxyl is oriented toward solvent in site 2, and not involved in hydrogen bonding to the protein. Glucose is the C_4 epimer of galactose and it does not bind to RTB, in accord with the specific bonds made to the C_4 alcohols. A line of nonpolar atoms of the galactose ring, C_4, C_5, and C_6, contact the plane of the aromatic ring found in each site. This juxtaposition of the hydrophobic face of a sugar against an aromatic ring is a common feature emerging in protein-carbohydrate interactions (28).

Crystallographic models of the two sugar-binding sites are shown in Figure 4. These graphically illustrate the residues involved in recognition

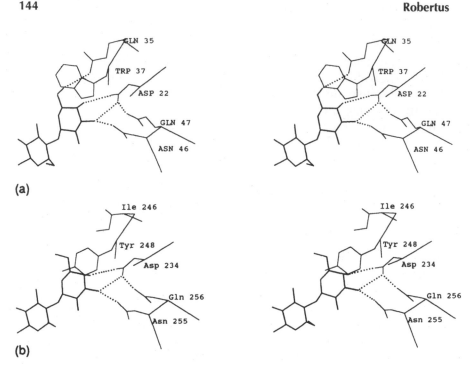

(a)

(b)

Figure 4   The lactose-binding sites of RTB. The sugar-binding sites of (a) domain 1 and (b) domain 2 are presented in a common orientation, as convergent stereo pairs. Key residues on the protein are labeled.

of cell surface galactosides and can be used to guide genetic engineering efforts aimed at exploring that protein function. Indeed, several studies have already been conducted.

The x-ray model can also explain why NAGal is bound only at site 2 and not at site 1. NAGal differs from galactose by being N-acetylated at C_2. The least squares superposition of the sugar-binding sites (not shown) reveals that the bound galactose residue of site 1 is rotated roughly 15 degrees relative to that of site 2. An N-acetyl group at the galactose C_2 position in site 2 extends freely into solution, but the sugar would encounter steric hindrance in site 1. In part, the N-acetyl group would collide with the side chain of Asp-44.

V. MODELS OF OTHER TOXINS

It is a generally accepted maxim of biochemistry that tertiary structure is more conservative than primary structure. As a result, the x-ray structure for one member of a family of proteins, like the plant toxins, can provide useful information about other members of the class. In our analysis of the structure of RTA (14), we made a detailed analysis of those residues which are conserved among the toxins. Active site residues directly involved in enzymatic catalysis are very conservative, as would be expected. A number of amino acid positions around the molecule are seen to be conserved as hydrophobic moieties. These cluster and pack together, controlling the folding and stability of the protein. One example of this is the clustering of four hydrophobic residues at positions 93, 117, 119, and 168. Each amino acid is donated to the cluster from a distinct topological element of the structure (14), and the clustering helps orient them in space.

We have used the RTA model as a template to construct models of several other toxins (6). The amino acid sequences of abrin A chain (ATA) and α-trichosanthin (α-TCS) were aligned with RTA to optimize identities. The linear sequence alignments were used to guide amino acid substitutions, insertions, and deletions made to the three-dimensional RTA model using an interactive graphics system. Stereochemical clashes resulting from the amino acid changes were initially adjusted by hand maintaining as much of the RTA geometry as possible.

The hand-adjusted models for ATA and α-TCS were each subjected to 40 computer cycles of energy minimization using the harmonic repulsive term option of program XPLOR (29) to reduce the number of unfavorable Van der Waals contacts. After this treatment, the root-mean-squared difference for corresponding alpha carbon positions between RTA and ATA is 0.30 Å; that between RTA and α-TCS is 0.85 Å. Positional differences between individual side chains were often larger, particularly for the less constrained residues on the molecular surface.

This modeling showed the overall folding of both ATA and α-TCS is probably very similar to that of RTA. Deletions of residues at the amino and carboxy-terminals, respectively, were easily accommodated. Internally, deletions and insertions tended to appear on the molecular surface, often at loops and turns which were readily accommodated.

We have used this same method to produce plausible models for a number of other toxins, including Shiga-like toxin A chain, SO-6 and *Mirabalis* antiviral protein (MAP). The latter protein is extraordinarily interesting because of its apparent ability to inhibit bacterial ribosomes (38). Its structure is also of interest because it appears to have a disulfide bond formed between cysteines 36 and 220 (33). Applying our usual strategy of model building to the MAP sequence placed the two Cys residues reasonably close in space, but not close enough to bond. A portion of the model had to be rebuilt with this chemical constraint in mind. A super-

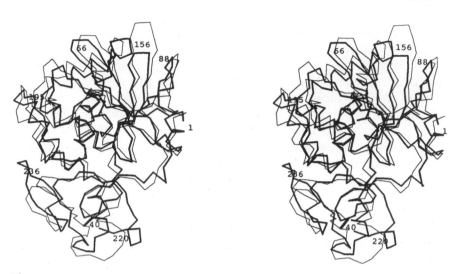

Figure 5 Superposition of RTA and MAP. The alpha carbon backbones are displayed in this convergent stereo pair. RTA is the darker model, whereas MAP is light. Several RTA positions are labeled with residue numbers. The disulfide bond of MAP is near position 40.

position of RTA and MAP is shown in Figure 5. The disulfide is shown as a dark bar connecting the light backbone lines of MAP, near the position labeled as RTA residue 40. Notice that peptide loops in this general area have been altered a fair bit to accommodate the crosslink. It is as yet unclear how these structural changes may be related to the unusual toxicity of MAP.

ACKNOWLEDGMENTS

We are grateful to Ed Collins and Michael Ready for continuing assistance and helpful discussions, and to Raquelle Keegan for her help in preparing the illustrations. This work was supported by grants GM 30048 and GM 35989 from the National Institutes of Health and by a grant from the Clayton Foundation for Research.

REFERENCES

1. Ready, M. P., Brown, D. T., and Robertus, J. D. Extracellular localization of pokeweed antiviral protein. Proc. Natl. Acad. Sci. U.S.A., *83*: 5053–5056.
2. References in: "Immunotoxins," Frankel, A. E. ed. Boston: Kluwer Academic Publishers, 1988.

3. McGrath, M. S., Hwang, K. M., Caldwell, S. E., Gaston, I., Luk, K. C., Wu, P., Ng, V. L., Crowe, S., Daniels, J., Marsh, J., Deinhart, Y., Lekas, P. V., Vennari, J. C., Yeung, H. W., and Lifson, J. D. GLQ223: An inhibitor of human immunodeficiency virus replication in acutely and chronically infected cells of lymphocyte and mononuclear phagocyte lineage. Proc. Natl. Acad. Sci. U.S.A., *86*: 2844–2848, 1989.

4. Ready, M., Wilson, K., Piatak, M., and Robertus, J. D. Ricin-like plant toxins are evolutionarily related to single-chain ribosome-inhibiting proteins from *Phytolacca*. J. Biol. Chem., *259*: 15252–15256, 1984.

5. Endo, Y., Tsurgi, K., and Lambert, J. M. The site of action of six different ribosome-inactivating proteins from plants on eukaryotic ribosomes: The RNA N-glycosidase activity of the proteins. Biochem. Biophys. Res. Commun., *150*: 1032–1036, 1988.

6. Collins, E. J., Robertus, J. D., LoPreti, M., Stone, K. L., Williams, K. R., Wu, P., Hwang, K., and Piatak, M. Primary amino acid sequence of α-trichosanthin and molecular models for abrin A-chain and α-trichosanthin. J. Biol. Chem., *26*: 8665–8669, 1990.

7. Olnes, S., and Pihl, A. The molecular action of toxins and viruses. *In*: P. Cohen and S. Van Heynigen (eds.), The Molecular Action of Toxins and Viruses, pp. 52–105. New York: Elsevier Biomedical Press, 1982.

8. Youle, R. J., and Colombatti, M. Hybridoma cells containing itracullular anti-ricin antibodies show ricin meets secretory antibody before entering the cytosol. J. Biol. Chem., *262*: 4676–4682, 1987.

9. Olsnes, S., Fernandez-Puentes, C., Carrasco, L., and Vasquez, D. Ribosome inactivation by the toxic lectins abrin and ricin: Kinetics of the enzymic activity of the toxin A-chains. Eur. J. Biochem., *60*: 281–288, 1975.

10. Endo, Y., and Tsurugi, K. RNA N-glycosidase activity of ricin A-chain. J. Biol. Chem., *262*: 8128–8130, 1987.

11. Endo, Y., Chan, Y. L., Lin, A., Tsurugi, K., and Wool, I. G. The cytotoxins α-sarcin and ricin retain their specificity when tested on a synthetic oligoribonucleotide (35-mer) that mimics a region of the 28 S ribosomal ribonucleic acid. J. Biol. Chem., *263*: 7917–7920, 1988.

12. Montfort, W., Villafranca, J. E., Monzingo, A. F., Ernst, S., Katzin, B., Rutenber, E., Xuong, N. H., Hamlin, R., and Robertus, J. D. The three-dimensional structure of ricin at 2.8 Å. J. Biol. Chem., *262*: 5398–5403, 1987.

13. Rutenber, E., Katzin, B. J., Collins, E. J., Mlsna, D., Ernst, S. E., Ready, M. P., and Robertus, J. D. The crystallographic refinement of ricin at 2.5 Å resolution. Proteins, *10*: 240–250, 1990.

14. Katzin, B. J., Collins, E. J., and Robertus, J. D. The structure of ricin A chain at 2.5 Å. Proteins, *10*: 251–259, 1991.

15. Rutenber, E., and Robertus, J. D. The structure of ricin B chain at 2.5 Å resolution, and its interaction with A chain. Proteins, *10*: 260–269, 1991.

16. Lewis, M. S., and Youle, R. J. Ricin subunit association: Thermodynamics and the role of the disulfide bond in toxicity. J. Biol. Chem., *261*: 11571–11577, 1986.

17. Mentch, F., Parkin, D. W., and Schramm, V. L. Transition-state structures for N-glycoside hydrolysis of AMP by acid and by AMP nucleosidase in the presence and absence of allosteric activator. Biochemistry, 26: 921–930, 1987.

18. Frankel, A., Welsh, P., Richardson, J., and Robertus, J. D. The role of arginine 180 and glutamic acid 177 of ricin toxin A chain in the enzymatic inactivation of ribosomes. Mol. Cell Biol., 10: 6257–6263, 1991.

19. Ready, M. P., Kim, Y., and Robertus, J. D. Directed alteration of active site residues in ricin A chain and implications for the mechanism of action. Proteins, 10: 270–278, 1991.

20. Schlossman, D., Withers, D., Welsh, P., Alexander, A., Robertus, J., and Frankel, A. Expression and characterization of mutants of ricin toxin A chain in Escherichia coli. Mol. Cell. Biol., 9: 5012–5021, 1989.

21. Rutenber, E., Ready, M., and Robertus, J. D. Structure and evolution of ricin B chain. Nature, 326: 624–626, 1987.

22. Zentz, C., Frenoy, J. P., and Bourrillon, R. Binding of galactose and lactose to ricin: Equilibrium studies. Biochim. Biophys. Acta, 536: 18–26, 1978.

23. Shimoda, T., and Funatsu, G. Binding of lactose and galactose to native and iodinated ricin D. Agr. Biol. Chem., 49: 2125–2130, 1985.

24. Houston, L., and Dooley, L. Binding of two molecules of 4-methylumbelliferyl galactose or 4-methylumbelliferyl N-acetylgalactosamine to the B chains of ricin and Ricinus communis agglutinin and to purified ricin B chain. J. Biol. Chem., 257: 4147–4151, 1982.

25. Villafranca, J. E. The Crystal Structure of Ricin at Low Resolution. Ph.D. Dissertation, University of Texas at Austin, 1980.

26. Yamasaki, N., Hatakeyama, T., and Funatsu, G. Ricin D-saccharide interaction as studied by ultraviolet difference spectroscopy. J. Biochem., 98: 1555–1560, 1985.

27. Hatakeyama, T., Yamasaki, N., and Funatsu, G. Identification of the tryptophan residue located at the low-affinity saccharide binding site of ricin D. J. Biochem., 100: 781–788, 1986.

28. Molecular recognition and protein-carbohydrate interactions. Transactions of the American Crystallographic Association, Vol. 25, (H. Einspahr and K. B. Ward, eds.) ACA, Buffalo, New York, 1989.

29. Brunger, A. T. Crystallographic refinement by simulated annealing. In: N. W. Issacs and M. R. Taylor (eds.), Crystallographic Computing 4: Techniques and New Technologies, Clarendon Press, Oxford, England, 1988.

30. Lamb, F. I., Roberts, L. M., and Lord, J. M. Nucleotide sequence of cloned cRNA coding for preproricin. Eur. J. Biochem., 148: 265–270, 1985.

31. Halling, K. C., Halling, A. C., Murray, E. E., Ladin, B. F., Houston, L. L., and Weaver, R. F. Genomic cloning and characterization of a ricin gene from Ricinus communis. Nucl. Acids Res., 13: 8019–8033, 1985.

33. Habuka, N., Murakami, Y., Noma, M., Kudo, T., and Horikoshi, K. Amino acid sequence of Mirabilis antiviral protein, total synthesis of its gene and expression in Escherichia coli. J. Biol. Chem., 264: 6629–6637, 1989.

34. Benatti, L., Saccardo, M. B., Dani, M., Nitti, G., Sassano, M., Lorenzetti, R., Lappi, D. A., and Soria, M. Nucleotide sedquence of cDNA coding for

saporin-6, a type-1 ribosome-inactivating protein from Saponaria officinalis. Eur. J. Biochem., *183*: 465–470, 1989.

35. Asano, K., Svensson, B., Svendsen, I., Poulsen, P., and Roepstorff, P. The complete primary structure of protein synthesis inhibitor II from barley seeds. Carlsberg Res. Commun., *51*: 129–141, 1986.

36. Calderwood, S. B., Auclair, F., Donohue-Rolfe, A., Keusch, G. T., and Mekalanos, J. J. The nucleotide sequence of the shiga-like toxin genes of *Escherichia coli*. Proc. Natl. Acad. Sci. U.S.A., *84*: 4364–4368, 1987.

37. Jackson, M. P., Neill, R. J., O'Brien, A. D., Holmes, R. K., and Newland, J. W. Nucleotide sequence analysis and comparison of the structural genes for Shiga-like toxin I and Shigl-like toxin II encoded by bacteriophages from Eschericia coli 933. Microbiol. Lett., *44*: 109–114, 1987.

38. Habuka, N., Akiyama, K., Tsuge, H., Miyano, M., Matsumoto, T., and Noma, M. Expression and secretion of *Mirabolis* antiviral protein by *Eschericia coli*, and its inhibition of *in vitro* eukaryotic and prokaryotic protein synthesis. J. Biol. Chem., *264*: 6629–6637, 1990.

39. Bradley, J. L., and McGuire, P. M. Site-directed mutagenesis of ricin A chain trp 211 to Phe. Int. J. Peptide Protein Res., *35*: 365–366, 1990.

41. Vitetta, E. S., and Yen, N. Expression and functional properties of genetically engineered ricin B chain lacking galactose binding activity. Biochim. Biopys. Acta, *1049*: 151–157, 1990.

8

Chemical and Genetic Characterization of the Enzymatic Activity Associated with Ricin A Chain

Lawrence Greenfield   *Roche Molecular Systems, Alameda, California*

I. INTRODUCTION

Ricin (M_r 58,000), a lectin from *Ricinus communis*, is a heterodimer consisting of a catalytically active A chain (M_r 32,000, 267 amino acids) and a binding B chain (M_r 34,700, 262 amino acids) which are linked by a single disulfide bond (1,2). The A chain is an N-glycosidase that inactivates eukaryotic ribosomes (3,4). When the A chain and B chains are disulfide-linked, the whole toxin is enzymatically inactive (5). The B chain binds to cell surface galactose residues and may play a role in the translocation of the A chain across the cell membrane (6). Both the A and B chains of native ricin are glycosylated. The B chain contains two high mannose-containing oligosaccharides of composition $(GlcNAc)_2(Man)_{4-7}$. There is heterogeneity in glycosylation of the A chain in that 64% (ricin A1) and 36% (ricin A2) contain one and two carbohydrate chains, respectively (7). One of the carbohydrate chains is similar in composition to that of the B chain, whereas the other has the composition $(GlcNAc)_2(Xyl)_1,(Fuc)_1,(Man)_{3-4}$ (7–10). These asparagine-linked sugars are attached through N-glycosylation to position 10 in the A chain (11,12) and positions 374 and 414 in the B chain (13). It has been postulated that Asn-236 is the second, variable glycosylation site in the A chain (14).

Correlation of the structure of ricin A chain with its catalytic activity will aid in the understanding of the mechanism by which it inactivates protein synthesis. The use of ricin A chain as a component of immunotoxins in the treatment of a number of human diseases may present new problems, including immuogenicity, toxicity, and localization to the desired target. Identification of portions of the molecule not necessary for its catalytic or translocating activities may aid in directing modifications of the protein to address issues of immunogenicity and nonspecific toxicity. Reducing the size of the toxin may decrease its immunogenicity and may aid in the passage into the extravascular space. Finally, an understanding of those portions of the molecule essential for its activity will help in the intelligent design of fusion proteins similar to those created with diphtheria toxin and *Pseudomonas* exotoxin A.

II. CHEMICAL MODIFICATION OF NATIVE RICIN A CHAIN

Prior to the cloning of ricin A chain, chemical modification of the protein helped delineate those residues important for various functions. By controlling reaction conditions, only a few amino acid residues were derivatized (Table 1).

Ricin A chain undergoes conformational changes when the holotoxin is reduced. The protease sensitivity of the isolated A chain is greater than when it is attached to the B chain (15). Ricin A chain epitopes detectable in the holotoxin are no longer evident on the free A chain (16). The conformation or accessibility of the catalytic site is altered upon removal of the B chain. Thus, the protein synthesis inhibitory activity of ricin A chain is increased 50- to 100-fold following reduction of the whole toxin (17). Cibacron blue F3GA, which is postulated to bind in the active site of the A chain, interacts with free ricin A chain but not whole toxin (18).

The pI values for intact ricin, ricin A chain, and ricin B chain are 7.1–7.3, 7.4–7.5, and 4.8–5.2, respectively (15,19). The B chain of ricin appears to stabilize the A chain against denaturation by heat, acid, and extremes of pH (15,20). In 5% SDS, ricin A chain retains 80% of its protein synthesis inhibitory activity (20). In contrast, the addition of SDS to ricin A chain–containing antibody conjugates reduces the in vitro cytotoxic activity of the conjugates (21), suggesting that the detergent alters the conformation of the portion of the A chain responsible for translocation across cell membranes. Conflicting results have been obtained for the effect of urea on the activity of the A chain (15,20).

Chemical modificaton of whole ricin and free A chain may give insight into amino acid residues important for the catalytic activity (Table 1). Results from trinitrophenylation of whole ricin D suggests that free amino groups exposed on the surface of the whole toxin may be involved in the enzymatic activity of the A chain (20). Modification of six of the available 11 amino groups (nine lysines plus two amino terminal ends) decreases the in vivo toxicity of whole ricin eightfold without effecting its cytoagglutinating activity. In contrast, modification of nine arginine residues of whole ricin with 1,2-cyclohexanedione has little effect on both the in vivo toxicity (50% of that of unmodified toxin) and cytoagglutinating activity (75% of that of unmodified toxin), suggesting that arginine residues on the surface of the whole toxin are not involved in its biological activity (20). Oxidation of tryptophan residues (7 of the 10) by N-bromosuccinimide decreases both the cytoagglutinating activity and in vivo toxicity to 14 and 13%, respectively, of that of unmodified toxin (20). Addition of hydroxynitrobenzyl groups to tryptophans (four to five groups per molecule) by treatment of whole toxin with 2-hydroxy-5-nitrobenzylbromide results in no reduction of the in vitro protein synthesis inhibitory activity of the toxin (22). Nitration of 7 of the 23 tyrosines of whole toxin by tetranitromethane affects the in vivo toxicity (1% of unmodified toxin) more than the cytoagglutinating activity (38% of unmodified toxin) (20). Modification of nine free whole toxin carboxyl groups by carbodiimide decreases both the in vivo toxicity and the cytoagglutinating activity to similar events (8–10% of unmodified toxin) (20). Succinylation of six whole ricin amino groups has no effect on the in vitro activity of the A chain (22). Acetylation of 15 whole ricin tyrosine residues by N-acetylimidazole reduces its in vitro protein synthesis inhibitory activity 1.4-fold and its in vivo toxicity to mice 17-fold (22). Most of the in vivo toxicity but none of the in vitro inhibitory activity can be restored by deacetylation with hydroxylamine. Nitration of whole ricin with tetranitromethane resulting in 3.2 nitrotyrosyl derivatives on the A chain (eight groups incorporated into the whole toxin), reduces the in vitro activity of the A chain by approximately sevenfold (23). With the reduction of these groups to aminotyrosines by sodium dithionite, full activity is restored. In contrast, the in vitro cytotoxicity and in vivo toxicity of both of these modified whole toxins is dramatically reduced. Reaction of whole ricin with 2,4,6-trinitrobenzenesulfonic acid under conditions which add eight trinitrophenyl groups per toxin modifies 3.4 amino groups in the A chain (24). The altered toxin retains full cytoagglutinating activity and greater than half of its in vitro protein synthesis inhibitory activity, but is tenfold less toxic to mice. The similar sensitivities of the modified toxin and unmodified toxin to digestion by trypsin, chymotrypsin, and nagarse suggested that the molecule was not significantly denatured.

Modification of arginine residues in isolated ricin A chain by either phenylglyoxal or 1,2-cyclohexanedione inactivates its inhibitor activity in a cell-free protein synthesis system (25). When three of the 20 ricin A chain arginine residues are altered, greater than 90% of the catalytic activity is lost, and there is a concomitant decrease in the ability to bind the dye Cibacron blue F3GA to the active site (18). Binding of dye to the active

Table 1 The Effect of Chemical Modification on the Activity of Ricin

Chemical	Amino acid[a]	No.[b]	Form[c]	% Wild-type activity			Reference
				toxicity[d]	agglut.[e]	prot. syn.[f]	
Sodium borohydride/formaldehyde	Amino	11	Whole	75	100	100	22
Sodium borohydride/formaldehyde	Amino	4	A chain			100	22
Succinic anhydride	Amino	6	Whole	100	100	100	22
Trinitrobenzene sulfonate	Amino	8	Whole	<10	100	>50	24
Trinitrobenzene sulfonate	Amino	6.0	Whole	12	100		20
Maleic anhydride	Amino	11	Whole	4	8		70
S-Acetylmercaptosuccinic anhydride	Amino	1.4	A chain			58	71
Carbodiimide	Carboxyl	9	Whole	8	10		20
Cystamine/1-ethyl-3-[3-dimethylaminopropyl]carbodiimide	Carboxyl	0.7	Whole			50	71
Phenylglyoxal	Arginine	2.9	A chain			6.0	25
1,2-Cyclohexanedione	Arginine	7.6	A chain			0.9	25
1,2-Cyclohexanedione	Arginine	9.4	Whole	50	75		20
N-Ethylmaleimide	Cysteine	1.0	A chain			100	71

	[a]	[b]	[c]	[d]	[e]	[f]	
Tetranitromethane	Tyrosine	8	Whole	<1		~15	23
Tetranitromethane/sodium dithionite	Tyrosine	8	Whole	<1		100	23
N-Acetylimidazole	Tyrosine	15	Whole	6	12	70	22
Tetranitromethane	Tyrosine	6.5	Whole	1	38		20
Potassium iodide	Tyrosine	5.3	Whole	2	6		70
Pyridyl dithiopropionate/ carbonyldiimidazole	Tyrosine	1.6	A chain			31	71
Maleimidocaroic acid/ carbonyldiimidazole	Tyrosine	1.0	A chain			11	71
N-Bromosuccinimide	Tryp	6.6	Whole	13	14		20
N-Bromosuccinimide	Tryp	1.9	Whole	3	100		22
2-Hydroxy-5-nitrobenzylbromide	Tryp	5	Whole	100	100	100	22

[a]Group predicted to be modified by the chemical treatment.
[b]Number of groups modified.
[c]Form of toxin subjected to chemical treatment.
[d]In vivo toxicity; applicable only to whole toxin; expressed as the percent of wild-type activity.
[e]Ability to agglutinate cells containing complex sugars on the surface; expressed as percent of wild-type activity.
[f]Percent of wild-type in vitro protein synthesis inhibitory activity.

site protects an important arginine residue in the A chain from modification by phenylglyoxal (18). Regeneration of the 1,2-cyclohexanedione–modified residues by reaction with hydroxylamine completely restores the enzymatic activity. Arginine residues 26, 29, 31, 39, 48, and 56 are particularly susceptible to derivatization by 1,2-cyclohexanedione. The addition of three to four methyl groups to ricin A by reductive methylation of either whole ricin or free ricin A chain with sodium borohydride and formaldehyde has no effect on the in vitro activity of the A chain or the toxicity of the whole toxin in vivo (22). This suggests that the exposed lysine residues do not play an important role in the protein synthesis inhibitory activity of ricin A chain.

Thus, the results of chemical modification of whole toxin and free ricin A chain suggests that an arginine residue plays an important role in the catalytic activity of the A chain. If tyrosine, tryptophan, or other amino groups are involved, they are protected by the B chain from chemical modification.

Ricin A chain has been deglycosylated both enzymatically (7) and chemically (8,9). There is no loss of protein synthesis inhibitory action of the A chain following oxidation of the carbohydrates with sodium metaperiodate with or without reduction by sodium cyanoborohydride (8,26). Similarly, oxidation of the A chain carbohydrates followed by blocking of the Schiff's base with excess amines results in an A chain which retains its in vitro protein synthesis inhibitory activity (71). The catalytic activity of the A chain is unaffected by removal of a portion of the carbohydrate with α-mannosidase or endoglycosidase H (27). Thus, the carbohydrates do not appear to play a role in catalysis by ricin A chain. In contrast, chemical destruction of whole ricin carbohydrate residues significantly decreases the potency of the toxin on cells in vitro without altering either the protein synthesis inhibitory activity of the A chain or the cell-agglutinating activity of the B chain (8,26). When enzymatically deglycosylated B chain is combined with unaltered A chain, a reduction in in vitro toxicity is observed, suggesting that the decrease in potency may involve alterations in B chain functions (27).

III. PROTEOLYTIC FRAGMENTS OF RICIN A CHAIN

In contrast to the relative protease resistance of intact ricin, the isolated A chain is sensitive to inactivation when digested with trypsin (5 μg/ml), pepsin (5 μg/ml), or pronase (5 μg/ml) (Table 2) (15,28,29). Under some denaturing conditions, whole ricin can be digested by trypsin (28). Whole ricin is not digested by trypsin in the presence of 0.3% of cetyltrimethylammonium bromide, 0.3% sodium deoxycholate, or 0.3% triton X-100,

Table 2 The Effect of Proteolysis on the Structure and Activity of Ricin

Enzyme	Conditions	Size range	In vivo toxicity	% wt protein synthesis inhibition	Reference
Trypsin	Heat-denatured whole ricin	"HA": 30,000	"Equivalent"	50	28
Trypsin	Guanidine-HCl–denatured whole ricin	"GC": 14,000–11,000	"Equivalent"	50	28
Trypsin	Guanidine-HCl–denatured whole ricin	II	10–47% (by weight)		29
Trypsin	Guanidine-HCl–denatured whole ricin	II_{tc} 22,000	8%	33	29
Trypsin	Guanidine-HCl–denatured whole ricin	II_{2a} 18,000.	5%	<28	29
Trypsin	Whole ricin, 0.3% SDS	RI:40,000 RII:23,500			30
Trypsin	Ricin A chain, 0.13% SDS	TI:23,500 TII:7,300		~100	30
Nagarase	Whole ricin		100%	100	34

but is digested in the presence of 0.3% SDS (30). The digestion pattern and in vivo toxicity of the resultant fragments differ depending on the denaturant used and the pH at which cleavage is performed. One fraction ("HA"), consisting of peptides of around 30,000 d, isolated from size fractionation of whole ricin digested with trypsin at pH 8.5 has in vivo toxicity to mice similar to that of the native toxin (28). Tryptic digestion of guanidine hydrochoride–denatured toxin and size fractionation yields a population of four to five fragments, ranging in size from 14,000 to 11,000 d, which partially retain their in vivo toxicity to mice (28,29). These peptides also retain 50% of the native toxin-inhibitory activity in a cell-free protein synthesis system. In contrast to intact toxin, which is inactive on the prokaryotic protein synthetic machinery (31,32), a limited tryptic hydrolysate can inhibit the in vitro protein synthesis activity of an *Esherichia coli* lysate (33).

A 22-kd (220 amino acids) and a 18-kd (181 amino acids) peptide can be purified by ion exchange chromatography of guanidine hydrocloride-denatured ricin digested with trypsin (29). Both fragments generate two peptides upon reduction, suggesting that they are contain portions of the A and B chains of native ricin. These purified peptides retain 5–8% of the in vivo toxicity of the native toxin, but are three to four times less active on whole cell in vitro and in a rabbit reticulocyte lysate cell-free extract.

Limited digestion of ricin D with nagarse removes a small amino-terminal peptide (34). The resulting large fragment, which begins with glutamine, retains the full cytoagglutinating activity and in vivo toxicity of whole toxin. Nagarse does not cleave between lysine and glutamine, suggesting that amino acids 1–18 of the native toxin were removed. Thus, the amino-terminal end of ricin A chain is not required for activity of the toxin.

IV. STRUCTURE OF RICIN A CHAIN

The structure of ricin A chain has been discussed in other chapters of this book. A summary of the major structural features is provided for reference (Table 3).

V. CLONING AND EXPRESSION OF RICIN A CHAIN

The cloning and expression of ricin has been covered elsewhere in this book and in recent papers (14,21,35–39). However, because some mutations result in the expression of less soluble products, there will be some discussion on the effect of expression on the solubility of the unmodified protein.

The solubility of the *E. coli*–expressed recombinant ricin A chain (rRA) is highly dependent on the method of induction (21,38,40). When

Table 3 Structure of Ricin A Chain (43)

Domain	Amino acid start	Amino acid end	Structure
1	7	13	β strand (a)
	18	32	α helix (A)
	59	64	β strand (b)
	68	73	β strand (c)
	82	86	β strand (d)
	89	93	β strand (e)
	99	104	α helix (B)
	115	117	β strand (f)
2	122	127	α helix (C)
	141	152	α helix (C)
	161	180	α helix (E)
	184	192	α helix (F)
	202	210	α helix (G)
3	211	219	α helix (H)

transcribed from the *trp*, *phoA*, and P_L promoters, the level of rRA represents 2, 6–8, and 6–8% of the total cell protein, respectively (21). The rRA expressed at 42°C is insoluble, and partially active. Synthesis at 37 or 30°C produces a soluble form of the protein with activity in a rabbit reticulocyte assay identical to that of native A chain (21,39). The soluble, active form produced at the lower temperatures can be converted into the insoluble form by incubation at 42°C in vivo but not in vitro, (21) suggesting that the conversion of soluble to insoluble forms is not an inherent thermal lability of the protein, but requires other *E. coli* environmental conditions (21). Similarly, expression from the coliphage T5 promoter P_{N25} produces a fully active protein when induced at 30°C but aggregated, partially active protein when induced at 37°C (38). The increased thermal lability of the latter protein may result from its fusion to 10 additional amino acids at the amino-terminal end of the mature sequence. The additional 10 amino acids does not alter the catalytic activity of the recombinant protein when expressed at the lower temperature (38).

In some studies, whole ricin reconstituted from recombinant A chain and native B chain has in vitro potency identical to whole ricin formed with native A chain, suggesting that the portion of the translocation domain present in rRA is properly folded (21,38). In another study, differences were reported in the in vitro cytotoxic activities of the two reconstituted forms (39). There has been some controversy as to the effect of substituting recombinant ricin A chain for native A chain in immunotoxins. In one report, conjugates made with the two forms of A chain had similar in vitro cytotoxicities (21), whereas in a second report, conjugates made with recombinant A chain were 100-fold less toxic than analogous conjugates made with native A chain (39). It is possible that subjecting the recombinant ricin A chain to pH 3 during purification may have contributed to the reduced potency in the latter study.

Fusion of ricin A chain sequences to a 60–amino acid collagen linker and β-galactosidase produces a soluble protein even at elevated temperatures (40). When the fused collagen linker and β-galactosidase are removed by digestion with collagenase, the recombinant A chain portion has 42% of the protein synthesis inhibitory activity of the native plant A chain. However, its ability to recombine with native ricin B chain is significantly reduced (40). When reconstituted with native B chain into whole ricin, both native and this recombinant ricin A chain construct show equivalent in vitro cytotoxicity to whole cells (40). The reduced ability to regenerate whole toxin with this ricin A chain construct may reflect differences in protein folding that affect the A chain–B chain interaction more than the deadenylation activity. Following cleavage with collagenase, this recombinant ricin A chain construct differs from the native sequence in several ways: The sequence MetAspProAsn is fused to the amino-terminal end of mature ricin A chain; Phe-267 is replaced by Leu-267; and the sequence AspProGlyProVal is fused to the carboxyl end of the molecule (40,41).

A ricin A chain–staphylococcal protein A fusion, consisting of 12 codons of λ–E. coli cro–lacZ gene fusion, 249 codons of the staphylococcal protein A gene, 5 codons from the multiple cloning site, 13 codons of the ricin A secretory leader, 267 codons acids of ricin A chain, the 12-amino acid linker peptide, 6 amino acids of ricin B chain, and 5 codons of the polylinker, has been expressed in E. coli by temperature induction (42). The expressed protein eluted from an IgG-Sepharose affinity column by low pH had approximately 20% of the in vitro protein synthesis inhibitory activity of the native A chain. The lower activity may result from the additional amino acids at either end of the fusion protein.

VI. MUTAGENESIS OF RICIN A CHAIN

A. Fusion of Additional Amino Acids to Amino- and Carboxy-terminal End of Ricin A Chain

When considering the effect of amino acid substitutions or mutations on the catalytic activity of ricin A chain, the background within which the mutations are made must be considered. Many of the constructs generated for the mutational analysis contain foreign amino acids fused to the ends of the polypeptide chain, which themselves may alter the enzymatic activity. The first six residues of ricin A chain interact loosely with the protein, and residues 220–267 appear to fold in a random coil (43), suggesting that addition of extraneous amino acids to either end of the peptide may have little effect on the folding of the remainder of the protein.

Addition of a Met to the amino-terminal end has no effect on the catalytic activity of ricin A chain (21). The amino-terminal end of the purified product from one such construct begins with Ile-Phe and Met-Ile-Phe in 60 and 40%, respectively, of the molecules (44). When the sequence MetGlySerSerArgValGluAspAsnAsnAla was fused to the amino-terminal end, the product had the same in vitro protein synthesis inhibitory activity and ability to reconstitute with native B chain as the native protein (38). However, the product from a construct consisting of 12 amino acids of λ–E. coli cro–lacZ fusion, 249 amino acids of staphylococcal protein A gene, 5 amino acids from the multiple cloning site, 13 amino acids of the ricin signal peptide, ricin A chain, 12 amino acids from the ricin junction peptide, 6 amino acids of ricin B chain, and 5 amino acids from the polylinker had 20% of the in vitro protein synthesis inhibitory activity of the native A chain (42).

Constructs in which the predicted translated sequence includes 22 amino acids fused to the amino-terminal end (MetProAlaGlyArgAspSerArgGly-SerThrSerGlyTrpSerPheThrLeuGluAspAsnAsn) and 26 amino acids (Ser-LeuLeuIleArgProValValProAsnPheAsnAlaAspValCysMet AspProProArg-ValProSerSerAsn) or 7 amino acids (SerLeuLeuIleArgProVal) fused to the carboxyl end, produce functional ricin A chain derivatives which both inactivate the reticulocyte lysate and depurinate the 28S rRNA (36). Removal of the nonnative amino-terminal acids moderately increases the activity of the in vitro transcription/translation product. The 10-fold difference in activity when the two carboxy-terminal fusions are compared may be due to differences in the level of expression of the two constructs in vitro, or to differences in protein folding. Many of the deletions to be discussed below were studied in this background. Similarly, constructs in which the predicted amino acid sequence contains the residues MetGlyLeuThrAla-HisAspArgProGlySerThrSerGlyTrpSerPheThrLeuGluAspAsnAsn at the amino-terminal end retain both protein synthesis inhibitory activity and depurinate the 28S rRNA (45).

Fusion of amino acids to the amino-terminal end of the mature ricin A chain sequence appears to alter the activity of the toxin in vivo in yeast cells (46). The structural gene for ricin A chain has been cloned into the low-copy-number plasmid, pBM150, and into the high-copy-number plasmid pRY131. In the latter case, the expressed protein was fused to 40 plasmid-encoded amino acids. Despite the higher copy number, induction of ricin A chain had less inhibitory effect on yeast cell growth from strains bearing the pRY131 ricin construct than the pBM150 ricin construct. In addition, higher levels of immunoreactive protein could be induced from the former construct before growth arrest (46). In another study, it was postulated that the addition of 24 amino acids to the amino-terminal of the mature ricin A chain may have lowered its catalytic activity in vivo in yeast cells (47).

B. Deletional Mutagenesis

A series of deletions have been made within the ricin A chain structural gene in an attempt to define amino acids important for the catalytic activity (Table 4). Removal of amino acids 1–18 by the proteolytic enzyme nagarse has little influence on the catalytic activity of the toxin (34). The product synthesized after removal of the codons encoding amino acids 1–9 has catalytic activity similar to that of the wild-type toxin (45). Removal of amino acids 1–12 abolishes the activity of the polypeptide. In contrast, the product synthesized after removal of the portions of the structural gene coding for amino acids 1–28 retains a protein synthesis–inhibitory activity that appears to be different from the deadenylation activity of the A chain (36).

A series of carboxyl-terminal deletion mutants have been generated using selection in yeast (48). The mutants contain in-phase termination codons, and therefore lack extraneous amino acids. Substantial deletions which remove amino acids 185–267, 194–267, 202–267, or 211–267, eliminating a number of the invariant residues implicated in the catalytic activity of the toxin, produce soluble, catalytically inactive muteins.

In a novel screening approach, the wild-type sequence and deletion mutants of the ricin A chain structural gene were cloned into an in vitro transcription vector, downstream from an SP6 promoter (45,36). The ability of a cell-free rabbit reticulocyte lysate system to translate the run-off transcription products was used as an assay for biological activity of the mutant proteins. The addition of toxin-encoded mRNA to the reticulocyte lysate completely blocked protein synthesis, suggesting that expression of ricin from the message inactivated the ribosomes before detectable protein synthesis could take place. In the final assay, wild-type or mutant toxin-encoded messages were translated for 5 min at 30°C and the remaining activity of the protein synthetic machinery was measured by the addition of a second mRNA. The system showed specificity in that antibodies to ricin A chain prevented the inhibition due to translation products from ricin-encoded message (36). In addition, the translation product were less effective in inhibiting wheat germ lysates (45,36); consistent with the finding that wheat germ ribosomes are more resistant to the action of ricin A chain (49,50). It was estimated that synthesis of approximately 10 molecules of ricin A chain from each mRNA molecule was sufficient to inhibit the translational activity of the reticulocyte lysate and produce detectable glycosylase activity (36). It would be interesting to know how this value compares to the number of ribosomes present in the system. These assays demonstrate qualitative and not quantitative alterations in ricin A chain activity. In addition, changes which reduce the inhibitory activity may result in higher levels of expression of the mutant toxin. Therefore, these assays may only detect those changes which result in dramatic inactivation of the catalytic activity of the toxin.

Expression of carboxyl deletions generated by DNA restriction of the structural gene may be complicated by the inability of ribosomes to release the nascent polypeptide chain when translating in vitro these messages lacking termination codons (51). Therefore, depending on the relative amounts of ribosomes to mRNA, the ability to translate a second message may be inhibited due to the lack of available ribosomes. That the protein inhibitory activity was due to the catalytic activity of ricin A chain was shown by the generation of depurinated 28S rRNA (36). Messenger RNA encoding ricin A from full-length wild-type constructs containing and lacking a termination codon had comparable protein synthesis inhibitory activity in this reticulocyte lysate system (36). It was concluded that the synthesized ricin A chain inactivates ribosomes other than those involved in the synthesis of the A chain (36). However, because further synthesis of ricin A chain may have occurred after the second addition reticulocyte lysate, it is unclear as to whether the synthesis product from transcripts lacking termination codons act on more than one ribosome per ricin product.

In constructs containing amino-terminal fusions (see Table 4), products lacking the carboxy-terminal 44 (224–267), 70 (198–267), 79 (189–267), and 154 (114–267) amino acids do not inhibit the reticulocyte lysate nor produce depurinated 28S rRNA (36). It is interesting that removal of amino acids 156–267 results in a product which inhibits the reticulocyte lysate without depurinating the 28S rRNA. This inhibition is blocked by the addition of antiricin antibodies before the addition of the second lysate (36).

Thus, the smallest deletion, removing amino acids 224–267, inactivates the catalytic activity of ricin A chain. This stretch is within the region of domain 3 which is mostly random coil (43). Although it contains a few conserved amino acids (Leu-232, Ile-252, and Met-255), it lacks the invariant amino acids and residues implicated in the catalytic activity. However, Ile-

Table 4 Deletional Analysis of the Ricin A Chain Gene

Deletion start	Deletion end	Additional amino acids	Solubility	Activity on protein synthesis[g]	Method[h]	Protein[i] purity	Aniline[j] product	Reference
1	9	No	N.D.[k]	Inhibited	Coupled	In vitro	N.D.	45
1	12	No	N.D.	No activity	Coupled	In vitro	N.D.	45
1	28	Yes[a,b]	N.D.	Inhibited	Coupled	In vitro	None	36
19	41	Yes[c]	Insoluble	>0.1% native	Reticulocyte	Crude	N.D.	52
45	267	No	N.D.	No activity	N.D.	N.D.	N.D.	46
56	56	Yes[d]	N.D.	Inhibited	Coupled	In vitro	N.D.	45
75	80	Yes[d]	N.D.	No activity	Coupled	In vitro	None	45
176	180	Yes[d]	N.D.	Inhibited	Coupled	In vitro	Present	45
114	267	Yes[e]	N.D.	No activity	Coupled	In vitro	None	36
156	267	Yes[f]	N.D.	Inhibited	Coupled	In vitro	None	36
156	267	Yes[e]	N.D.	Inhibited	Coupled	In vitro	None	36
185	267	No	Soluble	<0.3% native	Reticulocyte	Purified	N.D.	48
189	267	Yes[e]	N.D.	No activity	Coupled	In vitro	None	36
194	267	No	Soluble	<0.3% native	Reticulocyte	Purified	N.D.	48
198	267	Yes[e]	N.D.	No activity	Coupled	In vitro	N.D.	36
202	267	No	Soluble	<0.3% native	Reticulocyte	Purified	N.D.	48
211	267	No	Soluble	<0.3% native	Reticulocyte	Purified	N.D.	48
224	267	Yes[e]	N.D.	No activity	Coupled	In vitro	N.D.	36
1	28	Yes[a]	N.D.	No activity	Coupled	In vitro	N.D.	36
114	267							
1	28	Yes[a]	N.D.	Inhibited	Coupled	In vitro	None	36
156	267							

1 / 189	28 / 267	Yes[a]	N.D.	No activity	Coupled	In vitro	N.D.	36
1 / 198	28 / 267	Yes[a]	N.D.	No activity	Coupled	In vitro	N.D.	36
1 / 224	28 / 267	Yes[a]	N.D.	No activity	Coupled	In vitro	N.D.	36

[a] At amino-terminal end: MetProAlaGly—(ricin A amino acid 29)ArgGlyArg. . . . (pALR202 derivatives: deletions 1–28).

[b] At the carboxy-terminal end: SerLeuLeuIleArgProValValProAsnPheAsnAlaAspValCysMetAspProProArgValProSerSerAsn or SerLeuLeuIle-ArgProVal

[c] Glu-41Ile-42 modified to Ala-41Leu-42

[d] At amino-terminal end: MetGlyLeuThrAlaHisAspArgProGlySerThrSerGlyTrpSerPheThrLeuGluAspAsn.Asn.

[e] At amino-terminal end: MetProAlaGlyArgAspSerArgGlySerThrSerGlyTrpSerPheThrLeuGluAspAspAsnAsn (pALR201-derivatives).

[f] At amino-terminal end: MetGly (pRA derivatives).

[g] Activity on protein synthesis indicates the ability of the material to inhibit incorporation of radioactive amino acids in an in vitro system. Since many of the assays were nonquantitative, the ability of the test material to allow (no activity) or inhibit (inhibit) protein synthesis is given in qualitative terms Where the activity of purified protein was quantitated, the level of activity relative to the native protein (% native) is given.

[h] The assays testing the activity of the mutant toxins included a coupled in vitro transcription/translation system in which the mutated gene was transcribed, added to a reticulocyte lysate translation system and the ability of the translated product to inhibit further translation of added mRNA (coupled); or direct addition of either purified protein or total E. coli lysates from bacteria expressing the mutant protein.

[i] Different preparations were used to test the activity of the mutant proteins: in vitro indicates that the same in vitro reticulocyte lysate system was used to synthesize the mutein and to test for its catalytic activity; purified indicates that purified protein was tested; and crude indicates that total E. coli extract from bacteria expressing the mutein was tested.

[j] The mechanism for the protein synthesis inhibitory activity was measured by showing that aniline treatment of the 28S rRNA produced a product analogous to that produced by native ricin A chain. None indicates no cleavage product was generated; present indicates that a cleavage product similar to that of the toxin was generated.

[k] N.D., not reported in the reference.

252 and Met-255 contact Ile-42, and Leu-254 contacts Leu-45, which in turn contacts Trp-211. In addition, Leu-232 contacts G helix residue Leu-207 (43). It has been postulated that a number of the domain 3 hydrophobic residues (including Leu-232, Phe-240, Val-242, Ile-247, Leu-248) may play a role in governing folding of a portion of the ricin A chain protein (43).

Several internal deletions have been also investigated. The removal of amino acids 19–41 (also modifying Glu-41Ile-42 to Ala-41Leu-42) results in an enzymatically inactive, insoluble protein (52). This is not surprising in that this deletion removes residues Tyr-21 and Arg-29, two invariant amino acids which appear to be instrumental in creating the bend in helix E bringing Glu-177 and Arg-180 to the surface of the active site (43). In addition, it has been postulated that Arg-29 may be involved in ribosome recognition (43). This deletion also removes the invariant Phe-24 and conserved Ile-25 (43).

Ricin A chain and EF-2 may bind to similar sites within the ribosome (53,54). The amino acid sequence around ricin A chain residues 70–79 is homologous to a region of hamster EF-2 (ValThrLeuAlaLeuAspValThr-Asn in ricin, ValThrAlaAlaLeuArg-ValThrAsp in EF-2) (45). Deletion of residues 75–80 (AspValThrAsnAlaTyr), which removes this region of homology, inactivates the catalytic activity of ricin A chain (45). This stretch contains the conserved Asn-78 and the invariant Tyr-80, and lies in the loop between β strands c and d which contains several active site residues (43). Tyr-80 is important in the structure of the active site in that it makes a strong hydrogen bond to O_{121} (43). Asn-78 sits on the surface of the active site and may play a role in stabilizing the active site conformation (43). It has been postulated that Asn-78 may be involved with RNA base recognition. As discussed below, changing Tyr-80 to phenylalanine reduces the catalytic activity of ricin A chain sevenfold (55). Surprisingly, deletion of residues 176–180, which contain the invariant residues Glu-177, Ala-178, Ala-179, and Arg-180, produces a protein which both retains protein synthesis inhibitory activity and depurinates 28S rRNA (45). Glu-177 and Arg-180 lie in the active site (43), and, as discussed below, alteration of either amino acid inactivates the catalytic activity of the toxin (40,55,56). It is possible that the catalytic activity of this mutant may result from nearby residues substituting for the deleted amino acids, as has been reported for Glu-177 mutants (56). Deletion of Arg-56, which is conserved among many of the toxins and appears to lie in the active site (43), has no effect on the protein synthesis activity (45).

C. Site-Directed Mutagenesis

Comparisons of conserved amino acids among the members of the plant toxins with similar N-glycosidase activities (43,57–59) and the recent x-ray structure of the A chain (43,60) has pointed to several amino acids that may play a crucial role in the catalytic activity and folding of ricin A chain (Table 5). Conserved amino acids proposed to be involved in the catalytic activity of the toxin or participating in maintaining the conformation of the active site include Asn-78, Tyr-80, Tyr-123, Arg-134, Gln-173, Glu-177, Arg-180, Glu-208, Asn-209, and Trp-211 (43,55). Among these residues, Asn-78, Tyr-80, Tyr-123, Glu-177, Arg-180, and Asn-209 are conserved among many of the plant ribosome-inhibiting proteins (58,57). Glu-177 and Arg-180 are charged residues which reside on the surface of the active site as a result of the bend in the E helix (43). The mechanism of depurination of 28S rRNA could be ascertained through mutagenesis of a number of these residues. Because many alterations in the native sequence reduced the solubility of the mutant A chains, it is sometimes difficult to determine if the reduced protein synthesis inhibitory activity results from reduced catalytic activity of the toxin or reduced solubility of the polypeptide.

Conversion of Asn-209 to Ser reduced the activity of the toxin to threefold (55). The K_m of the mutein was reduced from 1.25 to 8 μM with no effect on the k_{cat}, suggesting that the amide group of asparagine, which is located on the surface of the active site cleft, is involved in recognition and binding of the substrate (55). The decrease in K_m corresponds to an increase in free energy for the enzyme substrate complex of 1 kcal/mol, implying that the serine residue can bind to the substrate less strongly than the asparagine (55).

Glu-177 has been implicated to have a key role in catalysis in that it may stabilize the positive charge on the ribose ring (oxycarbonium) (56). Glu-177 makes hydrogen bonds to Arg-180, Tyr-183, and an active site water (43). Substitution of Gln for Glu-177 results in a 180-fold decrease in catalytic activity (55). The more conservative change of Glu-177 to Asp reduces the activity 80-fold (40). The comparable mutation in Shiga-like toxin, in which Glu-167 is converted to Asp, reduces the protein synthesis inhibitor activity of the toxin by 1000-fold (61). Surprisingly, the mutein in which Glu-177 is converted to Ala has only a 20-fold loss in activity (40,56). It has been proposed that when the amino acid side chain at position 177 is small, Glu-208 can substitute for Glu-177 (56). Glu-208 forms an ion pair with Arg-134, forming one wall of the active site (43). Conversion of Glu-177 to Ala would allow rotation around the alpha-carbon resulting in the carboxylate group of Glu-208 to approach the position occupied by the wild-type Glu-177. However, an Asp at position 177 displaces the carboxylate function from the position occupied in the wild-type protein and sterically occludes Glu-208 from altering its position (56). Ala or Asp at positions 177 reduce the solubility of the protein.

Table 5 Mutational Analysis of Ricin A Chain

pos.	wt.	mut.	Added amino acids	Induction method for expression	Solubility[d]	% wild-type activity[e]	Method[f]	Protein purity[g]	Aniline product[h]	Reference
2	Phe	Ser	No	Low phosphate	wt	wt	Reticulocyte	Crude	N.D.	52
19	Gln	Ala	No	Low phosphate	wt	wt	Reticulocyte	Crude	N.D.	52
40	His	Gln	No	Low phosphate	Insoluble	wt	Reticulocyte	Crude	N.D.	52
41	Glu	Ala								
42	Ile	Leu								
46	Pro	Ser	No	Low phosphate	Insoluble	<0.3	Reticulocyte	Purified	N.D.	48
134	Arg	Gly								
48	Arg	Ala	Yes[a]	In vitro	N.D.	wt	Coupled	In vitro	Present	45
56	Arg	Ala	Yes[a]	In vitro	N.D.	wt	Coupled	In vitro	Present	45
80	Tyr	Phe	No	IPTG	N.D.	7	Artemia salina ribosomes	Purified	N.D.	55
83	Gly	Asp	No	Low phosphate	Insoluble	<0.3	Reticulocyte	Purified	N.D.	48
212	Gly	Glu								
120	Gly	Asp	No	Low phosphate	Insoluble	10	Reticulocyte	Purified	N.D.	48
121	Gly	Asp	No	Low phosphate	Soluble	7	Reticulocyte	Purified	N.D.	48
158	Gly	Asp								
123	Tyr	Phe	No	IPTG	Reduced	15	Artemia salina ribosomes	Purified	N.D.	55
134	Arg	Lys	No	Low phosphate	Insoluble	<0.3	Reticulocyte	Purified	N.D.	48
140	Gly	Glu								
135	Glu	Lys	No	Low phosphate	Soluble	<0.3	Reticulocyte	Purified	N.D.	48
212	Gly	Glu								
227	Ala	Ter								
140	Gly	Glu	No	Low phosphate	Insoluble	<0.3	Reticulocyte	Purified	N.D.	48

140	Gly	Arg	No	Low phosphate	Insoluble	<0.3	Reticulocyte	Purified	N.D.	48
140	Gly	Glu	No	Low phosphate	Insoluble	<0.3	Reticulocyte	Purified	N.D.	48
187	Glu	Lys								
212	Gly	Glu								
142	Gly	Asp	No	Low phosphate	Soluble	<0.3	Reticulocyte	Purified	N.D.	48
212	Gly	Glu								
177	Glu	Lys	No	Low phosphate	Insoluble	<0.3	Reticulocyte	Purified	N.D.	48
177	Glu	Lys	?[c]	Yeast, galactose	N.D.	Undetectable	Reticulocyte	Crude	N.D.	46
177	Glu	Gln	No	IPTG	Reduced	0.6	Artemia salina ribosomes	Purified	N.D.	55
177	Glu	Ala	Cut[b]	Thermal	Intermediate	6	Reticulocyte	Purified	N.D.	40,56
177	Glu	Asp	Cut[b]	Thermal	Reduced	1	Reticulocyte	Purified	N.D.	40
177	Glu	Ala	Cut[b]	Thermal	Intermediate	<0.1	Reticulocyte	Purified	N.D.	56
208	Glu	Asp								
180	Arg	Ala	Cut[b]	Thermal	Reduced	N.D.				56
180	Arg	Gln	Cut[b]	Thermal	Reduced	N.D.				56
180	Arg	Met	Cut[b]	Thermal	Reduced	N.D.				56
180	Arg	His	Cut[b]	Thermal	Soluble	<0.1	Reticulocyte	Purified	N.D.	56
180	Arg	Lys	Cut[b]	Thermal	Soluble	38	Reticulocyte	Purified	N.D.	56
203	Ser	Asn	Cut[b]	Thermal	Soluble	0.9	Reticulocyte	Purified	Present	47
208	Glu	Asp	Cut[b]	Thermal	Poor	70	Reticulocyte	Purified	N.D.	56
209	Asn	Ser	No	IPTG	Reduced	30	Artemia salina ribosomes	Purified	N.D.	?
211	Trp	Phe	No	Low phosphate	N.D.	11	Reticulocyte	Crude	N.D.	63
211	Trp	Arg	?[c]	Yeast, galactose	N.D.	Undetectable	Reticulocyte	Crude	N.D.	46
212	Gly	Trp	?[c]	Yeast, galactose	N.D.	Undetectable	Reticulocyte	Crude	N.D.	46
212	Gly	Glu	?[c]	Yeast, galactose	N.D.	Undetectable	Reticulocyte	Crude	N.D.	46
212	Gly	Glu	No	Low phosphate	Soluble	1	Reticulocyte	Purified	N.D.	48
215	Ser	Phe	No	Low phosphate	Insoluble	7	Reticulocyte	Purified	N.D.	48
215	Ser	Pro	?[c]	Yeast, galactose	N.D.	Undetectable	Reticulocyte	Crude	N.D.	46

Table 5 Continued

Mutation pos.	wt.	mut.	Added amino acids	Induction method for expression	Solubility[d]	% wild-type activity[e]	Method[f]	Protein purity[g]	Aniline product[h]	Reference
250	Pro	Leu	No	Low phosphate	Insoluble	1	Reticulocyte	Purified	N.D.	48
253	Ala	Val								
252	Ile	Arg	?[c]	Yeast, galactose	N.D.	Undetectable	Reticulocyte	Crude	N.D.	46

[a] At amino-terminal end: MetGlyLeuThrAlaHisAspArgProGlySerThrSerGlyTrpSerPheThrLeuGluAspAsnAsn.

[b] Expressed as a tripartate fusion consisting of sequence MetAspProAsn, ricin A chain amino acids 1–266, Leu (ricin A chain residue 267 is altered to a leucine), 60 amino acids of chicken proα-2 collagen, and amino acids 9 to the end of β-galactosidase. The fusion protein was partially purified, cut with collagenase, and further purified. Treatment of the tripartate fusion results in cleavage of the chicken proα-2 collagen at several sites, which potentially may result in heterogeneity at the carboxyl end of the protein. If digestion is complete, the minimal fusion to the carboxyl end of the ricin A chain sequence will be AspProGlyProVal (41). Heterogeneity of cleavage may result in molecules with 3, 6, 9, 21, and 24 additional amino acids.

[c] It is unclear from the paper whether these strains were derived from plasmid pBM150, in which there would be no additional amino acids, or plasmid pRY131, in which 40 extraneous amino acids would be fused to the amino terminal end of the mature ricin A chain sequence.

[d] Solubility of the expressed ricin A chain, given in terms relative to the native toxin.

[e] Percent of the wild-type ricin A chain in vitro protein synthesis inhibitory activity.

[f] Method by which the in vitro protein synthesis inhibitory activity measured: reticulocyte represents tested in a rabbit reticulocyte lysate; Artemia salina represents testing in an in vitro system containing Artemia salina ribosomes; and coupled represents testing in a coupled in vitro translation system in which the mutein was translated in vitro in the same system used to test the protein synthesis inhibitory activity.

[g] The purity of the mutein when tested for protein synthesis inhibitory activity; purified, purified to homogeneity; crude is a crude. E. coli extract from cells induced to express the mutein; and in vitro is when the mutein was translated in vitro in the same system used for testing its protein synthesis inhibitory activity.

[h] Testing for deadenylation of the 28S rRNA by the generation of a small molecular weight RNA product upon treatment of ribosomes with aniline.

The model explaining the activity of mutant Ala-177 was confirmed by examining modifications in position 208 alone and in conjunction with position 177 (56). Alteration of Glu-208 to Asp results in an enzyme with 70% of the wild-type activity (56). Simultaneous replacement of Glu-177 by Ala and Glu-208 by Asp reduces the enzymatic activity of the toxin by more than 1000-fold (56). It was proposed that the shorter length of the Asp-208 side chain may be insufficient for the carboxylated function to occupy a position near that of the wild-type Glu-177 (56).

Arg-180, another invariant residue which resides in the active site, lies parallel to Trp-211 and makes strong hydrogen bonds with O_{78}, Glu-177, and an active site water which may be involved in the N-glycosidase reaction (43,55). This amino acid may also bind the phosphate backbone of the 28S rRNA substrate (55). As discussed above, chemical modification of arginine residues in the A chain reduces its ability to inhibit protein synthesis by 20- to 100-fold (25). Conversion of Arg-180 to Lys has minimum effect on the catalytic activity of the toxin (56). Thus, the guanidinium moiety is not required catalysis mechanism (56). Like Arg-180, Lys-180 may efficiently hydrogen bond to the active site water and form an ion pair with Glu-177 (56). In contrast, conversion of Arg-180 to His dramatically reduces the catalytic activity (56). The His side chain is shorter than Arg and may not reach far enough into the active site to ion pair with Glu-177 or coordinate the active site water (56). Enzymatic activity and solubility of ricin A chain requires a positively charged residue at position 208 (56). It has been hypothesized that a positive charge at this position may be required for proper folding and stability of ricin A chain. Glu-177 and Arg-180 lie on the portion of helix E that is bent, and it has been suggested that only charged residues may have sufficient solvation attraction to pull the chain into the proper orientation (56).

Chemical modification suggests that arginine residues may be important in maintaining the activity of ricin A chain (18,25) (see Section II). One or a combination of arginine residues at positions 26, 29, 31, 39, 48, or 56 appear to be preferentially modified. Arginines at ricin A chain positions 29, 48, and 56 are conserved among several of the ribosome inhibitory proteins (45). Arg-48 and Arg-56 lie within the proposed active site (43). Conversion of Arg-48 or Arg-56 to Ala have no effect on the catalytic activity, suggesting that these are unlikely to be the arginine residues important in the catalysis reaction (45).

The invariant residue Tyr-80, which lies on the top of the proposed active site, makes a strong hydrogen bond with O_{121} and is thought to play an important role in stabilizing the conformation of the active site (43). Nitration of the homologous residue in barley protein synthesis inhibitor reduces its enzymatic activity 10-fold (62). In addition, nitration of tyrosine residues within whole ricin reduces its protein synthesis inhibitory activity by sevenfold (see Tables 1, 2, and 3). Modification of Tyr-80 to Phe reduces the catalytic activity by 15-fold (55). The K_m increases from 1.25 to 8 μM and the k_{cat} decreases from 300 to 100 min^{-1} (55).

The invariant Tyr-123 resides in the proposed catalytic cleft and is involved in a network of hydrogen bonds and may contribute to the binding of RTA to ribosomes (43,55). Conversion of Tyr-123 to Phe reduces the protein synthesis inhibitory activity by sevenfold (55). The modified protein has reduced solubility, suggesting alteration in folding of the protein.

In the process of generation deletions, several point mutations in the amino-terminal end of ricin A chain have been made (52). The conversion of Phe-2 to Ser was without effect on both solubility and activity. This is not surprising in that the first six residues are loosely connected with the protein molecule (43). The lack of effect on protein synthesis inhibitory activity when Gln-19 was changed to Ala may result from the surface location of the side chain in that region. Modification of His40Glu41Ile42 to GlnAlaLeu decreased the solubility of the mutein without effecting its catalytic activity (52). This stretch of amino acid residues lies on the surface of the molecule in the loop between helix A and β strand b, a region which is not conserved among the ribosome-inactivating proteins (52). The amino acid sequences within loops contain many differences among the ribosome-inactivating proteins, suggesting that the compositions of these loops may not be important in the folding of the protein or in the catalytic activity of the toxins.

Trp-211, another active site residue conserved among many of the ribosome inhibitory toxins (43), resides at the beginning of domain 3 and begins helix H (43,60). In the active site, Arg-180 lies parallel to the plane of Trp-211 and the guanidinium group passes over the indole ring to hydrogen bond to O_{78} (43). Preliminary studies of the interaction of a synthetic oligonucleotide analog of the ricin A chain's RNA depurination substrate indicates a fluorescence intensity quenching of Trp-211 along with a change in the polarity around the residue (63). The conservative substitution of a Phe for Trp-211 reduces the activity of the toxin by ninefold (63).

D. Random Mutagenesis and Expression in Yeast

Yeast ribosomes are sensitive to the action of ricin A chain. Upon expression of the toxin in *Saccharomyces cerevisiae* bearing plasmids carrying the wild-type gene, cell growth is immediately arrested (46). The effect ricin A chain expression yeast depends on a number of factors, including: number of copies of the ricin A chain structural gene; promoter transcription efficiency; translation efficiency; solubility of the product; specific activity of the expressed ricin A chain; expression kinetics; cell growth rate; and ribosomes concentration within the cell. The effect of mutein expression on the host yeast cell will be influenced by the inherent properties of the mutein, expression system, and host-related factors. The ability of yeast cells to grow when the catalytic activity of the toxin has been attenuated can be a powerful selection for mutant forms of the toxin.

The 35S rRNA precursor of the 25S and 5.8S RNAs makes up more than 60% of the total transcription in rapidly growing cells (64). Ribosomal proteins each represent 0.1 to 0.5% of the total cellular protein (64). The cellular content of ribosomes, which is proportional to the growth rate, can be altered by many factors effecting yeast growth (65). Under some conditions, ribosomes make up more than 15% of the cell mass (64). Yeast grown on ethanol as the carbon source have a 2.5-fold lower concentration of ribosomes than cells grown on glucose (65,66). Similarly, amino acid starvation reduces the cellular content of ribosomes (66). In all three situations, there is a coordinated decrease in the syntheses of both ribosomal RNA and ribosomal proteins (66). Following a temperature upshift, the cellular concentration of ribosomes decreases owing to inhibition in the synthesis of ribosomal proteins (66). Increasing the temperature from 23 to 36°C results in a decline of synthesis of ribosomal proteins to 20% of its original level within 20 min of the temperature change (67). Within the subsequent 90 min, the rate of synthesis returns to normal (67). The effect of temperature on the synthesis of ribosomal proteins is strongly dependent on the nutritional content of the medium (66).

In many studies, the ricin A chain structural gene was regulated by the gal1–10 promoter and induced by the simultaneous addition of galactose and removal of glucose from the growth medium (46). The kinetics of toxin induction were studied in a construct bearing the structural gene on a high-copy-number plasmid (46). Following induction, mRNA encoding ricin A chain was detectable by 15 min and reached maximum levels within 50 min (46). Toxin production was detectable by 1 hr when measured by protein synthesis inhibition and immunoradiometric analysis (46). By 6–8 hr, the levels of active toxin were maximum, reaching concentrations of 25–250 ng/mg of yeast protein (corresponding to 10^3–10^4 molecules per cell) (46). In contrast, the level of toxin produced within strains bearing the structural gene on a low-copy-number plasmid was less than 10 ng/mg of yeast protein (46).

Several approaches have been taken to obtain random, catalytically inactive mutants within the ricin A structural gene. In one study, plasmid bearing the ricin A chain wild-type sequence was transformed into yeast and colonies surviving induction of the ricin A chain selected (46). The mutation frequency was increased by passing shuttle vectors encoding the ricin A chain through the *E. coli* mutator strain *mut*D for 100 generations prior to transformation in yeast (46). This strain induced mutations at a rate 50–100 times higher than wild-type *E. coli* strains (68). Seventy-two colonies which grew following induction by galactose were obtained, 12 of which revealed immunoreactive RTA protein by Western blot analysis or radioimmunometric assay (46). Of these, four arose from strains bearing a low-copy-number plasmid and produced a protein of 28 kd on Western blot analysis. The remaining eight mutants arose from strains transformed with a high copy number and produced a 32-kd immunoreactive peptide (46). The level of immunoreactive protein produced by all mutants was similar to that produced by the wild-type gene borne on the high-copy-number plasmid. This approach yielded the following mutations: Glu-177 to Asp, Glu-177 to Lys, Trp-211 to Arg, Gly-212 to Trp, Gly-212 to Glu, Ser-215 to Pro, and Ile-252 to Arg. Crude lysates obtained from each strain bearing the mutant forms of the toxins did not inhibit protein synthesis in a rabbit reticulocyte lystate translation system (46).

In another approach, the ricin A chain was expressed from the *eno*1 promoter as an intracellular protein or as a secreted protein by fusion to the glucoamylase leader (48). It was hoped that by secretion, the ricin A chain would be sequestered away from the cytoplasmic compartment and result in less inhibitory effect on yeast growth. However, in both constructs, the number of transformants obtained with plasmids containing the structural gene for ricin A chain were 0.1% of that obtained in the absence of the toxin gene (48). Following nitrosoguanidine mutagenesis, a number of muteins were obtained, many of which contained multiple mutations. In general, muteins which were obtained from the cytoplasmically expressed toxin had lower specific activities than those produced from the secretion constructs (e.g., Asp-121Asp-158 and Glu-212). Many of the muteins obtained from these contructs contained mutations at positions similar to those obtained by others (e.g., Gly-212 to Glu, and Glu-177 to Lys).

It is not surprising that alteration of residues Glu-177 and Trp-211 result in ricin A chain of reduced activity (46). Substitution of proline for Ser-215 attenuated the catalytic activity (46). Although there has been no role assigned to this residue, it lies in the active site and is conserved in four out of five ribosome-inhibitory proteins (43). Exchange of an arginine for Ile-252 inactivated the activity of the toxin (46). Ile-252 contacts Ile-42 and many of the toxins contain a hydrophobic amino acid at this position (e.g. Ile, Leu, or Met) (43). Replacement with Arg may, therefore, disrupt hydrophobic interactions that may be important in the proper folding of the protein. The analogous position of ricin A chain Gly-212 is highly variable among the ribosome-inhibitory proteins, being occupied by aspartic acid, serine, alanine, or lysine (43). It is interesting that aspartic acid lies in the corresponding position of abrin A chain (69). In two double mutants, conversion of Arg-134 to either Gly or Lys results in an insoluble protein with undetectable protein synthesis inhibitory activity (48). Arg-134 is conserved in many of the ribosome inhibiting proteins (43). It sits on the active site and forms an ion pair with another highly conserved active site residue, Glu-208, where it may form one wall of the active site cleft (43).

By selection in yeast, a number of ricin A chain mutations in positions not present in the proposed active site were obtained which had reduced catalytic activity and solubility. Many of these mutations occurred in loops between alpha-helices or beta-strands. It is likely that some of these alterations in activity and solubility result from the effect the mutations have on overall protein folding. Gly-120 is not conserved among the toxins, and yet conversion to Asp decreases both the catalytic activity and solubility (48). A mutein in which both the partially conserved residues Gly-121 and Gly-158 were converted to Asp had 7% of the wild-type enzymatic activity (48). It has been proposed that the sharp loop between helices D and E (residues 152–161) feeds back to stabilize the carboxyl-terminal end of helix D (43). Substitution of a charged residue (Glu or Arg) for Gly-140 resulted in an inactive, insoluble protein (48). Gly-140 is conserved among many of the toxins and lies between α helices C and D, close to Arg-198 (43). Finally, substitution within the hydrophobic carboxyl-terminal at positions 250 and 253 reduced both the solubility and activity of the toxin.

The level of expression of plasmid-encoded proteins will reflect the multiple copies of plasmids present in the cell. Thus, selection of attenuated forms of ricin, when the toxin is expressed from plasmids, will result in mutations which dramatically reduce the catalytic activity of the toxin. However, when the gene for ricin A chain is integrated within the yeast chromosome, there will be no gene dosage effect on the expression level. In this case, mutations which have less of an inactivating effect on catalytic activity may permit survival of yeast expressing the mutant form of toxin. Yeast cells containing two integrated copies of the ricin A chain structural gene immediately stop growing upon toxin induction (47). Deadenylation of a proportion of the 26S rRNA is detectable by 1 hr following induction, and after 4 hr greater than 70% of the 26S rRNA has been modified (47). One ricin A chain mutein (with a Ser-203 to Asn modification) was obtained from mutagenesis with EMS (methane sulfonic acid, ethyl ester) of yeast containing two integrated copies of the ricin A chain gene. Although the growth rate of strains bearing this mutein are equivalent to that of strains lacking the ricin A chain gene, modification of a proportion of the 26S rRNA is still evident. In contrast to wild-type ricin A chain, muteins with an asparagine at position 203 appear to preferentially modify the 26S rRNA in free 60S ribosomal subunits (47). The amino acid residue at this position is not conserved among the ribosome inhibitory proteins and lies outside the active site (43). It has been proposed that this mutation may distort α helix G, alter the position of other residues involved in catalysis, decrease the solubility of the mutein, increase the turnover rate of the toxin in vivo, or impair the interaction of ricin A chain with the 60S ribosomal subunit (43).

Some of the mutant forms of ricin generated by site-directed mutagenesis were cloned and expressed in yeast to confirm in vivo some of the analyses that were obtained in vitro. Since yeast ribosomes are sensitive to ricin, expression of the toxin in the cell would be expected to inhibit cell growth. Mutants which have reduced activity may have intermediate effects on cell growth. It was hoped that monitoring of yeast growth may show some relation to the amount of residual N-glycosylase activity in the mutant toxins. Yeast cells transformed with plasmids containing regulated expression of ricin A chain are totally arrested in growth upon induction of the toxin (46). Expression of Asp-208, Lys-180, and Ala-177, which retain 70, 38, and 6%, respectively, of the residual activity of the wild-type sequence also arrest cell growth (40,56). However, expression of the mutant with Asp-177 in which the activity is reduced to 80-fold, retain a small, but measurable, residual growth (40). Expression of His-180 and the double mutation Ala-177–Asp-208 has no effect on yeast cell growth (56). The effect of the mutant proteins on growth are dependent on the expression of the ricin A chain–derived protein and are independent of growth in liquid medium or on solid medium (40).

The effect of expression of a number of the muteins in yeast is summarized in Table 6. On a low-copy number plasmid, expression of a mutein which retains 1% or greater of the wild-type protein synthesis activity is sufficient to decrease the growth rate of yeast. However, when the gene is integrated as a single copy into the host genome, expression of a mutant with reduced activity is better tolerated (e.g., conversion of Ser-203 to Asn-203).

VII. CONCLUSIONS

Site-directed mutagenesis of the structural gene for ricin A chain has complemented both comparison of the primary structures of several ribosome inhibitory proteins and the recent x-ray structure of the A chain in elucidating those amino acids important in both the catalytic activity and folding of ricin A chain. These results, along with future mutational analyses, will aid in the rational design of ricin A chain fusion proteins as well as provide

Table 6 Correlation Between Activity of the Ricin A Chain Mutein and Ability to Inhibit Growth of Yeast

Mutation				% Wild-type activity protein synthesis[b]	Protein purity[c]	Effect on yeast growth	Reference
pos.	wt.	mut.	Expression[a]				
0	All		1,2	100	Purified	Arrested	40
45	Leu	Stop	1 or 2	Undetectable	Crude	No effect	46
177	Glu	Asp	1	1	Purified	Residual	40
177	Glu	Ala	1	6	Purified	Arrested	40
180	Arg	His	1	<0.1	Purified	No effect	56
180	Arg	Lys	1	38	Purified	Arrested	56
203	Ser	Asn	3	0.9	Purified	No effect	47
208	Glu	Asp	1	70	Purified	Arrested	56
177	Glu	Ala	1	<0.1	Purified	No effect	56
208	Glu	Asp					
211	Trp	Arg	1 or 2	Undetectable	Crude	Slight	46

[a]1. Vector: pBM150, a low-copy-number CEN plasmid using the inducible *Gal*1–10 promoter.
2. Vector: pRY131, a high-copy-number 2-μm plasmid with an additional 120 nucleotides of *Gal*1 and polylinker sequence between the promoter and the gene for ricin A chain.
3. Integrated as a single copy into the host genome. Expression from the *Gal/Cyc*1 hybrid promoter. The expressed protein contains an additional 24 plasmid-encoded amino acid residues at the amino-terminal end.
[b]Percent of the protein synthesis inhibitory activity of wild-type ricin A chain.
[c]Purity of mutein preparation used to test the in vitro protein synthesis inhibitory activity.

substrates for the understanding of the mechanism by which ricin A chain functions in immunotoxins. In addition, these mutations may be helpful in understanding the toxicity of immunotoxins in vivo.

REFERENCES

1. Butterworth, A. G., and Lord, J. M. Ricin and *Ricinus communis* agglutinin subunits are all derived from a single-size polypeptide precursor. Eur. J. Biochem., *137*: 57–65, 1983.
2. Funatsu, G., and Funatsu, M. Separation of the two constituent polypeptide chains of ricin D. Agric. Biol. Chem., *41*: 1211–1215, 1977.
3. Endo, Y., Mitsui, K., Motizuki, M., and Tsurugi, K. The mechanism of action of ricin and related toxic lectins on eukaryotic ribosomes: The site and the characteristics of the modification in 28S ribosomal RNA caused by the toxins. J. Biol. Chem., *262*: 5908–5912, 1977.
4. Endo, Y., and Tsurugi, K. RNA N-glycosidase activity of ricin A-chain: Mechanisms of action of the toxic lectin ricin on eukaryotic ribosomes. J. Biol. Chem., *262*: 8128–8130, 1987.
5. Olsnes, S., and Pihl, A. Toxic lectins and related proteins, *In*: P. Cohen and S. Van Heyningen, (eds.), Molecular Action of Toxins and Viruses, pp. 51–105. New York: Elsevier Biomedical Press, 1982.
6. Vitetta, E. S. Synergy between immunotoxins prepared with native ricin A chains and chemically-modified ricin B chains. J. Immunol., *136*: 1880–1887, 1986.
7. Foxwell, B. M. J., Donovan, T. A., Thorpe, P. E., and Wilson, G. The removal of carbohydrates from ricin with endoglycosidases H, F and D and α-mannosidase. Biochim. Biophys. Acta, *840*: 193–203, 1985.
8. Thorpe, P. E., Detre, S. I., Foxwell, B. M. J., Brown, A. N. F., Skilleter, D. N., Wilson, G., Forrester, J. A., and Stirpe, F. Modifications of the carbohydrate in ricin with metaperiodate-cyanoborohydride mixtures: Effects on toxicity and in vivo distribution. Eur. J. Biochem., *147*: 197–206, 1985.
9. Blakey, D. C., and Thorpe, P. E. Effect of chemical deglycosylation on the in vivo fate of ricin A-chain. Cancer Drug Delivery, *3*: 189–196, 1986.
10. Kimuara, Y., Hase, S., Kobayashi, Y., Kyogoku, Y., Ikenaka, T., and Funatsu, G. Structures of sugar chains of ricin D. J. Biochem., *103*: 944–949, 1988.
11. Funatsu, G., Yoshitake, S., and Funatsu, M. Isolation of tryptic peptides from the Ile chain of ricin D and the sequences of some tryptic peptides including N- and C-terminal peptides. Agric. Biol. Chem., *41*: 1225–1231, 1977.
12. Funatsu, G., Yoshitake, S., and Funatsu, M. Primary strucutre of Ile chain of ricin D. Agric. Biol. Chem., *42*: 501–503, 1978.
13. Funatsu, G., Kimura, M., and Funatsu, M. Primary structure of Ala chain of ricin D. Agric. Biol. Chem., *43*: 2221–2224, 1979.
14. Lamb, F. I., Roberts, L. M., and Lord, J. M. Nucleotide sequence of cloned cDNA coding for preproricin. Eur. J. Biochem., *148*: 265–270, 1985.

15. Olsnes, S., Refsnes, K., Christensen, T. B., and Pihl, A. Studies on the structure and properties of the lectins from *Abrus precatorius* and *Ricinus communis*. Biochim. Biophys. Acta., *405*: 1–10, 1975.
16. Olsnes, S., and Saltvedt, E. Conformation-dependent antigenic determinants in the toxic lectin ricin. J. Immunol. *114*: 1743–1748, 1976.
17. Olsnes, S., and Pihl, A. Treatment of abrin and ricin with β-mercaptoethanol. Opposite effects on their toxicity in mice and their ability to inhibit protein synthesis in a cell-free system. FEBS Lett., *28*: 48–50, 1972.
18. Watanabe, K., and Funatsu, G. Interaction of Cibacron blue F3GA and polynucleotides with ricin A-chain, 60 S ribosomal subunit-inactivating protein. Biochim. Biophys. Acta, *914*: 177–184, 1987.
19. Funatsu, M. The structure and toxic function of ricin. *In*: M. Funatsu, K. Hiromi, K. Imahori, T. Murachi, and K. Narita (eds.), Proteins, Structure and Functions, pp. 103–139. New York: John Wiley & Sons, 1972.
20. Taira, E., Yochizuka, N., Funatsu, G., and Funatsu, M. Effects of physical and chemical treatments on the biological activity of ricin D. Agric. Biol. Chem., *42*: 1927–1932, 1978.
21. Piatak, M., Lane, J. A., Laird, W., Bjorn, M. J., Wang, A., and Williams, M. Expression of soluble and fully functional ricin A chain in *Escherichia coli* is temperature-sensitive. J. Biol. Chem., *263*: 4837–4843, 1988.
22. Sandvig, K., Olsnes, S., and Phil, A. Chemical modifications of the toxic lectins abrin and ricin. Eur. J. Biochem., *84*: 323–331, 1978.
23. Dalrymple, P. N., and Houston, L. L. Differential effects of nitrated ricin and nitrated and dithionite-reduced ricin or protein-synthesis inhibition and trans-membrane transport in eukaryotic cells. Biochem. J., *188*: 941–944, 1980.
24. Watanabe, K., and Funatsu, G. Effect of trinitrophenylation of amino groups on biological activities of ricin D. J. Fac. Agr., Kyushu Univ., *28*: 201–211, 1984.
25. Watanabe, K., and Funatsu, G. Involvement of arginine residues in inhibition of protein synthesis by ricin A-chain. FEBS Lett., *204*: 219–222, 1986.
26. Simeral, L. S., Kapmeyer, W., MacConnell, W. P., and Kaplan, N. O. On the role of the covalent carbohydrate in the action of ricin. J. Biol. Chem., *255*: 11098–11101, 1980.
27. Foxwell, B. M. J., Blakey, D. C., Brown, A. N. F., Donovan, T. A., and Thorpe, P. E. The preparation of deglycosylated ricin by recombination of glycosidase-treated A- and B-chains: Effects of deglycosylation on toxicity and in vivo distribution. Biochim. Biophys. Acta, *923*: 59–65, 1987.
28. Lugnier, A. A. J., Le Meur, M.-A., Gerlinger, P., and Dirheimer, G. Isolation and biochemical properties of toxic tryptic peptides of ricinotoxin from *Ricinus communis* seeds. Eur. J. Toxicol., *9*: 323–333, 1976.
29. Creppy, E.-E., Lugnier, A. A. J., and Dirheimer, G. Isolation and properties of two toxic tryptic peptides from ricin, the toxin of *Ricinus communis* L. (castor bean) seeds. Toxicon, *18*: 649–660, 1980.
30. Yoshitake, S., Watanabe, K.-i., and Funatsu, G. Limited hydrolysis of ricin D with trypsin in the presence of sodium dodecyl sulfate. Agric. Biol. Chem., *43*: 2193–2195, 1979.

31. Olsnes, S., Heiberg, R., and Pihl, A. Inactivation of eukaryotic ribosomes by the toxic plant proteins abrin and ricin. Mol. Biol. Rep., *1*: 15–20, 1973.

32. Lugnier, A. A. J., Küntzel, H., and Dirheimer, G. Inhibition of *Neurospora crassa* and yeast mitochondrial protein synthesis by ricin, A toxic protein inactive on *E. coli* protein synthesis. FEBS Lett., *66*: 202–205, 1976.

33. Haas-Kohn, L. J. G., Jugnier, A.-A. J., Tiboni, O., Ciferri, O., and Dirheimer, G. Inhibition of *Escherichia coli* protein synthesis by a limited tryptic digest of ricin, the toxin of *Ricinus communis* L. seeds. Biochem. Biophys. Res. Commun., *97*: 962–967, 1980.

34. Funatsu, G., and Funatsu, M. Limited hydrolysis of ricin D with alkaline protease from *Bacillus subtilis*. Agric. Biol. Chem., *41*: 1309–1310, 1977.

35. Halling, K. C., Halling, A. C., Murray, E. E., Ladin, B. F., Houston, L. L., and Weaver, R. F. Genomic cloning and characterization of a ricin gene from *Ricinus communis*. Nucl. Acids Res., *13*: 8019–8033, 1985.

36. Sundan, A., Evensen, G., Hornes, E., and Mathiesen, A. Isolation and *in vitro* expression of the ricin A-chain gene: Effect of deletions on biological activity. Nucl. Acids Res., *17*: 1717–1732, 1989.

37. Piatak, M., Lane, J. A., O'Rourke, E. Clark, R., Houston, L., and Apell, G. Expression of ricin and ricin B-chain in insect cells. ICSU Short Rep., *8*: 62, 1988.

38. O'Hare, M., Roberts, L. M., Thorpe, P. E., Watson, G. J., Prior, B., and Lord, J. M. Expression of ricin A chain in *Escherichia coli*. FEBS Lett., *216*: 73–78, 1987.

39. Shire, D., Bourrié, B. J. P., Carillon, C., Derocq, J.-M., Dousset, P., Dumont, X., Jansen, F. K., Kaghad, M., Legoux, R., Lelong, P., Pesséqué, B., and Vidal, H. Biologically active A-chain of the plant toxin ricin expressed from a synthetic gene in *Escherichia coli*. Gene, *93*: 183–188, 1990.

40. Schlossman, D., Withers, D., Welsh, P., Alexander, A., Robertus, J., and Frankel, A. Role of glutamic acid 177 of the ricin toxin A chain in enzymatic inactivation of ribosomes. Mol. Cell. Biol., *9*: 5012–5021, 1989.

41. Germino, J., and Bastia, D. Rapid purification of a cloned gene product by genetic fusion and site-specific proteolysis. Proc. Natl. Acad. Sci. U.S.A., *81*: 4692–4696, 1984.

42. Kim, J.-ho, and Weaver, R. F. Construction of a recombinant expression plasmid encoding a staphylococcal protein A-ricin A fusion protein. Gene, *68*: 315–321, 1988.

43. Katzin, B. J., Collins, E. J., and Robertus, J. D. The structure of ricin A-chain at 2.5 Å. Proteins, *10*: 251–259, 1991.

44. Ben-Bassat, A., Bauer, K., Chang, S.-Y., Myambo, K., Boosman, A., and Chang, S. Processing of the initiation methionine from proteins: Properties of the *Escherichia coli* methionine aminopeptidase and its gene structure. J. Bacteriol., *169*: 751–757, 1987.

45. May, M. J., Hartley, M. R., Roberts, L. M., Krieg, P. A., Osborn, R. W., and Lord, J. M. Ribosome inactivation by ricin A chain: A sensitive method to assess the activity of wild-type and mutant polypeptides. EMBO J. *8*: 301–308, 1989.

46. Frankel, A., Schlossman, D., Welsh, P., Hertler, A., Withers, D., and Johnston, S. Selection and characterization of mutants of ricin toxin A-chain in *Saccharomyces cerevesiae*. Mol. Cell. Biol., *9*: 415–420, 1989.

47. Gould, J. H., Hartley, M. R. Welsh, P. C., Hoshizaki, D. K., Frankel, A., Roberts, L. M., and Lord, J. M. Alteration of an amino acid residue outside the active site of ricin A chain reduces its toxicity toward yeast ribosomes. in press.

48. Emerick, A. E., Gates, C. A., Long, C. M., Bolonick, J., and Greenfield, L. Ricin A-chain mutants that are reduced in enzymatic activity. Manuscript in preparation.

49. Harley, S. M., and Beevers, H. Ricin Inhibition of *in vitro* protein synthesis by plant ribosomes. Proc. Natl. Acad. Sci. U.S.A., *79*: 5935–5938, 1982.

50. Battelli, M. G., Lorenzoni, E., Stirpe, F., Cella, R., and Parisi, B. Differential effect of ribosome-inactivating proteins on plant ribosome activity and plant cells growth. J. Exp. Botany, *35*: 882–889, 1984.

51. Perara, E., Rothman, R. E., Lingappa, V. R. Uncoupling translocation from translation: Implications for transport of proteins across membranes. Science, *232*: 348–352, 1986.

52. Bradley, J. L., Piatak, M., Lane, J. A., and McGuire, P. M. Site-directed mutagenesis at amino terminus of recombinant ricin A chain. Int. J. Peptide Protein Res., *34*: 2–5, 1989.

53. Olsnes, S., Fernandez-Puentes, C., Carrasco, L., and Vazquez, D. Ribosome inactivation by the toxic lectins abrin and ricin: Kinetics of the enzyme activity of the toxin A-chains. Eur. J. Biochem., *60*: 281–288, 1975.

54. Fernandez-puentes, C., Benson, S., Olsnes, S., and Pihl, A. Protective effect of elongation factor 2 on the inactivation of ribosomes by the toxic lectins abrin and ricin. Eur. J. Biochem., *64*: 437–443, 1976.

55. Ready, M. P., Kim, Y., and Robertus, J. D. Site-directed mutagenesis of ricin A-chain and implications for the mechanism of action. Proteins, *10*: 270–278, 1991.

56. Frankel, A., Welsh, P., Richardson, J., and Robertus, J. D. Role of arginine 180 and glutamic acid 177 of ricin toxin A chain in enzymatic inactivation of ribosomes. Mol. Cell. Biol., *10*: 6257–6263, 1990.

57. Xuejun, Z., and Jiahuai, W. Homology of trichosanthin and ricin A chain. Nature, *321*: 477–478, 1986.

58. Ready, M. P., Katzin, B. J., and Robertus, J. D. Ribosome-inhibiting proteins, retroviral reverse transcriptases, and RNase H share common structural elements. Proteins, *3*: 53–59, 1988.

59. Ready, M., Wilson, K., Piatak, M., and Robertus, J. D. Ricin-like plant toxins are evoluntionarily related to single-chain ribosome-inhibiting proteins from *Phytolacca*. J. Biol. Chem., *259*: 15252–15256, 1984.

60. Montfort, W., Willafranca, J. E., Monzingo, A. F., Ernst, S. R., Katzin, B., Rutenber, E., Xuong, N. H., Hamlin, R., and Robertus, J. D. The three-dimensional structure of ricin at 2.8 Å. J. Biol. Chem., *262*: 5398–5403, 1987.

61. Hovde, C. J., Calderwood, S. B., Mekalanos, J. J., and Collier, R. J. Evidence

that glutamic acid 167 is an active-site residue of shiga-like toxin I. Proc. Natl. Acad. Sci. U.S.A., *85*: 2568–2572, 1988.

62. Asano, K., and Svensson, B. Chemical modification Studies on protein synthesis inhibitor II from barley seeds. Identification of an essential tyrosyl residue. Carlsberg Res. Commun. *51*: 501–507, 1984.

63. Bradley, J. L., and McQuire, P. M. Site-directed mutagenesis of ricin A chain Trp 211 to Phe. Int. J. Peptide Protein Res. *35*: 365–366, 1990.

64. Warner, J. R. Synthesis of Ribosomes in *Saccharomyces cerevisiae*. Microbiol. Rev. *53*: 256–271, 1989.

65. Kief, D. R., and Warner, J. R. Coordinate control of synthesis of ribosomal ribonucleic acid and ribosomal proteins during nutritional shift-up in *Saccharomyces cerevisiae*. Mol. Cell. Biol. *1*: 1007–1015, 1981.

66. Kief, D. R., and Warner, J. R. Hierarchy of elements regulating synthesis of ribosomal proteins in *Saccharomyces cerevisiae*. Mol. Cell. Biol. *1*: 1016–1023, 1981.

67. Kim, C. H., and Warner, J. R. Mild temperature shock alters the transcription of a discrete class of *Saccharomyces cerevisiae* genes. Mol. Cell. Biol. *3*: 457–465, 1983.

68. Degnen, G. E., and Cox, E. C. Conditional Mutator gene in *Escherichia coli*: Isolation, mapping, and effector studies. J. Bacteriol. *117*: 477–487, 1974.

69. Funatsu, G., Taguchi, Y., Kamenosono, M., and Yanaka, M. The complete amino acid sequence of the A-chain and abrin-a, a toxic protein from the seeds of *Abrus precatorius*. Agric. Biol. Chem. *52*: 1095–1097, 1988.

70. Funatsu, G., Miyauchi, S., Yoshizuka, N., and Funatsu, M. Hybridization between the heterologous chains of native ricin D and iodinated or maleyl ricin D. Agric. Biol. Chem. *41*: 1217–1223, 1977.

71. Jansen, F. K., Bourrie, B., Casellas, P., Dussossoy, D., Gros, O., Vic, P., Vidal, H., and Gros, P. Toxin selection and modification: Utilization of the A chain of ricin, *In* A. E. Frankel, (ed.), Immunotoxins, pp. 97–111. Boston: Kluwer Academic Publishers, 1988.

9

Chimeric Proteins Containing Ricin A Chain

J. Michael Lord and Lynne M. Roberts *University of Warwick, Coventry, England*

Philip E. Thorpe *Cancer Immunobiology Center, University of Texas Southwestern Medical Center, Dallas, Texas*

I. INTRODUCTION

Bacterial toxins such as diphtheria toxin and *Pseudomonas* exotoxin A and plant toxins such as ricin and abrin are able to kill their target cells by delivering a catalytically active polypeptide into the cell cytosol (1). The toxic polypeptides inhibit host cell protein synthesis either by the ADP-ribosylation of elongation factor 2 (the bacterial toxins) or by the specific depurination of 28S or 26S ribosomal RNA (the plant toxins). In addition to the toxic polypeptide (sometimes referred to as the A fragment or A chain), these proteins also contain a cell-binding polypeptide (the B fragment or B chain). This second peptide domain is responsible for binding the holotoxin to cell surface receptors leading to internalization by the classic endocytic route of coated pits and coated vesicles (1). In the bacterial toxins, the ADP-ribosylating peptide and the cell-binding peptide are initially synthesized as distinct domains within a single chain protein molecule (2,3). These two domains are separated in linear amino acid sequence by an arginine-rich, serine protease–sensitive cleavage site. The cleavage site lies within a disulfide loop. This region in diphtheria toxin (4) is shown in Figure 1. During cellular entry, the bacterial toxins are processed by a cellular, possibly endosomal, protease which generates two polypeptide fragments, one containing the ADP-ribosylating domain, or A fragment, and the other the cell-binding domain, or B fragment (5,6). These fragments remain covalently joined by the single disulfide bond which initially formed the intrachain disulfide loop surrounding the cleavage site. At some stage after the proteolytic processing step, the disulfide bond is reductively broken and the ADP-ribosylating polypeptide is translocated across the membrane of an intracellular compartment into the cytosol.

II. CHIMERIC PROTEINS CONTAINING BACTERIAL TOXINS

Recombinant chimeric bacterial toxins with novel cell-binding properties have been created by deleting the portion of the genes encoding the cell-binding domain of diphtheria toxin and *Pseudomonas* exotoxin A and replacing them with DNA encoding an alternative cell-binding peptide such as a growth factor (7), lymphokine (8,9), antibody fragment, or single chain Fv (10). Provided that the new cell-binding domain remains bound to any necessary membrane translocation sequences and remains attached to the ADP-ribosylating domain by a disulfide-bonded loop containing a protease-sensitive site, the chimeras are potently cytotoxic, although their target cell recognition has been altered (11). The chimeric proteins are endocytosed by their target cells and are converted through the action of intracellular proteases into their cytotoxic heterodimeric forms.

III. CHIMERIC PROTEINS CONTAINING RICIN

The approach of making functionally cytotoxic, single-chain chimeric proteins by the generation and expression of novel DNA fusions cannot be directly extended to the plant toxins. Although the plant toxins also consist of a toxic polypeptide and a cell-binding polypeptide, both of which initially occur within a single proprotein, the proteolytic processing step that separates the two polypeptides occurs during toxin biogenesis within the plant (12). Processing does not involve cleavage at a serine protease–sensitive site, but rather the removal of a short linker peptide from within a disulfide loop (13). The endoprotease(s) responsible for this processing, which in many instances requires cleavage after asparagine residues, is commonly found in plant cells but is apparently absent from mammalian cells (14). Such proteolytic processing during biogenesis generated mature heterodimeric toxin molecules linked by a disulfide bond. Intoxication of mam-

Figure 1   The amino acid sequence in the disulfide loop containing a serine protease-sensitive region between the ADP-ribosylating and cell-binding domains of diphtheria toxin. The single letter amino acid symbols are used.

malian cells with mature toxins of this type does not, therefore, require target cell proteolytic activity to release the toxic polypeptide for its translocation into the cytosol. Plant toxin genes can be fused to DNA encoding an appropriate ligand and, after expression in *Escherichia coli* or some other appropriate host, a recombinant chimeric molecule can be generated which possesses the biological activities of both component parts. For example, we have described fusions between DNA encoding staphylococcal protein A (PA) and that encoding ricin A chain (RTA) (15). The recombinant fusion protein produced in *E. coli* was able to bind to immunoglobulin G (via PA) and was able to specifically depurinate 28S ribosomal RNA (via RTA). This bifunctional biological activity was observed regardless of whether the fusion protein contained N-terminal PA and C-terminal RTA (PA–RTA) or vice-versa (RTA–PA). However, neither fusion protein was cytotoxic to antigen-bearing target cells in the presence of an appropriate monoclonal antibody. Presumably the RTA moiety could not be released from the fusion protein by reduction intracellularly and was, therefore, unable to enter the cell cytosol where its ribosomal substrate is located. We overcame this limitation by inserting DNA encoding a short amino acid sequence from diphtheria toxin between the RTA and PA to allow production of a fusion protein with an internal disulfide linked loop containing a trypsin-sensitive cleavage site (RTA–DT–PA; Figure 2).

Recombinant RTA–PA and PA–RTA premixed with rabbit antihuman kappa light chain (RAHk) were not cytotoxic to Daudi cells. Their IC_{50} values were greater than that of free RTA (10^{-7} M or greater [Figure 3]). In contrast, the fusion protein containing the diphtheria toxin loop (RTA–DT–PA) premixed with RAHk was cytotoxic to Daudi cells ($IC_{50} = 8 \times 10^{-11}$ M). Trypsin pretreatment of RTA–DT–PA increased its toxicity fourfold ($IC_{50} = 2 \times 10^{-11}$ M). Thus, inclusion of the diphtheria toxin sequence and cleavage site within the recombinant fusion protein significantly enhances cytotoxicity, with or without trypsin pretreatment. In the presence of antibody, the IC_{50} for the trypsin-treated chimeric protein approached that of whole ricin (6×10^{-12} M). The toxicity of RTA–DT–PA premixed with RAHk was specific since RTA–DT–PA applied alone was several hundredfold less toxic ($IC_{50} = 3 \times 10^{-8}$ M) for noncleaved RTA–DT–PA and 6×10^{-9} M for trypsin-treated RTA–DT–PA) and RTA–DT–PA mixed with a control mouse immunoglobulin G_{2a} antibody (which was also able to bind PA) was no more cytotoxic than RTA–DT–PA added alone.

Because the inclusion of a proteolytic cleavage site into a chimeric protein also requires that the two derived fragments remain covalently linked by a disulfide bond after cleavage, formation of the relevant disulfide bond is clearly crucial. To facilitate this in recombinant chimeras produced

(a)

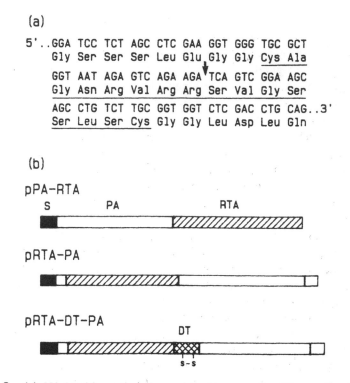

(b)

Figure 2 (a) Nucleotide and amino acid sequence of a 90-mer *Bam*HI–*Pst*I fragment, which includes the trypsin cleavage site (arrowed) of the DT disulfide loop (underlines), and (b) schematic illustration of the fusion proteins encoded by the expression plasmids generated.

in *E. coli*, it is necessary to include a signal sequence in the construct to allow the expressed product to be directed to the bacterial periplasmic space. Likewise, expression in a eukaryotic host would also require an N-terminal signal sequence to ensure the recombinant product enters the endoplasmic reticulum lumen.

One advantage of recombinant chimeras containing the diphtheria toxin loop is that target cell protease(s) can process the molecule intracellularly to a cytotoxic heterodimeric form. This precludes the need to treat the chimera with trypsin before use. Trypsin treatment may also be avoided by creating an alternative cleavage site. One strategy we have used is illustrated schematically in Figure 4. DNA encoding preproricin is modified so that most of the DNA encoding RTB is deleted. This generates a construct comprising the ricin signal sequence, RTA, the linker, and a small

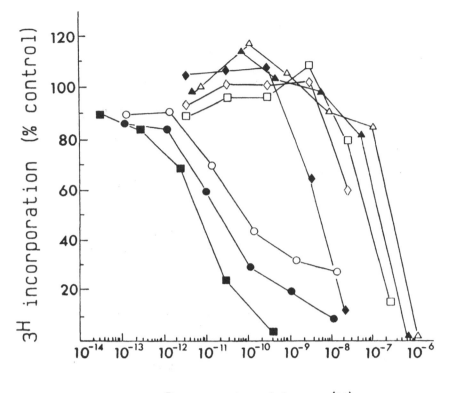

Figure 3 Inhibition of protein synthesis in Daudi cells. A range of concentrations of purified recombinant fusion proteins were mixed with equimolar quantities of rabbit antihuman kappa light chain antibody and incubated with Daudi cells before determining the ability of the cells to incorporate [^{14}C]leucine into protein. Cells were incubated with RTA-PA (▲), PA-RTA (△), RTA-DT-PA without trypsin cleavage (○), RTA-DT-PA after trypsin cleavage (●). In other cultures, cells were incubated with RTA (□), whole ricin (■), RTA-DT-PA without trypsin cleavage (◇) or RTA-DT-PA after trypsin cleavage (◆), all of which were not preincubated with rabbit and antihuman kappa light chain.

portion of RTB. A short N-terminal RTB sequence is retained in this case because it contains the cysteine residue which is involved in the formation of a disulfide bond with RTA. Once again, the signal sequence is included to ensure correct targeting and concommitant disulfide bond formation during expression. DNA encoding an alternative cell-binding function, such as human interleukin-2 (IL-2), is included in place of deleted RTB. Having

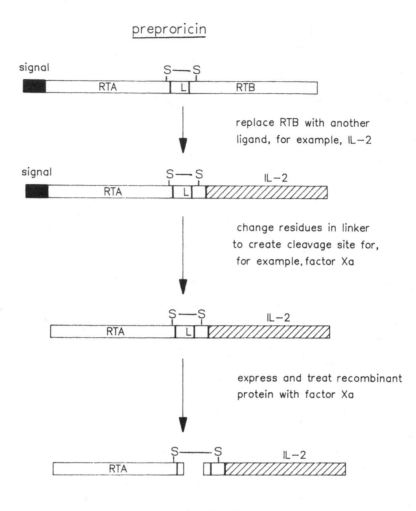

Figure 4   Schematic illustration of a strategy for replacing most of the preproricin B chain with an alternative cell-binding protein such as interleukin-2, and for introducing a factor X cleavage site into to the linker region.

produced the recombinant chimera, the RTA and IL-2 moieties can be cleaved in one of two ways. The first would employ the purified plant endoprotease(s) which naturally processes proricin to the heterodimeric toxin. In practice, this obvious approach is not yet feasible because of difficulties in purifying the processing enzyme. The alternative approach we have successfully used is to employ site-directed mutagenesis to alter residues within the ricin linker sequence to generate a specific protease recognition site. In this way, we have inserted a factor X cleavage site and, following expression of the RTA–factor X–IL-2 fusion, the resulting chimeric molecule can be efficiently cleaved to produce a biologically active, disulfide-linked heterodimer by treatment with commercially available factor Xa.

REFERENCES

1. Olsnes, S., and Sandvig, K. How protein toxins enter and kill cells. In: Immunotoxins, A. E. Frankel (ed.), pp. 39–73. New York: Kluwer Academic Publishers, 1988.
2. Pappenheimer, A. M. Diphtheria toxin. Ann. Rev. Biochem., 46: 64–94, 1977.
3. Wick, M. J., Hamood, A. N., and Iglewski, B. H. Analysis of the structure-function relationship of Pseudomonas aeruginosa exotoxin A. Mol. Microbiol., 4: 527–535, 1990.
4. Greenfield, L., Bjorn, M. J., Horn, G., Fong, D., Buck, G. A., Collier, R. J., and Kaplan, D. Nucleotide sequence of the structural gene for diphtheria toxin carried by corynebacteriophage β. Proc. Natl. Acad. Sci U.S.A., 80: 6853–6857, 1983.
5. Moskaug, J. O., Sandvig, K., and Olsnes, S. Low pH-induced release of diphtheria toxin A fragment in Vero cells. Biochemical evidence for transfer to the cytosol. J. Biol. Chem., 263: 2518–2525, 1988.
6. Ogata, M., Chaudhary, V. K., Pastan, I., and FitzGerald, D. J. Processing of Pseudomonas exotoxin by a cellular protease results in the generation of a 37,000-Da toxin fragment that is translocated into the cytosol. J. Biol. Chem., 265: 20678–20685, 1990.
7. Siegall, C. B., Xu, Y.-H., Chaudhary, V. J., Adhya, S., FitzGerald, D., and Pastan, I. Cytotoxic activities of a fusion protein comprised of TGFα and Pseudomonas exotoxin. FASEB J., 3: 2647–2652, 1989.
8. Lorberboum-Galski, H., FitzGerald, D., Chaudhary, V. J., Adhya, S., and Pastan, I. Cytotoxic activity of an interleukin 2–Pseudomonas exotoxin chimeric protein produced in Escherichia coli. Proc. Natl. Acad. Sci. U.S.A., 85: 1922–1926, 1988.
9. Kelley, V. E., Bacha, P., Pankewycz, D., Nichols, J. C., Murphy, J. R., and Strom, T. B. Interleukin 2–diphtheria toxin fusion protein can abolish cell-mediated immunity in vivo. Proc. Natl. Acad. Sci. U.S.A., 85: 3980–3984, 1988.
10. Chaudhary, V. K., Queen, C., Junghans, R. P., Waldmann, T. A., FitzGerald, D. J., and Pastan, I. A recombinant immunotoxin consisting of two antibody variable domains fused to Pseudomonas exotoxin. Nature, 339: 394–397, 1989.
11. Pastan, I., and FitzGerald, D. Pseudomonas exotoxin: Chimeric toxins. J. Biol. Chem., 264: 15157–15160, 1989.

12. Lord, J. M. Precursors of ricin and *Ricinus communis* agglutinin. Glycosylation and processing during synthesis and intracellular transport. Eur. J. Biochem., *146*: 411–416, 1985.

13. Lamb, F. I., Roberts, L. M., and Lord, J. M. Nucleotide sequence of cloned cDNA coding for preproricin. Eur. J. Biochem., *148*: 265–270, 1985.

14. Lord, J. M., and Robinson, C. Role of proteolytic enzymes in the posttranslational modification of proteins. *In*: M. J. Dalling (ed.), Plant Proteolytic Enzymes, pp. 69–80. Boca Raton, Florida: CRC Press, 1986.

15. O'Hare, M., Brown, A. N., Hussain, K., Gebhardt, A., Watson, G., Roberts, L. M., Vitetta, E. S., Thorpe, P. E., and Lord, J. M. Cytotoxicity of a recombinant ricin A chain fusion protein containing a proteolytically-cleavable spacer sequence. FEBS Lett., *273*: 200–204, 1990.

III
Plant Hemitoxins

10
Studies on Ribosome-Inactivating Proteins from *Saponaria officinalis*

Marco R. Soria   *San Raffaele Research Institute, Milan, Italy*

Luca Benatti and Gianpaolo Nitti   *Farmitalia Carlo Erba, Milan, Italy*

Aldo Ceriotti and Michela Solinas   *Istituto Biosintesi Vegetali del CNR, Milan, Italy*

Douglas A. Lappi   *The Whittier Institute for Diabetes and Endocrinology, La Jolla, California*

Rolando Lorenzetti   *Marion Merrell Dow Research Institute—Lepetit Research Center, Gerenzano, Italy*

I. INTRODUCTION: SEQUENCING STUDIES

The widespread occurrence of single-chain ribosome-inactivating proteins (RIPs) throughout the plant kingdom is now an established fact (1). These are designated type I RIPs as opposed to RIPs consisting of two nonidentical subunits (A and B chains) that are joined by a disulfide bond (2). Complete amino acid sequences are now known for some type I and Type II RIPs. For several plants and plant tissue, the names of the corresponding type I and type II RIPs are reported in Table 1.

Complete sequences of type I RIPs from *Trichosanthes kirilowii*, from barley, and from *Mirabilis jalapa*, are now known (3–6). Partial N-terminal sequences of type I RIPs from plants of the Phytolaccaceae family have been reported (7–9). Several type I RIPs were isolated from *Saponaria officinalis* (10). Among these, saporin-6 (SO6) had 40% amino acid sequence similarity with the RIPs from *Phytolacca americana* and *P. dodecandra* at its NH_2-terminal sequence, though immunologically distinct from them and several other RIPs (11).

Table 1 RIP Nonmenclature

Plant	Tissue	RIP	Abbrev. in Fig. 1
		a. Type I:	
Saponaria officinalis	Seeds	Saporin-6 (SO6)	SO6
	Leaves	Saporin-1 (SO1, SO4)	SO4
Momordica charantia	Seeds	Momordin	Momo
Bryonia dioica	Seeds	Bryodin	Bryo
Trichosanthes kirilowii	Roots	Trichosanthin	Tric
	Seeds	Trichokirin	Trik
Gelonium multiflorum	Seeds	Gelonin	Gelo
Phytolacca americana	Leaves	Pokeweed Antiviral Protein (PAP)	PAP
	Summer leaves	PAP-II	PAP2
	Seeds	PAP-S	PAPS
Phytolacca dodecandra	Leaves	Dodecandrin	Dode
Mirabilis jalapa	Leaves	*Mirabilis* antiviral	Mira
	Roots	protein (MAP)	
		b. Type II:	
Ricinus communis	Seeds	Ricin	Rici
Abrus precatorius	Seeds	Abrin	Abri
Adenia digitata	Roots	Modeccin	Mode

In a previous publication, we reported the characterization and the N-terminal sequences of RIPs isolated from the seeds of a plant of the family Euphorbiaceae, *Gelonium multiflorum*, of two plants of Cucurbitaceae, *Momordica charantia* and *Bryonia dioica*, and from the seeds and leaves of a plant of the Caryophyllaceae, *Saponaria officinalis* (Table I). Purified gelonin, momordin, bryodin, and saporin-1 eluted as single peaks in reverse phase chromatography prior to sequencing, whereas purified saporin-6 yielded two peaks with similar molecular weight, identical NH$_2$-terminal sequences, and cross-reactivity to an anti–saporin-6 antibody. Also, Bjorn et al. (8) reported that the reverse phase chromatography of pokeweed antiviral protein (PAP), previously considered to be a homogeneous protein, yielded two peaks with the same N-terminal sequence. These authors were unable to explain the basis of the observed heterogeneity. The existence of a multigene family (see below) might possibly explain the presence of multiple peaks in samples that are nonglycosylated and quite homogeneous in different conditions of chromatography, electrophoresis, and isoelectric focusing (11,12). Saporin-1 had only minor cross-reactivity to anti–saporin-6 antibody. Indeed, saporin-1 and saporin-6 showed some differences at their NH_2-terminal extremities (13) (Fig. 1).

The sequence analysis program GENALIGN (Intelligenetics Inc., Mountain View, CA) was used to assist in aligning the sequenced regions of type I and type II RIPs by similarity (Figure 1). Gaps were inserted as needed to maximize the total similarity of key amino acid residues, and positions across the sequences occupied by the same amino acids were grouped by vertical lines. The sequenced portions of momordin and bryodin presented high similarity with trichosanthin. All three are members of the Cucurbitaceae family. The N-terminals of bryodin and momordin showed remarkable sequence similarity with gelonin and with ricin A chain, both RIPs from the Euphorbiaceae family. This is a possible consequence of a common evolutionary origin for the Cucurbitaceae and Euphrobiaceae. There is also remarkable similarity between RIPs of the taxonomically closely related families of Caryophyllaceae and Phytolaccaceae, as represented by identities between the PAP group and the saporin seed and leaf proteins. Lower sequence similarity was found between the RIPs of the Cucurbitaceae and Euphorbiaceae families and those of the Phytolaccaceae or Caryophyllaceae familys. The published amino acid sequence of barley protein synthesis inhibitor II (5) was excluded from the comparison owing to an even lower overall similarity, except for the region of the putative active site (4).

The different effects of different RIPs on ribosomes from various plants or protozoa, described by Battelli et al. (14) and by Villemez et al. (15) suggest that RIPs, while sharing a common mechanism of action, might be quite different from each other with respect to their interaction with their substrates. Cenini et al. (16) have reported that RIPs that are members of the Phytolaccaceae and Caryophyllaceae families are both extremely effective in the inhibition of ribosomes from trypanosomes and *Leishmania*. On the other hand, RIPs from the families Euphorbiaceae and Cucurbitaceae showed little effect against these parasitic ribosomes. Thus, there might be a functional difference in the activities of RIPs from different families reflected by the sequence of the proteins. On the basis of sequence one might be able to predict the ability of a RIP to inhibit trypanosome or *Leishmania* ribosomes and that other RIPs from members of the Phytolaccaceae and Carophyllaceae families would be able to inhibit these ribosomes. In contrast, one does not see a similiar correspondence between sequence similarity and the ability to inhibit ribosomes from the protozoans *Tetrahymena* and *Acanthamoeba* (17). In the case of *Tetrahymena* ribosomes, momordin was one-tenth as active as bryodin, whereas the two proteins have notable sequence similarity (70%). Likewise, saporin-6 (Car-

```
Gelo  1   glDtVSFstkGAtyitYvnFlneLRv-kLk--pEgnshgIPLLrkkkddd....
          I III    II   I   I  II   I    I   IIII
Momo  1   D-VSFRLSGAdprSYGmFIKdLRn-ALPf-REKVY-NIPLLLpsvsgagxy...
          I IIIIIIIII  III III II  III  I III IIIIII
Bryo  1   D-VSFRLSGATttSYGVFIKNLRe-ALPyER-KVY-NIPLLL-Rhxig....
          I IIIIIIIII  IIIIII III III II I I IIII I
Tric  1   D-VSFRLSGATssSYGVFIsNLRk-ALPnER-KlY-DIPLL--RssLPgsQRya
          III III   II  II   I  I        III I  I  II   II
Trik  1   D-VSFsLSGggtaSYek.... I   I        IIII I  I  II   II
                          II II   I   I      III I  I  II   II
Mode  1   FPKvteddtr----ATVeSYTt.... I   I      III I  I  II   II
          III          III III II   I   I    III I  I   II  II
Rici      iFPKqyPIinFtTaGATVQSYTnFIrAvRg-RLttGadvr-HDIPVLPnRvgLPinQRfI
          II  I I III III  II I I   II    I  IIIIIII   I   I I
Abri  1   qdrPI-kFsTeGATsQSYkqFIeAlRe-RLrgGli---HDIPVLPdptTLqeRnRYI
          I   I  I  I II  I   II      III I II
Mira  1   aptletIasLDLnNPT--tYlSFItnIRt---Kvad-KteqcT-IqkIskTftqRysYI
          I  III III  I  II   II    I    I  I I        I
SO6   1   VtsI-tLDLvNPTagQYSSFv-dIRnn-vKDPNLK-YGgTDIavigppskekflrI
          I I  I I  I  III    I       IIIII II  I
SO4   1   Vi-IyeLNLqgTTkaQYSTiLkqLRdd-iKDPNLx-YGxxDys...
          I I   I   II  I I I II    III I  II
PAP   1   VNTI-IYNVGSTTISkYATFLndLRNEA-KDPSLkxYG-ipmlpnt....
          IIII IIIIIIIIII IIII   IIIII IIIII
Dode  1   VNTI-IYNVGSTTISnYATFmdnLRNEA-KDPSL....
          I I    I  I  I   I  II II II  I
PAP2  1   N-I-vFDVenATpetYsnFltSLR-EAvKDk-Lt....
          I I  II   II   I  I  II III II
PAPS  1   iNtI-tFDaghATinkYatFyeSLxnEA-KD....
```

(a)

Figure 1 Alignment of sequenced portions of type I and type II RIPs. The GENALIGN program by Intelligenetics, Inc., was used to assist in the alignments. Vertical bars between the aligned protein sequences indicate identical amino acids. Dashes indicate gaps introduced to highlight similarity between sequences. (a) NH$_2$-terminal region; (b) remaining part of sequenced RIPs. RIP nomenclature is described in Table 1.

yophyllaceae) had almost a hundredfold higher activity against *Acanthamoeba* ribosomes than PAP-S (Phytolaccaceae). Thus, there was no apparent relation in this case between the activities of members of the Phytolaccaceae and Caryophyllaceae families. However, whether these differences are due to selective action on ribosomes and not to some form of differential susceptibility of the various RIPs to the experimental conditions employed by the authors (e.g., the presence of proteases degrading the various RIPs to different extents) remains to be ascertained.

Little overall sequence similarity is present among the sequenced N-terminals of type I RIPs and the corresponding segments of ricin and abrin A chains. Though there is a good, if expected, similarity between RIPs from the same family, often it is difficult to find significant similarity when one compares RIPs from different families (Figure 1). However, despite the surprising lack of similarity between proteins that have similar catalytic

```
Tric  49   LihLtNyAdetisvAiDVTNvYimGY---------RAGdtsYFF--neasatEAakyvFkDamrkvTlpysGNYeRL----
            I I I I    I I IIII I  II        III  II    I   I I         I  I   III II
Rici  59   LVELSNhAelSvtlAlDVTNAYVVGY---------RAGnsaYFFhpdnqedaEAithLFTDvqnryTfaFgGNYdRLE---
            IIIII   I  IIIIIIII I            III  II      I  III        I I  II
Abri  54   tVELSNsdTeSIevgIDVTNAYVVaY---------RAGtqsYF---lrdApssAsdyLFT-GTdqhsLpFyGtYGDLE---
            I I  I   II   II I        II        I     I      I II  I I I IIIII
Mira  53   dlivSS--TqkItLaIDmadLYVlgY-sdianNkgRA----fFfkdvTeAvannfFPgAT-GTNrikLtFTGsYGDLEKN-
            II I   I  I I I   I I I  I  I I         I  I   I I     IIII II   I I I  III
SO6   54   -nfqSSrgTvslgLkrD--nLYVvaYlamdntNvnRAy---yFrseiTsAestalFPeAT-taNqkaLeyTedYqsiEKNa
```

```
Tric  115  Q-taAGkiRENIpLGlpaLdsAIttLfYY------nansAasalmvlIQstSEAARykfIE----qqIgk
            I  II IIIII II  I  II  I II          I          II IIIII   II     I
Rici  128  Ql--AGnlRENIeLGngpLeeAISaLyYYstGGtqlptlA-RsfIicIQMiSEAARFqYIEgemRtrI--
            I  I III  I  I I      II      I  I III IIII I    I
Abri  118  --rwAhqsRqqIPLGlqaLthgIS---ffrsGGnDneekA-RtLIviIQMVaEAARFrYISnrvRvsI--
            I  III  I  I I           II II  I   III IIII  IIIII   I
Mira  125  ----Ggl-RKdnPLGIfrLenSI---vniygKAgDVKkqAkfFLL-AIQMVsEAARFkYIS----dkIps
            I  II III I  I          II II  I  III IIII IIIII I     I
SO6   128  qitqGdqsRKelgLGIdlLstS---meavnkKArvVKdeA-rFLLiAIQMtaEAARFrYIqn---lvI-k
```

```
Tric  174  RvdktflPslaiIsLE-NSWsaLSkqIQiaStNnGqFetPvvLinaqnqrvmitnvdagvvtsn-----IALllnrnnma
            I   I IIIII  II II I I I I I   I                               III
Rici  193  RynrrsaPDpsvItLE-NSWgrLStaIQE-S-NQGaFasPiqLqrrNgskfsVy--DvSiliPi-----IAL-MvyrCapPPssqf
            II  I IIII I   IIIII   II  I  I  I    I    I   I  I  I I II     I I  I  II
Abri  180  qtgtafqPDaaMIsLE-NNWdnLrg-vQE-S-vQdtFpnqvtL--TNirnePV-IvD-SlshP--TVavlaL-MlfvCn-PPn
            II  I IIIII  II I  I  I II   I       III   I II   II   I  I
Mira  182  ekyeevtvDeyMtaLE-NNWaKLSTAvYn-SKpstttatkcqLa-Ts----PVtIspwifK----TVeeikLvM--GllKss
            I I I IIII I                           I I I       I I
SO6   190  nfpnkfnsenkviqfEvN-WkKiSTAiYgdaK--------------ngvfnkdydfgfgkvrqvkdlqmgLlMylGkpK
```

(b)

functions, there could possibly be a strong retention of three-dimensional structure. Hovde et al. (18), in comparing the active sites of ricin A chain and Shiga-like toxin, note that among the Shiga and ricin toxin families, only 10 amino acids are conserved, but that seven of those surrounded a cleft proposed as the active site (19). This different in sequence between proteins that have the same function can be explained if the three-dimensional structures are retained. Efforts to crystalize the various RIPs are thus needed in order to compare them with the known crystal structures of ricin A chain (19) and of trichosanthin (20). Energy-minimized molecular models of abrin A chain and of trichosanthin were recently derived by fitting their primary sequences to the backbone structure of ricin A chain (4).

III. CLONING OF THE GENE FOR SAPORIN-6

Gene families are the result of the duplication and divergence of old genes arising during evolution. Some of the genes have related function and some are not functional, for example, pseudogenes. Examples of gene families include the globin, histone, and structural RNA genes. Most of the toxin cloning studies have revealed multiple related toxin genes in plants. Their role in normal plant physiology is still unknown.

To date, very few studies have been reported on the genomic organization and sequence of type I RIPs and their transcripts. As a first step toward this goal, we isolated and characterized bacterial clones containing portions of a cDNA coding for the saporin-6 RIP of *Saponaria officinalis*.

Cloning and expressing a plant toxin gene in microorganisms is advantageous because contaminating components conferring toxicity (as could be the case with contaminating ricin, abrin, and modeccin B chains) can be totally eliminated by this approach. Moreover, chemical conjugation between a monoclonal antibody or ligand and the toxic moiety for targeting purposes might result in unpredictable loss of activity of both components, while removal of unreacted toxin or ligand from the chemical conjugate is not a trivial matter. Instead, the assembly of chimeric toxins by gene fusion techniques results in a precise peptidic linkage between the toxic moiety and the ligand. Finally, components of the conjugate and the conjugate itself can be obtained by cost-effective fermentation of recombinant microorganisms.

The gene coding for saporin-6 was recently cloned in our laboratory (21). A cDNA library from the leaves of *S. officinalis* was prepared. The library was constructed using λ gt10 and poly(A)$^+$ RNA from summer leaves, yielding 1.2×10^6 independent clones after background subtraction. Inserts ranged in size between 600 and several kilobases.

Initial efforts at screening the library with short mixtures of oligonucleotides derived from sequences at the NH$_2$-terminal end were unsuccessful. Therefore, a 111-bp oligonucleotide was assembled since complementary stretches in this probe should be sufficiently long for specific hybridization even in the presence of mismatches. Other genes were successfully isolated using this approach (22).

Two hundred thousand plaques from insert-containing clones were screened with the labeled 111-bp oligonucleotide probe, yielding positive plaques that were isolated and successively rehybridized to an oligonucleotide mixture to confirm the first selection. Two clones were selected for further analysis and were characterized by subcloning into *Eco*RI–digested M13 mp8 in both directions of insertion. The translated sequence of the cDNA clones thus identified showed the signal peptide and the coding region for saporin-6. Comparison of the NH$_2$-terminal sequences and of internal sequences from peptide fragments of saporin-6 purified from seeds (23) to the predicted amino acid sequence from the leaf cDNA (Fig. 2a) showed complete identity between these sequences at all but a few amino acid residues along the molecule. The differences between amino acid residues predicted by cDNA cloning and those determined by direct sequencing (Figure 2b) might be due to different forms of saporin-6 in the same plants as in the case of PAP (7), of ricin (24), and of trichosanthin

```
          V   T   S   I   T   L   D   L   V   N   P   T   A   G   Q   Y
         GTC ACA TCA ATC ACA TTA GAT CTA GTA AAT CCG ACC GCG GGT CAA TAC
                  9          18          27          36          45

  S   S   F   V   D   K   I   R   N   N   V   K   D   P   N   L   K   Y
 TCA TCT TTT GTG GAT AAA ATC CGA AAC AAC GTA AAG GAT CCA AAC CTG AAA TAC
          57          66          75          84          93         102

  G   G   T   D   I   A   V   I   G   P   P   S   K   E   K   F   L   R
 GGT GGT ACC GAC ATA GCC GTG ATA GGC CCA CCT TCT AAA GAA AAA TTC CTT AGA
         111         120         129         138         147         156

  I   N   F   Q   S   S   R   G   T   V   S   L   G   L   K   R   D   N
 ATT AAT TTC CAA AGT TCC CGA GGA ACG GTC TCA CTT GGC CTA AAA CGC GAT AAC
         165         174         183         192         201         210

  L   Y   V   V   A   Y   L   A   M   D   N   T   N   V   N   R   A   Y
 TTG TAT GTG GTC GCG TAT CTT GCA ATG GAT AAC ACG AAT GTT AAT CGG GCA TAT
         219         228         237         246         255         264

  Y   F   R   S   E   I   T   S   A   E   S   T   A   L   F   P   E   A
 TAC TTC AGA TCA GAA ATT ACT TCC GCC GAG TCA ACC GCC CTT TTC CCA GAG GCC
         273         282         291         300         309         318

  T   T   A   N   Q   K   A   L   E   Y   T   E   D   Y   Q   S   I   E
 ACA ACT GCA AAT CAG AAA GCT TTA GAA TAC ACA GAA GAT TAT CAG TCG ATT GAA
         327         336         345         354         363         372

  K   N   A   Q   I   T   Q   G   D   Q   S   R   K   E   L   G   L   G
 AAG AAT GCC CAG ATA ACA CAA GGA GAT CAA AGT AGA AAA GAA CTC GGG TTG GGG
         381         390         399         408         417         426

  I   D   L   L   S   T   S   M   E   A   V   N   K   K   A   R   V   V
 ATT GAC TTA CTT TCA ACG TCC ATG GAA GCA GTG AAC AAG AAG GCA CGT GTG GTT
         435         444         453         462         471         480

  K   D   E   A   R   F   L   L   I   A   I   Q   M   T   A   E   A   A
 AAA GAC GAA GCT AGA TTC CTT CTT ATC GCT ATT CAG ATG ACG GCT GAG GCA GCG
         489         498         507         516         525         534

  R   F   R   Y   I   Q   N   L   V   I   K   N   F   P   N   K   F   N
 CGA TTT AGG TAC ATA CAA AAC TTG GTA ATC AAG AAC TTT CCC AAC AAG TTC AAC
         543         552         561         570         579         588

  S   E   N   K   V   I   Q   F   E   V   N   W   K   K   I   S   T   A
 TCG GAA AAC AAA GTG ATT CAG TTT GAG GTT AAC TGG AAA AAA ATT TCT ACG GCA
         597         606         615         624         633         642

  I   Y   G   D   A   K   N   G   V   F   N   K   D   Y   D   F   G   F
 ATA TAC GGG GAT GCC AAA AAC GGC GTG TTT AAT AAA GAT TAT GAT TTC GGG TTT
         651         660         669         678         687         696

  G   K   V   R   Q   V   K   D   L   Q   M   G   L   L   M   Y   L   G
 GGA AAA GTT AGG CAG GTG AAG GAC TTG CAA ATG GGA CTC CTT ATG TAT TTG GGC
         705         714         723         732         741         750

  K   P   K      <COOH-terminal end of mature protein
 AAA CCA AAG
         759
                                        COOH-terminal extension>

  S   S   N   E   A   N   S   T   V   R   H   Y   G   P   L   K   P   T   L   L   I   T stop
 TCGTCAAACGAGGCGAATTCTACCGTACGCCACTACGGTCCTCTGAAGCCTACTTTACTAATCACATGA
```

(a)

Figure 2 (a) DNA and deduced amino acid sequence of saporin-6 cDNA. The 3' end of the cDNA coding for the COOH-terminal propeptide extension (27) is also indicated. (b) Amino acid sequence of saporin-6 purified from seeds (23).

```
           10         20         30         40         50
     VTSITLDLVN PTAGQYSSFV DKIRNNVKDP NLKYGGTDIA VIGPPSKEKF

           60         70         80         90        100
     LRINFQSSRG TVSLGLKRDN LYVVAYLAMD NTNVNRAYYF RSEITSAELT

          110        120        130        140        150
     ALFPEATTAN QKALEYTEDY QSIEKNAQIT QGDKSRKELG LGIDLLLTFM

          160        170        180        190        200
     EAVNKKARVV KNEARFLLIA IQMTAEVARF RYIQNLVTKN FPNKFDSDNK

          210        220        230        240        250
     VIQFEVSWRK ISTAIYGDAK NGVFNKDYDF GFGKVRQVKD LQMGLLMYLG

          253
     KPK
```

(b)

Figure 2   Continued

(25). The existence of several genes coding for various forms of saporin-6 is indicated by genomic amplification experiments on leaf DNA using the polymerase chain reaction (26).

Both cDNA clones coding for saporin-6 ended with a "natural" *Eco*RI site at their 3' end; that is, a site not resulting from the addition of linkers to the cDNA. Thus, we could not identify a translation termination codon at the 3' end of these clones. Initially, positive identification of the COOH-terminal end of saporin-6 was hindered by the resistance of this protein to treatment with some proteases, including carboxypeptidases (10) Table 2. In this context, it is worth noting that resistance to various proteases was also observed in the case of other type I RIPs, including those of *Mirabilis jalapa*, which were recently sequenced by biochemical procedures (6).

The missing portion of the cDNA for saporin-6 codes for a 22-amino acid COOH-terminal extension that is not found in the mature protein, followed by the stop codon and polyadenylation signal (27). Thus, saporin-6 derives by a processing mechanism from a longer precursor whose cDNA extends beyond the 3' end of our cDNA clones. Therefore, all the coding portion for the COOH-terminal of mature saporin-6 is contained in the cDNA clones. This is evidenced by the positive identification of the COOH-terminal residues as . . . Pro-Lys by hydrazinolysis and by treating saporin-6 with carboxypeptidase P after thermal denaturation (27). Similar post-translation processing mechanisms of a COOH-terminal extension were described in the precursors to the hole-forming toxin from *Aeromonas hydrophila*, aerolysin (28), to interferon gamma (29), to tobacco glucanase, a plant defense-related enzyme (30), to barley lectin (31), and to another type I RIP, trichosanthin (25).

Table 2   Protease Treatment of Saporin-6

	Conditions				
Enzyme	Time	Temp. (°C)	Urea	Prior denat.	Results
Trypsin	24 hr	37			−
Trypsin	24 hr	37	2 M		+ + +
Chymotrypsin	24 hr	37			−
Chymotrypsin	24 hr	37	2 M		+
Staph. V8 protease	24 hr	37			−
Staph. V8 protease	24 hr	37	2 M		−
Clostripain	24 hr	37			−
Clostripain	24 hr	37	4 M		+ + + +
Pepsin	5 min/4 hr	37			+ + + + +
Thermolysin	24 hr	56			+ + + + +
Carboxypeptidases:					
A, B, Y	2 hr	37	2–6 M		−
Y	2 hr	37	4 M	100° C 10 min	
P	2 hr	37	4 M	−	+ +
P	2 hr	25	4 M	100° C 10 min	+ + + +

The biosynthesis of ricin has been investigated in considerable detail, showing a close relationship between its biosynthetic pathway and that of several storage proteins and lectins accumulating in the seeds of *Ricinus communis* and of other plants. These proteins are synthesized on the membranes of the rough endoplasmic reticulum, in which they are cotranslationally segregated. This segregation step is likely to be common to all plant RIPs, allowing a topological separation between the toxin and the translation machinery. The newly synthesized proteins are transported out of the endoplasmic reticulum and across the Golgi complex, where modification of the oligosaccharide chains can occur. The final site of accumulation are the protein bodies, membrane-bound organelles derived from the central vacuole.

Early processing events of plant RIPs consist of the removal of an N-terminal signal peptide and, in the case of glycoproteins, in the addition of oligosaccharide side chains. The protein body is the subcellular compartment in which further processing of the oligosaccharide chains and/or of the polypeptide backbone takes place. In the case of ricin, a proteolytic step consisting of the removal of 12 linker amino acids produces two disulfide-linked subunits from a single precursor polypeptide. Since the ricin precursor seems to be inactive, it is tempting to speculate the post-translational, vacuole-localized modifications might activate the enzymatic activity of the RIP, ensuring that any precursor molecule escaping the

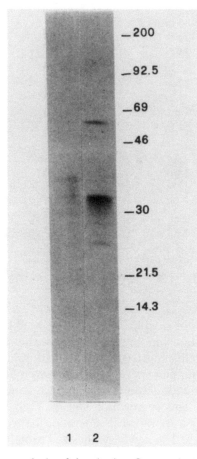

Figure 3 Pulse-chase analysis of developing *Saponaria officinalis* seeds. Developing seeds were incubated for 3 hr in nutrient medium containing 2% sucrose and 0.01 mg/L of abscissic acid, supplemented with 5 μCi/seed of [^{3}H]leucine and 5 μCi/seed of [^{35}S]methionine, and then homogenated either directly (lane 1) or after an 18-hr chase in nutrient medium supplemented with 10 mM leucine, methionine, and glutamine (lane 2). Homogenates were immunoprecipitated using an antiserum raised against purified saporin-6 (11), and the immunoprecipitates were analyzed by SDS-PAGE and fluorography.

initial segregation step does not cause any damage to the cell protein-synthesizing machinery. Interestingly, it has been shown that extracts from the soluble matrix of castor bean protein bodies (but not from castor bean leaves) and from *Phaseolus multiflorum* seeds catalyze in vitro the processing of ricin precursor (32). These findings suggest that processing activities related to those involved in ricin maturation are ubiquitous in seeds, whose protein reserves are subjected to proteolytic maturation.

That proteolytic processing events affect the carboxy-terminal of saporin-6 is consistent with results of pulse-chase experiments performed on developing seeds of *Saponaria officinalis*. Saporin-6 was immunoprecipitated from pulse-labeled seeds, and presented an SDS-PAGE pattern consisting of at least four polypeptides in the 30- to 35-kd range (Figure 3, lane 1). The slower-migrating bands were absent in the immunoprecipitate from seeds that had been labeled and then chased in cold medium to allow processing of the newly synthesized proteins, and a corresponding increase in the intensity of the faster-migrating polypeptides was observed (Figure 3, lane 2), suggesting the existence of a precursor-product relationship. Since saporin-6 is not glycosylated, it is likely that this change reflects a proteolytical processing of the polypeptide chain with a reduction in molecular weight of about 400 d. In the case of trichosanthin, a 19-residue fragment is removed from the COOH-terminal end of the precursor (25).

Preliminary attempts at expressing recombinant saporin-6 in *E. coli* have been performed by fusing the coding region of the gene to that of β-galactosidase. Starting from the saporin-6 gene carried by one of the cDNA clones, a 900-bp *Eco*RI fragment (Figure 4) was subcloned into the *Eco*RI site of the vector pUC8 (33). The plasmid thus obtained was then cut with *Bgl*II and *PST*I, allowing retrieval of a fragment containing the saporin-6 gene but lacking the 5′ region encoding the signal peptide and the first 6 amino acid residues (Figure 5). At the 3′ end of the fragment, the saporin-6 cDNA was followed by an extra piece of DNA deriving from the pUC8 polylinker between the *Eco*RI and the *Pst*I site. Ligation of the *Bgl*II–*Pst*

Figure 4 Restriction map of cDNA coding for saporin-6. Enzymes cutting at single sites are shown on the map.

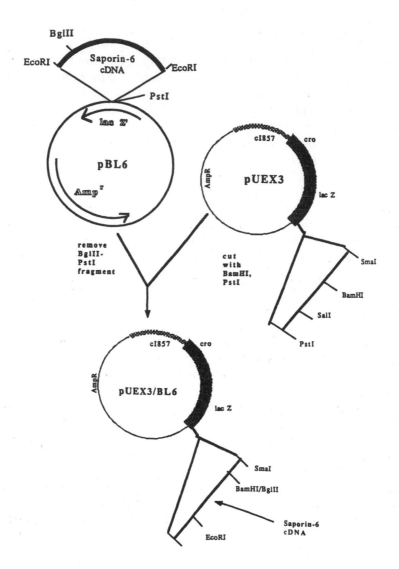

Figure 5   Steps in the in-frame positioning of a fragment containing the cDNA sequence coding for saporin-6 into the expression vector pUEX3.

fragment carrying the gene for saporin-6 to the expression vector pUEX3 (34) cut with *Bam*HI and *Pst*I, resulted in an in-frame fusion between the gene coding for β-galactosidase (35) and the gene containing most of the coding sequence for saporin-6. Downstream from the saporin-6 gene, translation termination was provided by stop codons present in the pUEX3 vector.

The recombinant plasmid thus obtained was transformed into the bacterial host *E. coli* JM105 (33). The hybrid gene, under the control of the λ P_R promoter, was expressed essentially in accordance with Zabeau and Stanley (36). The molecular weight of the hybrid β-galactosidase–SO6 protein was found to be, as expected, 145 kd on SDS-PAGE. The band migrating at this position on the gel was specifically recognized by an anti-SO6 antiserum as shown by immunoblot analysis (37). JM105 cells harboring plasmids coding for the chimeric protein were grown at 30°C to an OD_{600} of 0.9. The temperature was quickly raised to 42°C by addition of an equal volume of broth preheated at 54°C. Cultures were incubated at 42°C for 2 hr before harvest. Total cell lysates were loaded on polyacrylamide slab gels at the appropriate concentration and gels were then transferred on nitrocellulose filters. Filters were then washed and distilled water, incubated with buffered solution containing bovine serum albumin, and incubated with an anti–saporin-6 antiserum raised in rabbits. After further washing, the filters were incubated with a goat antirabbit IgG conjugated with horseradish peroxidase and stained with 4-chloro-1-napthol. Purification of the hybrid protein was obtained by resuspending bacterial pellets in buffer containing 7 M urea, followed by extensive dialysis and affinity chromatography of p-aminophenyl thiogalactoside–Sepharose as described by Ullmann (22). SDS-PAGE of the purified material was run in parallel to the unpurified material as described above, and revealed a prominent band of the expected molecular weight that was recognized by the specific anti–saporin-6 antiserum after immunoblotting by the procedure already described. The purified recombinant protein corresponded in migration to the hybrid β-galactosidase–saporin-6 present in extracts of the induced *E. coli* strain (37).

IV. CONJUGATION STUDIES OF SAPORIN-6

Progress with plant RIPs as partners for totally recombinant conjugates has been slower than with their bacterial counterparts; i.e., those obtained with *Pseudomonas* exotoxin or diphtheria toxin derivatives (reviewed in Ref. 38). Therefore, the reported *E. coli* expression of ricin and abrin A chain, and of a synthetic gene coding for a *Mirabilis jalapa* type I RIP, is an important preliminary step in this direction (6, 39). *E. coli* expression of recombinant saporin-6 has now been achieved (40), and an active RIP

Table 3 Some Recent Growth Factor–Based Saporin Conjugates

Effector	Cytotoxicity	Reference
Antirat nerve Growth factor Receptor monoclonal antibodies	Superior cervical ganglion neurons (immunolesioning)	43
Recombinant human basic fibroblast growth factor	Contaminating fibroblasts for pancreatic islet cell purification	44, 45
"	Neurons of the CA3 region of the hippocampus	46
"	Smooth muscle cells after balloon catheter injury	47
"	Cells derived from Dupuytren's contracture	48
"	Human melanoma cells in nude mice	49
Antihuman interleukin-2 receptor (CD25) monoclonal antibodies	Activated T lymphocytes	50

has been obtained (I. Barthélemy and D. A. Lappi, unpublished). In the meantime, mitotoxins made of "natural" saporin-6 linked to growth factors have been made and shown to be effective (Table 3).

V. CONCLUSIONS

Explosive growth is now taking place in the characterization of protein domains and of signals involved in molecular aspects of cellular adhesion mechanisms, of intracellular membrane traffic and topogenic signals, of cell surface receptors and their ligands, and of autocrine/paracrine growth factors (41). The accumulated knowledge about the structural and functional aspects of interacting molecules will be increasingly used to design molecular strategies for recombinant drug delivery in order to interfere specifically with these structures for a wide variety of therapeutic approaches. These new tools allow molecular biologists to focus on mechanisms that could be exploited for selective delivery of substances possessing biological and/or pharmacological activity, and allow molecular pharmacologists to deliver therapeutic agents in innovative ways (42).

ACKNOWLEDGMENTS

This work was supported in part by the Ministero della Sanità–Istituto Superiore di Sanità—IV, V Progetto AIDS 1991, 1992.

REFERENCES

1. Stirpe, F., and Barbieri, L. Ribosome-inactiviating proteins up to date. FEBS Lett. *195*; 1–8, 1986.
2. Lappi, D. A., Kapmeyer, W., Beglan, J. M., Kaplan, N. O. The disulfide bond connecting the chains of ricin Proc. Natl. Acad. Sci. U.S.A., *75*; 1096–1100, 1978.
3. Zhang, X., Wang, J. Homology of trichosanthin and ricin A-chain. Nature, *321*; 477–478, 1986.
4. Collins, E. J., Robertus, J. D., LoPresti, M., Stone, K. L., Williams, K. R., Wu, P., Hwang, K., and Piatak, M. Primary amino acid sequence of alpha-trichosanthin and molecular models for abrin A-chain and alpha-trichosanthin. J. Biol. Chem., *265*; 8665–8669, 1990.
5. Asano, K., Svensson, B., Svendsen, I., Poulsen, F. M., Roepstorff, P. The complete primary structure of protein synthesis inhibitor II from barley seeds, Carlsberg Res. Commun., *51*; 129–141, 1986.
6. Habuka, N., Murakami, Y., Noma, M., Kudo, T., and Horikoshi, K. Amino acid sequence of Mirabilis antiviral protein, total synthesis of its gene and expression of E. coli. J. Biol. Chem., *264*; 6629–6637, 1989.
7. Houston, L. L., Ramakrishnan, S., and Hermodson, M. A. Seasonal variations in different forms of pokeweed antiviral protein, a potent inactivator of ribosomes. J. Biol. Chem., *258*; 9601–9604, 1983
8. Bjorn, M. J., Larrick, J., Piatak, M., Wilson, K. J. Characterization of translation inhibitors from *Phytolacca americana*: Amino-terminal sequence determinations and antibody-inhibitor conjugates. Biochim. Biophys. Acta, *790*: 154–163, 1984.
9. Ready, M., Wilson, K., Piatak, M., and Robertus, J. D. Ricin-like plant toxins are evolutionarily related to single-chain ribosome-inhibiting proteins from *Phytolacca*. J. Biol. Chem., *259*: 15252–15255, 1985.
10. Stirpe, F., Gasperi-Campani, A., Barbieri, L., Falasca, A., Abbondanza, A., and Stevens, W. A. Ribosome-inactivating proteins from the seeds of Saponaria officinalis L. (soapwort), of Agrostemma githago L. (corn cockle) and Asparagus officinalis L. (asparagus), and from the latex of Hura crepitans L. (sandbox tree). Biochem. J., *216*: 617–625, 1983.
11. Lappi, D. A., Esch, F., Barbieri, L., Stirpe, F., and Soria, M. Characterization of a ribosomal inactivating protein from seeds of Saponaria officinalis (soapworth): Immunoreactivities and sequence homologies. Biochem. Biophys. Res. Commun., *129*: 934–942, 1985.

12. Gelfi, C., Bossi, M. L., Bjellqvist, B., and Righetti, P. G. Isoelectric focusing in immobilized pH gradients in the pH 10–11 range. J. Biochem. Biophys. Methods, *15*: 41–48, 1987.

13. Montecucchi, P. C., Lazzarini, A. M., Barbieri, L., Stirpe, F., Soria, M., and Lappi, D. A. The N-terminal sequence of some ribosome inactivating proteins. Int. J. Pept. Protein Res., *33*: 263–267, 1989.

14. Battelli, M. G., Lorenzoni, E., Stirpe, F., Cella, R., Parisi, B. Differential effect of RIPs on plant ribosome activity and plant cell growth. J. Exp. Bot. *35*: 882–889, 1984.

15. Villemez, C. L., Russell, M. A., Barbieri, L., Stirpe, F., Irvin, J. D., Robertus, J. D. Toxins for protozoan immunotoxins J. Cell Biochem. Suppl. *10b*: 78, 1986.

16. Cenini, P., Bolognesi, A., Stirpe, F. Ribosome-inactivating proteins from plants inhibit ribosome activity of Trypanosoma and Leishmania. J. Protozool., *35*: 384–387, 1988.

17. Cenini, P., Battelli, M. G., Bolognesi, A., Stirpe, F., Villemez, C. Effect of ribosome-inactivating proteins on ribosomes from Tetrahymena pyriformis and Acanthamoeba castellanii. Biochem. Biophys. Res. Commun., *148*: 521–526, 1987.

18. Hovde, C. J., Calderwood, S. B., Mekalanos, J. J., and Collier, R. J. Evidence that glutamic acid 167 is an active-site residue of Shiga-like toxin I. Proc. Natl. Acad. Sci. U.S.A., *85*: 2568–2572, 1988.

19. Monfort, W., Villafranca, J. E., Monzingo, A. F., Ernst, S. R., Katzin, B., Rutenber, E., Xuong, N. H., Hamlin, R., Robertus, J. D. The three-dimensional structure of ricin at 2.8 A. J. Biol. Chem., *262*: 5398–5403, 1987.

20. Pan, K. Z., Zhang, Y. M., Lin, Y. J., Wu, C. W., Zheng, A., Chen, X. Z., Dong, Y. C., Chen, S. Z., Wu, S., Ma, X. Q., Wang, Y. P., Zhang, M. G., Xia, Z. X., Tian, G. Y., Fan, Z. C., Ni, C. Z., Ma, Y. L., Sun, X. X. Tertiary structure of trichosanthin. *In*: H. M. Chang, H. W. Yeung, W. W. Tso, and A. Kos (eds.), Advances in Chinese Medicinal Materials Research, pp. 297–303. Singapore: World Scientific, 1985.

21. Benatti, L., Dani, M., Lorenzetti, R., Nitti, G., Saccardo, B., Lappi, D. A., and Soria, M. Molecular cloning and characterization of the gene coding for a ribosome inactivating protein from the leaves of Saponaria officinalis. Eur. J. Biochem. *183*: 465–470, 1989.

22. Ullman, A. One-step purification of hybrid proteins which have Beta-galactosidase activity, Gene, *29*: 27–31, 1984.

23. Maras, B., Ippoliti, R., De Luca, E., Lendaro, E., Bellelli, A., Barra, D., Bossa, F., and Brunori, M. The amino acid sequence of a ribosome-inactivating protein from Saponaria officinalis seeds. Biochem. Inc., *21*: 631–638, 1990.

24. Lamb, F. I., Roberts, L. M., and Lord, J. M. Nucleotide sequence of cloned cDNA coding for preproricin. Eur. J. Biochem., *148*: 265–270, 1985.

25. Chow, T. P., Feldman, R. A., Lovett, M., and Piatak, M. Isolation and DNA sequence of a gene encoding alpha-trichosanthin, a type I ribosome-inactivating protein. J. Biol. Chem., *265*: 8670–8674, 1990.

26. Soria, M. Protein structure and gene organization of saporin-6, a type I ri-

bosome Inactivating Protein with unusual resistance to proteases. 2nd Int. Symp. on Immunotoxins. Orlando, Florida, 1990.

27. Benatti, L., Nitti, G., Solinas, M., Valsasina, B., Vitale, A., Ceriotti, A., and Soria, M. R. A saporin-6 cDNA containing a precursor sequence coding for a carboxyl-terminal extension. FEBS Lett., *291*: 285–288, 1991.

28. Howard, S. P., Buckley, J. T. Activation of the hole-forming toxin aerolysin by extracellular processing. J. Bacteriol., *163*: 336–340, 1985.

29. Pan, Y. E., Stern, A. S., Familletti, P. C., Khan, R. F., Chizzonite, R. Structural characterization of human interferon gamma. Heterogeneity of the carboxyl terminus. Eur. J. Biochem., *166*: 145–149, 1987.

30. Shinshi, H., Wenzler, H., Neuhaus, J. M., Felix, G., Hofsteenge, J., Mein, F. Jr. Evidence for N- and C-terminal processing of a plant defense-related enzyme: primary structure of tobacco prepro-beta-1,3-glucanase Proc. Natl. Acad. Sci. U.S.A., *85*: 5541–5545, 1988.

31. Wilkins, T. A., Bednarek, S. Y., and Raikhel, N. V. Role of propeptide glycan in post-translational processing and transport of barley lectin to vacuoles in transgenic tobacco. Plant Cell, *2*: 301–313, 1990.

32. Harley, S., Lord, J. M. In vitro endoproteolytic cleavage of castor bean lectin precursors. Plant Sci., *41*: 111–116, 1985.

33. Messing, J. New vectors for cloning. Methods Enzymol. *101C*: 20–78, 1985.

34. Bressan, G. M., and Stanley, K. K. pUEX, a bacterial expression vector related to pEX with universal host specificity. Nucl. Acids Res., *15*: 10056, 1987.

35. Stanley, K. K., and Luzio, P. Construction of a new family of high efficiency bacterial expression vectors: Identification of cDNA clones coding for human liver proteins. EMBO J. *3*: 1429–1434, 1984.

36. Zabeau, M., and Stanley, K. K. Enhanced expression of cro-beta galactosidase fusion proteins under the control of the Pr promoter of bacteriophage lambda. EMBO J., *1*: 1217–1224, 1982.

37. Lorenzetti, R., Benatti, L., Dani, M., Lappi, D. A., Saccardo, B. M., and Soria, M. Nucleotide sequence encoding plant ribosome-inactivating protein. Brit. Pat. Appl. n. 8801877, 1988.

38. Soria, M. Immunotoxins, ligand-toxin conjugates and molecular targeting. Pharmacol. Res., *21*: 35–46, 1989.

39. O'Hare, M., Roberts, L. M., Thorpe, P. E., Watson, G. J., Prior, B., and Lord, J. M. Expression of ricin A chain in E. coli. FEBS Lett., *216*: 73–78.

40. Prieto, I., Lappi, D. A., Ong, M., Matsunami, R., Benatti, L., Villares, R., Soria, M., Sarmientos, P., and Baird, A. Expression and characterization of a basic fibroblast growth factor-saporin-6 fusion protein in E. coli. Ann. N. Y. Acad. Sci, in press, 1991.

41. Soria, M., and Martini, D. Recombinant strategies in the search for targeted pharmaceuticals. *In*: O. M. Neijssel, R. R. Van der Meer, and K. Ch. A. M. Luyben (eds.) Proc. 4th European Congress on Biotechnology, Vol. 4, pp. 135–148. Amsterdam: Elsevier, 1987.

42. Soria, M. Molecular targeting and delivery: applications of recombinant DNA technology. Biotechnol. Appl. Biochem., *11*: 527–551, 1989.

43. Wiley, R. G., Oeltmann, T. N., and Lappi, D. A. Immunolesioning: selective

destruction of neurons using immunotoxin to rat NGF receptor. Brain Res., *562*: 149–153, 1991.

44. Beattie, G. M., Lappi, D. A., Baird, A., and Hayek, A. Selective elimination of fibroblasts from pancreatic islet monolayers by basic fibroblast growth factor-saporin mitotoxin. Diabetes, *39*: 1002–1005, 1990.

45. Beattie, G. M., Lappi, D. A., Baird, A., and Hayek, A. Functional impact of attachment and purification in the short-term culture of human pancreatic islets. J. Clin. Endocrinol. Metab., *73*: 93–98, 1991.

46. Gonzalez, A., Lappi, D. A., Buscaglia, M. L., Carman, L. S., Gage, F. H., and Baird, A. Basic FGF-SAP mitotoxin in the hippocampus: specific lethal effect on cells expressing the basic FGF receptor. Ann. NY Acad. Sci., *638*: 442–444, 1991.

47. Lindner, V., Lappi, D. A., Baird, A., Majack, R. A., and Reidy, M. A. Role of basic fibroblast growth factor in vascular lesion formation. Cir. Res., *68*: 106–113, 1991.

48. Lappi, D. A., Martineau, D., Maher, P. A., Florkiewicz, R. Z., Buscaglia, M., Gonzalez, A. M., Fox, R., and Baird, A. Basic growth factor in cells derived from Dupuytren's contracture: Synthesis, presence and implications for the therapy of the disease. J. Hand Surg., in press, 1992.

49. Beitz, J., Davol-Lewis, P., Clark, J., Kato, J., Medina, M., Frackelton, A. R., Lappi, D. A., Baird, A., and Calabresi, P. Inhibitory effects of the mitotoxin FGF-saporin on human melanoma growth in vitro and in vivo. Clin. Res., *38*, 777A, 1990.

50. Tazzari, P. L., Bolognesi, A., De Totero, D., Pileri, S., Conte, R., Wijdenes, J., Hervè, P., Soria, M. R., Stirpe, F., and Gobbi, M. BB-10 (AntiCD25)—saporin immunotoxin: a possible tool in graft versus host disease treatment. Transplantation, in press, 1992.

BIBLIOGRAPHY

Barbieri, L., Dinota, A., Lappi, D. A., Soria, M., Stirpe, F., and Tazzari, P. G. Selective killing of CD4+ and CD8+ cells by immunotoxins containing saporin. Scand. J. Immunol., *30*: 369–372, 1989.

Bergamaschi, G., Cazzola, M., Dezza, L., Savino, E., Consonni, L., and Lappi, D. A. Killing of K562 cells with conjugates between human transferrin and a ribosome-inactivating protein (SO-6). Br. J. Haematol., *68*: 379–384, 1989.

Bergonzoni, L., Isacchi, A., Cauet, G., Caccia, P., Sarmientos, P., and Soria, M. Expression and characterization of recombinant human basic Fibroblast Growth Factor and its molecular variants in *E. coli*. EMBL Conference "Oncogenes and Growth Control," Heidelberg, 1988, p. 21.

Bregni, M., Lappi, D. A., Siena, S., Villa, S., Formosa, A., Soria, M., Bonadonna, G., and Gianni, A. M. Activity of a monoclonal antibody-saporin 6 conjugate against B-lymphoma cells. J. Natl. Cancer Inst., *80*: 511–517, 1988.

Bregni, M., Siena, S., Formosa, A., Lappi, D. A., Martineau, D., Malavasi, F., Dorken, B., Bonadonna, G., and Gianni, A. M. B-Cell Restricted Saporin Immunotoxins: Activity against B-Cell Lines and Chronic Lymphocyte Leukemia Cells. Blood, *73*: 753–762, 1989.

Caligaris-Cappio, F., Schena, M., Bergui, L., Tesio, L., Riva, M., Rege-Cambrin, G., Funaro, A., and Malavasi, F. C3b receptors mediate the growth factor-induced proliferation of malignant B-chronic lymphocytic leukemia lymphocytes. Leukemia, *1*: 746–752, 1987.

Lappi, D. A., Martineau, D., and Baird, A. Biological and chemical characterization of basic FGF-saporin mitotoxin. Biochem. Biophys. Res. Commun., *160*: 917–923, 1989.

Lappi, D. A., and Baird, A. Mitotoxins: Growth factor-targeted cytotoxic molecules. Prog. Growth Factor Res., *2*: 223–226, 1990.

Letvin, N. L., Goldmacher, V. S., Ritz, J., Schlossman, S. F., and Lambert, J. M. In vivo administration of lymphocyte-specific monoclonal antibodies in nonhuman primates: In vivo stability of disulphide-linked immunotoxin conjugates. J. Clin. Invest., *77*: 977–984, 1986.

Marcucci, F., Lappi, D. A., Ghislieri, M., Martineau, D., Formosa, A., Siena, S., Bregni, M., Soria, M., and Gianni, A. M. In vivo effects in mice of an anti-T cell immunotoxin. J. Immunol., *142*: 2955–2960, 1989.

Reimann, K. A., Goldmacher, V. S., Lambert, J. M., Chalifoux, L. V., Schlossman, S. F., and Letvin, N. L. In vivo administration of lymphocyte-specific monoclonal antibodies in nonhuman primates: IV. Cytotoxic effect of an anti-T11-gelonin immunotoxin. J. Clin. Invest., *82*: 128, 1988.

Richardson, N. E., Chang, H. C., Brown, N. R., Hussey, R. E., Sayre, P. H., and Reinherz, E. L. Adhesion domain of human T11 (CD2) is encoded by a single exon. Proc. Natl. Acad. Sci. U.S.A., *85*: 5176–5180, 1988.

Sarmientos, P., Isacchi, A., Bergonzoni, L., Cauet, G., Caccia, P., and Soria, M. Basic Fibroblast Growth Factor: Expression, characterization and site-directed mutagenesis of the recombinant molecule. Proc. 34th Cong. Ital. Soc. Biochem., Padova, p. 277, 1988.

Siena, S., Villa, S., Bregni, M., Bonadonna, G., and Gianni, A. M. Amantadine Potentiates T Lymphocyte Killing by an Anti-Pan-T Cell (CD5) Ricin A-Chain Immunotoxin. Blood, *69*: 345–348, 1987a.

Siena, S., Villa, S., Bonadonna, G., Bregni, M., and Gianni, A. M. Specific ex-vivo depletion of Human Bone Marrow T Lymphocytes by an anti–Pan-T cell (CD5) ricin A-chain immunotoxin. Transplantation, *43*: 421–426, 1987b.

Siena, S., Lappi, D. A., Bregni, M., Formosa, A., Villa, S., Soria, M., Bonadonna, G., and Gianni, A. M. Synthesis and characterization of an anti-human T-lymphocyte saporin immunotoxin (OKT1-SAP) with in vivo stability into nonhuman primates. Blood, *72*: 756–765, 1988a.

Siena, S., Bregni, M., Formosa, A., Martineau, D., Lappi, D. A., Bonadonna, G., and Gianni, A. M. Evaluation of antihuman T lymphocyte saporin immunotoxins potentially useful in human transplantation. Transplantation, *46*: 747–753, 1988b.

Siena, S., Bregni, M., Formosa, A., Brando, B., Lappi, D. A., Bonadonna, G., and Gianni, A. M. Immunotoxin mediated killing of clonable T-lymphocytes infiltrating an irreversibly rejected human renal allograft. Int. J. Biol. Reg. Homeost. Agents, *3*: 84–88, 1989a.

Siena, S., Bregni, M., Formosa, A., Brando, B., Marenco, P., Lappi, D. A., Bonadonna, G., and Gianni, M. Immunotoxin mediated inhibition of chronic lymphocytic leukemia cell proliferation. Cancer Res., *49*: 3328–3332, 1989b.

Thorpe, P. E., Brown, A. N., Bremmer, J. A. G., Jr., Foxwell, B. M. J., and Stirpe, F. An immunotoxin composed of monoclonal anti-Thy 1.1 antibody and a ribosome-inactivating protein from Saponaria officinalis: potent anti-tumor effects in-vitro and in-vivo. J. Natl. Cancer Inst., *75*: 151–159, 1985.

11
Cloning and Expression of Trichosanthin and α-Momorcharin cDNA

Pang-Chui Shaw, Rong-Huan Zhu,* Mei-Hing Yung, Hin-Wing Yeung, and Walter K.-K. Ho   *The Chinese University of Hong Kong, Shatin, Hong Kong*

I. INTRODUCTION

Trichosanthin (TCS) and α-momorcharin (α-MMC) are plant proteins isolated from *Trichosanthes kirilowii* and *Momordica charantia*, respectively. In China, TCS has been administered for abortion and the treatment of ectopic pregnancy, hydatidiform mole, invasive mole, and choriocarcinoma (1–3). Work in Hong Kong and elsewhere has shown that these proteins contain different pharmacological activities. Besides possessing abortifacient properties (4,5), they also inhibit tumor growth (6) and suppress immune responses (7,8). McGrath et al. (9) also discovered that these proteins can selectively inhibit the replication of type 1 human immunodeficiency virus in acutely infected lymphoblastoid cells or chronically infected macrophages.

The exact mechanism contributing to these diversified properties is not known. However, TCS and α-MMC belong to single-chain ribosome-inactivating proteins (RIP) (10), and our preliminary study has shown that like diphtheria toxin (11), TCS also possesses nuclease activity.

In order to elucidate the structure-function relationship of these two proteins and to improve their pharmacological usage, we cloned and expressed the cDNA of these proteins.

---

*Present affiliation: Institute of Genetics, Academia Sinica, Beijing, People's Republic of China

II. CLONING AND SEQUENCING OF α-MMC AND TCS

Messenger RNA for α-MMC was extracted from young seeds of *M. charantia* and a cDNA library was constructed using λgt11 as vector. The α-MMC cDNA was screened by immunoscreening the library with polyclonal antibodies against succinylated α-MMC (12).

Messenger RNA for TCS was extracted from root tubers of *T. kirilowii* using an elaborate procedure owing to the presence of a high level of contaminants (13). Since the cDNA for α-MMC had been cloned by us and its deduced amino acid sequence is highly homologous to that of TCS, we decided to use the heterologous probe hybridization strategy to screen the cDNA of TCS. Therefore, the cDNA of TCS was cloned into λgt10 and a 120-bp α-*MMC* DNA probe was used to screen the gene.

A complete α-*MMC* and two partial clones containing the C-terminal region have been sequenced. The complete α-*MMC* cDNA encodes a polypeptide of 286 amino acids. The N-terminal 23-amino acid is a signal peptide as it is hydrophobic and does not exist in the mature α-MMC protein. There is a potential glycosylation site at amino acid 227 with the sequence NVT not found in TCS.

Two *TCS* cDNAs have been sequenced. These cDNAs encode a polypeptide of 289 amino acids. Comparison to the amino acid sequence deduced from the natural product (14) shows that there is a signal peptide of 23 amino acids at the N-terminus and an extra 19 amino acids at the C-terminus. The signal peptide of TCS is 56% homologous to that of α-MMC.

There are three silent nucleotide changes in our two *TCS* cDNAs. These together with the evidence that at least four bands have been found from the hybridization of a *TCS* probe with the genomic DNA (15) and the existence of a RIP, trichokirin (16) with related N-terminal sequence in the leaves of *T. kirilowii*, reveals the existence of *TCS* as a multigenic family.

Our *TCS* sequence in pTCS48210 has six nucleotide differences as compared to the genomic sequence obtained by Chow et al. (15). Although this may be due to polymorphism of *TCS*, it is also likely that the root tubers we obtained from Southern China were of a slightly different variety to the leaves collected by Chow's group from South Korea.

Although the three α-*MMC* cDNAs have identical sequence, *MMC*, probably, like ricin A and *TCS* also exists in a multigenic family. Another 23-kd RIP has been found in seeds of *M. charantia* (17). Our group has also found a 28-kd RIP, β-MMC, which has immunochemical properties distinct from MMC but has similar activities for inducing abortion and immunosuppression (18,19).

It is not clear why polymorphism exists in TCS, MMC, and other RIPs. Hypotheses such as spatial distribution of different variants (may not be applied to α-MMC, β-MMC, and the 23-kd RIP mentioned previously as they all exist in abundance in seeds of *M. charantia*) and antibody analogous function to defend the plant against different foreign attacks have been proposed.

III. EXPRESSION OF TCS

Strategies for the synthesis of recombinant TCS (rTCS) in *Escherichia coli* under the control of a modified *trp–lac* promoter have been described (13). This rTCS contains a propeptide of 19 amino acids at the C-terminus and a Met codon at the −1 region to replace the N-terminal signal peptide. However, induction of this clone, RB791 (pTRC48210), only produces about 0.01% rTCS.

To improve expression, we then removed the nonessential 0.4 kb from the encoded propeptide of TCS up to the transcriptional terminal signal of the vector to generate the expression vector pTRC58210 (Figure 1). Soluble protein with the right side of 27 kd was synthesized. Nevertheless, the level of expression did not improve significantly (Figure 2).

Other expression systems are being exploited for high-level expression of TCS.

IV. COMPARISON OF AMINO ACID SEQUENCE BETWEEN TCS, α-MMC, AND OTHER PLANT RIPs

The percentage identity of amino acid sequences between TCS and α-MMC, luffin A, ricin A, abrin A, barley protein synthesis inhibitor, and mirabilis are 66, 58, 36, 35, 16, and 21, respectively. Trichosanthin, α-MMC, and luffin A are more closely related because they are from plants of the Curcubitaceae family. Nevertheless, energy-minimization models for TCS and abrin A show overall similarity to ricin A (14).

Sequence alignment of the above RIPs shows at least five highly conserved regions (Figure 3): amino acids 4–25, 60–74, 109–136, 148–168, and 186–200. The latter four conserved regions contain amino acids lining the putative active site cleft. Mutagenesis of Glu-160 to Asp and Arg-163 to Lys in TCS has led to drastic decrease in RIP activities (20).

The less conserved regions at amino acids 26–35, 98–102, and 169–180 and the C-terminus contain loops lying on the surface of the TCS molecule and some of them may act as epitopes to elicit antibody production. As TCS and ricin A are being used as pharmacological agents, it may be useful to shuttle some of these regions among their homologous species

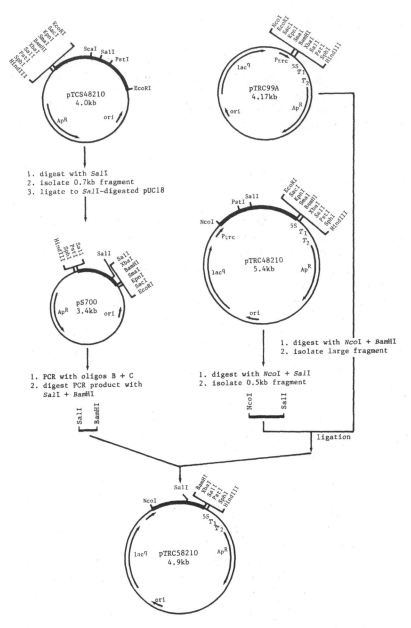

Figure 1 Construction of expression vector pTRC58210. The pTCS48210, a *TCS* cDNA cloned in pUC18, was digested by *Sal*I. The 0.7-kb fragment was removed and ligated to pUC18 to form pS700. A 17-nucleotide oligodeoxyribonucleotide (oligo) B with sequence 5'GTTTTCCCAGTCACGAC and a 35-nucleotide oligo C with sequence 5'CGCGGATCCTATGCCATATTGTTTCTATTCAGCAG were made. One hundred nanograms of pS700 and 50 pmol of oligos B and C were mixed in a reaction mixture of 50 μl (21) and 21 cycles of polymerase chain reaction

216

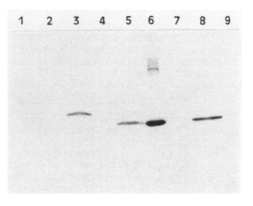

Figure 2 Induction of RB791 (pTRC58210) for rTCS. *E. coli* RB791 (23) was transformed by pTRC58210. Procedures for induction, sample collection and immunological detection of rTCS are according to Shaw et al. (13). Lanes: 1. pTRC99A, uninduced, total protein; 2. pTRC99A, induced, total protein; 3. pTRC48210, induced, total protein; 4. pTRC48210, uninduced, total protein; 5. pTRC58210, induced, total protein; 6. TCS, 500 ng; 7. pTRC58210, uninduced, total protein; 8. pTRC58210, induced, soluble fraction; 9. pTRC58210, induced, sedimented fraction.

to create a set of immunological non–cross-reacting variants for multiple dosage administration.

ACKNOWLEDGMENTS

The α-*MMC*s cDNA were cloned and sequenced by Suk-Ching Liu. Thanks are due to Tsz-Kwong Man and Wai-Hai Tsang for technical assistance.

(PCR) were performed. The PCR product was then extracted twice with chloroform, digested by *Sal*I + *Bam*HI, and purified by the Geneclean Kit (Bio 101). The vector pTRC99A was digested by *Nco*I + *Bam*HI and the large fragment was recovered. This was ligated to the 0.5-kb *Nco*I + *Sal*I TCS N-terminal fragment isolated from pTRC48210 (13). The product was further ligated to the PCR fragment to form pTRC58210. The PCR-generated region was examined by nucleotide sequencing to ensure no mutation had occurred. DH5α cells (21) were used as the host for vector construction. Symbols in the maps are according to Shaw et al. (13).

```
                      1                    20
TCS             DVSFRLSGATSSSYGVFISNLRKALPNERKL YDIPLL    RS
α-MMC           DVSFRLSGADPRSYGMFIKDLRNALPFREKV YNIPLL    LP
LUFFIN A        DVRFSLSGSSSTSYSKFIGDLRKALPSNGTVYNLTILL    S
RICIN A     IFPKQYPI INFTTAGATVQSYTNFIRAVRGRLTTGADVRHDIPVLPNRV
ABRIN A       EDRPI KFSTEGATSQSYKQFIEALRERLRGGL I HDIPVLPDPT
BPSI     AAKMAKNVDKPLFTATF NVQASSADYATFIAGIRNKLRNPAHFSHNEPVLPPVE
MIRA          APTLETIASLDLNNPTTYLSFITNIRTKVADKTEQ    CTIQKI

         40                      60                      80
TCS      SLPGSQR YALIHLTNYAD ETISVAIDVTNVYIMGYRAGD    TSYFFNEASATEA
α-MMC    SVSGAGR YLLMHLFNYDG KTITVAVDVTNVYIMGYLADT    TSYFFNEPAAELA
LUFFIN A SASGASR YTLMTLSNYDG KAITVAVDVSQLYIMGYLVNS    TSYFFNESDAKLA
RICIN A  GLPINQR FILVELSNHAE LSVTLALDVTNAYVVGYRAGN    SAYFFHPDNQEDA
ABRIN A  TLQERNR YITVELSNS DTESIEVGIDVTNAYVVAYRAGT    QSYFLRDAPSS A
BPSI     PNVPPSRWF HVVLKASPTSAGLTLAIRADNIYLEGFKSSD    GTWWELTPGLIPG
MIRA     SKTFTQR YSYIDLIVS STQKITLAIDMADLYVLGYSDIANNKGRAFFFKDVTEAVA

              100                     120                     140
TCS      AK   YV FKDAMRKVTLPYSGNYERLQTAAGKIRENIPLGLPALDSAITTLFYY
α-MMC    SQ   YV FRDARRKITLPYSGNYERLQIAAGKPREKIPIGLPALDSAISTLLHY
LUFFIN A SQ   YV FKGSTI VTLPYSGNYEKLQTAAGKIREKIPLGFPALDSALTTIFHY
RICIN A  EAITHL FTDVQNRYTFAFGGNYDRLEQLAGNLRENIELGNGPLEEAISALYY
ABRIN A  SD   YL FTGTDQH SLPFYGTYGDLERWAHQSRQQIPLGLQALTHGIS  FFR
BPSI     AT   YVGFGGTYRDLL  GDTDKLTNVALG RQQLEDAVTAL HGRTKADKA
MIRA     NNF FPGATGTNR IKLTFTGSYGDLEKN GGLRKDNPLGIFRLENSIVNIYGK

                           160                     180
TCS      NANSA ASALMVLIQSTSEAARYKFIEQQIGKRVDKTF    LPSLAI
α-MMC    DSTAA AGALLVLIQTTAEAARFKYIEQQIQERAYRDE    VPSLAT
LUFFIN A DSTAA AAAFLVILQTTAEASRFKYIEGQIIERISKNQ    VPSLAT
RICIN A  STGGTQLPTLARSFIICIQMISEAARFQYIEGEMRTRIRYN  RRSAPDPSV
ABRIN A  S GGNDNEEKARTLIVIIQMVAEAARFRYISNRVRVSIQTGTAFQ   PDAAM
BPSI     SGPKQQQAREAVTTLLL  MVNEATRFQTVSGFVAGLLHPKAVEKKSGKIGNE
MIRA     AGDVKKQAK  FFLLAIQMVSEAARFKYISDKIPSEKYEE    VTVDEYM

                   200                     220
TCS      ISLENS WSALSKQIQIASTNNGQFESPVVLINAQNQRVTITNVDAGVVTSNIA
α-MMC    ISLENS WSGLSKQIQLAQGNNGIFRTPIVLVDNKGNRVQITNVTSKVVTSNIQ
LUFFIN A ISLENSLWSALSKQIQLAQTNNGTFKTPVVITDDKGQRVEITNVTSKVVTKNIQ
RICIN A  ITLENS WGRLSTAIQE S NQGAFASPIQLQRRNGSKFSVYDVS  ILIPIIA
ABRIN A  ISLENN WDNLRG VQE SV QDTFPNQVTLTNIRNEPVIVDSLSH PTVAVLA
BPSI     MKAQVNGWQDLSAALLKTDVKPPPGKSPAKFTPIEKMGVRTAEQAAA   TLG
MIRA     TALENN WAKLSTAVYNSKPSTTTATKCQLATSPVTISPWIFKTVEEI KLVMG

         240
TCS      LLLNRNNMA
α-MMC    LLLNTRNIAEGDNGDVSTTHGFSSY
LUFFIN A LLLNYKQNVA
RICIN A  LMVYRCAPPPSSQF
ABRIN A  LMLFVCNPPN
BPSI     ILLFVEVPGGLTVAKALFLFHASGGK
MIRA     LLKSS
```

Figure 3 Alignment of RIP sequences to show maximum homologies. The sequence of TCS is numbered for reference. Amino acids found in all seven sequences are boxed. The number of lines beneath the amino acid sequences corresponds to the following degree of homology: ≡ : 6 out of 7 sequences; = : 5 out of 7 sequences;

This work was supported by a grant from the University and Polytechnic Grants Committee of Hong Kong.

NOTE ADDED IN PROOF

Recently, we have succeeded in expressing *TCS* to a high level in *E. coli* and the work of expression and purification has been reported in Int. J. Protein Peptide Res. *39*:77–81, 1992.

REFERENCES

1. Liu, G., Liu, F., Li, Y., Yu, S., et al. A summary of 402 cases of termination of early pregnancy with crystalline preparations of trichosanthin. *In*: H.-M. Chang, H.-W. Yeung, W.-W. Tso, and A. Koo (eds.), Advances in Chinese Medicinal Materials Research, pp. 327–333. World Science Publishing Co., Singapore, 1985.

2. Jin, C.-Y. Clinical study of trichosanthin. *In*: H.-M. Chang, H.-W. Yeung, W.-W. Tso, and A. Koo (eds.), Advances in Chinese Medicinal Materials Research, pp. 319–325. World Science Publishing Co., Singapore, 1985.

3. Huang, Y. A clinical study on treatment of malignant trophoblastic neoplasia with trichosanthin [in Chinese]. Chin. J. Integrat. Tradit. West. Med., 7: 154–155, 1987.

4. Law, L. K., Tam, P. P. L., and Yeung, H. W. Effects of α-trichosanthin and α-momorcharin on the development of peri-implantation embryos. J. Reprod. Fertil., *69*: 597–604, 1983.

5. Chan, W. Y., Tam, P. P. L., and Yeung, H. W. The termination of early pregnancy in the mouse by β-momorcharin. Contraception, *29*: 91–100, 1984.

6. Tsao, S. W., Yan, K. T., and Yeung, H. W. Selective killing of choriocarcinoma cells *in vitro* by trichosanthin, a plant protein purified from root tubers of the Chinese medicinal herb *Trichosanthes kirilowii*. Toxicon, *24*: 831–840, 1986.

7. Leung, K.-N., Yeung, H.-W., and Leung, S.-O. The immunomodulatory and antitumor activities of trichosanthin—an abortifacient protein isolated from Tian-hua-fen (*Trichosanthes kirilowii*). Asian Pac. J. Allergy Immunol., *4*: 111–120, 1986.

8. Yeung, H.-W., Poon, S.-P., Ng, T.-B., and Li, W.-W. Isolation and characterization of an immunosuppressive protein from *Trichosanthes kirilowii* root tubers. Immunopharmacol. Immunotoxicol., *9*: 25–46, 1987.

− : 4 out of 7 sequences. TCS from Shaw et al. (13); α-MMC from Ho et al. (12); luffin A, from Islam et al. (22). Sequences for ricin A, abrin A, BPSI (barley protein synthesis inhibitor), and MIRA (mirabilis) are taken from Chow et al. (15).

9. McGrath, M. S., Hwang, K. M., Caldwell, S. E., Gaston, I., Luk, K.-C., Wu, P., Ng, V. L., Crowe, S., Daniels, J., Marsh, J., Deinhart, T., Lekas, P. V., Vennari, J. C., Yeung, H.-W., and Lifson, J. D. GLQ223: An inhibitor of human immunodeficiency virus replication in acutely and chronically infected cells of lymphocyte and mononuclear phagocyte lineage. Proc. Natl. Acad. Sci. U.S.A., *86*: 2844–2848, 1989.

10. Yeung, H.-W., Li, W.-W., Feng, Z., Barbieri, L., and Stirpe, F. Trichosanthin, α-momorcharin and β-momorcharin: Identity of abortifacient and ribosome-inactivating proteins. Int. J. Peptide Protein Res., *31*: 265–268, 1988.

11. Bruce, C., Baldwin, R. L., Lessnick, S. L., and Wisnieski, B. J. Diphtheria toxin and its ADP-ribosyltransferase-defective homologue CRM197 possess deoxyribonuclease activity. Proc. Natl. Acad. Sci. U.S.A., *87*: 2995–2998, 1990.

12. Ho, W. K.-K., Liu, S.-C., Shaw, P.-C., Yeung, H.-W., Ng, T.-B., and Chan, W.-Y. Cloning of the cDNA of alpha-momorcharin: a ribosome inactivating protein. Biochim. Biophys. Acta, *1088*: 311–314, 1991.

13. Shaw, P.-C., Yung, M.-H., Zhu, R.-H., Ho, W. K.-K., Ng, T.-B., and Yeung, H.-W. Cloning of trichosanthin cDNA and its expression in *Escherichia coli*. Gene, *97*: 267–272, 1991.

14. Collins, E. J., Robertus, J. D., LoPresti, M., Stone, K. L., Williams, K. R., Wu, P., Hwang, K., and Piatak, M. Primary amino acid sequence of α-trichosanthin and molecular models for abrin A-chain and α-trichosanthin. J. Biol. Chem., *265*: 8665–8669, 1990.

15. Chow, P. T., Feldman, R. A., Lovett, M., and Piatak, M. Isolation and DNA sequence of a gene encoding α-trichosanthin, a type I ribosome-inactivating protein. J. Biol. Chem., *265*: 8670–8674, 1990.

16. Casellas, P., Dussossoy, D., Falasca, A. I., Barbieri, L., Guillemot, J. C., Ferrara, P., Bolognesi, A., Cenini, P., and Stirpe, F. Trichokirin, a ribosome-inactivating protein from the seeds of *Trichosanthes kirilowii* Maximowicz. Purification, partial characterization and use for preparation of immunotoxins. Eur. J. Biochem., *176*: 581–588, 1988.

17. Barbieri, L., Zamboni, M., Lorenzoni, E., Montanaro, L., Sperti, S., and Stirpe, F. Inhibition of protein synthesis *in vitro* by proteins from the seeds of *Momordica charantia* (Bitter pear melon). Biochem. J., *186*: 443–452, 1980.

18. Yeung, H.-W., Li, W.-W., Chan, W.-Y., Law, L.-K., and Ng, T.-B. Alpha and beta momorcharins. Int. J. Peptide Protein Res., *28*: 518–524, 1986.

19. Leung, S. O., Yeung, H. W., and Leung, K. N. The immunosuppressive activities of two abortifacient proteins isolated from the seeds of bitter melon (*Momordica charantia*). Immunopharmacology, *13*: 159–171, 1987.

20. Hwang, K., Lifson, J., McGrath, M., Yeung, H., Piatak, M., Chow, T., Luk, K., Wu, P., Caldwell, S., Robertus, J., and Collins, E. Anti-HIV activity of trichosanthin (Abstract, S5A.2). Third SCBA International Symposium and Workshop, June 24–30, 1990, Hong Kong, 1990.

21. Sambrook, J., Fritsch, E. F., and Maniatis, T. Molecular Cloning, A Laboratory Manual. 2nd ed., Cold Spring Harbor, New York: Cold Spring Harbor Laboratories Press, 1989.

22. Islam, M. R., Nishida, H., and Funatsu, G. Complete amino acid sequence of luffin-a, ribosome-inactivating protein from the seeds of *Luffa cylindrica*. Agric. Biol. Chem., *54*: 1343–1345, 1990.

23. Amann, E., Ochs, B., and Abel, K.-J. Tightly regulated *tac* promoter vectors useful for the expression of unfused and fused proteins in *Escherichia coli*. Gene, *69*: 301–315, 1988.

12
Cloning and Expression of a *Luffa* Ribosome-Inactivating Protein–Related Protein

Bi-Yu Li and S. Ramakrishnan *University of Minnesota, Minneapolis, Minnesota*

I. INTRODUCTION

Two types of ribosome-inactivating proteins have been purified from plants: type I, which is made of a single polypeptide chain, and type II, which consists of two nonhomologous subunits (1–4). Both varieties of proteins inactivate eukaryotic ribosome irreversibly. Current evidence indicates that at least some of these proteins possess N-glycosidase activity and depurinate a single adenine residue from 28S rRNA (5). Removal of adenine from position 4324 functionally inactivates the ribosomes. Type II ribosome-inactivating proteins (RIPs) bind to the target cell surface by the B chain, which has specific binding sites. Type I proteins, in contrast, do not have a cell surface recognition moiety and, therefore, are not toxic to eukaryotic cells by themselves. The lack of a binding subunit has some advantage in targeting these polypeptides to tumor cells via carrier molecules such as monoclonal antibodies. A variety of single-chain RIPs have been isolated from plants (6) and chemically linked to antibodies to form highly toxic, specific immunotoxins. In the recent past we isolated a class of proteins belonging to type I RIPs from the seeds of *Luffa*. In this chapter, we would describe some of the properties of these proteins, named *Luffa* ribosome-inactivating proteins (LRIPs), and summarize the cloning and expression of a related protein.

The RIP isolated from the seeds of *Luffa cylindriaca* (LRIP) has a molecular weight of about 30,000 kd. Like many other type I RIPs, LRIP is basic with a pI of over 9.6. Initial characterization indicated that LRIP did not have any free thiol groups, but at least two SH groups could be

```
        1        5        10              15
LRIP I:  D V R F S L S G L A/V  –  T  –  Y K/S K F I
LRIP II: D V S F S L S G S  S C/D T  –  Y Q/S K F I


        20
        G D
                  25
        G D L L K A L P S
```

Figure 1 Amino terminal sequence of LRIP I and LRIP II.

introduced by the modification of free amino groups. The amino-terminal sequence suggests that LRIP is not homologous to other RIPs. Very recently, we isolated another RIP with similar characteristics from the seeds of *Luffa aegyptiaca*. The amino-terminal sequence of both the proteins are shown in Figure 1. Although LRIP II was closely related to LRIP I, the former was found to be relatively more potent in inhibiting protein synthesis in cell-free translation. The total amino acid composition of LRIP II is listed in Table 1, and the DNA and amino acid sequence of LRIP-related protein is listed in Table 2.

Table 1 Total Amino Acid Composition of LRIP II

Amino acid	Amount (%)
D	9.87
T	9.68
S	9.87
Q	8.13
P	4.05
G	11.69
A	8.20
V	6.04
I	7.07
L	9.30
Y	2.84
F	3.62
H	0.60
K	7.18
R	1.85

Total amount injected was 26.15 nmols. Amino acid composition was determined in a Beckman analyzer.

Table 2 DNA and Amino Acid Sequence of LRIP-Related Protein

```
          10              20              30              40              50
           *               *               *               *               *
TTC TTC GGC AAG GAT CCG CGC AAG GAC GTG AAC CCC GAC GAA GCC GTC GCC GCT
 F   F   G   K   D   P   R   K   D   V   N   P   D   E   A   V   A   A

          60              70              80              90             100             110
           *               *               *               *               *               *
GGC GCC GCC ATC CAG GGT TCC GTG CTG TCG GGC GAC CGC AAG GAC GTG CTG CTG
 G   A   A   I   Q   G   S   V   L   S   G   D   R   K   D   V   L   L

                 120             130             140             150             160
                  *               *               *               *               *
CTG GAC GTC ACC CCC CTG TCC CTG GGT ATT GAA ACC CTG GGC GGC GTG ATG ACC
 L   D   V   T   P   L   S   L   G   I   E   T   L   G   G   V   M   T

         170             180             190             200             210
          *               *               *               *               *
AAG ATG ATC CAG AAG AAC ACG ACC ATC CCG ACC CGG TTC TCG CAG ACC TTC TCG
 K   M   I   Q   K   N   T  .T   I   P   T   R   F   S   Q   T   F   S

220             230             240             250             260             270
 *               *               *               *               *               *
ACC GCT GAT GAC AAC CAG CCG GCC GTG ACG ATC AAG GTG TTC CAG GGC GAG CGC
 T   A   D   D   N   Q   P   A   V   T   I   K   V   F   Q   G   E   R

         280             290             300             310             320
          *               *               *               *               *
GAA ATC GCC GCG GGC AAC AAG GCC CTG GGC GAG TTC AAC CTC GAA GGG ATC CCG
 E   I   A   A   G   N   K   A   L   G   E   F   N   L   E   G   I   P

330             340             350             360             370             380
 *               *               *               *               *               *
CCG TCG CCG CGC GGC ATG CCG CAG ATC GAA GTC ACG TTC GAC ATC GAT GCC AAC
 P   S   P   R   G   M   P   Q   I   E   V   T   F   D   I   D   A   N

         390             400             410             420             430
          *               *               *               *               *
GGC ATC CTG CAC GTG TCG GCC AAG GAC AAG GGC ACT GGC AAG GAA AAC AAG ATC
 G   I   L   H   V   S   A   K   D   K   G   T   G   K   E   N   K   I

         440             450             460             470             480
          *               *               *               *               *
ACC ATC AAG GCC AAC TCG GGT CTG TCG GAA GAC GAG ATC CAG CGC ATG GTC AAG
 T   I   K   A   N   S   G   L   S   E   D   E   I   Q   R   M   V   K

490             500             510             520             530             540
 *               *               *               *               *               *
GAT GCC GAG GCC AAT GCC GAG GAA GAT CAC CGC GTC GCC GAG CTG GCC CAG GCC
 D   A   E   A   N   A   E   E   D   H   R   V   A   E   L   A   Q   A

         550             560             570             580             590
          *               *               *               *               *
CGC AAC CAG GCC GAT GCG CTG GTG CAC GCC ACC CGC AAG TCG CTG ACC GAA TAC
 R   N   Q   A   D   A   L   V   H   A   T   R   K   S   L   T   E   Y

         600             610             620             630             640             650
          *               *               *               *               *               *
GGC GAC AAG CTC GAG GCG TCC GAG AAG GAA AGC ATC GAG GCG GCC ATC AAG GCG
 G   D   K   L   E   A   S   E   K   E   S   I   E   A   A   I   K   A

                 660             670             680             690             700
                  *               *               *               *               *
CTG GAA GAC ACG CTG AAG GAC GGC GAC AAG GCC GCC ATC GAC GCC AAG GTC GAA
 L   E   D   T   L   K   D   G   D   K   A   A   I   D   A   K   V   E

         710             720             730             740             750
          *               *               *               *               *
GCC CTG TCG ACC GCC TCG CAG AAG CTC GGC GAA AAG ATG TAC GCC GAC ATG CAG
 A   L   S   T   A   S   Q   K   L   G   E   K   M   Y   A   D   M   Q

760             770             780             790             800             810
 *               *               *               *               *               *
GCG CAG CAA GCG GCC GGC CAG CAG CAG GCC GCC GAC AAC GCG AAG CCG GTG GAC
 A   Q   Q   A   A   G   Q   Q   Q   A   A   D   N   A   K   P   V   D

         820             930             940             950             960
          *               *               *               *               *
GAC AAC GTC GTC GAG GCC GAC TTC AAG GAA GTC AAG CGC GAC CAA TAA
 D   N   V   V   E   A   D   F   K   E   V   K   R   D   Q   *
```

DNA sequence of the open reading frame is shown in the upper lane as triplets. Macvector (IBI) program was used to deduce the amino acid sequence which are represented in single letter code.

Amino acid sequences were determined in a Beckman automated peptide sequencer. Single letter codes are used for individual amino acid residues and the information is compiled from three independent experiments. Unidentified residues are indicated as " $-$ " and positions 10 and 15 of LRIP I need further confirmation.

II. CONSTRUCTION OF GENOMIC LIBRARY

A. Isolation of Genomic DNA from *Luffa* leaves

Since most of the RIP genes are intronless, it was convenient to clone fragments of genomic DNA into expression vectors directly. One of the advantages of this strategy is to select clones in the right reading frame in a single step. However, this strategy has some drawbacks in screening and characterization of the cloned fragment. Since the expression of the LRIP gene during the developmental stages of the plant was not known, DNA was isolated from the leaves. *Luffa aegyptiaca* seeds were sprouted in normal lighting conditions for 3 days before transferring the seedlings to dark. This method is to reduce the carbohydrate content in the leaves, which would otherwise interfere with the purification of DNA. Briefly, leaves were snap frozen and ground in a coffee mill in the presence of a small amount of liquid nitrogen. The powder was immediately transferred to a mortar and ground further in the presence of lysis buffer (0.15 M NaCl, 0.1 M EDTA and 1% SDS) and 10 ml of TE- (10 mM Tris, 1 mM EDTA, pH 8.0) saturated phenol. The lysate was extracted with phenol-chloroform mixture to remove proteins and lipids. The aqueous phase was removed and the DNA was precipitated with cold ethanol. The pellet was dissolved in TE and incubated with RNase and proteinase K to further purify the DNA. Finally, the DNA was subjected to CsCl density gradient centrifugation. The quality of the DNA was checked in a low-concentration agarose gel (0.25%) using lambda DNA as a marker.

B. Partial Digestion of the Genomic DNA

About 100 ug of the genomic DNA was digested with *Eco*RI enzyme at a ratio of 1 unit of enzyme to 1 ug of DNA at 37°C for 1 hr in a total volume of 3 ml. At the end, EDTA at a final concentration of 20 mM was added to stop further digestion. After extracting the digest with phenol, the fragments were precipitated with ethanol and resuspended in TE. DNA fragments thus generated were size fractioned in a 10-ml sucrose gradient (10–40%) at 28,000 rpm for 16 hrs using a SW 41 rotor. One milliliter fractions were collected and an aliquot from each fraction was analyzed on a 0.5% agarose gel. Fractions containing 2–7 kb size were pooled and used for the construction of the library.

C. Ligation to an Expression Vector

With the assumption that the LRIP gene is intronless, the genomic DNA fragments were directly cloned into an expression vector, λgt11. *Eco*RI digested, dephosphorylated λgt11 arms were obtained from Strategene Inc. In this vector, there is an *Eco*RI site at the C-terminal end of the *lac*Z gene, 53 base-pairs upstream from the stop codon. Insertion of foreign DNA fragments at the right orientation at this site would result in a fusion protein and the expression of which will be under the control of *lac* promoter. About 0.5 μg of DNA fragment (2–7 kb) was ligated with 1 μg of the vector DNA. The ligated DNA was then packaged in vitro using a kit from Stratagene Inc., La Jolla, CA. An aliquot of the packaged DNA was titrated against the host, *Escherichia coli* strain Y1088. In this process, a library containing about 6.34 million recombinant plaque-forming units were obtained.

D. Amplification and Screening

For amplification, strain Y1088 (hsdR$^-$, hsdM$^+$) was used. The amplified library had 2×10^{12} plaque-forming units (PFUs) at a concentration of 9×10^{10}/ml. For screening, *E. coli* strain Y1090 was used since this strain is deficient in *lon* protease, which increases the stability of the recombinant fusion protein. About 2 million recombinant PFUs were plated in 150-mm Petri dishes at a density of 1×10^5 PFU/dish. Fusion proteins are sometimes toxic to host cells and this could complicate the screening procedures. To avoid this problems, plaque formation was initiated without inducing the expression of the *lac*Z gene. After the number of infected cells surrounding the plaques is large, *lac*Z-directed gene expression was induced by placing isopropyl-β-D-thiogalactopyranoside (IPTG) saturated nitrocelluose membranes over the lawn of bacteria and incubated for a few more hours at 37°C. The plaque lifts were blocked with a solution containing 1% BSA and then probed with antibodies specific for LRIP. A polyclonal antiserum against purified LRIP was prepared by hyperimmunization of rabbits with multiple intradermal injections in the presence of complete Freund's adjuvant. Antibodies generated in this manner had a high titer (binding capacity 50% binding of radioiodinated LRIP at 1:30,000 dilution). To minimize nonspecific binding of the antibodies to *E. coli* proteins, the antiserum was preabsorbed with an *E. coli* lysate (BNN97, which is a λgt11-harboring lysogen). For preabsorption, nitrocellulose membranes were soaked in *E. coli* lysate at room temperature for 30 mins. The membranes were than washed in TBS solution (50 mM Tris and 150 mM NaCl) and used for absorbing the antiserum. Typically, the antiserum (1:10 dilution) was incubated with the lysate-coated nitrocellulose membranes for about 2 hr with constant shaking. The original plaque lifts after induction with IPTG were then incubated with the preabsorbed antiserum for an hour at room temperature. After thorough washing in PBS containing 0.05% Tween 20, the membranes were probed with ^{125}I-labeled goat anti-rabbit immunoglobulin (IgG) (second antibody). Unbound radioactivity was removed by washing in Tween 20 containing PBS and then the membranes were exposed to x-ray films. Plaques showing a positive signal were then repurified by plating at a lower density until all plaques in the dish were positive for the expression of fusion protein. In our cloning procedure, four rounds of plaque purification were necessary to reach that stage.

E. Characterization of the Cloned Fragment

Positive clones were expanded and the phage was isolated by precipitation with polyethyleneglycol/NaCl and subsequently DNA was purified by CsCl gradient centrifugation. When digested with *Eco*RI enzyme, two of the representative clone (#15 and #18) had an insert of about 2 kb. This fragment was gel purified and subcloned into pBluescript KSII (+) vector (Stratagene Inc.) and digested with various restriction enzymes either alone or in combination to obtain a restriction map. A simplified version of the restriction map is shown in Figure 1. Suitable subfragments of the recombinant DNA was then cloned into M13mp18 vector and sequenced by the dideoxy chain termination method (7). Sequence information was subsequently analyzed by the MacVector gene analysis program (IBI) to identify additional restriction sites and open reading frames (ORFs). These studies revealed that the cloned fragment had a single ORF (bp 3–857) in the right reading frame with *lacZ* gene.

III. PREPARATION OF FUSION PROTEIN

Y1089 host cells were lysogenized with recombinant phages. This host strain has a mutation (hflA50) which enhances the frequency of lysogeny. Infected cells were selected for positive growth at 32°C but not at 42°C. Lysates were prepared from 250-ml culture of the Y1089 lysogen with thermal (42°C) and IPTG induction. As a control, lysates were prepared from the BNN97 strain. An aliquot of the lysates were analyzed on SDS-PAGE. Figure 2A shows the comparative electrophoretic mobility of the proteins. BNN97 cells had a protein band corresponding to β-galactosidase (M_r 116,000). Y1089 cells harboring the recombinant λgt11 showed a pro-

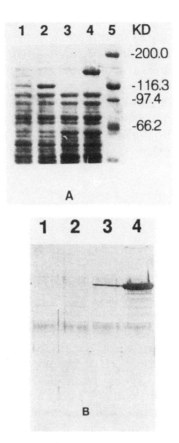

Figure 2   Expression of fusion protein. (A) About 10 μg of lysate from control, BNN97 cultures (lanes 1 and 2) and recombinant cultures (lanes 3 and 4) were run on a 7.5% SDS-PAGE. Lane 5 shows molecular weight markers (BIORAD, high molecular weight standards). Cultures treated with IPTG were shown in lanes 2 and 4. (B) Western blot of the bacterial lysates. Lanes 1 and 2 were from control cultures, BNN97 in the presence (lane 2) and absence (lane 1) of the inducer, IPTG. Lysates from recombinants were shown in lanes 3 and 4. Lane 3, in the absence of IPTG, and lane 4, treated with IPTG.

tein with a molecular weight of about 145,000. In a subsequent experiment, the proteins were transferred to nitrocellulose membranes by electroblotting and probed with antibodies specific to LRIP (Fig. 2B). The fusion protein had a distinct cross-reactive band at a position corresponding to M_r 145,000. Normal β-galactosidase enzyme expressed by the control cultures did not react with anti-LRIP antibodies. Based on the size of the open reading frame, the recombinants should code for a protein of the size of 31,350 contiguous with β-galactosidase protein. The fusion protein with β-galactosidase did not show significant inhibition of protein synthesis in a cell-free system.

IV. EXPRESSION OF LRIP-RELATED PROTEIN IN pET VECTOR

Fusion proteins have limited utility in the biological characterization of the recombinants. Therefore, it was necessary to engineer the molecule in an expression vector wherein the cloned fragments could be expressed as a single polypeptide. For this purpose we utilized the T7 RNA polymerase-based expression vector developed by Studier (Brookhaven National Laboratory, Upton, New York). In this system, the host cells [BL21 (DE3)] carry a lysis-defective phage which contains the T7 polymerase gene under the control of IPTG-inducible *lac*UV-5 promoter. The vector system has T7 promoter and T0 transcription terminator flanking a multiple cloning site. By inducing the expression of T7 polymerase, cloned fragments can be transcribed very efficiently. This system had been previously used in our laboratory to produce ricin A chain at high levels. A simplified version of the cloning strategy is shown in Figure 3. After selecting the clones containing recombinant DNA at the right orientation (*Eco*RI digestion was used to verify the orientation), expression of LRIP-related protein was achieved by inducing the cultures with IPTG. Induced proteins were then electrophoresed on SDS-PAGE and electroblotted. Western blot analysis indicate a cross-reactive band corresponding to 35,000 (Figure 4). The size difference is due to the presence of additional amino acid residues introduced from the expression vector. This protein is being currently characterized. Preliminary studies indicate that this proteins is possibly a truncated version of a protein related to LRIP. The amino-terminal sequence had some homology to positions 17–20 (last four residues of the amino-terminal sequence, KFIGD) of LRIP. Although, the cloned protein is immunologically cross-reactive to native LRIP I and LRIP II, the subsequent residues (positions 21–27 of LRIP II) did not show any homology. Further studies are in progress to complete the sequence of LRIP-related protein at the 5' end and assess the relationship of this protein to RIPs.

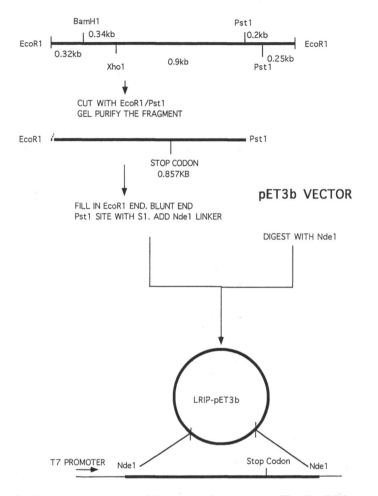

Figure 3 Schematic diagram of the expression strategy. The *Eco*RI fragment (2 kb) was excised from the clone #15. A partial restriction map of the clone is shown in the top line. Subsequent treatment with *Pst*I gave a fragment of the size of 1.56 kb. This fragment was engineered into a pET3b vector at the *Nde*I site (LRIP–pET3b) to obtain recombinant plasmids with the right reading frame. HMS 174 cells were transformed with recombinant plasmids and the transformants were selected on ampicillin medium. Plasmid containing the cloned fragment in the right orientation was finally introduced into BL21 (DE3) host cells for expression.

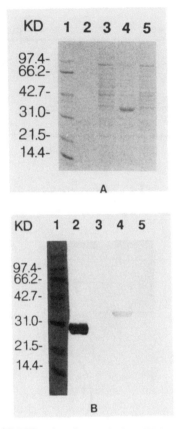

Figure 4   Expression of LRIP-related protein in pET3b vector under the control of a T7 promoter. (A) Bacterial lysates were run on a 12.5% SDS-PAGE. Lane 1, molecular Weight markers (BIORAD, low molecular weight standards). Lane 2, native LRIP II isolated from the seeds. BL21 (DE3) cells transformed with recombinant pET3b plasmid in the absence (lane 3) and the presence (lane 4) of IPTG. Lane 5, lysate from BL21 (DE3) as control. (B) Western blot analysis. Lanes 1–5 as above.

ACKNOWLEDGMENTS

This work was supported in part by a grant from the National Cancer Institute, CA 48608.

REFERENCES

1. Frankel, A. E. (ed.), Immunotoxins, Boston: Kluwer Academic Publishers, 1988.

2. Ready, M., Wilson, K., Piatak, M., and Robertus, J. D. Ricin-like plant toxins are evolutionarily related to single-chain ribosome-inhibiting proteins from phytolacca. J. Biol. Chem. *295*: 15252–15256, 1984.
3. Houston, L. L., Ramakrishnan, S., and Hermodson, M. A. Seasonal variations in different forms of pokeweed antiviral protein, a potent inactivator of ribosomes. J. Biol. Chem. *258*: 9601–9604, 1983.
4. Olsnes, S., Stirpe, F., Sandvig, K., and Pihl, A. Isolation and characterization of vis cumin, a toxic lectin from *Viseum album*. J. Biol. Chem. *257*: 13263–13270, 1982.
5. Endo, Y., Mitsui, K., Motizuki, M., and Tsurugi, K. The mechanism of action of ricin and related toxic lectins on eukaryotic ribosomes. J. Biol. Chem. *262*: 5098–5912, 1987.
6. Stirpe, F., and Barbieri, L. Ribosome-inactivating proteins up to date. FEBS Lett., *195*: 1–8, 1986.
7. Sanger, F., Nicklen, S., and Coulson, A. R. DNA sequencing with chain-terminating inhibitors. Proc. Natl. Acad. Sci. U.S.A. *74*: 5463–5467, 1977.

IV
Fungal Ribotoxins

13
The *Aspergillus* Ribonucleolytic Toxins (Ribotoxins)

Bernard Lamy and Julian Davies   *Institut Pasteur, Paris, France*

Daniel Schindler   *Weizmann Institute of Science, Rehovoth, Israel*

I. INTRODUCTION

Following the isolation of α-sarcin, restrictocin, and mitogillin in the late 1950s in the course of a screening program for naturally occurring antineoplastic activities (1), there were many tests of their efficacy against a variety of different tumors in animal models and in mammalian cell cultures. However, in spite of some promising indications, the antitumor activity of the fungal protein toxins was considered to be too limited and the toxicity too great to merit continued investigation. In general, inhibition was most marked against transformed or virus-infected cell lines. Nonetheless, these toxins (we propose they be named ribotoxins) have proved to be of considerable biochemical value, particularly with reference to their activity as specific intraribosomal rRNA nucleases. Their formal relationship to the plant-derived ribosome-inactivating proteins (RIPs) and site-specific ribonucleases has stimulated additional studies. In view of their potency and their highly specific mode of action, the fungal ribotoxins are worthy of consideration as components of targeted immunotoxins, and some preliminary success has been noted in animal models.

II. BIOCHEMICAL STUDIES: MODE OF ACTION

The mode of action of the fungal ribotoxins is of theoretical and practical interest. There were only limited biochemical studies until the early 1970s when Schindler (2) tested these toxins for their ability to inhibit protein synthesis in wheat germ extracts. α-Sarcin inhibited amino acid incorporation by 50% at 0.1 nM (Figure 1), a concentration which was several

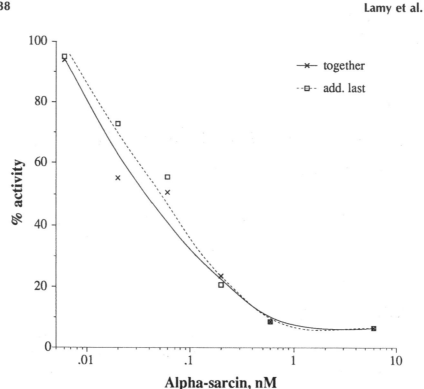

Figure 1 Inhibition of wheat germ protein-synthesizing system by α-sarcin. The rate of protein synthesis is measured by incorporation of [^{14}C-]leucine into acid precipitable material by wheat germ extracts (incubation of 30 min at 30°C with Brome mosaic virus RNA). 100% activity is 33 pmol of [^{14}C-]leucine. x: wheat germ extract and α-sarcin added together. □: wheat germ extract added last.

orders of magnitude lower than the concentration of ribosomes in the incubation mixture; this identified α-sarcin and its close analogs, restrictocin and mitogillin, among the most potent inhibitors of translation to be isolated. In principle, a single molecule of one these inhibitors inside a susceptible cell would suffice to kill the cell! It was immediately obvious that such potency could be due either to the fact that the toxins had a catalytic action (enzymatic), or that to inhibit translation in the cell-free system it sufficed to block only a small proportion of the ribosomes. The latter is a well-known phenomenon in mode of action studies of antibiotic inhibitors of ribosome function since inhibition of elongating ribosomes results in blockade of polyribosome development. Identical inhibitory effects with α-sarcin, restrictocin, and mitogillin were demonstrated in cell-free extracts of the yeast *Saccharomyces cerevisiae* and against translation on rat liver polysomes. By contrast, only with concentrations of toxin several orders of magnitude higher was it possible to inhibit *Escherichia coli* cell-free translation (Table 1). It should be noted that α-sarcin had no effect on growing *E. coli* or yeast cells. HeLa cells proved also to be refractory, presumably due to failure of the ribotoxin to penetrate cells; as we shall see later, the fungal ribotoxins are single-chain protein toxins and do not possess a specific entry component as is found in the dimeric plant RIPs. The mode of action of α-sarcin was examined in more detail by testing its effect on discrete steps in protein synthesis; restrictocin and mitogillin have identical modes of action to α-sarcin (the molecules do not differ significantly in amino acid sequence). Since the drug inhibited both polyuridylic acid (poly-U) and natural mRNA-directed polypeptide synthesis (Figure 2), it was concluded that α-sarcin was not a specific inhibitor of steps involved in initiation. When wheat germ protein synthesis was allowed to initiate by preincubation of the complete system in the absence of toxin, the onset of inhibition was not delayed, as occurs with the initiation inhibitor, T-2 toxin (3) (Figure 3). The rapid inhibition of translation strongly suggested that α-sarcin must inhibit protein synthesis during the elongation steps. The toxin has no effect on the puromycin-induced release of peptidyl-tRNA from rat liver polysomes, which eliminates the possibility of inhibition of peptidyl transferase activity or some of the steps known to be involved in the translocation stage of protein synthesis. In addition, when protein synthesis in rabbit reticulocytes was blocked by α-sarcin, the peptidyl-tRNA is bound to the acceptor site and does not react with puromycin. Preliminary experiments using partially purified preparations of

Table 1   Effect of α-Sarcin on Various Protein-Synthesizing Systems

		α-Sarcin concentration (ng/ml)	% Inhibition
Wheat germ extract	+ BMV	10	96
	+ Poly U	10	83
Rat liver polysomes + yeast supernatant fraction		10	57
Yeast ribisomes + yeast supernatant fraction		500	84
	+ poly U	10	49
E. coli S-30	+ R 17	50,000	85
	+ RNA	5,000	20

The rate of protein synthesis is measured by incorporation of ^{14}C–amino acids into acid-precipitable material, as in Figure 1, 100% inhibition is 0% activity.

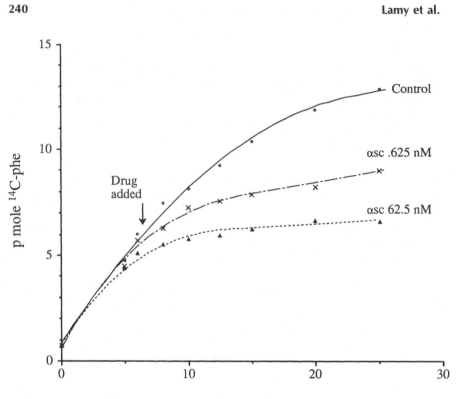

Figure 2 Inhibition of poly U-dependent polyphenylalanine synthesis by α-sarcin. The polyphenylalanine synthesis activity is measured by kinetics of incorporation of [^{14}C-]phenylalanine by yeast ribosomes and supernatant fraction in TCA-precipitable material. α-Sarcin (αsc) was added at the indicated concentration after 5.5 min of incubation at 30°C. Control is without toxin.

EF-1 showed that EF-1–dependent aminoacyl-tRNA binding to the ribosome was inhibited by α-sarcin; however, the inhibition was not complete even at very high concentrations. Furthermore, binding of aminoacyl-tRNA in the presence of GMPCP in place of GTP was not inhibited by α-sarcin (Table 2).

One interpretation of this partial inhibition would be that one of the steps implicated in the binding of aminoacyl-tRNA to the ribosome is resistant to inhibition by α-sarcin. Such a reaction would be the formation of the ternary complex (aminoacyl tRNA–EF-1–GTP) and the binding of this complex to the ribosome. An alternative step for inhibition would be EF-2 catalyzed GTP hydrolysis and translocation, which releases EF-1 from the ribosome. The finding that the nonhydrolyzable GTP analog, GMPPCP,

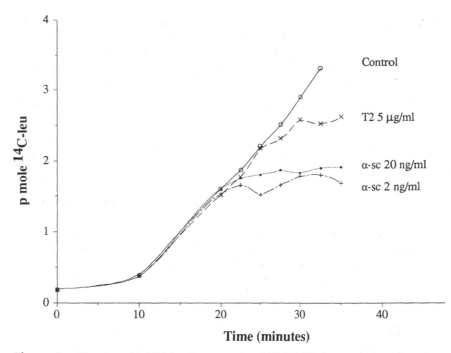

Figure 3 Kinetics of inhibition by α-sarcin of BMV RNA translation. Incorporation of leucine is measured as described in Figure 1. In the control, yeast extracts were incubated without toxin. α-Sarcin (αsc) or T-2 toxin (T2) were added at the indicated concentrations.

Table 2 α-Sarcin Inhibition of EF-1–Dependent Binding of Aminoacyl tRNA

Nucleotide			cpm
GTP	−	α-sarcin	1008
	+	α-sarcin	338
GMPPCP	−	α-sarcin	98
	+	α-sarcin	146

Activity is expressed as count per minute of [^{14}C]-leucyl tRNA bound to ribosomes in the presence of EF-1. The binding in the absence of EF1 has been subtracted to give the above values. α-Sarcin was used at 50 μg/ml.

supported the binding of aminoacyl tRNA to the ribosome, and that this binding was resistant to α-sarcin was consistent with the idea that the *catalytic* reuse of EF-1 on the ribosome is blocked. The binding of aminoacyl tRNA to the ribosome is not affected by α-sarcin; this mode of action has not been confirmed in molecular detail, but this interpretation is consistent with studies carried out with other inhibitors of protein synthesis.

α-Sarcin and the other *Aspergillus* protein toxins are active in inhibiting protein synthesis in vitro in the same concentration range as pancreatic ribonuclease. This raised the following question: Since α-sarcin appears to act enzymatically, is it also a ribonuclease? Presumably, the fact that α-sarcin is not an effective inhibitor of protein synthesis in cell-free extracts of *E. coli* implies that it is not a generalized ribonuclease such as RNAse A (4). In addition, when α-sarcin was tested against Moloney viral messenger RNA as substrate, the toxin was found to have nuclease activity, but was 10,000 times less active than pancreatic ribonuclease at the same concentration. Thus, it would appear that random nuclease digestion is not an explanation for the inhibition of protein synthesis by α-sarcin.

Attempts were then made to examine if the toxins inactivated a component of the protein synthesis machinery. When treated with α-sarcin, purified ribosomes from eukaryotic cells were found to be inactive. On more detailed analysis it was found that when the ribosomal RNA species were separated by gel electrophoresis after incubation of yeast 80S ribosomes with α-sarcin, a new RNA fragment was detected. Incubation of isolated 40S and 60S subunits from yeast with α-sarcin showed that the fragment produced was from the 60S subunit, the 40S subunit remaining unaffected after this treatment; similar results were obtained with ribosome subunits from wheat germ. The rRNA cleavage product from *Saccharomyces cerevisiae* was found to be about 300 nucleotides long and was derived from the 3′ end of the rRNA (2). Was the cleavage of 28S RNA the result of α-sarcin action, and did this cleavage cause the absolute loss of protein synthetic activity on the ribosome? In other words, do α-sarcin and restrictocin (which acted in the same way) possess novel mechanisms of action for protein synthesis inhibitors? A crucial question concerned the purity of the toxin preparations. Could the nucleolytic cleavage have been due to a contaminating nuclease?

To resolve this uncertainty, α-sarcin was repurified by chromatography to provide a fraction containing a single band on analytical electrophoresis which, alone, inhibited and caused the production of the RNA fragment. In a separate experiment, it was shown that treatment of yeast ribosomes with pancreatic RNAse produced a variety of rRNA cleavage products, none of which corresponded to the RNA fragment obtained by incubation with α-sarcin. Both the ability to generate the RNA fragment and to inhibit protein synthesis were destroyed by heating α-sarcin at 90°C for 10 min in 3 mM KOH. When heated at 90°C for 10 min in 3 mM HCl, conditions which do not affect the activity of the ribotoxin to inhibit protein synthesis, α-sarcin still produced the small RNA fragment on incubation with ribosomes (Table 3). This supported the notion that α-sarcin (and not a contaminant) both inhibits protein synthesis and induces the cleavage of the large ribosomal RNA (it should be mentioned that this is not a trivial consideration; recent debate on the mode of action of diphtheria toxin revolves on this issue) (5). It is now generally believed that protein synthesis inhibition by the ribotoxins involves both prevention of EF-1–dependent binding of aminoacyl-tRNA and the GTP-dependent binding of EF-2 to their ribosomal sites (2). It appears that ribosomes are most susceptible to inhibition by the ribotoxins when peptidyl-tRNA is in the ribosome A site prior to translocation. In confirmation of this, it has been shown that cleavage of *E. coli* 32S rRNA with α-sarcin interferes with the binding of the elongation factors EF-Tu and EF-G (6), and footprinting studies with these factors have shown that they interact directly with the α-sarcin domain (7).

While the ribotoxins are extremely potent inhibitors of protein synthesis in cell-free extracts, they have substantially limited activity against whole cells. This is almost certainly due to the fact that they penetrate most cells very poorly. It has been found that a variety of transformed cell lines are more susceptible to the ribotoxins, and cells are also much more sensitive following infection with a virus. Carrasco and his coworkers (8,9)

Table 3   Inactivation of α-Sarcin's Ability to Inhibit Protein Synthesis

Concentration of α-sarcin	Experiment 1		Experiment 2	
	4 ng/ml (%)	1 ng/ml (%)	4 ng/ml (%)	1 ng/ml (%)
Treatment of α-sarcin				
None	87	59	59	29
Heat in TKM	76	31	80	47
Heat in 3 mM HCl	74	25	72	41
Heat in 3 mM KOH	5	4	6	4
Heat in H$_2$O	58	20	56	25

Values are expressed as percentage inhibition obtained with the preparation of α-sarcin which was preheated in the indicated buffer 10 min at 90° C, and the rate of protein synthesis measured by incorporation of [^{14}C]-leucine into acid-precipitable material as in Table 1.

have capitalized on this fact to develop a convenient and sensitive system of testing effects on protein synthesis by adding the ribotoxins at fixed times after infection of Vero cells with viruses such as Semliki Forest virus. The permeabilized cells permit the entry of the ribotoxins and detailed analyses of protein synthesis inhibition were possible. This work indicated the potential of the ribotoxins as "suicide" systems for the study of cell specificity in development.

Although a specific endonucleolytic cleavage had not previously been shown to be the mode of action of any eukaryotic protein synthesis inhibitor, it is in this way that colicin E3 inhibits bacterial protein synthesis, causing the cleavage of the 16S rRNA about 50 nucleotides from the 3′ end. The concentrations needed to obtain this cleavage are comparable to those necessary for cleavage of eukaryotic 25S RNA with α-sarcin (10,11). Colicin E3 has been reported to inactivate mouse ascites ribosomes for the EF-1–dependent binding of aminoacyl-tRNA, the same step inhibited by α-sarcin, although other steps in eukaryotic protein synthesis were not tested for their sensitivity to the colicin (12). However a highly purified preparation of colicin E3 did not inhibit the wheat germ protein–synthesizing system; it is possible that contaminating nuclease in crude colicin E3 preparations was the cause of inhibition in the ascites extract. This example serves to emphasize the importance of demonstrating the purity of microbial products before their use in sensitive cell-free systems. It should be mentioned that subsequent studies on the ribotoxins has confirmed their highly specific ribonucleolytic activity (13).

There are several other eukaryotic protein synthesis inhibitors which are known to act catalytically. These include the plant protein toxins ricin and abrin, which inhibit the GTP-dependent binding of EF-2 to eukaryotic ribosomes; pokeweed antiviral protein (PAP) (14), and the well-known diphtheria and *Pseudomonas* toxins (15). The latter two inhibitors do not act on the ribosome per se, but inactivate unbound elongation factor II by ADP ribosylation. Pokeweed antiviral protein has a mechanism of inhibitory action very similar to that proposed for α-sarcin. It prevents aminoacyl tRNA binding to the ribosome and inhibits elongation factor 1–dependent GTPase activity. The properties of such inhibitors are discussed in more detail elsewhere in this volume.

III. REDISCOVERY OF THE *ASPERGILLUS* PROTEIN TOXINS

Following the initial mode of action studies described above, which left a number of important questions unresolved, little work was carried out on α-sarcin and its close relatives (except certain protein sequence determination studies) until Ira Wool and his collaborators at the University of Chicago began their definitive work on the structure, function, and applications of α-sarcin. In summary, in a remarkable series of papers (13,24,31) these authors showed that:

1. α-Sarcin and the other ribotoxins are highly specific ribonucleases acting on the RNA of the eukaryotic large ribosomal subunit. Cleavage of one single phosphodiesterase bond accounts for the cytotoxic effects of the ribotoxins.
2. The site of cleavage was identified and shown to occur between a G and an A residue in a 15-base sequence which is highly conserved among the large subunit rRNAs of all living species (the α-sarcin domain, according to Wool) (Figures 4 and 5).
3. α-Sarcin is a potent ribonuclease that cleaves naked RNA molecules at all phosphodiester bonds 3′ to the purine bases. This activity has been used to advantage in employing α-sarcin for footprinting analyses of protein/RNA interactions.
4. α-Sarcin, restrictocin, and mitogillin show sequence similarities with RNAse U_2 (Table 4). However, only the ribotoxins demonstrate spe-

		ric. α-s.
Rat	AAUCCUGCUC	**AGUACGAGAGGAAC** CGCAGGUUCA
X. laevis	AAUCCUGCUC	**AGUACGAGAGGAAC** CGCAGGUUCA
Yeast	AAUUGAACUU	**AGUACGAGAGGAAC** AGUUCAUUCG
E.coli	GGCUGCUCCU	**AGUACGAGAGGAcC** GGAGUGGACG
Z. mays chloro.	ACCUUUCACU	**AGUACGAGAGGAcC** GGGAAGGACG
Yeast mito.	UCUCAUUAUU	**AGUACGcaAGGAcC** AUAAUGAATC
Mouse mito.	GAUUUCUCCC	**AGUACGA-AaGgAC** AAGAGAAAUA
Human mito.	AAUUCCUCCU	**uGUACGA-AaGgAC** AAGAGAAAUA

Figure 4   Comparison of the nucleotide sequence at the α-sarcin site of rat 28S rRNA with analogous regions from other cytoplasmic and organelle ribosomes. The bold capital letters designate bases that are identical to the rat rRNA, and the bold letters indicate the purine-rich segment of 14 nucleotides that is nearly universal. The arrows show the site of cleavage by α-sarcin and the adenine removed by ricin.

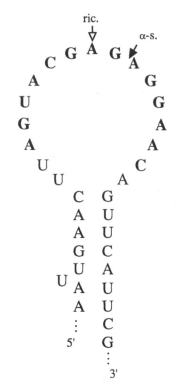

Figure 5 Structure of the α-sarcin loop from yeast 23S rRNA. The bold letters indicate the purine-rich segment of 14 nucleotides that is nearly universal. The arrows show the phosphodiester bond between G3025 and A3026 cleaved by α-sarcin and the A3024 removed by ricin.

cific 28S RNA cleavage on complete ribosomes; the structural differences responsible for this are not presently known. It is clearly of interest to understand how the ribotoxins bind (and where) on the large ribosomal subunit and what specific protein/RNA interactions are involved in the cleavage of a single phosphodiester bond.

5. α-Sarcin has nucleolytic activity on DNA but this is not efficient and requires specific buffer conditions.

6. Finally, in a series of typical ''Ira Wool'' experiments, the structure of the α-sarcin domain was analyzed in detail. The results have been reviewed recently and it is not necessary to repeat them here (16). Suffice it to say that, the cleavage of rat liver robosomes, next to G-4325 in the 28S rRNA, appears to require recognition of a *structure* within the RNA, rather than simply a conserved sequence of nucleo-

Table 4 Comparison of Amino Acid Sequences of Guanyl Ribonucleases and Ribotoxins

```
Barnase   QVINTFDGVADYLQTYHKL.PDNYITKSEAQA....LGWVASKGNLADVAPGK...SIGG
Binase    AVINTFDGVADYLIRYKRL.PNDYITKS..QASA..LGWVASKGDLAEVAPGK...SIGG
Sa                    DV..SG..TVCLS...ALPPEATDTLN.LIA.SD...GPFP..YSQ
Se            E...APCGDT......SGFEQVRLA..DLPPEATDTYE.LIEKG....GPYP..YPE
F1            E.SATTCGST....NYS.ASQVRAA..ANAAC.QYYQND.DSAGSTT.......YPH
F11           E..ASTCGST....P.YS.ASQVRAA..ANAAC.QYYQSD.DTAGSTT.......YPH
N1            ACMYICGSV....CYS.SSAISAA...LNKGYSYYE.DGATAGS........SSYPH
Ms            ESCEYTCGST....CYW.SSDVSAA...KAKGYSLYES.GDTI.DD.........YPH
C2            DCDYTCGSH....CYS.ASAVSDA...QSAGYQL.ESAG...QS.V..GR..SRYPH
T1            ACDYTCGSN....CYS.SSDVSTA...QAAGYQLHE.DGETVGSNS.......YPH
Pb            ACAATCGTV...CYT.SSAISSA...QAAGYNLYSTN.DDV.SN.........YPH
U2                C.NI...PESTNCGGN......VYSNDDINT.AIQGALDDVAR.PDGDNYPH
aS            AVTWTCLNDQKNPK.TNKYETKRLLYNQNKAESNSHHAP...LSD...GKTGSSYPH
rest          A.TWTCINQQLNPK.TNKWEDKRLLYSQAKAESNSHHAP...LSD...GKTGSSYPH
consensus ...ac.ytcgs....pcys.ss.vsaa...qaagy.lye.agdtalsdv..gk..ssyph
                10        20        30        40        50        60

Barnase   D..IFSNR.EGK.LPGKSGRT.....................W.READINYTSGFRN
Binase    D..VFSNR.EGR.LPSAGSRT.....................W.READINYVSGFRN
Sa        DGVVFQNR.ESV.LPTQSYGCC..................YYHEYTVI.TPGAR.
Se        DGTVFENR.EGI.LPDCAEG...................YYHEYTVK.TPSGD.
F1        T...YNNY.EGFDFPVDGP.....................YQEFPIKS.GG.VY
F11       T...YNNY.EGFDFAVNGP.....................YQEFPIR.TGG.VY
N1        R...YNNY.EGFDFPTAKP.....................WY.EFPILS.SGRVY
Ms        G...YHDY.EGFDFPVGTS.....................YY.EYPIMS.DYDVY
C2        Q...YRNY.EGFNFPVSGN.....................YY.EWPILS.SGSTY
T1        K...YNNY.EGFDFSVSSP.....................YY.EWPILS.SGDVY
Pb        E...YHNYDEGFDFPVSGT.....................YY.EFPILK.SGKVY
U2        ...QY..YDEASD.......QITLC...CG.PGSWS..........EFPLVY.NGP.Y
aS        .W.FTNGYD.GDGKLPKGRTPIKFGKSDCDRPPKHSKDGNGKTDHYLLEFPTFP.DGHDY
rest      .W.FTNGYD.GNGKLIKGRTPIKFGKADCDRPPKHSQNGMGKDDHYLLEFPTFP.DGHDY
consensus d...ynny.egfdfpv..p.....................yy.Efpils.sg.vy
              70        80        90        100       110       120

Barnase   ..........SDRILYSSDW.LIY..KTTDHYQ.....TFTKIR
Binase    ..........ADRLVYSSDW.LIY..KTTDHY.A....TFTRIR
Sa        TR........GTRRIICGEATQEDY..YTGDHY.A....TESLIDQTC
Se        DR........GARREVVGDGG.E.YF.YTEDHYESFRLCTIVN
F1        TGGS......PGADRVVINTNC.E.Y.AGAITHTGA.SGNNFVGCSGTN
F11       SGGS......PGADRVIINTSC..QY.AGAITHTGA.SGNNFVGCSNST
N1        TGGS......PGADRVIFDSHGN...LDMLITHNGA.SGNNEVACN
Ms        TGGS......PGADRVIFDGDD.EC.LAGVITHTGAAGGDDVFACSSSC
C2        NGGG......PGADRVVFNDND.E..LAGLITHTGA.SGDGFVACY
T1        SGGS......PGADRVVFNENN..Q.LAGVITHTGA.SGNNFVECT
Pb        TGSS......PGADRVIFNDDD.E..LAGVITHTGA.SGNNFVACT
U2        .YSSRDNYVSPGPDRVIYQTNTGE..FCATVTHTGAASYDGFTQCS
aS        KFDSKKPKENPGPARVIY.TYPNKV.FCGGITAHTKENQGE.LKLCSH
rest      KFDSKKPKEDPGPARVIY.TYPNKV.FCGIVAHQRGNQGD.LRLCSH
consensus tggs......pgadRvi.nt...e.ylagtitHtga.sgnnfvacs...
              130       140       150       160       170
```

Ribonucleases: Barnase (*Bacillus amyloliquefaciens*); Binase (*Bacillus intermedius*); Sa (*Streptomyces auerofaciens*); Se (*Saccharopolyspora erythreus*); F1 (*Fusarium moniliforme*); N1 (*Neurospora crassa*); Ms (*Aspergillus phoenicis*); C2 (*Aspergillus clavatus*); T1 (*Aspergillus oryziae*); Pb (*Penicillium brevicompactum*); U2 (*Ustillago sphaerogena*); ribotoxins: aS, α-sarcin (*Aspergillus giganteus*); rest, restrictocin (*Aspergillus restrictus*). Residues that are homologous in eight or more sequences are boxed.

tides. In addition, ribosomes are sensitive to the ribotoxins only when peptidyl-tRNA is in the ribosome A site before translocation.

IV. TOXIN SUBSTRATES

Elongation factors apparently interact with the stem-loop structure of the 60S ribosomal subunit that is the enzymatic target of both the fungal toxins (α-sarcin, mitogillin, restrictocin) and plant and bacterial toxins (ricin, gelonin, abrin, *Shiga* toxin, among others). Further, diphtheria and *Pseudomonas* exotoxin modify EF-2 by ADP-ribosylation of the dipthamide residue so that binding to the ribosome is blocked. This key stem-loop rRNA structure may have a role in mediating transfer of tRNAs from A site to P site on the ribosome. Recently, Arthur Pardi and associates have defined a solution structure to tetranucleotide GNRA loops which reveal an adenine or cytosine base stacked on top of the loop (32). To form a GNRA loop, the conserved stem loop in the rRNA must form a 4-base substem composed of

$$
\begin{array}{cccc}
\text{C} & \text{A} & \text{U} & \text{G} \\
| & | & | & | \\
\text{G} & \text{G} & \text{A} & \text{A}
\end{array}
$$

Experiments done previously by Endo et al. (33) would support the conservation of these nucleotides in addition to the GAGA. Further experiments should better define the conserved stem-loop structure and cloned expressed toxins should help identify the mechanism of interaction of toxins with the ribosome and how that inhibits protein synthesis.

V. MOLECULAR STUDIES OF THE FUNGAL PROTEIN TOXINS

Since the fungal ribotoxins are secreted in substantial amounts from their *Aspergillus* hosts, it has proved relatively easy to obtain quantities sufficient for structure analysis and the complete amino acid sequences for α-sarcin, restrictocin, and mitogillin have been determined. It was originally thought that α-sarcin contained an unusual amino acid, sarcinine, but this is no longer tenable, based on amino acid and nucleic acid sequencing.

Furthermore, in the interest of understanding certain fundamental aspects of the synthesis and function of α-sarcin, restrictocin, etc., efforts have been made to isolate and characterize the genes for these proteins. The restrictocin gene has been cloned from *Aspergillus restrictus* by probing a bacteriophage lambda library with synthetic oligonucleotides derived from the amino acid sequence. Degenerate probes were used and several positive signals were obtained; the entire restrictocin gene (*res*) complete with putative 5′ and 3′ nontranslated sequences was obtained on a single

*Sal*I fragment (17). Attempts to isolate the gene using expression or epitope libraries were not successful. The gene and flanking regions were sequenced and potential regulatory functions identified as indicated in Figure 6. The *res* gene contains a single intron of 52 bp which essentially separates the structural gene sequence from the signal sequence. Other aspects of this gene sequence are indicated. It should be mentioned that restrictocin and mitogillin differ, apparently, by only a single amino acid. They are produced by different strains of *A. restrictus*, and on probing several different restriction enzyme digests of *A. restrictus* DNA with different fragments of the cloned *res* gene, evidence for only one sequence could be obtained. Therefore, it is highly probable that restrictocin and mitogillin are two different alleles of the same gene and they may be members of a large family of related proteins.

An oligonucleotide probe derived from the restrictocin gene was used to examine the presence of the α-sarcin gene in DNA preparations of *Aspergillus giganteus* (the α-sarcin producer). The results are shown in Figure 7; the α-sarcin gene is organized in a different manner than the restrictocin and mitogillin genes. The less intense hybridization signal obtained with *A. giganteus* compared to *A. restrictus* fragments using a restrictocin gene probe is consistent with the limited degree of homology between the two genes. However, it is evident that by using the restrictocin probe it will be possible to clone the α-sarcin gene of *A. giganteus* and analyze its structure in detail.

A cDNA clone for α-sarcin has been isolated and sequenced (18), and Figure 6 shows a comparison of the two ribotoxins, α-sarcin and restrictocin, and their leader sequences. There is a high degree of similarity between the two nucleic acid sequences. Since the two toxins are produced in limited quantities in exponential phase, but in larger amounts in stationary phase, it will be of interest to examine the regulatory sequences involved in the control of expression of the two genes. The availability of the complete sequences of the α-sarcin and restrictocin genes will permit the production of specific gene fusions that may be more effective immunotoxins. In particular, it will be possible to carry out site-directed mutagenesis of the two ribotoxin genes to provide information on the functions necessary for the specific interaction of the toxins with eukaryotic ribosomes and to understand the mechanisms of rRNA cleavage. Concomitant with these genetic studies, the crystallization and analysis of three-dimensional structure of one of these toxins is absolutely imperative, and supplies of pure protein should no longer be a negative consideration.

Promising preliminary results have been obtained by Martinez and Smith (personal communication) in work with mitogillin. It will be interesting to study the structural interaction between these 17 kd toxins and

```
r   GCA ACA GCG AGG CAC TTT CTA ACT AGC TGT GTC ACC TTG AAA ATG CTC AAT GTC TCA CCC  60
r   GCA AGG GTC AGC TAG TAC GCT TTT CAG CTC TTT ACA TGT TGC AAT CCA GGA TCT AGT CTT  120
r   CGG TGT TGA TAT GCA TGA AGC TTG GGT TGA GTA TCC ACT CCA TCA GCT TAC AGT CAT CTC  180
r   TTC TGT CTC CAT TAG GCA ACT GGG GCA GAA CAT CCA ATT TCT CTT AGC TTA CAA TGT TAC  240
r   ATT GAC TCC AAA GCT CGC ACT CCA CTT GTC AGT TTG AGG CAT GAT CTA CCC TGT GAG GGG  300
                                                                 ------->         <------
r   CGT CTT ATG GCC C̲C̲A̲ A̲T̲A̲ ATT ACC TCT ACA GAC AGA GGT ACT CAC C̲C̲A̲ A̲T̲A̲ TCA TCT GAG  360
r   ATG TGC AGA GGA CAA AAG ACG CAT CTT CGC GA̲C̲ A̲A̲T̲ CCC ATG TCC TGG ACG AAT AAT GTC  420
r   AGG T̲A̲T̲ A̲A̲A̲ AGC GGT TGA ATT CTC TCG TTT GTC GCA ACC AAA TAG GAA GAC ATC GTC ACA  480
α                                                                               cC ACA
            1 met val ala ile lys asn leu phe leu leu ala ala thr
r   ATT GCC CTG ACT ACG TCC AAG ATG GTT GCA ATC AAA AAT CTT TTC CTG CTG GCT GCC ACA  540
α   Agc atC tcc atc tca TtC AAa ATG GTT GCA ATC AAA AAc CTT gTC CTG gTG GCc ctC ACg
                                                        val ... val ... leu
         14 ala val ser val leu ala ala pro ser pro leu asp ala arg ala ==============
r   GCC GTG TCT GTT CTA GCT GCT CCC TCG CCC CTC GAC GCT CGT GCG GTA AGA GTC ACA TCG  600
α   GCC GTG aCc Gcc CTt GCa Gtg CCC TCG CCt CTC GAg GCg CGc GCG GTg
         ... ... thr ala ... ... val ... ... ... glu ... ... ... val
         29 ============ intron ============================ thr trp thr cys ile asn gln gln
r   AAA GGC CTT CGA AAG GAT GAC TGA CAT GACCCCC TAG ACC TGG ACA TGC ATC AAC CAA CAG  661
α                       "" "" ""                    ACC TGG ACC TGC tTg AAC gAc CAG
                                                                    leu ... asp ...
         37 leu asn pro lys thr asn lys trp glu asp lys arg leu leu tyr ser gln ala lys ala
r   CTG AAT CCC AAG ACA AAC AAA TGG GAA GAC AAG CGG CTT CTA TAC AGT CAA GCC AAA GCC  721
α   aaG AAc CCC AAG ACC AAc AAg Tat GAg acC AAa CGc CTc CTc TAC AAc CAg aaC AAg GCC
         lys ... ... ... ... ... ... tyr ... thr ... ... ... ... ... asn ... asn ...
         57 glu ser asn ser his his ala pro leu ser asp gly lys thr gly ser ser tyr pro his
r   GAA AGC AAC TCC CAC CAC GCA CCT CTT TCC GAC GGC AAG ACC GGT AGC AGC TAC CCG CAC  781
α   GAg AGC AAC TCg CAC CAt GCg CCT CTc TCC GAC GGC AAG ACC GGg AGC AGC TAt CCt CAC
         ... ... ... ... ... ... ... ... ... ... ... ... ... ... ... ... ... ... ...
         77 trp phe thr asn gly tyr asp gly asn gly lys leu ile lys gly arg thr pro ile lys
r   TGG TTC ACT AAC GGC TAC GAC GGG AAT GGC AAG CTC ATC AAG GGT CGC ACG CCC ATC AAA  841
α   TGG TTC ACC AAC GGt TAt GAt GGc gAT GGa AAG CTC ccC AAG GGc CGC ACG CCC ATC AAg
         ... ... ... ... ... ... asp ... pro ... ... ... ... ... ... ... ... ... ...
         97 phe gly lys ala asp cys asp arg pro pro lys his ser gln asn gly met gly lys asp
r   TTC GGA AAA GCC GAC TGT GAC CGT CCC CCG AAG CAC AGC CAG AAC GGC ATG GGC AAG GAT  901
α   TTC GGA AAA tCC GAC TGT GAC CGT CCt CCc AAG CAC AGC aAg gAC GGa Aac GGC AAG acT
         ... ... ... ser ... ... ... ... ... ... ... ... ... lys asp ... asn ... thr
         117 asp his tyr leu leu glu phe pro thr phe pro asp gly his asp tyr lys phe asp ser
r   GAC CAC TAC CTG CTG GAG TTC CCG ACT TTT CCA GAT GGC CAC GAC TAT AAG TTT GAC TCG  961
α   GAt CAC TAC CTG CTG GAG TTC CCa ACc TTc CCt GAT GGC CAt GAC TAc AAG TTT GAt TCG
         ... ... ... ... ... ... ... ... ... ... ... ... ... ... ... ... ... ... ...
         137 lys lys pro lys glu asp pro gly pro ala arg val ile tyr thr tyr pro asn lys val
r   AAG AAA CCC AAG GAA GAC CCG GGC CCA GCG AGG GTC ATC TAT ACT TAT CCC AAC AAG GTG  1021
α   AAG AAG CCC AAG GAA aAt CCt GGC CCg GCG cGG GTC ATC TAc ACc TAT CCt AAC AAG GTG
         ... ... ... ... ... asn ... ... ... ... ... ... ... ... ... ... ... ... ...
         157 phe cys gly ile val ala his gln arg gly asn gln gly asp leu arg leu cys ser his
r   TTT TGC GGC ATT GTG GCC CAT CAG CGG GGG AAT CAG GGA GAC TTG AGA CTG TGT TCT CAT  1081
α   TTc TGt GGt ATc aTt GCt CAT act aaG GaG GAa cTt AAG CTc TGc TCT CAT
         ... ... ... ile ... ... thr lys glu ... ... glu ... lys ... ... ... ... ...
         AMB
r   TAG TTA TGT GGG TAT TAT GCC TTA GGA GGG CGT TGC TGC CAC TCC TTT CTT CGA CCA CAA  1141
α   TAG aag gGc ttG cAg aAg aag aaA GGt GGt tcg aGg ccC ttt Ttt Tgg CTg CGg ttg atg
         AMB
r   TAT GCG CCT TAT TTA CTG ATC GGG TAG ACT TGG GCT CCA GTG GTA TCT GGA CAA TAT GCT  1201
α   cta aat CGc acc Tgt tct ggt aca gtG caa gGt Gaa tgt tat cTA gtc cac Cct aga ttT
r   TGT TTA TCA TTA TTT GAG ATG CAA ATT GCC GCA GAC ATT CGC TCC GCT CAC TGT TAG ATA  1261
α   cta gcT Ttg aTA cca cAc gTt tgg Aac tat tTA tAc ATc TgG
r   TTT GAA TAG ACA TTT TAA GCC CTT tCT AAG TAT CAG TTT GTC TAG AGG TAA CAT TGC CTT  1321
r   CAT ATT TGC AAC CGA GAT CAA GAA GCT TCT AGT AGT CAG TAG ACA GGA CAG TCG AGC ACT  1381
```

Figure 6 Alignment of the nucleotide sequences of the *res* gene (r) and cDNA for α-sarcin (α). Lower case letters denote bases not homologous with the restrictocin sequences. The restrictocin intron is underlined: (=). The lariat signal is indicated: ("). The proposed TATA box and the three CAAT boxes are underlined, one CAAT box is flanked by inverted repeats (arrows). The deduced amino acid sequence of restrictocin is shown over, and the amino-acid differences in α-sarcin under the nucleotide sequences (Dots: ⋯ indicate identical amino acids). The predicted signal peptide is underlined (17).

250

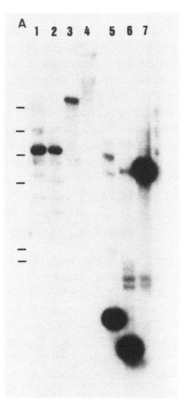

Figure 7 Southern analysis of the restrictocin and of the α-sarcin genes. Genomic DNA from *A. giganteus* (lanes 1–3), *A. nidulans* (lane 4), or *A. restrictus* (lanes 5–7) was digested with *Hin*dIII (Lanes 1,4,5), *Eco*RI (lanes 3,7), and *Hin*dIII plus *Eco*RI (lanes 2,6). The probe used is an oligonucleotide of 26 bp derived from the gene *res* (position 671 to 696, Figure 6). *Hin*dIII fragments from λ phage DNA as size markers are shown on the left.

the 60S ribosomal subunit, which is of the order of 200 times larger; do the ribotoxins have specific domains for ribosome binding and nuclease activity? Where does the specificity lie? For the latter, we know already that there is a region with strong homology to a known ribonuclease.

VI. INTRONS IN A TOXIN GENE

Interrupted genes were discovered in 1977 in human adenovirus DNA and subsequently in ovalbumin and β-globin genes. Sequences present in the DNA but omitted from the mRNA are called intron sequences and those present in the mRNA are called exon sequences. Introns are not present in prokaryotes (hence, their absence from diphtheria toxin, *Pseudomonas* exotoxin, or Shiga toxin genes), and have not been reported for a number of plant toxin genes (ricin toxin, abrin toxin, trichosanthin, or momorcharin). Hence, their presence in the fungal toxin restrictocin gene was unexpected. The primary RNA transcript is a faithful copy of the gene containing intron and exon sequences, and the intron sequences are removed from the mRNA to produce a transcript coding for a protein. RNA splicing occurs in the cell nucleus, and the mRNA is exported to the cytoplasm when processing is complete. Small nuclear ribonucleoproteins (SnRNPs) complex to the primary RNA transcript by recognizing specific sequences at the 5' and 3' cleavage sites:

$$\frac{C}{A} \text{ AG GT } \frac{A}{G} \text{ AGT} \quad \text{and} \quad \frac{T}{C}\frac{T}{C}\frac{T}{C}\frac{T}{C}\frac{T}{C}\frac{T}{C}\frac{T}{C}\frac{T}{C}\frac{T}{C}\frac{T}{C}\frac{T}{C} \text{ N } \frac{C}{T} \text{ AG } \frac{G}{A}$$

The introns can be from 50–10,000 nucleotides long and only the end sequences (needed for splicing) are conserved. The bound SnRNPs form a spliceosome. First, a special A nucleotide in the 3' splice site attacks the 5' splice site, causing cleavage and joining to the A nucleotide, forming a branched nucleotide. Then the 3'-OH from the first exon cleaves the RNA at the 3' splice site joining the exons. The intron sequence is released as a lariat. The restrictocin donor and acceptor intron splice sites are similar but not identical to the consensus sequences for higher eukaryotes: G C G G T A A G A and C C C C C T A G A. Interestingly, the exons for restrictocin are different functional domains—leader peptide and toxin.

VII. HOW CAN MICROBES MAKE RIBOTOXINS AND SURVIVE?

The production of ribotoxins as potent as α-sarcin and restrictocin by aspergilli raises questions about the ability of the microbes to protect themselves during production. This problem has been addressed for other toxic metabolities such an antibiotics, for which it is known that several different mechanisms can operate to prevent the producing microbe from committing suicide (19).

In the case of α-sarcin, Hobden has shown that the translation system of the producing organism *A. giganteus* is sensitive (20); we can assume that the same situation pertains for restrictocin. Thus, the cell would die if one molecule of the active toxin was released into the cytoplasm. There are several models for a self-protection system in this case. The first is that the ribotoxins are inactive when in their precursor (nonsecreted) form; this assumes that the pretoxins are folded in a conformation which suppresses all ribonucleolytic activity, and that the molecules, produced on membrane-bound ribosomes, are secreted immediately through the Golgi system, being matured during the process. All three ribotoxins contain disulfide bridges, which are probably formed during maturation and secretion from the cell. Once secreted, the toxin cannot reenter the cell; it is known that the fungi are resistant to high concentrations of exogenous toxin. Another possibility is that the producing organism coproduces an inhibitor of the nuclease activity of the ribotoxin; an immunity protein. There is a well-studied example of such a combination; the extracellular ribonuclease of *Bacillus amyloliquefaciens*, barnase, is specifically inhibited by a stochiometric interaction with another small protein, barstar (21). These two proteins must be produced coincidentally during growth; in the absence of cytoplasmic bastar, the gene for barnase has been cloned only when converted into an inactivated form (by point mutation or transposon insertion). The production of an "immunity" protein of this type is reminiscent of colicin immunity and may be a general phenomenon for such toxic molecules. Having presented the reader with these various possibilities, we can state that production of the ribotoxins is not accompanied with an immunity protein; they are produced apparently in an inactive preprotein form which is activated on processing through the Golgi. This conclusion was drawn from the finding that when the isolated gene for restrictocin was cloned onto an *Aspergillus* vector and introduced into *A. nidulans*, restrictocin was secreted in active form and there no lethal effect on the new host was apparent.

VIII. USES FOR THE *ASPERGILLUS* PROTEIN TOXINS

There are several potential applications for ribosomal ribonucleases such as α-sarcin and restrictocin; we have mentioned their practicality in footprinting studies as described by Wool's group. Now that the genes have been cloned, not only should it be possible to produce the corresponding toxins in quantity, employing appropriate high-level expression systems, but it should be possible to obtain constructions in which the synthesis of the toxin is controlled by a tightly regulated promoter. Such a suicide-cassette would be useful in studying the role and function of specific cells in multicellular systems. Activating the expression of the regulated toxin gene under appropriate conditions and at specific times would block protein synthesis completely in specific subsets of cells or tissues during studies of cell development.

Another application of the fungal protein toxins would be to employ them as components of immunotoxins; given their great specificity of action and their high degree of resistance to inactivation by heat or acid treatment, ribotoxin derivatives targeted at discrete cell types would seem to offer a promising therapeutic development. They are extremely cytotoxic once inside the cell cytoplasm. This approach has already been investigated by Conde and collaborators (22), who demonstrated that immunoconjugates of restrictocin with cell-specific monoclonal antibodies were effective antitumor agents in an animal model.

Studies of activity, although encouraging, were limited; for example, it is known that increased cation concentrations or the presence of certain ionophores may potentiate the activity of immunotoxins. With the cloned ribotoxin genes now available, it will be possible to make appropriate recombinants that provide translational fusions between the toxin and appropriate antibodies or antibody heavy chain fragments (23,23a). Alternatively, any appropriate ligand function could be genetically fused to the toxin (or part of it), an approach which should provide more flexibility in the design of targeted therapeutic agents based on the ribotoxins. Better knowledge of active sites or motifs in the ribotoxin proteins would be of great value in this respect.

As mentioned above, the potential for studying structure-function relationships in the ribotoxins is now feasible and three-dimensional structure analyses and epitope mapping would be of considerable value at this stage of development.

IX. CONCLUSIONS

The ribotoxins α-sarcin, restrictocin, and mitogillin form a specific class of ribosome-inactivating protein (RIPs), they differ in structure and mode of action from the plant RIPs. The plant RIPs exist as single chain (type I) or double chain (type II); the latter are substantially more potent in vivo since one of the chains (B chain) is a cell surface binding protein which targets the toxin. Although formally analogous to the type I RIPs of plants, the ribotoxins are smaller: Restrictocin and α-sarcin have molecular weights of ~17,000, whereas the plant type I RIPs have molecular weights of the order of 30,000. The type I RIPs block eukaryotic protein synthesis, like the ribotoxins, by catalytic inactivation of the 60S ribosomal subunit. However, unlike α-sarcin and restrictocin, the plant RIPs modify but do not cleave ribosomal RNA. It has been shown by Endo and his collaborators (24) and adequately confirmed (13) that the plant RIPs are specific N-glycosidases that release a nucleic acid base from the phosphodiester backbone of rRNA. Subsequently, the backbone can be cleaved by the addition of an aniline in vitro, but inhibition of protein synthesis is due to loss of a single base without cleavage of the rRNA. As we have described here, the ribotoxins have a specific phosphodiesterase activity; the ribosome structure remains essentially intact after α-sarcin or restrictocin treatment, but the larger rRNA is in two pieces inside the structure. Comparison of the sequences of ribonucleases from a variety of sources with the ribotoxins and RIPs shows some common domains; it is tempting to speculate that additional sequences found in the translation inhibitors represent regions that determine substrate binding specificity (Table 4).

The mechanism of action of the plant RIPs at the level of protein synthesis differs from that of the fungal toxins. As mentioned previously, the latter interfere with both binding of EF-1 and EF-2 to eukaryotic ribosomes. The plant toxins (e.g., ricin) act by blocking the binding of EF-2; however, it must be stated that detailed molecular studies of the inhibition reactions remain to be performed. The target base in rRNA of the two types of inhibitors also differs; in yeast 26S RNA, α-sarcin and restrictocin cleave between G3025 and A3026, whereas the plant RIPs remove the base from A3024 (Figures 4 and 5). The specificity of these toxins in identifying one phosphodiester bond in 7000 in the ribosome is nothing less than striking and is surely worthy of further investigation! It is intriguing that both the fungal and plant toxins act by modifying different sites within the same universally conserved sequence (pro- and eukaryotes and archaebacteria) of large eukaryotic rRNAs; the structure of the ribosome must play a role in the resistance of prokaryotes. The value of these powerful inhibitors as therapeutic agents has yet to be demonstrated, but if at all, it is likely to come from incorporation into some form of targeted hybrid. Their usefulness as tools in analyzing translation and protein/nucleic interactions is evident.

As mentioned previously, the mode of action of the fungal protein toxins has clear analogies with the activity of colicin E3 (a bacterial ribonuclease) and similar inhibitors. A more distant, but nonetheless interesting, comparison can be made between the ribotoxins and angiogenin, potent stimulator of blood vessel formation. Angiogenin, a protein of molecular weight 14.4 kd has a distinct ribonuclease activity (25) and possesses 35% sequence identity with ribonuclease A. Angiogenin at low concentrations has been shown to inhibit protein synthesis in rabbit reticulocyte lysates (26) and this inhibition has been shown to be due to specific cleavage of ribosomal RNA! The action of angiogenin differs from the ribotoxins in that its inhibited target is the 40S ribosomal subunit, in which the rRNA is cleaved. Studies of the ribonucleolytic activity of angiogenin indicate that both isolated 18S and 28S RNA are digested.

At present, there is no evidence that α-sarcin and restrictocin have any essential function in their producing hosts. However, the formal similarity to the activity of angiogenin suggests that the ribotoxins have a range of biological effects. The roles of cellular ribonucleases have been reconsidered (27,28); proteins with this type of activity are usually considered to be scavenging or processing enzymes required to provide appropriate precursors for cell metabolism. However, the fact that ribonucleolytic activity and sequence homology with "typical" ribonucleases can be found associated with proteins that have diverse biological roles not (apparently) related to this activity suggests that such proteins may play specific roles in cell function. Given our current notions on the role of RNA in evolution, it is conceivable that proteins or their domains implicated in the processing of RNA molecules were among the earliest polypeptides to evolve.

One final point is worthy of mention with respect to the ribotoxins; there are a number of life-threatening fungal infections which are considerably enhanced in immunocompromised patients. *Aspergillus fumigatus*, in particular, is responsible for high mortality in systemic fungal infections and it is not known what, if any, virulence factors may contribute to the pathogenicity of fungal diseases. However, restrictocin (or its analogs) have been detected in the urine of patients suffering from severe aspergillosis and the question of the role of ribotoxins as factors of virulence in these infections has become significant (29,30). In addition, restrictocin has been shown to be an allergen in individuals hypersensitive to *Aspergillus funigatus* (R. Crameri, personal communication). The availability of the cloned restrictocin and α-sarcin genes will permit detail molecular analyses of their role in the pathogenicity of fungal infections of man and animals.

REFERENCES

1. Olson, B. H., and Goerner, G. L. Alpha-sarcin, a new antitumor agent I. Isolation, purification, chemical composition, and the identity of a new amino acid. Appl. Microbiol., *13:* 314–321, 1965.
2. Schindler, D. G., and Davies, J. E. Specific cleavage of ribosomal RNA caused by alpha-sarcin. Nucl. Acids Res., *4:* 1097–1110, 1977.
3. Cundliffe, E., Cannon, M., and Davies, Mechanism of inhibition of eukaryotic protein synthesis by trichothecene fungal toxins. Proc. Natl. Acad. Sci. U.S.A., *71:* 30–34, 1974.
4. Schindler, D. G. Use of Protein Synthesis Inhibitors to Study the Biochemistry and Genetics of the Yeast Ribosome. Ph.D. Thesis, University of Wisconsin—Madison, 1976.
5. Bodley, J. W., Johnson, V. G., and Wilson, B. A. Does diphteria toxin have nuclease activity? Science, *250:* 832–836, 1990.
6. Hausner, T. P., Atmadja, J., and Nierhaus, K. H. Evidence that the G2661 region of 23S rRNA is located at the ribosomal binding sites of both elongation factors. Biochimie, *69:* 911–923, 1987.

7. Moazed, D., Robertson, J. M., and Noller, H. F. Interaction of elongation factors EF-Tu with a conserved loop in 23S RNA. Nature, *334:* 362–364, 1988.

8. Munoz, A., Castrillo, J. L., and Carrasco, L. Modification of membrane permeability during Semliki Forest Virus infection. Virology, *146:* 203–212, 1985.

9. Otero, M. J., and Carrasco, L. Proteins are cointernalized with virion particles during early infection. Virology, *160:* 75–80, 1987.

10. Boon, T. Inactivation of ribosomes *in vitro* by colicin E3 and its mechanism of action. Proc. Natl. Acad. Sci. U.S.A., *69:* 549–552, 1972.

11. Bowman, C. M., Sidikaro, J., and Nomura, M. Specific inactivation of ribosomes by colicin E3 *in vitro* and mechanism of immunity in colicinogenic cells. Nature New Biol., *234:* 133–137, 1971.

12. Turnowsky, F., Drews, J., Eich, F., and Hogenauer, G. In vitro inactivation of ascites ribosomes by colicin E3. Biochem. Biophys. Res. Commun., *52:* 327–334, 1973.

13. Endo, Y., Huber, P. W., and Wool, I. G. The ribonuclease activity of the cytotoxin α-sarcin. J. Biol. Chem., 258: 2662–2667, 1983.

14. Obrig, T. G., Irvin, J. D., and Hardesty, B. The effect of an antiviral peptide on ribosomal reactions of the peptide elongation enzymes, EF-I and EF-II. Arch. Biochem. Biophys., *155:* 278–289, 1973.

15. Chen, J. Y., and Bodley, J. W. Biosynthesis of diphthamine in *Saccharomyces cerevisiae.* J. Biol. Chem., *263:* 11692–11696, 1988.

16. Wool, I. G., Endo, Y., Chan, Y. L., and Glück, A. Structure, function, and evolution of mammalian ribosomes. *In:* W. E. Hill, A. Dahlberg, R. A. Garrett, P. B. Moore, D. Schlessinger, and J. R. Warner, (eds.), The Ribosome Structure, Function, & Evolution, pp. 203–214. American Society of Microbiology, Washington, D.C., 1990.

17. Lamy, B., and Davies, J. Isolation and nucleotide sequence of the *Aspergillus restrictus* gene coding for the ribonucleolytic toxin restrictocin and its expression in *Aspergillus nidulans.* Nucl. Acids Res., *19:* 1001–1005, 1991.

18. Oka, T., Natori, Y., Tanaka, S., Tsurugi, K., and Endo, Y. Complete nucleotide sequence of cDNA for the cytotoxin alpha-sarcin. Nucl. Acids Res., *18:* 1897, 1990.

19. Cundliffe, E. How antibiotic producing organisms avoid suicide. Ann. Rev. Microbiol., *43:* 207–233, 1989.

20. Hobden, A. N. Ph.D. dissertation, Thesis, University of Leicester, Leicester, England, 1978.

21. Hartley, R. W. Barnase and barnstar: Two small proteins to fold and fit together. Trends Biochem. Sci., *14:* 450–454, 1989.

22. Conde, F. P., Orlandi, R., Canevari, S., Mezzanzanica, D., Ripamonti, M., Munoz, S. M., Jorge, P., and Colnaghi, M. I. The *Aspergillus* toxin restrictocin is a suitable cytotoxic agent for generation of immunoconjugates with monoclonal antibodies directed against human carcinoma cells. Eur. J. Biochem., *178:* 795–802, 1989.

23. Orlandi, R., Canevari, S., Conde, F. P., Leoni, F., Mezzanzanica, D., Ripamonti, M., and Colnaghi, M. I. Immunoconjugate generation between the ribosome inactivating protein restrictocin and an anti-human breast carcinoma MAB. Cancer Immunol. Immunother., *26:* 114–120, 1988.

23a. Wawrzynczak, E. J., Henry, R. V., Cumber, A. J., Parnell, G. D., Derbyshire, E. J., and Ulbrich, N. Eur. J. Biochem. *196:* 203–209, 1991.

24. Endo, Y., and Wool, I. G. The site of action of α-sarcin on eukaryotic ribosomes. J. Biol. Chem., *257:* 9054–9060, 1982.

25. Shapiro, R., Riodan, J. F., and Vallee, B. L. Characteristic ribonucleolytic activity of human angiogenin. Biochemistry, *25:* 3527–3532, 1986.

26. St. Clair, D. K. S., Rybak, S. M., Riodan, J. F., and Vallee, B. L. Angiogenin abolishes cell-free protein synthesis by specific ribonucleolytic inactivation of 40S ribosomes. Biochemistry, *27:* 7263–7268, 1988.

27. Hill, C., Dodson, G., Heinemann, U., Saenger, W., Mitsui, Y., Nakamura, K., Borisov, S., Tischenko, G., Polyakov, K., and Pavlovsky, S. The structural and sequence homology of a family of microbial ribonucleases. Trends Biochem. Sci., *8:* 364–369, 1983.

28. Benner, S. A., and Allemann, R. K. The return of pancreatic ribonucleases. Trends Biochem. Sci., *14:* 396–397, 1989.

29. Arruda, L. K., Platts-Mills, A. E., Fox, J. W., and Chapman, M. D. *Aspergillus fumigatus* allergen I, a major IgE-binding protein, is a member of the mitogillin family of cytotoxins. J. Exp. Med., *172:* 1529–1532, 1990.

30. Lamy, B., Moutaouakil, M., Latge, J. P., and Davies, J. Secretion of a potential virulence factor, a fungal ribonucleotoxin, during human aspergillosis infection. Mol. Microbiol., *7:* 811–1815, 1991.

31. Chan, Y. L., Endo, Y., and Wool, I. G. The sequence of the nucleotides at the α-sarcin cleavage site in rat 28S ribosomal ribonucleic acid. J. Biol. Chem., *21:* 12768–12770, 1983.

32. Heus, H., Pardi, A. Structural features that give rise to the unusual stability of RNA hairpins containing GNRA loops. Science, *253:* 191–194, 1991.

14

An Efficient Expression System for α-Sarcin in *Escherichia coli*

Yaeta Endo*   *Yamanashi Medical College, Tamaho, Nakakoma, Yamanashi, Japan*

Tatsuzo Oka and Yuji Aoyama   *Tokushima University School of Medicine, Kuramoto-cho, Tokushima, Japan*

I. INTRODUCTION

α-Sarcin is a small, basic, cytotoxic protein (molecular weight 16,987) produced by the mold *Aspergillus giganteus* (1,2) that inhibits protein synthesis by inactivating ribosomes (3–5). The molecular basis of the inhibition is the hydrolysis of a phosphodiester bond (6) on the 3' side of G4325, which is in a single-stranded loop 459 residues from the 3' end of 28S rRNA (7,8). The cleavage site is embedded in a purine-rich single-stranded segment of 17 nucleotides that is nearly universal (7). This is one of the most strongly conserved regions of rRNA, and, indeed, the ribosomes of all the organisms that have been tested, including the producing fungus, are sensitive to the toxin (9). The finding that cleavage of a single phosphodiester bond in the α-sarcin domain inactivates the ribosomes implies that this sequence is crucial for ribosome function since ribosomes survive treatment with nucleases despite many nicks in their RNA (10). This presumption has gained considerable reinforcement from the elucidation of the mechanism of action of ricin (11–13). Ricin, which is among the most toxic substance known, is an RNA N-glycosidase and the single base that is depurinated is A4324 in 28S rRNA; i.e., the nucleotide adjacent to the α-sarcin cut site (11–13). There is an evidence that the α-sarcin/ricin domain is involved in elongation factor– (EF-) 1–dependent binding of aminoacyl-

---

*Present affiliation:
Department of Applied Chemistry, Faculty of Engineering, Ehime University, Bunkyo-cho, Matsuyama, Japan

tRNA to the ribosomal A site and EF-2 catalyzed GTP hydrolysis and translocation of peptidyl-tRNA to the P site (14,15).

A definition of the chemistry of the interaction of the toxin with ribosomes requires specification of the nucleotides in 28S rRNA and amino acid side chains in α-sarcin that contribute to recognition and catalysis. The identity element, i.e., the nucleotide sequence and the conformation, of the α-sarcin/ricin domain stem and loop have been defined by analyzing the effect of α-sarcin on mutants of a small RNA that reproduces this structure (16). What is needed is complementary information on the identity of the amino acids residues in the protein that interact with the RNA and that are necessary for catalysis. To achieve this goal, it is an essential step to develop a gene expression system which would enable us to have a substantial amount of each one of a series of mutant α-sarcins.

Here we describe an *Escherichia coli* system which secretes α-sarcin into the periplasmic space efficiently, and a convenient procedure for the purification of the protein (17).

II. RESULTS

A. Construction of the *E. coli* Expression Vector

Outline of the construction of pUSARTB15-1 is shown in Figure 1A. A plasmid, pUSAR15-5, was constructed by inserting cDNA of α-sarcin (18) into pUC19. The expression vector, pKTN2-2, contained the *tac* promoter, the bla signal sequence, and a Shine-Dalgarno sequence that is under the control of isopropyl-β-D-thiogalactopyranoside (IPTG) (Figure 1A). The pUSAR15-5 was digested with *Bst* EII and the ends of the linearized plasmid were filled with DNA polymerase (Figure 1B [a] and [b]), then the *Bst* EII–*Sal*I fragment which contained the α-sarcin coding sequence was excised by the subsequent digestion with *Sal*I. The fragment (822 bp) was ligated with the 456-bp *Eco*RI–*Nae*I fragment which contained the *tac* promoter, the bla signal peptide, and a Shine-Dalgarno sequence from pTKN2-2. The pUSARTB15-5 was constructed by ligating this sequence with the large *Eco*RI–*Sal*I fragment of pBR322 (Figure 1A). The nucleotide sequence of the insert was confirmed by the dideoxy chain termination procedure (Figure 1B [b]). The resulting construct, pUSARTB15-5, encoded a fusion of the bla signal peptide and α-sarcin. Since we lost one codon at the step of the excision of α-sarcin cDNA (GCG code for alanine, as shown in Figure 1B [a]), a mature product should have valine at its NH$_2$-terminal, which corresponds to the second amino acid residue of the authentic protein when the signal peptidase cleaves a peptide bond at correct site in the primary translation product (Figure 1B [b]).

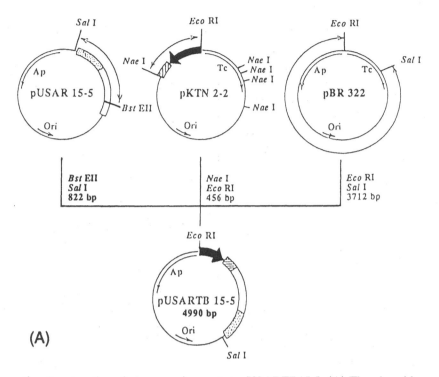

Figure 1 Construction of an expression vector, pUSARTB15-5. (A) The plasmid of approximately 5 kb contains the following genetic loci: ⬚ ; α-sarcin coding sequence, ▨ ; α-sarcin noncoding sequence, ▨ ; bla signal sequence, ◀ ; ribosome-binding site and *tac* promoter. (B) The nucleotide sequence at around the ligation site between the bla signal peptide sequence and α-sarcin cDNA. The number indicates the position of amino acid residues from the NH₂-terminal of the two peptides. The nucleotides in the shadowed box are filled with DNA polymerase.

B. Expression of Recombinant α-Sarcin

Figure 2A shows the time course of α-sarcin production in the *E. coli* transformant JM109/pUSARTB15-5. The extracts obtained by a brief sonication of the cells were prepared and were analyzed by SDS-PAGE. A faint band was detected in the expected position which corresponds to authentic α-sarcin before the induction by IPTG (Figure 2A, lane 1). One hour after the induction, the intensity of the band was slightly increased (Figure 2A, lane 2). A preliminary immunostaining experiment revealed that the increase of the band intensity could be due to an expression of recombinant α-sarcin (data not shown). The amount of expressed α-sarcin

Nae I   Bst EII
↓   ↓

```
          15   16   17   18   19   20   21   22        1    2    3    4    5
         ·Ala  Phe  Cys  Leu  Pro  Val  Phe  Ala       Ala  Val  Thr  Trp  Thr·
(a)  5'-GCC  TTT  TGC  CTT  CCT  GTC  TTC  GCC    5' GCG GTG  ACC  TGG  ACC-3'
     3'-CGG  AAA  ACG  GAA  GGA  CAG  AAG  CGG    3' CGC CAC  TGG ACC  TGG-5'
```

bla signal peptide   α-sarcin

```
          15   16   17   18   19   20   21   22   2    3    4    5
         ·Ala  Phe  Cys  Leu  Pro  Val  Phe  Ala  Val  Thr  Trp  Thr·
(b)  5'-GCC  TTT  TGC  CTT  CCT  GTC  TTC  GCC GTG  ACC  TGG  ACC-3'
     3'-CGG  AAA  ACG  CAA  GGA  CAG  AAG  CGG CAC  TGG  ACC  TGG-5'
```

(B)

Figure 1 Continued

in the cells was increased until 7 hr, then decreased at 12 hr (evidenced by the intensity of the bands) (Figure 2A, lanes 5 and 6). The immunoblot analysis established an identity of the band (Figure 3B). The results indicated that the recombinant fusion protein had synthesized and processed into apparently mature size. The analysis of samples obtained by cell fractionation indicated that the mature type of recombinant α-sarcin was accumulated exclusively in the periplasmic space (Figure 2B, lane 2; Figure 3C, lanes 1 and 2). Furthermore, the immunoblot analysis revealed the absence of a precursor protein in the cell extract, indicating an efficient segregation of the product took places in the cells (Figure 3C, lane 3). Using ELISA, the amount of the recombinant α-sarcin produced in the cells 7 hr after the induction was estimated to be 1.2 mg/L of culture.

C. Purification of Recombinant α-Sarcin

Cells (from 6 L of culture) were harvested by centrifugation at 10,000 × g for 10 min and the cells (5 g by wet weight) were suspended in 10 volumes of water and disrupted by the sonication described in Figure 2. The resultant supernatant fluid was brought to 40% saturation of ammonium sulfate at 0°C, stirred 1 hr, and centrifuged at 10,000 × g for 10 min. The supernatant fluid was dialyzed against Tris-NaCl (TN) (10 mM Tris-HCl, pH 7.4, 10 mM NaCl). The sample was applied to a column of Affi-Gel Blue (20 × 100 mm) previously equilibrated with TN, and the column was washed well with TN. The recombinant α-sarcin was eluted with 50 ml of 10 mM Tris-HCl, pH 7.4, 500 mM NaCl. The eluate dialyzed against TN was

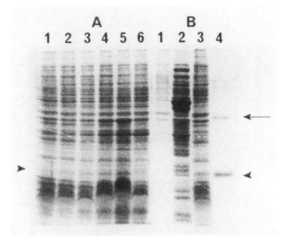

Figure 2   Time course and localization of α-sarcin production after induction with IPTG. Recombinant plasmid pUSARTB15-5 was propagated in *E. coli* strain JM109. Bacteria were precultured in LB-broth until an A_{600} value of 0.3, then the fusion protein was induced by adding IPTG (250 μg/ml). The cells were harvested by centrifugation at the indicated period. The cell pellet was resuspended in 10 volumes of water and the product was solubilized by sonication (Branson Sonifier model 250, Danbury, Connecticut; microtip, 80–90 W; 20 × 20 sec 1-min interval) at 0°C. The supernatant fluid was recovered by a centrifugation of the lysate at 16,000 × g for 15 min and was submitted for analyses. A, each extract containing 30 μg protein was separated on 15% polyacrylamide gels containing 0.1% SDS according to Laemmli (19) and the protein bands were visualized by Coomassie brilliant blue. Lane 1, start of the induction; lanes 2–6, 1, 3, 5, 7, and 12 hr after the induction, respectively. B, Localization of the recombinant product. Samples from 7 hr after the induction were analyzed as above. The isolation of a protein fraction from the periplasmic space was done according to the standard method (20). The cell extract was prepared by a prolonged sonication of the precipitate followed by centrifugation of the lysate with the same conditions as above. Lane 1, culture medium; lane 2, periplasmic fraction; lane 3, cell extract; lane 4, authentic α-sarcin as a standard marker. Arrowheads and an arrow indicate positions of σ-sarcin, monomer, and its dimer (22), respectively.

rechromatographed on Affi-Gel Blue with the same conditions as above. The eluate was dialyzed against TN and loaded onto a column of Bio-Rex 70 (20 × 100 mm) previously equilibrated with the same buffer. The column was washed with TN thoroughly and then the recombinant α-sarcin was eluted with 50 ml of 10 mM Tris-HCl, pH 7.4, 500 mM NaCl. The sample was dialyzed against water and lyophilized. The amount of the protein

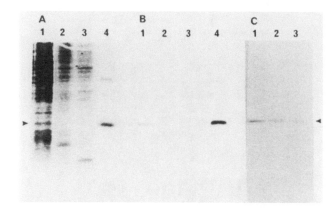

Figure 3 Immunoblot analysis of the cell extracts. Extracts were prepared 7 hr after the induction. The samples were separated on 15% SPS-PAGE. Gels were stained with Coomassie brilliant blue (A) (as in Figure 2B) or separated and immunoblotted (B) (17). Lane 1, pUSARTB15-5/JM 109 cells; lane 2, pKTN2-2/JM109 cells; lane 3, JM109 cells nontransformed; lane 4, an authentic α-sarcin as a standard marker (arrowhead). Localization of the product (C). Lane 1, total cell lysate; lane 2, periplasmic fraction; lane 3, cell extract.

recovered was 3.9 mg from 6 L of the culture; thus the recovery of the protein was estimated to be 54%. Analysis of the purified protein by SDS-PAGE revealed a single band having no appreciable mobility difference to the authentic α-sarcin (Figure 4, lanes 1 and 2). By an analysis of the purified protein using ELISA, we confirmed that the preparation had an equivalent antigenticity to the authentic one (data not shown). The first five amino acid residues from the NH_2-terminal of the purified protein were determined to be Val-Thr-Trp-Thr-Lys. An NH_2-terminal amino acid residue, valine, corresponds to the second amino acid residue of authentic α-sarcin (22). The result confirmed that the cleavage of a peptide bond in the fusion protein by signal peptidase between the bla signal peptide and the designed α-sarcin took place at the expected site in the bacterial cells.

D. Protein Synthesis Inhibition and the Formation of α-Fragment

The inhibitory effect of both secreted and authentic α-sarcin was assessed on in vitro protein synthesis from rabbit reticulocyte lysates. The recombinant α-sarcin inhibited protein synthesis in a similar manner with that of authentic α-sarcin (Figure 5). The recombinant protein, however, was twice as active as the authentic one; IC_{50} for recombinant and authentic α-sarcin

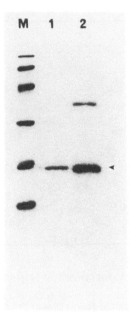

Figure 4 SDS-PAGE analysis of purified α-sarcin. Three micrograms of the purified recombinant α-sarcin (lane 1) or 10 μg of authentic protein (lane 2) was separated and stained as in Figure 2. Arrowhead points α-sarcin (monomer). Molecular weight markers used are phosphorylase b 94,000; bovine serum albumin 67,000; ovalbumin 43,000; carbonic anhydrase 30,000; trypsin inhibitor 20,000; α-lactoalbumin 14,000 (from top to bottom).

were 7.5 ng/ml (0.44 nM) and 15 ng/ml (0.88 nM), respectively. An analysis of rRNAs extracted from similar reaction mixtures revealed the formation of a characteristic RNA α-fragment, the oligonucleotide containing 498 bases that is cleaned from the 3′ end of 285 rRNA by the action of α-sarcin, indicating that the recombinant α-sarcin has the same ribonuclease activity as authentic α-sarcin does (Figure 6, lanes 2, 3 and 4).

III. DISCUSSION

We developed an efficient system for the expression of α-sarcin in *E. coli* using cDNA encoding α-sarcin and the expression vector pKTN2-2 (17). The resulting construct in the expression plasmid loses three nucleotides (GCG) between the bla signal sequence and the sequence for α-sarcin that encodes alanine of authentic α-sarcin. Thus, it is expected that the first amino acid in the mature recombinant protein would be valine, which

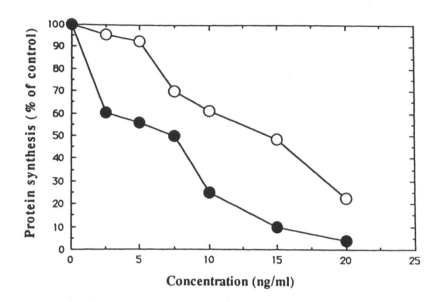

Figure 5 Inhibitory effect of the recombinant product on in vitro protein synthesis system from rabbit reticulocyte lysate. The activity of the purified recombinant product was assessed using a protein-synthesizing system derived from rabbit reticulocyte lysate (21). Inhibition of protein synthesis was assayed by reduction in the incorporation of L-[U-C^{14}]leucine into globin in the reaction mixture (60 μl) containing 0.5 μCi of L-[U-C^{14}]leucine, 15 μM hemin, and indicated amounts of purified recombinant protein (●) or authentic α-sarcin (○). The incubations were carried out for 30 min at 30°C. After the incubation, 30 μl of the reaction mixture was added to 0.5 ml of 1M NaOH, 5% H$_2$O$_2$ to hydrolyze aminoacyl tRNA complex and then 2 ml of 25% trichloroacetic acid was added to precipitate proteins. The precipitated proteins were trapped onto glass-fiber filters (Whatman GF/C) and washed three times with 20 ml of 8% trichloroacetic acid. The filters were dried and then counted by liquid scintillation counter (Aloka, LSC-3000). An incorporation of L-[U-C^{14}]leucine into the trichloroacetic acid insoluble activity of 46046 dpm/assay was taken as 100% of protein synthesis.

corresponds to the second amino acid residue of authentic protein. A purified recombinant α-sarcin showed a single band having an indistinguishable mobility difference to that of authentic α-sarcin on the gel. The results of amino acid sequencing of the protein showed that the NH$_2$-terminal amino acid of the product is valine, confirming the construction. The recombinant protein was twice as active as authentic α-sarcin on the inhibition of protein synthesis. This fact can be ascribed to a difference in the purification procedures. The original method developed by Olson et

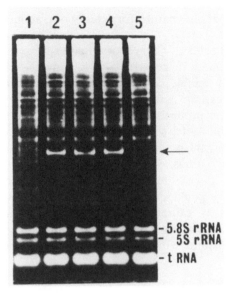

Figure 6 Formation of α-fragment. The unique ribonuclease activity of the recombinant α-sarcin was tested by analyzing the production of α-fragment during the translation reaction according to the method described previously (7). The half volume of the reaction mixture described in Figure 5 was added 5 volumes of 1% SDS in 50 mM Tris-HCl (pH 7.4) to terminate the translation and ribosomal RNAs were extracted by phenol and then precipitated by ethanol. The RNAs were separated by electrophoresis on 3.5% polyacrylamide gels and RNA bands were visualized with ethidium bromide. The reactions were carried in the absence (lane 1), or in the presence of the purified recombinant protein, 50 ng/ml (lane 2), 100 ng/ml (lane 3), 100 ng/ml of authentic protein (lane 4), and 100 ng/ml of extracts from *E. coli* transformed by the vector pKTN2-2 (lane 5). Arrow indicates α-fragment.

al. (1) includes many steps, during which partial inactivation of the protein may occur. Our purification procedure has fewer steps, reducing the risks of protein modification or denaturation. Alternatively, the *E. coli* product may be intrinsically more active. We also showed that production of α-fragment in ribosomes by the action of the recombinant protein. The result indicated that the molecular basis of the action of the product was the hydrolysis of a single phosphodiester bond in 28S rRNA at the specific site. All of these results provided evidence that the expressed fusion protein was translocated through the inner membrane of *E. coli* and was processed precisely in the periplasmic space and then folded into the active conformation. It is worthwhile to note that the recombinant product is fully active despite the fact that the mature form lacks one amino acid residue at its NH$_2$-terminal.

Recently, Henze et al. (23) reported an expression system of α-sarcin in *E. coli*. They made a construct by ligating the cDNA and a DNA sequence of *ompA* peptide, a major outer transmembranous protein of *E. coli* (24), so as to express α-sarcin with *ompA* signal peptide as a secretory signal. Their system produced only small amount of mature protein identified by immunoblot analysis. This phenomenon can easily be explained since the *ompA* signal peptide directs the protein to stay in the outer membrane (24). In contrast, we designed our plasmid to secrete the product into the periplasmic space in a soluble form (using the bla signal peptide sequence) (17). We could show a definitive band on the SDS-PAGE gel of the cell extract by a direct staining with Coomassie brilliant blue, indicating a massive production of the protein. Thus, the yield of recombinant α-sarcin in our system is much higher than that reported by Henze et al. (23).

CONCLUSIONS

We have developed an efficient *E. coli* system and a convenient procedure for its purification that enable us to have preparative amounts of fully active α-sarcin. The system can provide a useful took for further study in the field—such as the molecular mechanism of action of this unique enzyme or the mechanism of internalization of the protein into virus-infected cells.

REFERENCES

1. Olson, B. H., and Goerner, G. L. Appl. Microbiol., *13*: 314–321, 1965.
2. Olson, B. H., Jennings, J. C., Roga, V., Junek, A. J., and Schuurmans, D. M. Appl. Microbiol., *13*: 322–326, 1965.
3. Fernandez-Puentes, C., and Vazquez, D. FEBS Lett., *78*: 143–146, 1977.
4. Conde, F. P., Fernandez-Puentes, C., Montero, M. T. V., and Vazquez, D. FEMS Microbiol. Lett., *4*: 349–355, 1978.
5. Hobden, A. N., and Cundliffe, E. Biochem. J., *170*: 57–61, 1978.
6. Schindler, D. G., and Davies, J. E. Nucleic Acids Res., *4*: 1097–1110, 1977.
7. Endo, Y., and Wool, I. G. J. Biol. Chem., *257*: 9054–9060, 1982.
8. Endo, Y., Huber, P. W., and Wool, I. G. J. Bio. Chem., *258*: 2662–2667, 1983.
9. Wool, I. G. TIBS, *9*: 14–17, 1984.
10. Cahn, F., Schachter, E. M., and Rich, A. Biochim. Biophys. Acta, *209*: 512–520, 1970.
11. Endo, Y., Mitsui, K., Motizuki, M., and Tsurugi, K. J. Biol. Chem., *262*: 5908–5912, 1987.

12. Endo, Y., and Tsurugi, K. J. Biol. Chem., *263*: 7917–7920, 1988.
13. Endo, Y., and Tsurugi, K. J. Biol. Chem., *263*: 8735–8739, 1988.
14. Hausner, T-P., Atmadja, J., and Nierhaus, K. H. Biochimie, *69*: 911–923, 1987.
15. Moazed, D., Robertson, J. M., and Noller, H. F. Nature, *334*: 362–364, 1988.
16. Endo, Y., Glück, A., Chan, Y-L., Tsurugi, K., and Wool, I. G. J. Biol. Chem., *265*: 2216–2222, 1990.
17. Oka, T., Aoyama, Y., Natori, Y., Katano, T., and Endo, Y. Biochim. Biophys. Acta, in press, 1992.
18. Oka, T., Natori, Y., Tanaka, S., Tsurugi, K., and Endo, Y. Nucleic Acids Res., *18*: 1897, 1990.
19. Laemmli, U. K. Nature, *227*: 680–685, 1979.
20. Neu, H. C., and Heppel, L. A. J. Biol. Chem., *240*: 3685–3692, 1965.
21. Hunt, T., Vanderhoff, G., and London, I. M. J. Mol. Biol., *66*: 471–481, 1972.
22. Sacco, G., Drickamer, K., and Wool, I. G. J. Biol. Chem., *285*: 5811–5818, 1983.
23. Henze, P-P., Hahn, U., Erdmann, V. A., and Ulbrich, N. Eur. J. Biochem., *192*: 127–131, 1990.
24. Bremer, E., Cole, S. T., Hindennach, I., Henning, U., Beck, E., Kurz, C., and Schaller, H. Eur. J. Biochem., *122*: 223–231, 1982.

V
Diphtheria Toxin

15
Diphtheria Toxin Cloning and Expression in *Corynebacterium diphtheriae*

Lawrence Greenfield   *Roche Molecular Systems, Alameda, California*

I. INTRODUCTION

In 1888, the bacterium *Corynebacterium diphtheriae* was shown to secrete a protein toxin which contributed to the pathogenicity of diphtheria (1). It took over 65 years to demonstrate an association between bacteria producing diphtheria toxin and lysogenization by bacteriophage. Another 20 years ensued before it was shown conclusively that the gene for the toxin was encoded by the bacteriophage. Since that time, knowledge about diphtheria toxin has grown enormously. The application of current molecular biological techniques has permitted dissection and manipulation of the gene, thus permitting a detailed analysis of the protein and a way to wield the extreme potency of the toxin for targeting to discrete cell populations.

II. DIPHTHERIA TOXIN

Diphtheria toxin is a protein of 535 amino acids with an estimated molecular weight of 58,342 which is highly lethal to most animals (reviewed in Refs. 2 and 3). It consists of two major functional domains: the amino-terminal A domain (fragment A, or DTA, 193 amino acids) is the effector moiety which transfers the ADP-ribose group from nicotinamide adenine dinucleotide to eukaryotic elongation factor 2; and the carboxyl-terminal B domain (fragment B, or DTB, 342 amino acids) is the targeting moiety which both binds to most eukaryotic cells and aids in the translocation of fragment A across the cytoplasmic membrane. The two fragments can be separated by trypsin cleavage followed by reduction. Following binding to the cell surface, the toxin is taken up by receptor-mediated endocytosis.

During the acidification of the receptosome, the toxin is believed to undergo conformational changes which are required for the translocation of the A chain into the cytosolic compartment. The A chain then ADP-ribosylates the histamine of elongation factor-2 (EF-2); thus inhibiting protein synthesis.

III. IDENTIFICATION OF THE GENE ENCODING DIPHTHERIA TOXIN

A. Corynebacteriophages

The bacterium *Corynebacterium diphtheriae* serves as a host for a wide variety of closely related bacteriophages (including α, β, P, γ, π, K, ρ, L, δ, and ω), which are similar in morphology and physiology to the *E. coli* bacteriophage lambda (4). In the lytic cycle, 30–60 plaque-forming units are released per cell after a latent period of approximately 1 hr. In the lysogenic cycle, the phage DNA is inserted into the host chromosome at an attachment site. The lysogen bearing the integrated phage genome becomes resistant to superinfection by phages of the same immunity specificity (encoded by *imm*). With lambda, the packaged linear chromosome once injected into the cell is annealed and ligated at the cohesive ends (*cos*) (5). It can then replicate in the productive, lytic cycle, or integrate into the bacterial chromosome in the prophage state by recombination between the phage attachment site (*att*P) with the bacterial attachment site (*att*B). The physiological similarity of the corynebacteriophages and lambda (6) suggested that the former goes through similar stages (Figure 1).

B. The Gene for Diphtheria Toxin Is Carried on the Genome of *Corynebacterium diphtheriae* Bacteriophages

Non-toxin-producing strains of *C. diphtheriae* can be converted to toxinogenicity (the ability to produce diphtheria toxin) following infection by phage which carry the specific genetic marker designated *tox*$^+$ (reviewed in Refs. 7 and 8). Thus, infection of a bacterium by phage B (9,10) or phage 444V/A (11) resulted in strains which produced necrotic lesions and death in guinea pigs. That the in vivo toxicity was due to diphtheria toxin was demonstrated by the filterable nature of the toxic factor and its neutralization by diphtheria antitoxin (9). Not all clinical isolates of *Corynebacterium* could be infected by phage carrying *tox*$^+$ (7,9,12,13), and some corynebacteriophages (e.g., phage A, corynebacteriophage γ) were incapable of converting strains to toxinogenicity despite infection (9,10,14).

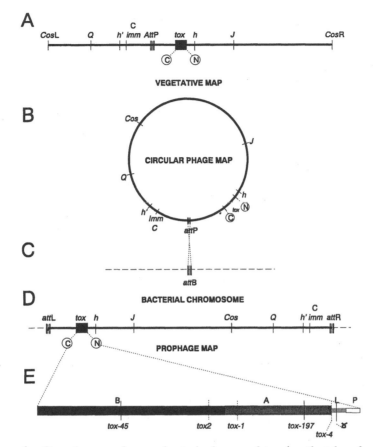

Figure 1 Genetic map of corynebacteriophage and *tox* (partly taken from Ref. 35). The conversion of the vegetative phage (A) to the integrated prophage map (D) by annealing of the cohesive ends to generate the circular map (B) followed by reciprocal recombination between the phage attachment site (*att*P) with the bacterial attachment site (*att*B) (C). As result of the recombination, the *att*L (leftward attachment site) and the *att*R (rightward attachment site) are generated. An expanded view of the *tox* genetic map including several missense and nonsense mutations is shown in E. Those mutations which are exactly known from sequence analysis (*tox*-45, *tox*-97, and γ, are shown by the solid line, whereas those which are imprecisely known (*tox*-2, *tox*-1, and *tox*-4) are indicated by the dotted lines. the placement of the genetic loci in A–D are only approximate. N and C represents the amino- and carboxy-terminal ends of the diphtheria toxin gene, respectively.

As might be expected of a lysogen, the toxinogenic strain resulting from infection of a susceptible, avirulent strain of *C. diphtheriae* by phage B was resistant to superinfection with phage B, had associated phage particles, and was stable to repeated stock culture passage as well as passage in vivo. However, because the filtrate of a naturally virulent strain of *C. diphtheriae* could not convert an avirulent strain to toxinogenicity, the direct association of *tox* with the bacteriophage was not made. Instead, it was hypothesized that phage lysis selected for spontaneous toxinogenic mutants (9), or that bacterial metabolism was altered owing to the formation of a lysogenic complex (10). We now know that this lack of conversion can be explained by restriction modifications systems (15), lysogeny by a nontoxinogenic strain of bacteriophage (e.g., γ [16]), or host range incompatibility (4,17).

The lack of spontaneous conversion of nontoxinogenic strains to toxinogenicity in the absence of infection by bacteriophage (11,18), and the rapid appearance of toxinogenic cells following infection, argued against the selection of a "mutant" population in favor of an induced change associated with the phage (11). In addition, conversion to toxinogenicity occurred at a high frequency following infection of a sensitive strain of *C. diphtheriae* by a *tox*$^+$ phage (11,19). An exact correlation existed between the acquisition of toxinogenicity and of lysogeny by *tox*$^+$ phage (9,20); loss of the lysogen resulted in reversion from the toxinogenic to the nontoxinogenic state (11,16). The observation that recombination between the nontoxinogenic bacteriophage γ and the toxinogenic bacteriophage β resulted in a toxinogenic bacteriophage γ′ and a nontoxinogenic bacteriophage β′ (14,17,21) indicated that *tox*$^+$ resided on the phage genome. Furthermore, nontoxinogenic strains of *C. diphtheriae* could be converted to toxinogenicity by exposure to lysate from a second nontoxinogenic strain, suggesting that recombination of two nonconverting phages resulted in converting phage progeny (12,13).

C. Genetic Localization of *tox* on the Bacteriophage Genome

As described below, classic genetic analysis localized the *tox* locus on the corynebacteriophage genome. *C. diphtheriae* is susceptible to infection by several genetically related phages: α^{tox^+}, β^{tox^+}, ω^{tox^+}, L^{tox^+}, P^{tox^+}, π^{tox^+}, ω^{tox^+}, and δ^{tox^+}, which result in toxinogenic lysogens; and γ^{tox^-}, K^{tox^-}, and ρ^{tox^-}, which result in nontoxinogenic lysogens. The bacteriophages β^{tox^+}, γ^{tox^-}, and L^{tox^+} are heteroimmune; i.e., each phage can infect lysogens of the other bacteriophage but are unable to infect lysogens carrying its own phage (4). In contrast, the phages β^{tox^+}, α^{tox^+}, P^{tox^+}, and ω^{tox^+} are homoimmune; i.e., each phage cannot infect lysogens of the other bacteriophage (4). The genetic similarity among the phages allowed for recombination between markers during matings (8,14); e.g., by heteroduplex analysis, over 99% of the β phage genome is homologous to the γ genome (22).

The closely linked regions determining the immunity specificity (*imm*), a locus required for lysogenization (*c*), a locus which allows the bacteriophage to infect lysogens (*vir*), and two loci determining extended host ranges, *h* and *h'*, were identified and mapped on the phage genome (17,23) (see Figure 1). A more extensive map of the vegetative phage genome, ordering 19 temperature-sensitive markers, the *h* and *h'* host range markers, the *c* gene, the *v1* virulence maker, and *tox* was made from two- and three-factor crosses between wild-type β phage and nitrosoguanidine temperature-sensitive mutants (24). The *tox* locus was localized between the cistrons involved in capsid (*Q, R, S, T, U, C, D, E,* and *F*) and tail (*W, X, G, H, I,* and *J*) morphogenesis on the vegetative map (24,25). Three-factor crosses among the heteroimmune temperate bacteriophage β^{tox^+}, γ^{tox^-}, and L^{tox^+}, defined the tox^+ region as a single phage-encoded genetic locus which resided between *h* and *imm* (17). Thus, genetic analysis suggested that the gene order was *h-tox-attP-imm-h'* (17,26) (see Figure 1). However, recent data aligning the physical and genetic maps have suggested that the order may be *h-attP-tox-imm-h'* (27). These data could not determine whether the tox^+ locus encoded diphtheria toxin itself.

Identification of *tox* as the structural gene for diphtheria toxin followed from studies of toxin cross-reacting mutants (28). Mutant phage (e.g., β_{45}) encoding truncated forms of the toxin (*tox*-45) resulted from nitrosoguanidine mutagenesis of a C7(β) lysogen. Strains lysogenized with β_{45}, (C7[β_{45}]), produced a 45,000-d peptide (CRM45, by SDS-PAGE; known as Crm45), possessing ADP-ribosyltransferase activity, which cross-reacted with wild-type diphtheria toxin but was nontoxic to guinea pigs. The findings that the production of the mutant protein required lysogenization of sensitive strains by β_{45}, and that the regulation of synthesis of CRM45 by C7(β_{45}) was identical to that of wild-type toxin by C7(β) suggested that the structural gene for diphtheria toxin was carried on the phage genome and was therefore the *tox* locus. This initial observation was confirmed and extended by the isolation of additional structural mutations within *tox* (29–32). Furthermore, complementation did not generate wild-type toxin when C7(β_{197}) containing a missense mutation within fragment A (see Figure 1E) was superinfected with the mutant β_{45}, which contains a nonsense mutation within fragment B (see Figure 1E) (31). In contrast, wild-type toxin could be generated in vitro when reduced, nicked CRM45 and CRM197 were mixed (33,34). This indicated that the toxin gene was coded by a single phage gene (31).

Localization of *tox* relative to identifiable physical structures on the phage genome followed. The prophage map was found to be a circular permutation of the vegetative map with *tox* at one end and *imm* at the other end of the prophage map (26) (see Figure 1). In contrast, the *tox* and *imm* genes were closely linked on the vegetative map (17). This could be explained if a phage attachment site (*att*P) was present between the toxin and immunity genes. Thus, in the vegetative map, the *att*P site would reside between *tox* and *imm*; in the prophage map, the *att*P locus would be divided into two halves, each located adjacent to the host chromosome, with the *imm* locus at one end of the phage genome and the *tox* locus at the other (see Figure 1).

Analysis of recombination among nontoxinogenic mutants oriented the *tox* gene within the prophage (35) and vegetative (29) maps. The carboxyl-terminal of the diphtheria toxin gene is adjacent to the phage attachment site (*att*P) with transcription proceeding toward *att*P (see Figure 1). By correlating the presence of enzymatic activity with the length of toxin peptides produced by a series of nonsense mutant CRMs, fragment A containing the ADP-ribosyltransferase activity was localized to the amino-terminus of the toxin molecule (31).

Naturally occurring nontoxinogenic corynebacteriophage may contain all or a portion of the *tox* gene cryptically. Genetic analysis indicated that corynebacteriophage γ contains at least a portion of the *tox* gene (30); in fact, the mutation is due to the presence of a 26-bp inverted repeat sequence beginning at base + 54 in the *tox* structural gene (36). Similarly, many nontoxinogenic strains of *C. diphtheriae* contain within their chromosome sequences homologous to the diphtheria toxin gene; 14 of 43 clinical nontoxinogenic isolates were found to contain chromosomal DNA homologous to the diphtheria toxin gene, with 12 of the 14 expressing some ADP-ribosyltransferase activity, but not producing intact diphtheria toxin (37).

D. Identification of Phage Genome Restriction Fragments Containing *tox*

In order to clone the diphtheria toxin gene, phage genome restriction fragments bearing *tox* were first identified so that the smallest fragment containing the entire diphtheria toxin gene could be cloned with a minimum amount of extraneous phage DNA. Because of the National Institutes of Health guidelines at the time, the intact structural gene for diphtheria toxin could be cloned only under high-level containment and with special permission (38). Therefore, a detailed restriction map of the toxin gene was required in order that hypotoxic fragments could be cloned using low-level containment. The localization of *tox* within restriction fragments was facilitated by the previous determination that *tox* was located close to *att*P (see Section III.C above, and see Figure 1).

The vegetative corynebacteriophage β and γ genomes are linear and have molecular weights of $22-23 \times 10^6$ d and 25×10^6 d, respectively, by electron microscopy and restriction analysis (22, Wolfson and Dressler quoted in Refs. 24,27, and 39).

The presence of the *cos* locus was used to help orient and order restriction fragments. Restriction digest analysis of both the vegetative phage and prophage indicated that during lysogenization, the vegetative DNA circularizes through the *cos* sites and integrates into the bacterial chromosome by recombination between the phage *att*P site and the bacterial *att*B site (39) (see Figure 1). In fact, circular forms of phage DNA were visualized by electron microscopy (22). Therefore, the vegetative linear DNA could be circularized in vitro by reannealing of the *cos* site. In this way, restriction fragments containing *cos* were identified as a single fragment when *cos* was reannealed, or as two fragments (each containing half of the *cos* sequence) when *cos* was denatured (39,40). The localization of the *cos* site served as a reference for the construction of a detailed restriction map (including *Bam*HI,*Eco*RI, *Hin*dIII, *Hpa*I, *Kpn*I, *Sal*I, and *Sma*I restriction sites) of the bacteriophage β (39,40) and γ (39) genomes.

Similarly, comparison of restriction digest patterns of β phage DNA with γ phage DNA simplified the construction of restriction maps and localization of *tox*. Heteroduplex analysis between β-converting (β-*tsr*-3) and γ-nonconverting (γ-*tsr*-1) bacteriophage demonstrated greater than 99% DNA homology with three regions of nonhomology: two deletion-insertion loops (DI-1 and DI-2) and one substitution bubble (S) (see Figure 2B) (22,39). In contrast, a single deletion-insertion loop (DI-2) and substitution bubble (S) was observed in heteroduplexes between DNA from β-*tsr*-3 and γ-*tsr*-2 (39). The *att*P locus lies near the DI-1 loop of β/γ-*tsr*-1 DNA heteroduplexes (41), whereas *tox* resides near the DI-2 loop of β/γ DNA heteroduplexes (42). Restriction fragments containing these altered sequences therefore differ between the two phages (27,39,41,42).

Because of the close genetic linkage of *tox* and *att*P (26,35), *tox* was expected to be located within the same restriction fragment as *att*P, or on a neighboring restriction fragment. As a result of integration of the corynebacteriophage genome into the *C. diphtheriae* chromosome through reciprocal recombination between *att*P and *att*B (see Figure 1), restriction fragments containing *att*P would differ between the two states: in the vegetative state, *att*P would be expected to reside on one restriction fragment (ATTP); whereas in the prophage state *att*P would be present on two restriction fragments (ATTL and ATTR). This was confirmed by hybridizing radioactively labeled vegetative phage DNA to Southern blots of restricted bacterial chromosomal DNA containing integrated β or γ phage DNA (27,39). Identification of the *att*P-bearing fragment was confirmed by hybridizing purified radioactively ^{32}P-labeled, identified *att*P-containing fragment to restriction digests of the prophage and vegetative phage gen-

A. GENETIC MAP

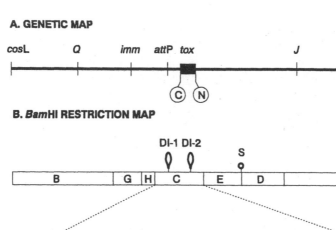

B. *Bam*HI **RESTRICTION MAP**

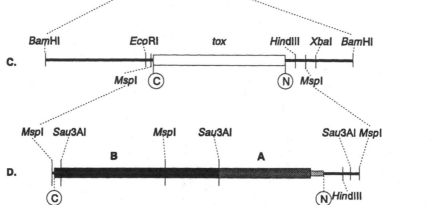

Figure 2 Restriction map of β phage and the diphtheria toxin gene. In A, the linear vegetative map is repeated for reference. The letters in the *Bam*HI restriction map of β phage (B) indicated decreasing size of fragments as indicated on an agarose gel and is taken from Ref. 39. DI-1 and DI-2 indicates the insertion loops associated with *att*P and *tox*, respectively; and s represents the substitution bubble (39,41,42). An expanded restriction map of the *Bam*HI fragment C shows the position of restriction fragments used in subsequent cloning (C). C and N represent the carboxy- and amino-terminal ends of the diphtheria toxin gene (*tox*). The position of *tox* within the fragment is approximate, whereas the position of the *Msp*I, *Eco*RI, and *Hin*dIII sites are taken from the DNA sequence and are therefore exact. Finally, a more detailed restriction map encompassing *tox* is shown indicating fragments expected to produce hypotoxic fragments to be used in cloning. The position of the *Msp*I, Sau3AI, and *Hin*dIII sites as well as the portions of the diphtheria toxin gene were derived from the sequence.

ome. The 3.9-kb *BamH*I band C fragment (Figure 2) of β-*tsr*-3 phage genome was thus identified as carrying *att*P (39). The *att*P site was further localized to a 50-bp region adjacent to an *Eco*RI site (Figure 2C) on the $β^{tox^+}$ and $γ^{tox^-}$ genomes (27).

Hybridization of toxin-encoding mRNA to restriction digests of corynebacteriophage DNA localized the *tox* gene to specific restriction fragments (Table 1) (42,40). As will be discussed below, diphtheria toxin synthesis is induced by growth of cultures in iron-deficient medium and the protein is secreted in the medium. Thus *tox* mRNA is associated with membrane-bound polysomes (43–45) and is present only in bacteria grown in media containing low concentrations of iron (46,47). Specific probes were made by isolating from lysogenic bacterial strains induced under low-iron conditions, either whole cell mRNA (42) or RNA isolated from membrane-bound polysomes (40). As a control, mRNA was also isolated from nonlysogenic bacterial strains (42) or lysogenic strains grown under conditions which repress diphtheria toxin production (42,40). Thus, the specific *tox*-containing restriction fragments were identified as those bands which hybridized to mRNA extracted from toxin-producing strains and not to mRNA isolated from nonlysogenic controls (42) or lysogenic controls grown under iron repression (42,40). Hybridization of mRNA from toxin-producing cultures to nitrocellulose filter blots of restricted vegetative β-*tsr*-3 phage DNA localized *tox* within the *Sal*I band B and the *BamH*I band C (see Table 1 and Figure 2B) (42); these same fragments had been previously identified as containing the *att*P locus (39). Similarly, the *tox* gene was found to lie within *BamH*I band 4 (3.9 kb), *Eco*RI band 1 (7.4 kb), *Hind*III band 1 (14.0 kb), and *Kpn*I bands 1 and 3 (23.0 kb and 3.5 kb, respectively) of the corynebacteriophage $β_c$ genome (see Table I) (40), again in a region close to the *att*P site. Hybridization of the *tox*-enriched mRNA to a single 2.1-kb region between *Eco*RI and *Hind*III restriction sites indicated that the *tox* message was no larger than 2.1 kb, suggesting that the *tox* operon was monocistronic; i.e., the *tox* operon encodes only for the diphtheria toxin gene and its control elements (40).

Localization and orientation of *tox* within the corynebacteriophage genome also resulted from combining both physical and genetic analysis of β phage and γ phage (42). Although the genetically similar β and γ phages can recombine within the *h-h′* segment containing *tox*, there are enough differences in the DNA to give slightly different restriction patterns. Therefore, restriction analysis of both parental and recombinant phage can identify the origins of parts of the genome. By restriction analysis of recombinants between the parental $β^{tox^+}$, $β^{tox^{-45}}$, and $γ^{tox^-}$, recombination within the 3.9-kb *BamH*I band C fragment was correlated with

Table 1   Size of Restriction Fragments Containing the Diphtheria Toxin Gene

Restriction fragment	Size (kb)					
	β	β$_c$	β$_{vir}$	γ	γ$_c$	ω$_c$
*Bam*HI	3.9	3.9	3.9	5.3	5.6	7.8
*Bam*HI–*Eco*RI			2.6			
*Eco*RI		7.4	3.9		2.6	6.2
*Xba*I–*Eco*RI			2.0			
*Sal*I	5.7					
*Hin*DIII	15.0	14.0	12.5		8.6	15.0
*Eco*RI–*Hin*dIII		2.1	2.1		1.6	
*Kpn*I		23.0				
		3.5				
*Hpa*I	13.0					
*Sma*I	26.0					

The fragment sizes were determined by agarose gel electrophoresis of restricted β (41,42), β$_c$ (40), β$_{vir}$ (40,48), γ (41), γ$_c$ (27), and ω$_c$ (15) phage DNA.

genetic alterations of the diphtheria toxin gene product. Furthermore, the γ$^{tox^-}$ genome was found to have a 1.4–1.7 kb insert within this *Bam*HI fragment which was closely linked (if not identical to) the *tox$^-$* mutation and resided close to the amino terminal end of the diphtheria toxin gene (27,39,41).

Finally, *tox* was localized to restriction fragments within the genome of the corynebacteriophage ω$_c^{tox^+}$, the phage isolated from the hyper-toxinogenic Park-Williams No. 8 (PW8) strain of *C. diphtheriae* (see Table 1) (15). In a similar manner, the *cos* site was localized by comparing restriction digests of ligated and heated phage DNA, and was used as a point of reference for constructing a detailed restriction map of the phage genome. The *tox* gene was located within the *Bam*HI band 3 and within a 1.7-kb region between *Hin*dIII and *Eco*RI restriction sites. In addition, an insertion-deletion region was identified adjacent to the diphtheria toxin promoter (Bouquet, quoted in Ref. 15).

IV. MAPPING AND SEQUENCING OF THE DIPHTHERIA TOXIN GENE

A. Mapping of *tox*

National Institutes of Health guidelines required that the cloning of the intact structural gene for diphtheria toxin be carried out under high-level containment (38). Therefore, either inactive mutants or hypotoxic fragments of the structural gene were cloned into *E. coli*. Restriction mapping of the corynebacteriophage genomes identified restriction fragments containing most, if not all, of the *tox* gene (see Table 1). In addition, the gene was found to have restriction sites for *Kpn*I (40), *Hinc*II (42), *Hae*II (42), *Mbo*I (48), and *Msp*I (48), thereby identifying fragments containing only portions of the gene.

The entire *tox* genes from β^{tox+} (49), $\beta^{tox+228}$ (50), β^{tox-45} (51), $\beta^{tox-197}$ (51), phage B (52), phage $\varphi984^+$ (52), phage $\varphi9tox^+$ (52), and ω^{tox+} (53) have been cloned and most have been sequenced. The DNA sequence (Figure 3) contains a single open reading frame encoding a protein sequence consistent with that determined for diphtheria toxin isolated from PW8(ω^{tox+}) (54–56). The mature toxin consists of 535 residues, with a calculated molecular weight of 58,342. Fragment A (bases 76–654) contains 193 amino acids with a molecular weight of 21,164, and fragment B (bases 655–1680) consists of 342 amino acid residues with a molecular weight of 37,194. The start of translation is predicted to lie between the amino-terminal glycine (codon GGC at position 76) of the mature protein and the first in-phase termination codon residing 99 bases upstream from the structural gene (codon TGA at position -24). In the absence of an ATG start codon within this upstream region, translation initiation is predicted to begin at the GTG codon 75 bases upstream (position -75).

B. Features of the Sequence

Secretory Leader

The sequence for the mature protein is preceded by a leader sequence of 25 amino acids, with a molecular weight of 2,460, suggesting that a 60,782-d proprotein is first synthesized. This amino-terminal extension has several features common to many secretory leaders from *Escherichia coli* (57–59): (1) The amino-terminal end is basic, containing two basic residues (Arg and Lys at positions -23 and -22, respectively. (2) The basic portion is followed by an extremely hydrophobic stretch of amino acids (positions -21 to -3, 19 residues). (3) A basic residue follows the hydrophobic region (His at -2). (4) There are proline residues at positions -5 and -6. (5) A serine is located at position -4. (6) No acidic residues are present. (7) Ala-His-Ala, the most frequent amino acid triplet adjacent to the signal peptidase cleavage site, is present immediately prior to the amino-terminal glycine of the mature protein. As is found with other signal peptides from gram-positive bacteria (60), this predicted leader is longer and contains more positively charged residues at the amino-terminal region than those from *E. coli*.

```
      -230              -210                -190                -170                -150                -130
        .       .         .       .          .       .          .       .          .       .          .       .
GGCGTTGCGTATCCAGTGGCTACACTCAGGTTGTAATGATTGGGATGATGTACCTGATCTGAGAGCGATTAAAAACTCATTGAGGAGTAGGTCCCGATTGGTTTTTGCTAGTGA


              -110                -90                 -70                 -50         mRNA    -30                 -10
        .       .          .       .          .       .          .       .    vv    .       .          .       .
AGCTTAGCTAGCTTTCCCCATGTAACCAATCTATCAAAAAAGGGCATTGATTTCAGAGCACCCTTATAATTAGGATAGCTTTACCTAATTATTTTATGAGTCCTGGTAAGGGGA
                                              "  -35"                "  -10"
                                                          ————————→              ←————————


               10                  30                  50                  70                  90
        .       .          .       .          .       .          .       .          .       .          .
TACGTTGTGTGAGCAGAAAACTGTTTGCGTCAATCTTAATAGGGGCGCTACTGGGGATAGGGGCCCCACCTTCAGCCCATGCAGGCGCTGATGATGTTGTTGATTCTTCTAAATCT
           ValSerArgLysLeuPheAlaSerIleLeuIleGlyAlaLeuLeuGlyIleGlyAlaProProSerAlaHisAlaGlyAlaAspAspValValAspSerSerLysSer
                -20                     -10                    -1  1                   10
 110                 130                 150                 170                 190                 210
  .       .          .       .          .       .          .       .          .       .          .       .
TTTGTGATGGAAAACTTTTCTTCGTACCACGGGACTAAACCTGGTTATGTAGATTCCATTCAAAAAGGTATACAAAAGCCAAATCTGGTACACAAGGAAATTATGACGATGAT
PheValMetGluAsnPheSerSerTyrHisGlyThrLysProGlyTyrValAspSerIleGlnLysGlyIleGlnLysProLysSerGlyThrGlnGlyAsnTyrAspAspAsp
            20                  30                  40


              230                 250                 270                 290                 310                 330
        .       .          .       .          .       .          .       .          .       .          .       .
TGGAAAGGGTTTTATAGTACCGACAATAAATACGACGCTGCGGGATACTCTGTAGATAATGAAAACCCGCTCTCTGGAAAAGCTGGAGGCGTGGTCAAAGTGACGTATCCAGGA
TrpLysGlyPheTyrSerThrAspAsnLysTyrAspAlaAlaGlyTyrSerValAspAsnGluAsnProLeuSerGlyLysAlaGlyGlyValValLysValThrTyrProGly
50                  60                  70                  80


              350                 370                 390                 410                 430                 450
        .       .          .       .          .       .          .       .          .       .          .       .
CTGACGAAGGTTCTCGCACTAAAAGTGGATAATGCCGAAACTATTAAGAAAGAGTTAGGTTTAAGTCTCACTGAACCGTTGATGGAGCAAGTCGGAACGGAAGAGTTTATCAAA
LeuThrLysValLeuAlaLeuLysValAspAsnAlaGluThrIleLysLysGluLeuGlyLeuSerLeuThrGluProLeuMetGluGlnValGlyThrGluGluPheIleLys
            90                  100                 110                 120


              470                 490                 510                 530                 550
        .       .          .       .          .       .          .       .          .       .          .
AGGTTCGGTGATGGTGCTTCGCGTGTAGTGCTCAGCCTTCCCTTCGCTGAGGGGAGTTCTAGCGTTGAATATATTAATAACTGGGAACAGGCGAAAGCGTTAAGCGTAGAA.TT
ArgPheGlyAspGlyAlaSerArgValValLeuSerLeuProPheAlaGluGlySerSerSerValGluTyrIleAsnAsnTrpGluGlnAlaLysAlaLeuSerValGluLeu
            130                 140                 150                 160


              570                 590                 610                 630                 650                 670
        .       .          .       .          .       .          .       .          .       .          .       .
GAGATTAATTTTGAAACCCGTGGAAAACGTGGCCAAGATGCGATGTATGAGTATATGGCTCAAGCCTGTGCAGGAAATCGTGTCAGGCGATCAGTAGGTAGCTCATTGTCATGC
GluIleAsnPheGluThrArgGlyLysArgGlyGlnAspAlaMetTyrGluTyrMetAlaGlnAlaCysAlaGlyAsnArgValArgArgSerValGlySerSerLeuSerCys
            170                 180                 190                 200


              690                 710                 730                 750                 770                 790
        .       .          .       .          .       .          .       .          .       .          .       .
ATAAATCTTGATTGGGATGTCATAAGGGATAAAACTAAGACAAAGATAGAGTCTTTGAAAGAGCATGCGCCTATCAAAAATAAAATGAGCGAAAGTCCCAATAAAACAGTATCT
IleAsnLeuAspTrpAspValIleArgAspLysThrLysThrLysIleGluSerLeuLysGluHisGlyProIleLysAsnLysMetSerGluSerProAsnLysThrValSer
            210                 220                 230


              810                 830                 850                 870                 890
        .       .          .       .          .       .          .       .          .       .          .
GAGGAAAAAGCTAAACAATACCTAGAAGAATTTCATCAAACGGCATTAGAGCATCCTGAATTGTCAGAACTTAAAACCGTTACTGGGACCAATCCTGTATTCGCTGGGGCTAAC
GluGluLysAlaLysGlnTyrLeuGluGluPheHisGlnThrAlaLeuGluHisProGluLeuSerGluLeuLysThrValThrGlyThrAsnProValPheAlaGlyAlaAsn
240                 250                 260                 270


              910                 930                 950                 970                 990                 1010
        .       .          .       .          .       .          .       .          .       .          .       .
TATGCGGCGTGGGCAGTAAACGTTGCGCAAGTTATCGATAGCGAAACAGCTGATAATTTGGAAAAGACAACTGCTGCTCTTTCGATACTTCCTGGTATCGGTAGCGTAATGGGC
TyrAlaAlaTrpAlaValAsnValAlaGlnValIleAspSerGluThrAlaAspAsnLeuGluLysThrThrAlaAlaLeuSerIleLeuProGlyIleGlySerValMetGly
280                 290                 300                 310


             1030                1050                1070                1090                1110                1130
        .       .          .       .          .       .          .       .          .       .          .       .
ATTGCAGACGGTGCCGTTCACCACAATACAGAAGAGATATGGCACAATCAATAGCTTTATCGTCTTTAATGGTTGCTCAAGCTATTCCATTGGTAGGAGAGCTAGTTGATATT
IleAlaAspGlyAlaValHisHisAsnThrGluGluIleValAlaGlnSerIleAlaLeuSerSerLeuMetValAlaGlnAlaIleProLeuValGlyGlyLeuValAspIle
            320                 330                 340                 350


             1150                1170                1190                1210                1230
        .       .          .       .          .       .          .       .          .       .          .
GGTTTCGCTGCATATAATTTTGTAGAGAGTATTATCAATTTTATTTCAAGTAGTTCATAATTCGTATAATCGTCCCGCGTATTCTCCGGGGCATAAAACGCAACCATTTCTTCAT
GlyPheAlaAlaTyrAsnPheValGluSerIleIleAsnLeuPheGlnValValHisAsnSerTyrAsnArgProAlaTyrSerProGlyHisLysThrGlnProPheLeuHis
            360                 370                 380                 390


             1250                1270                1290                1310                1330                1350
        .       .          .       .          .       .          .       .          .       .          .       .
GACGGGTATGCTGTCAGTTGGAACACTGTTGAAGATTCGAATAATCCGAACTGGTTTTCAAGGGGAGAGTGGGCACGACATAAAAATTACTGCTGAAAATACCCCGCTTCCAATC
AspGlyTyrAlaValSerTrpAsnThrValGluAspSerIleIleArgThrGlyPheGlnGlyGluSerGlyHisAspIleLysIleThrAlaGluAsnThrProLeuProIle
            400                 410                 420


             1370                1390                1410                1430                1450                1470
        .       .          .       .          .       .          .       .          .       .          .       .
GCGGGTGTCCTACTACCGACTATTCCTGGAAAGCTGGACGTTAATAAGTCCAAGACTCATATTTCCGTAAATGGTCGGAAAATAAGGATGCGTTGCAGAGCTATAGACGGTGAT
AlaGlyValLeuLeuProThrIleProGlyLysLeuAspValAsnLysSerLysThrHisIleSerValAsnGlyArgLysIleArgMetArgCysArgAlaIleAspGlyAsp
430                 440                 450                 460


             1490                1510                1530                1550                1570                1590
        .       .          .       .          .       .          .       .          .       .          .       .
GTAACTTTTTGTGCCCTAAATCTCCTGTTTATGTTGGTAATGGTGTGCATGCGAATCTTCACGTGGCATTTCACAGAAGCAGCTCGGAGAAAATTCATTCTAATGAAATTTCG
ValThrPheCysArgProLysSerProValTyrValGlyAsnGlyValHisAlaAsnLeuHisValAlaPheHisArgSerSerSerGluLysIleHisSerAsnGluIleSer
            470                 480                 490                 500


             1610                1630                1650                1670                1690
        .       .          .       .          .       .          .       .          .       .          .
TCGGATTCCATAGGCGTTCTTGGGTACCAGAAAACAGTAGATCACACCAAGGTTAATTCTAAGCTATCGCTATTTTTTGAAATCAAAAGCTGAAAGGTAGTGGGGTCGTGTGC
SerAspSerIleGlyValLeuGlyTyrGlnLysThrValAspHisThrLysValAsnSerLysLeuSerLeuPhePheGluIleLysSerEnd
            510                 520                 530


             1710                1730                1750                1770                1790                1810
        .       .          .       .          .       .          .       .          .       .          .       .
CGGTAAGCGGAACGGTTCCGGAATGGCGCTATAGTATGCACAGGTAGAGCAGAATTCGAATCTGACTACGGATCAGAAGGTTGGGGGTTCGAATCCCTCCGGGCGCACAAGTG


             1830                1850                1870                1890                1910
        .       .          .       .          .       .          .       .          .       .
AAACCCCAGCTCATAGCATGTTTGAGCTGGGGTTTCTCATGGCGTGTGGGTTGTCTGACTGTTGGCTGTTGTTGCGGTGGTTGGTGCTCGTACCGAACCGAACG
```

Promoter Localization Within The DNA Sequence

A promoter can be defined as a sequence that "signals the start of transcription" (61). Because most of the work has been done in *E. coli*, the initial description will focus on what is known in that bacterium. The core enzyme of *E. coli* RNA polymerase has the subunit composition $\alpha_2\beta\beta'$ (62). The formation of holoenzyme by the association of σ factor with core enzyme increases the specificity of the enzyme for promoters by 10^4-fold (62). In addition, σ may be involved in DNA strand separation (62). Four σ factors have been identified in *E. coli*: σ^{70}, σ^{32}, σ^{54}, and σ^{28}. The predominant sigma factor of *E. coli* is σ^{70}.

RNA polymerase protects from nuclease degradation a region of 40 to 60 bp, extending from 40–50 bases upstream to 10–20 bases downstream from the initiation of transcription (63,64). Promoters recognized by $\alpha_2\beta\beta'\sigma^{70}$ holoenzyme contain two highly conserved sequences: the "−35" region (TTGACA) and the "−10" region (TATAAT) (65). The 17 ± 1 bp spacing between these two regions is highly conserved and the mRNA is initiated 7 ± 1 bases downstream from the "−10" region (65). There appears to be some correlation between the degree of homology of a sequence with the consensus and its strength as a promoter in vitro (66). However, in promoters that are positively regulated, the "−35" region is not well conserved, and the activator protein's binding site generally resides within the promoter sequence (62). In addition, other sequences may be important with defining promoters: Some strong promoters contain (A/T)-rich sequences upstream from the "−35" region; and sequences within the 5' end of the mRNA from +1 to +25 may contribute to promoter strength (62).

The *tox* sequence stream upstream of the presumed GTG translational start site has homology with the promoter sequence for σ^{70} holoenzyme of *E. coli* (62) (see Figures 3 and 4A). The sequence TTGATT (extending from −74 to −69) contains the four of the five (TTGA) most highly

Figure 3 DNA sequence of *tox*. The DNA sequence of the β phage *tox* gene is shown. The numbers above the sequence identify the bases with +1 indicating the first nucleotide of the presumed GTG start codon. The numbers below the sequence position the amino acid residues with amino acid 1 indicating the Gly at the amino-terminal end of the mature protein. Amino acids −1 to −25 are the presumed secretory leader. Also indicated are sequences homologous to the "−35" and "−10" sequences of *E. coli* promoters, the 5' start of the messenger RNA (mRNA), and the diad symmetry (facing arrows) believed to be involved with binding of repressor. Bases −234 to 1707 are taken from Ref. 49 and bases 1708 to 1921 from Ref. 50.

conserved bases of the "−35" *E. coli* consensus sequence TTGACA (67). As with some *E. coli* promoters, an (A/T)-rich sequence is located upstream from this sequence and an A at position "−45."

Two possible "−10" sequences are present. The sequence TATAAT, extending from −56 to −51, is identical to the *E. coli* consensus sequence but has suboptimal spacing from both the "−35" region (12 compared to the conserved 17 ± 1 base pairs) and the 5' start of the message (see below). In contrast, the sequence TAGGAT, extending from −50 to −45, spaced 18 bp from the suggested "−35" region and 4–5 bases from the transcriptional start site. Although it is less homologous to the "−10" *E. coli* consensus sequence, it retains the most conserved bases (TA . . . T) (67).

Several other *Corynebacterium* genes have been sequenced. The *meso*-diaminopimelate D-dehydrogenase gene from *C. glutamicum* contains regions of homology to *E. coli* promoters, including a "−35" region (TCATCA) and a "−10" region (TAAGCT) (68). Similarly, the erythromycin-resistance gene from the *C. diphtheriae* plasmid pNG2 contains homologous "−35" (TTCACA) and "−10" (TATAAT) regions (69). In contrast, the *lys*A gene of C. glutamicum, which expresses at low levels in *E. coli*, does not show homology with either of these two regions (70). However, it has been suggested that this gene may be positively regulated (70).

Promoters from other gram-positive organisms frequently contain regions of homology with known *E. coli* promoters. In *Bacillus subtilis*, the promoter recognized by σ^{55} contains the conserved "−35" sequence TTGACA and "−10" sequence TATAAT, both similar to the *E. coli* promoters involving σ^{70} (71). The initiation region of *Staphylococcus aureus* β-lactamase gene contains the sequences TTGACA and TATTAT separated by 18 bp (72). Therefore, it is not surprising to find homology between the *C. diphtheriae* and *E. coli* promoter sequences.

Presumptive Ribosomal Binding Site

Translation initiation is a major mechanism of controlling the rate of protein synthesis (73). The region of mRNA within an initiation complex that is protected from nuclease digestion has been referred to as the ribosome-binding site (RBS) (74). The sequence, approximately 35 bases long, contains the initiation codon (usually AUG or GUG, but sometimes UUG or AUU). Within the sequence is the Shine-Dalgarno (SD)-sequence, which complements the highly conserved region (CUCCU) near the 3' end of 16S rRNA during translation initiation (75). This sequence is within 7 ± 2 bases from the start codon (76). The secondary structure of mRNA appears to play a major role in translation efficiency (74). In addition, the presence of other sequences appear to play a role in the efficiency of translation initiation (73). In gram-positive bacteria, the length of complementarity around the SD sequence is longer than found in *E. coli* (76).

The first amino acid of the protoxin is unknown. The GTG codon at position +1 is a good candidate for the start of translation of the secretory leader. The Gly codon (GGC) at position +76 (see Figure 3) encodes the first amino acid of mature diphtheria toxin. Although there are no in-phase ATG start codons between this GGC codon and the first in-phase termination codon (TGA at position −24), a GTG is present at position +1. The sequence TAAGGGG (positions −14 to −8) has a high degree of complementarity with the 3'-terminus of 16S *E. coli* rRNA (3'OH-AU-UCCUCCACUAG-5') and is 7 bases from this predicted start codon.

Several other features of the sequence are typical of *E. coli* ribosomal binding sites (77). Nucleotides 10–13 are rich in A and U (AAAC) (76). The sequence 5'-UGAUCC-3', which can base-pair to the *E. coli* 16S rRNA around position 1529, has been noted in the mRNA for highly expressed genes (73). A similar sequence (TGAGTCC) is present at positions −24 to −18 in *tox* (Figure 4). Most *E. coli* translational initiation regions contain, within the region +4 to +21, at least three consecutive nucleotides complementary to the first 16 nucleotides from the 5' end of the 16S rRNA sequence (5'-AAAUUGAAGAGUUUGA-3') (78). The sequence 5'-AAACTGTT-3' at positions +9 to +17 is complementary to 5 of the 8 bases of the sequence 5'-AAGAGTTT. Although the sequence of the 16S rRNA of *C. diphtheriae* is unavailable, the sequence at the 5' end of the *C. variabilis* 16S rRNA contains the sequence AGTTT (79). In addition, the sequences neighboring the translational start codon frequently show extended complementarity with the tRNA[fmet] A loop such that they contain a subset of the sequence GUUAUGAGC surrounding the AUG start codon: these additional 6 bases encompass the proposed GTG start codon of *tox* (see Figure 3).

Sequences Downstream from the tox Structural Gene

The 3' end of the *tox* message has been mapped close to the *Eco*RI site (positions 1755–1760). A 7-bp inverted repeat (CGGTAAG . . . GAATGGC) extending between 1704 and 1730 may represent a transcription termination sequence (50). It is interesting that the sequence GGGTTTCT (positions 1846–1853) is present as a palindromic structure following the *C. glutamicum lys*A, where it has been implicated in transcription termination (70).

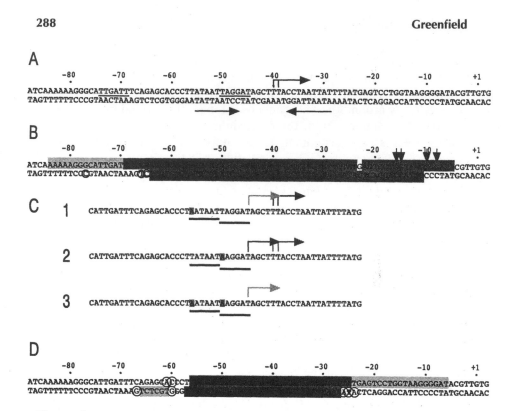

Figure 4   tox promoter and operator region. The DNA sequence of the *tox* regulatory region (A) indicates the "−35" (TTGATT) and "−10" (TAGGAT) consensus sequences, the 5′ end of the mRNA (T at positions −40 and −39) and the diad symmetry believed to be involved in repressor binding (inward pointing arrows). The sequence protected from DNAse I digestion by binding of extracts from *C. diphtheriae* (B), referred to as "F.exp" (taken from Ref. 113) shows both strongly protected (dark box) and weakly protected (light box) regions. Methylation at the G residues at positions −16, −15, −10, and −8 (arrows) prevent the extract from protecting the region from DNAse I. The circles bases on the bottom strand (−65, −66, and −77) are positions of enhanced cleavage. Mutational analysis of the two possible "−10" consensus sequences (C) indicate that changing the sequence TATAAT to AATAAT (C1) decreases the amount of mRNA beginning at position −45, whereas substitution of the sequence AAGGAT for TAGGAT (C2) decreases the amount of the mRNAs initiated at position −40 and −39. Modification of both sequences decreases all *tox* mRNA (taken from 115). The DNA sequence protected from DNAse I digestion by binding of an *C. diphtheriae* aporepressor in the presence of Fe²⁺ (F.rep.) (D) results in regions that are maximally protected (dark box) and variably protected (light box) as well as bases whose cleavage is enhanced (circled bases). Taken from Ref. 113.

The *att*P site has been mapped to within 50 bp from the *Eco*RI site (27). The perfect 14-bp inverted repeat interrupted by 9 bp extends from 1816 to 1852 (50) is a candidate for this locus. Recently, the bacterial *att*B site has been cloned and sequenced (80). Within a 120-nucleotide stretch homologous among three *Corynebacterium* species lies a segment of 96 nucleotides (core sequence) that is highly homologous to the γ phage sequence, with only 6 base mismatches. With the exception of the deletion of one base, the β phage genome contains the exact sequence between positions 1763–1867 (50) (see Figure 3).

IV. EXPRESSION OF DIPHTHERIA TOXIN IN *CORYNEBACTERIUM DIPHTHERIAE*

A. General

Diphtheria toxin is synthesized and secreted as a single polypeptide chain (protoxin) of around 62,000 d by *C. diphtheriae* lysogenized with a family of corynebacteriophages (81). As is true of most secreted proteins, the protoxin is produced by membrane-bound polysomes as a larger precursor (molecular weight of 68,000) which is cotranslationally processed to the 62,000-d mature protein and secreted (44,45). The removal of the secretory leader requires membrane enzymes in vitro, following complete synthesis of the precursor form (44). In vivo synthesis is associated with membrane-bound polysomes (43,44); no toxin is made on free polysomes (44). As with other secreted proteins, the intracellular concentration of toxin is very low (43,82).

Expression of *tox* under conditions of iron limitation (see below) is independent of the state of the bacteriophage, and is therefore under a separate control from that which regulates the genes involved in phage replication: it is expressed from the β-phage genome whether it is present in the prophage state (20,83–86), as a nonreplicating exogenote (87), or during the vegetative cycle (88,89); *tox* expression is not linked to expression of other phage genes. In addition, nontoxigenic mutants of β phage are able to lysogenize normally, have normal burst sizes, and produce phage with the same kinetics as that of wild-type phage, indicating that the toxin gene has no role in phage replication or lysogenization (29–32). Although most β phage genes are not expressed in the prophage state, low levels of both *tox* and the phage repressor genes are expressed (90).

The average time required for synthesis and extracellular secretion of toxin is 3 min (91). The rate of toxin production is a function of the cell concentration and can occur in the absence of bacterial growth. Thus, the higher production of diphtheria toxin by PW8 is partly due to its ability to continue growing at a linear rate in the absence of exogenous iron, thereby increasing cellular mass five- to sixfold (92); during this time, 5% of the total cell protein synthesized is diphtheria toxin (92,93). Under suitable conditions, diphtheria toxin can be produced by nongrowing cultures at a rate similar to that of growing cultures (94). When succinate is used as the sole energy source in iron-deficient medium, toxin is synthesized at a maximum rate of 1 $\mu g/10^9$ bacteria per hour (94), or 10^4 molecules per bacteria per hour. Under these circumstances, the toxin accounts for 35% of the total protein synthesized, and most of the extracellular protein produced.

The dependence of diphtheria toxin production on growth cycle and cell concentration seen with *C. diphtheriae* CN2000 (PW8) relates to the intracellular concentration of iron (93) (see below). In both glucose-limited and glucose-excess media, toxin production is undetectable until the cell concentration reaches 1.6 g bacteria protein/L, at which time the intracellular concentration of iron reaches 2.5–3.0 $\mu g/g$ bacterial protein (93). Toxin production is greater in slow-growing glucose-limited cultures probably because of a longer period of growth after the bacterial concentration reached 1.6 g/L. In chemostat cultures in which growth rate is controlled to 0.051 hr^{-1} by limited feeding of glucose, toxin production begins when the cell concentration reaches a level in which the intracellular iron concentration is 2.7 $\mu g/g$ bacterial protein (93). (See Section IV.C for discussion on the role of iron in diphtheria toxin synthesis.)

B. Role of the Host in Diphtheria Toxin Expression

The role of the bacterial genome in diphtheria toxin expression has been well documented in the strain PW8. The PW8 is the most toxigenic strain of *C. diphtheriae* owing to a defective cytochrome system (95) which allows the bacteria to grow for three doublings after iron becomes the rate-limiting substrate and because the ω^{tox^+} phage is integrated into its genome in two nontandem phage attachment sites (15). In contrast, the growth of $C7_s(\beta^{tox^+})$, and hence the level of diphtheria toxin production, is coupled to its cytochrome-mediated electron transport chain. Thus, under optimal conditions, the yield of diphtheria toxin from $C7_s(\beta^{tox^+})$ is approximately 30 $\mu g/ml$ and from PW8 is approximately 450 $\mu g/ml$ (15,96), representing 5% of the total bacterial proteins synthesized or over 75% of the protein secreted by the latter strain (96).

The influence of the host on diphtheria toxin expression is demonstrated in strains other than PW8: The yield of toxin can vary greatly when the same corynebacteriophage is present in different naturally occurring host strains of *C. diphtheriae* (7,97). In addition, mutagenesis of lysogenized C7 with N-methyl-N'-nitro-N-nitrosoguanidine (NTG) resulted in a mutant host strain, C7-262(β), which produced 10–12% of the levels of diphtheria toxin as the parent, despite possessing a normal β phage genome (31). The finding that the mutant host strain, C7-262, grew slower in broth and formed smaller colonies on plates than did the wild-type strain suggested that the mutation affected genes involved in regulation of host metabolism in addition to diphtheria toxin production (31). In fact, most bacterial mutations which result in low toxin yield from normal β phage genomes also result in poor growth of the mutant strain (98).

Finally, the bacterial host plays an important role on the well documented effect of iron on inhibiting expression of the toxin.

C. Iron Regulation of Diphtheria Toxin Synthesis in C. diphtheriae

Production of diphtheria toxin by *C. diphtheriae* is influenced by the concentration of metals in the medium (99,100). Addition of iron (II), manganese (II), copper (II), nickel (II), and cobalt (II) inhibits the production of diphtheria toxin by lysogenic strains at concentrations that have no effect on bacterial growth (101). The inhibitory effect is in the order $Fe^{2+} >$ $Cu^{2+} > Co^{2+} > Ni^{2+}$, with iron being from 60 to 750 times as effective on a weight basis as cobalt, copper, or nickel (101) when the extracellular concentrations are considered. In contrast, cobalt effects toxin production more dramatically than iron when intracellular concentrations of the metals in considered (102). Addition of strontium, barium, thorium, aluminum (III), and vanadium salts has no effect (101,103), whereas the addition of cadmium (II) and zinc (II) inhibited both toxin production and strain growth (103).

Although both Fe(II) and Fe(III) are taken up equally well by bacteria, the ferrous iron inhibits toxin synthesis much earlier than the ferric form, suggesting that Fe(III) requires conversion to Fe(II) in order to exert an inhibitory effect (104). Optimal production of diphtheria toxin occurs within a very narrow range of extracellular iron concentration, with maximum toxin yield at 140 μg/L (2.6 μM) of iron (100). When considering the intracellular levels of iron, diphtheria toxin production occurs only when the intracellular concentration of iron falls below 1.4–3.0 μg Fe/g cell protein (93,104). The level of toxin production (from 0 to 0.12 g toxin/g bacterial protein) is inversely related to the intracellular iron concentration (from 2.7 to 2.0 μg Fe/g bacterial protein) (93). Lower amounts of iron in the media results in poor cell growth, whereas a larger amount results in low toxin yields despite good cell growth. The limitation in cell growth at low iron concentrations likely results from reduced synthesis of iron-containing enzymes necessary for bacterial growth (105,93). Even at extremely high iron concentrations (e.g., 4000 μg/L iron), diphtheria toxin production is still detectable (105).

This correlation between extracellular iron levels and toxin production is true of all toxinogenic strains (101,105). Phage production and cell lysis is unaffected by the presence of iron (84). As a result of the inhibition of cell growth resulting from the low iron concentration, toxin yield is determined by the extent of bacterial growth following depletion of the exogenous iron supply (92).

D. Involvement of the *C. diphtheriae* Host Genes in Iron-Regulated Control of Diphtheria Toxin Synthesis

The role of *C. diphtheriae* in mediating the effect of iron on diphtheria toxin production is clearly shown by the isolation of host mutants in which toxin synthesis continues in the presence of high extracellular iron concentrations. The NTG mutagenesis of C7(β) produced five strains (C7*hm*722, C7*hm*723, C7*hm*726, C7*hm*728, and C7*hm*729) which produced toxin at normal rates even when the extracellular iron concentration was 3000 μg/L (83). In contrast, diphtheria toxin production by the unmutagenized strain, C7(β), is completely inhibited at 300 μg/L Fe^{2+}. All strains rapidly took up iron intracellularly. A mutant strain of C7*hm*723 which lost its prophage, C7*hm*723(−), was isolated by UV irradiation. In this strain, production of a series of toxin-related peptides following lysogenization were also unaffected by the addition of iron to the medium.

Another series of host mutants (HC1, HC3, HC4, and HC5), capable of producing diphtheria toxin in the presence of high iron concentrations, were obtained through NTG and ethyl membrane sulfonic acid ester (EMS) mutagenesis of C7(β) (106). In contrast to the previously describe mutants, all mutants had extremely defective iron transport systems.

Similarly, a mutant strain of PW8 which produces diphtheria toxin in medium containing high iron concentrations was obtained by mutagenesis (107). Three mutants of PW8 (PW8-3819, PW1901, and PW801907) which produced diphtheria toxin in the presence of high iron concentrations were obtained by NTG mutagenesis. The PW8-1901 and PW8-1907 mutants produced greater amounts of toxin in medium containing 1000 μg/L Fe^{2+} than did the parent PW8 strain in medium containing 100 μg/L Fe^{2+} (107).

The role of a host-encoded factor can also be demonstrated in vitro. Protein-synthesizing extracts isolated from *E. coli* and nontoxinogenic *C. diphtheriae* produce similar amounts of β phage proteins when programmed with purified corynebacteriophage DNA (108). In contrast, diphtheria toxin is produced only in the *E. coli* extracts. In addition, extracts from $C7_s(-)^{tox-}$ inhibited the toxin synthesis directed by the *E. coli* system, suggesting that both toxinogenic and nontoxinogenic strains of *C. diphtheriae* contain a factor which blocks diphtheria toxin production (108). The ability of *E. coli* extracts to synthesize diphtheria toxin indicates that expression involves a negative rather than a positive regulation by *C. diphtheriae*.

Therefore, a host factor, altered in the c7hm723 strain, is involved in iron-regulated inhibition of diphtheria toxin. This factor is not essential for cell growth in that each mutant grows well at all iron concentrations (83). Acquisition of resistance to iron-regulated production of diphtheria toxin results in a simultaneous insensitivity to inhibition by high concentrations of copper, cobalt, nickel, and magnesium, suggesting that the mechanism of inhibition by these metals is the same (101).

E. Bacteriophage Factors Influencing Iron Regulation of Diphtheria Toxin Synthesis

The role of the bacteriophage in regulating iron repression of *tox* is directly evident from mutants which constitutively express diphtheria toxin. Expression of toxin by C7 lysogenized with the mutant β phage (β_{ct1}^{tox+}), isolated by NTG mutagenesis, is partially insensitive to iron-mediated repression (109). Diphtheria toxin production by the mutant was 25 times that of the wild-type parent under conditions of high iron concentrations (200 μM FeCl$_3$). The mutation is *cis* acting, in that the double lysogen C7(β_{45}^{crm+}/β_{ct1}^{tox+}) produced both CRM45 and wild-type toxin under conditions of iron starvation, but predominantly wild-type toxin in the presence of inhibitory concentrations of iron. These data are consistent with a constitutive operator mutation (O^c) (109,110).

In contrast to the host mutant C7hm723, in which there is a simultaneous acquisition of resistance to suppression by iron, copper, cobalt, nickel, and manganese (101), the strain C7(β_{ct1}^{tox+}) demonstrates no increase resistance to copper, nickel, and manganese. The pattern of metal-induced inhibition by the lysogen C7(β_{ct1}^{tox+})hm723 is similar to that of C7hm723, suggesting that the host mutation is dominant (101).

A second series of bacteriophage mutants capable of producing diphtheria toxin under varying iron conditions were generate by mutagenesis with EMS, nitrous acid, or NTG (111). Toxin gene expression in the mutants is inhibited at different concentrations of extracellular iron: $\beta^{tox-201}$ by 1200 μg/L, $\beta^{tox-202}$ by 400 μg/L, and $\beta^{tox-203}$ by 200 μg/L. At 75 μg/L Fe^{2+}, toxin production is maximal for each strain and the levels are higher than that of the parent strain. As is true for the parental strain, the accumulation of toxin occurs predominantly during the final phases of bacterial growth (111). This suggests that physiological changes occurring during the terminal stages of growth are important for regulation of diphtheria toxin production by *C. diphtheriae* (111). The *tox*-201 mutation results in a 30- to 45-fold suppression of toxin synthesis when the iron concentration is raised from 75 μg/L to 1000 μg/L; in contrast, parental strain is suppressed by 210- to 475-fold (111). Thus, under optimum low iron conditions, C7($\beta^{tox-201}$) produced four times the amount of diphtheria toxin as C7(β); in contrast, under high-iron conditions, the mutant produced 200 times as much toxin as the wild-type strain. As with β_{ct1}^{tox+}, the mutations are *cis* acting.

F. Mechanism of the Iron-Regulated Synthesis of Diphtheria Toxin

Addition of iron to the medium results in a rapid cessation of diphtheria toxin production (93). The level of toxin expression is increases as a function of iron concentration when the iron concentration in the medium ranges between 16 nM and 16 μM (46,104), and when the intracellular concentration of iron ranges between 1.0 and 2.7 μg Fe/g bacterial protein (93). Addition of 16 μM iron to the medium results in the maximum inhibition of diphtheria toxin production without effecting total protein synthesis (46). Extracellular iron is rapidly taken up by *C. diphtheriae* (104). Following the addition of iron, diphtheria toxin production decreases with a half-life of 6 min (31,46). The kinetics of inhibition is similar to that produced following the addition of transcriptional inhibitors such as rifampin (46), suggesting that the mechanism of iron-induced inhibition is at the level of transcription.

More direct evidence that the inhibition is at the level of transcription comes from hybridization studies. Total RNA or mRNA from membrane-bound polysomes extracted from *C. diphtheria* PW8 during periods of maximal toxin production hybridizes to immobilized β phage DNA to a greater extent than that extracted prior to induction (40,46). Similarly, total RNA isolated from C7(β^{tox+}) grown in the presence of excess iron does not hybridize to any phage DNA restriction fragments (42). Quantitative hybridization of total RNA to DNA fragments containing parts of the *tox* gene demonstrated directly that the presence of iron significantly reduced the level of *tox* mRNA in *C. diphtheriae* C7($\beta^{tox-228}$) (47). As might be expected if the mechanism of iron regulation is at the level of transcription, *tox*-containing message is detectable in the iron-insensitive lysogen C7(β)^{tox+}hm723, but not in the nonlysogenic control C7($-$)$^{tox-}$ or in the iron-sensitive lysogenic C7(β)$^{tox+}$ grown in the presence of high iron concentrations (42).

Initial localization of the region of *tox* responsible for iron regulation comes from mapping several mutations. The *tox*-201 mutation is closely linked to the amino-terminal end of the *tox* structural gene, suggesting that is may be a regulatory region mutation (112). The fact that expression of *tox* message from C7(γ^{tox-}) is regulated by iron despite the insertion sequence at position -54 (see Figure 1) suggests that the region responsible for regulation is upstream of position -54 (36).

In order to explain the above findings, the following model was proposed (109,110) (Figure 5). *C. diphtheriae* produces a diphtheria *tox* aporepressor regardless of its lysogenic state. In the presence of iron, the repressor forms a complex with iron, and the complex binds to the *tox* operator. When the concentration of iron is low, the repressor-iron complex disassociates, resulting in a low affinity for the operator, thereby derepressing *tox* expression. Thus, the repressor-iron complex acts as a negative controlling element for transcription of the *tox* gene. If the number of toxin aporepressor molecules per bacterial cell were limited, this model could explain why toxin is produced by *C. diphtheriae* infected with the hypervirulent phage $\beta hv64^{tox+}$ in the presence of 500 μM iron (88).

G. Identification of the Promoter and Operator Regions of *tox*

The region of the *tox* sequence involved in regulation of transcription in *C. diphtheriae* has been identified through protein-DNA binding, and DNA footprinting (113) (see Figure 4). Crude protein extracts isolated from lysogenic C7($\beta^{tox-228}$) grown in both iron-containing and iron-deficient medium contain proteins which bind to the sequence extending between -114 and $+49$ (see Figure 3), resulting in a shift in mobility when the complex is analyzed by gel electrophoresis (113). Crude extract from cells grown in iron-deficient medium contains proteins which complexed to the DNA protecting the bases extending between nucleotides -84 and -5 on the top strand from digestion by DNase I, and enhancing cleavage of guanidine at position -23 (see Figure 4B). Methylation of the upper strand guanidine residues at position -16, -15, -10, and -8 inhibited this protection (113). This pattern, referred to as "F.exp," is specific to extracts derived from cells grown in the absence of iron, and suggests that it represents interaction between a *C. diphtheriae* RNA polymerase and *tox* promoter (113).

In contrast to the specific protection of the upper stand by extracts derived from cells grown in iron-deficient medium, the lower strand is protected when extracts are isolated from cells grown in both iron-containing and iron-deficient medium (113) (see Figure 4D). This protection, referred to as "F.rep," is increased by the addition of ferrous but not ferric ions to the binding reaction. This sequence overlaps the 27-bp interrupted palin-

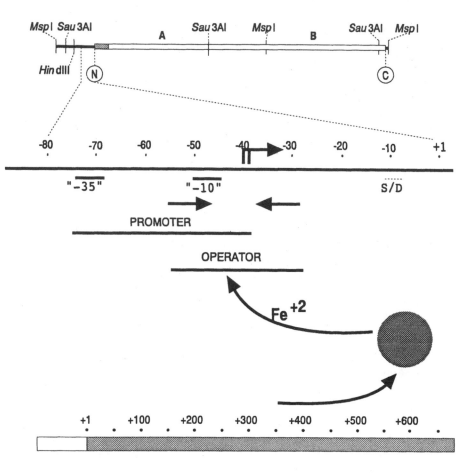

Figure 5 Model for the Fe^{2+}-induced repression of *tox*. The upper line represents the restriction map of *tox* between bases -236 and 1708 given for reference. Line 2 is a map of the regulatory region (see Figure 4) indicating the theoretical promoter and operator regions. For purposes of this figure, the promoter region is defined as the sequence extending from the start of the "-35" consensus sequence and the 5′ end of the mRNA; the operator region is defined as the sequence covering the diad symmetry. The bottom line represents the map of *dtx*R, the presumed aporepressor (115). In the absence of iron, the RNA polymerase can bind to the promoter and initiate transcription of *tox*. In the presence of iron, the constitutively expressed aporepressor is converted to an active repressor and bind to the operator sequence, thereby interferring with transcription initiation.

drome located from -55 to -29, as well as the start site for transcription located at bases -40 and -41. The palindromic nature of the sequence suggests, as with *Fur* protein (114), that *DtoxR*, a corynebacterial Fe^{2+}-sensitive protein, may interact as a dimer.

These results are consistent with a model in which the binding of a repressor to the lower strand inhibits binding of RNA polymerase to the upper strand. This factor, termed *DtoxR*, is present in both DT-expressing and DT-nonexpressing cells, and is activated in the presence of Fe^{2+}, suggesting that it is constitutively expressed (113).

H. The Diphtheria Toxin Gene Aporepressor

A *C. diphtheriae* protein, named *DtoxR* or *dtxR*, has been identified which can bind to the *tox* promoter region (109,113,115). The ability of the iron-dependent repressor to inhibit expression from the *tox* promoter/operator region was utilized in cloning the structural gene for the repressor (115). The *tox* promoter/operator region (extending from positions -80 to -16) was substituted for the *lac* promoter in front of *lacZ*, thereby placing β-galactosidase expression under the control of the *tox* regulatory elements. The construct was placed into the *E. coli* chromosome as a single copy, and the resultant strain was used to screen genomic libraries of nontoxinogenic, nonlysogenic *C. diphtheriae*. A *tox* regulatory element, *dtxR*, was identified which regulated the expression of the *tox*PO–*lacZ* fusion in an iron-sensitive manner.

The deduced amino acid sequence of the presumed *dtxR* protein has an expected molecular weight of 25,316 and has 66% nucleic acid and 25% amino acid homology to the *E. coli fur* protein, the *E. coli* analog protein which represses iron-controlled operons. In addition, the protein sequence has regions of homology to other DNA-binding proteins. Hybridization studies demonstrated that the sequence is also present in C7($-$), PW8($-$), and 1030($-$). Although this protein inhibits expression from the *tox*PO regulatory sequences, it has little effect on expression of *E. coli* proteins regulated by *fur* (115).

V. CONCLUSIONS

The purpose of this chapter was to give a historical review of the genetics of the gene for diphtheria toxin from the perspective of what is currently known. In so doing, it attempted to remind current investigators of the many mutants which were analyzed by the then current genetic techniques. These mutations, when analyzed by current molecular biological tools (e.g., DNA sequencing) may produce a wealth of new information. It also attempts to show how classic genetics can be applied to the understanding of new toxin genes.

ACKNOWLEDGMENTS

I would like to thank Shing Chang and Jon Raymond for careful reading of this manuscript.

REFERENCES

1. Roux, E., and Yersin, A. Contribution A l'etude de la diphterie. Ann. Inst. Pasteur, 2: 629–661, 1988.
2. Eidels, L., and Draper, R. K. Diphtheria toxin. p 217–247. *In*: M. C. Hardegree and A. T. Tu (eds.), Handbook of Natural Toxins, Vol. 4. Bacterial Toxins, pp. 217–247. New York: Marcel Dekker, 1988.
3. Collier, R. J. Diphtheria toxin: Mode of action and structure. Bacteriol. Rev. *39*: 54–85, 1975.
4. Holmes, R. K., and Barksdale, L. Comparative studies with *tox*+ and *tox*+ corynebacteriophages. J. Virol., *5*: 783–794, 1970.
5. Hendrix, R. W., Robers, J. W., Stahl, F. W., and Weisberg, R. A. (eds.). Lambda II. Cold Spring Harbor Laboratory, New York, 1983.
6. Singer, R. A. Lysogeny and toxinogeny in *Corynebacterium diphtheriae*, p. 1–30. *In*: A. W. Bernheimer (ed.), Mechanisms in Bacterial Toxinology, pp. 1–30. New York: John Wiley & Sons, 1976.
7. Barksdale, L. Lysogenic conversion in bacteria. Bacteriol. Rev., *23*: 202–212, 1959.
8. Hewitt, L. F. Mechanism of virulence transfer by bacterial viruses. J. Gen. Microbiol., *11*: 272–287, 1954.
9. Freeman, V. J. Studies on the virulence of bacteriophage-infected strains of *Corynebacterium diphtheriae*. J. Bacteriol., *61*: 675–188, 1951.
10. Freeman, V. J., and Morse, I. U. Further observations on the change to virulence of bacteriophage-infected avirulent strains of *Corynebacterium diphtheriae*. J. Bacteriol., *63*: 407–414, 1952.
11. Groman, N. B. Evidence for the induced nature of the change from nontoxigenicity to toxigenicity in *Corynebacterium diphtheriae* as a result of exposure to specific bacteriophage. J. Bacteriol., *66*: 184–191, 1953.
12. Groman, N. B. Conversion in *Corynebacterium diphtheriae* with phages originating from nontoxigenic strains. Virology, *2*: 843–844, 1959.
13. Parsons, E. I. Induction of toxigenicity in nontoxigenic strains of *C. diphtheriae* with bacteriophages derived from nontoxigenic strains. Proc. Soc. Exp. Biol. Med., *90*: 91–93, 1955.
14. Groman, N. B., and Eaton, M. Genetic factors in Corynebacterium diphtheriae conversion. J. Bacteriol., *70*: 637–640, 1955.

15. Rappuoli, R., Michel, J. L., and Murphy, J. R. Restriction-endonuclease map of Corynebacteriophage ω_c^{tox+} isolated from the Park-Williams No. 8 strain of Corynebacterium diphtheriae. J. Virol., *45*: 524–530, 1983.

16. Groman, N. B. Evidence for the active role of bacteriophage in the conversion of nontoxigenic Corynebacterium diphtheriae to toxin production. J. Bacteriol. *69*: 9–15, 1955.

17. Holmes, R. K., and Barksdale, L. Genetic analysis of *tox*$^+$ and *tox*$^-$ bacteriophages of Corynebacterium diphtheriae. J. Virol., *3*: 586–598, 1969.

18. Crowell, M. J. Morphological and psysiological variations in the descendants of a single diphtheria bacillus. J. Bacteriol., *11*: 65–74, 1926.

19. Groman, N. B. The relation of bacteriophage to the change of Corynebacterium diphtheriae from avirulence to virulence. Science, *117*: 297–299, 1953.

20. Barksdale, W. L., and Pappenheimer, A. M., Jr. Phage-host relationships in nontoxigenic and toxigenic diphtheria bacilli. J. Bacteriol., *67*: 220–232, 1953.

21. Groman, N. B., Eaton, M., and Booher, Z. K. Studies of mono- and poly-lysogenic *Corynebacterium diphtheriae*. J. Bacteriol., *75*: 320–325, 1958.

22. Buck, G., Groman, N., and Falkow, S. Relationship between β converting and γ non-converting corynebacteriophage DNA. Nature, *271*: 683–685, 1978.

23. Holmes, R. K. and Barksdale, L. Recombinational analysis of *Tox*$^+$ phages of *Corynebacterium diphtheriae* C7$_s$(-)$^{tox-}$. Bacteriological Proceedings: Abstracts of the 67th Annual Meeting, Abstract 155, 1967.

24. Singer, R. A. Temperature-sensitive mutants of toxinogenic corynebacteriophage beta: I. Genetics. Virology, *55*: 347–356, 1973.

25. Matsuda, M., Kanei, C., and Yoneda, M. Temperature-sensitive mutants of nonlysogenizing corynebacteriophage β$_{vir}$: Their isolation, characterization and relation to toxinogenesis. Biken J., *14*: 119–130, 1971.

26. Laird, W., and Groman, N. Prophage map of converting corynebacteriophage beta. J. Virol., *19*: 208–219, 1976.

27. Michel, J. L., Rappuoli, R., Murphy, J. R., and Pappenheimer, A. M., Jr. Restriction endonuclease map of the nontoxigenic corynephage γ$_c$ and its relationship to the toxigenic corynephage β$_c$. J. Virol., *42*: 510–518, 1982.

28. Uchida, T., Gill, D. M., and Pappenheimer, A. J., Jr. Mutation in the structural gene for diphtheria toxin carried by temperate phage β. Nature New Biol., *233*: 8–11, 1971.

29. Holmes, R. K. Characterization and genetic mapping of nontoxinogenic (*tox*) mutants of Corynebacteriophage beta. J. Virol., *19*: 195–207, 1976.

30. Laird, W., and Groman, N. Isolation and characterization of *tox* mutants of corynebacteriophage beta. J. Virol., *19*: 220–227, 1976.

31. Uchida, T., Pappenheimer, A. M., Jr., and Greany, R. Diphtheria toxin and related proteins. I. Isolation and properties of mutant proteins serologically related to diphtheria toxin. J. Biol. Chem., *248*: 3838–3844, 1973.

32. Matsuda, M., Kanei, C., and Yoneda, M. A phage-mutant directed synthesis of a fragment of diphtheria toxin protein. Biochem. Biophys. Res. Commun., *46*: 43–49, 1972.

33. Uchida, T., Pappenheimer, A. M., Jr., and Harper, A. A. Reconstitution of diphtheria toxin from two nontoxic cross-reacting mutant proteins. Science, *175*: 901–903, 1972.
34. Uchida, T., Pappenheimer, A. M., Jr., and Harper, A. A. Diphtheria toxin and related proteins. III. Reconstitution of hybrid "diphtheria toxin" from nontoxic mutant proteins. J. Biol. Chem., *248*: 3851–3854, 1973.
35. Laird, W., and Groman, N. Orientation of the *tox* gene in the prophage of corynebacteriophage beta. J. Virol., *19*: 228–231, 1976.
36. Leong, D., and Murphy, J. R. Characterization of the diphtheria *tox* transcript in *Corynebacterium diphtheriae* and *Escherichia coli*. J. Bacteriol., *163*: 1114–1119, 1985.
37. Groman, N., Cianciotto, N., Bjorn, M., and Rabin, M. Detection and expression of DNA homologous to the *tox* gene in nontoxinogenic isolates of *Corynebacterium diphtheriae*. Infection and Immunity *42*: 48–56.
38. National Institutes of Health. Guidelines for research involving recombinant DNA molecules. Fed. Reg., *51*: 16958–16985, 1986.
39. Buck, G. A., and Groman, N. B. Physical mapping of β-converting and γ-nonconverting corynebacteriophage genomes. J. Bacteriol., *148*: 131–142, 1981.
40. Costa, J. J., Michel, J. L., Rappuoli, R., and Murphy, J. R. Restriction map of corynebacteriophage β$_c$ and β$_{vir}$ and physical localization of the diphtheria *tox* operon. J. Bacteriol., *148*: 124–130, 1981.
41. Buck, G. A., and Groman, N. B. Genetic elements novel for *Corynebacterium diphtheriae*: Specialized transducing elements and transposons. J. Bacteriol., *148*: 143–152, 1981.
42. Buck, G. A., and Groman, N. B. Identification of deoxyribonucleic acid restriction fragments of β-converting corynebacteriophages that carry the gene for diphtheria toxin. J. Bacteriol., *148*: 153–162, 1981.
43. Uchida, T., and Yoneda, M. Evidence for the association of membrane with the site of toxin synthesis in *Corynebacterium diphtheriae*. Biochim. Biophys. Acta, *145*: 210–213, 1967.
44. Smith, W. P., Tai, P.-C., Murphy, J. R., and Davis, B. D. Precursor in cotranslational secretion of diphtheria toxin. J. Bacteriol., *141*: 184–189, 1980.
45. Smith, W. P. Cotranslational secretion of diphtheria toxin and alkaline phosphatase in vitro: Involvement of membrane protein(s). J. Bacteriol., *141*: 1142–1147, 1980.
46. Murphy, J. R., Michel, J. L., and Teng, M. Evidence that the regulation of diphtheria toxin production is directed at the level of transcription. J. of Bacteriol., *135*: 511–516, 1978.
47. Kaczorek, M., Zettlmeissl, G., Delpeyroux, F., and Streeck, R. E. Diphtheria toxin promoter function in *Corynebacterium diphtheriae* and *Escherichia coli*. Nucl. Acids Res., *13*: 3147–3159, 1985.
48. Bjorn, M. J., Kaplan, D. A., and Collier, R. J. Identification of DNA restriction fragments of corynebacteriophage β corresponding to hypotoxic peptides of diphtheria toxin. FEMS Microbiol. Lett., *20*: 177–180.

49. Greenfield, L., Bjorn, M. J., Horn, G., Fong, D., Buck, G. A., Collier, R. J., and Kaplan, D. A. Nucleotide sequence of the structural gene for diphtheria toxin carried by corynebacteriophage β. Proc. Natl. Acad. Sci. U.S.A., *80*: 6853–6857, 1983.

50. Kaczorek, M., Delpeyroux, F., Chenciner, N., Streek, R. E., Murphy, J. R., Boquet, P., and Tiollais, P. Nucleotide sequence and expression of the diphtheria *tox*228 gene in *Escherichia coli*. Science, *221*: 855–858, 1983.

51. Giannini, G., Rappuoli, R., and Ratti, G. The amino-acid sequence of two non-toxic mutants of diphtheria toxin: CRM45 and CRM197. Nucl. Acids Res., *12*: 4063–4069, 1984.

52. Kovgan, A. A., and Zhdanov, V. M. Comparative analysis of the DNA of *Corynebacterium diphtheriae* phages and the cloning of the gene which determines the synthesis of diphtheria toxin. Zh. Mikrobiol. Epidemiol. Immunol., *8*: 23–31, 1988.

53. Ratti, G., Rappuoli, R., and Giannini, G. The complete nucleotide sequence of the gene coding for diphtheria toxin in the Corynephage omega (tox$^+$) genome. Nucl. Acids Res., *11*: 6589–6595, 1983.

54. DeLange, R. J., Drazin, R. E., and Collier, R. J. Amino-acid sequence of fragment A, an enzymatically active fragment from diphtheria toxin. Proc. Natl. Acad. Sci. U.S.A., *73*: 69–72, 1976.

55. DeLange, R. J., Williams, L. C., Drazin, R. E., and Collier, R. J. The amino acid sequence of fragment A, an enzymatically active fragment of diphtheria toxin III. The chymotryptic peptides, the peptides derived by cleavage at tryptophan residues, and the complete sequence of the protein. 1979. J. Biol. Chem., *254*: 5838–5842, 1978.

56. Falmagne, P., Capiau, C., Lambotte, P., Zanen, J., Cabiaux, V., and Ruysschaert, J.-M. The complete amino acid sequence of diphtheria toxin fragment B. Correlation with its lipid-binding properties. Biochem. Biophys. Acta, *827*: 45–50, 1985.

57. Inouye, M., and Halegoua, S. Secretion and membrane localization of proteins in *Escherichia coli*. CRC Crit. Rev. Biochem., *7*: 339–371, 1980.

58. Perlman, D., Halvorson, H. O. A putative signal peptidase recognition site and sequence in eukaryotic and prokaryotic signal peptides. J. Mol. Biol., *167*: 391–409, 1983.

59. Michaelis, S., and Beckwith, J. Mechanism of incorporation of cell envelope proteins in *Escherichia coli*. Ann. Rev. Microbiol., *36*: 435–465, 1982.

60. Chang, S. Engineering for protein secretion in gram-positive bacteria. Methods Enzymol., *153*: 507–516, 1987.

61. Reznikoff, W. S., and McClure, W. R. *E. coli* promotors. *In*: W. Reznikoff and L. Gold (eds.), pp. 1–33, Maximizing Gene Expression. Stoneham, Massachusetts: Butterworth Publishers, 1986.

62. Horwitz, M. S. Z., and Loeb, L. Structure-function relationships in *Escherichia coli* promoter DNA. Prog. Nucl. Acid Res., *38*: 137–164, 1990.

63. Schmitz, A., and Galas, D. J. The interaction of RNA polymerase and lac repressor with the lac control region. Nucl. Acids Res., *6*: 111–136, 1979.

64. Metzger, W., Schickor, P., and Heumann, H. A cinematographic view of *Escherichia coli* RNA polymerase translocation. EMBO J., *8*: 2745–2754, 1989.

65. Harley, C. B., and Reynolds, R. P. Analysis of *E. coli* promoter sequences. Nucl. Acids Res., *15*: 2343–2361, 1987.

66. Mulligan, M. E., Hawley, D. K., Entriken, R., and McClure, W. R. *Escherichia coli* promotor sequences predict in vitro RNA polymerase selectivity. Nucl. Acids Res., *12*: 789–800, 1984.

67. McClure, W. R. Mechanism and control of transcription initiation in prokaryotes. Ann. Rev. Biochem., *54*: 171–204, 1985.

68. Ishino, S., Mizukami, T., Yamaguchi, K., Katsumata, R., and Araki, K. Nucleotide sequence of the *meso*-diminopimelate D-dehydrogenase gene from *Corynebacterium glutamicum*. Nucl. Acids Res., *15*: 3917, 1987.

69. Hodgson, A. L. M., Krywult, J., and Radford, A. J. Nucleotide sequence of the erythromycin resistance gene from the *Corynebacterium* plasmid pNG2. Nucl. Acids Res., *18*: 1891, 1990.

70. Yeh, P., Sicard, A. M., and Sinskey, A. J. Nucleotide sequence of the *lys*A gene of *Corynebacterium glutamicum* and possible mechanisms for modulation of its expression. Mol. Gen. Genet., *212*: 112–119, 1988.

71. Losick, R., and Pero, J. Cascade of sigma factors. Cell, *25*: 582–584, 1981.

72. McLaughlin, J. R., Murray, C. L., and Rabinowitz, J. C. Unique features in the ribosome binding site sequence of the gram-positive *Staphylococcus aureus* β-lactamase gene. J. Biol. Chem., *256*: 11283–11291, 1981.

73. Thanaraj, T. A., and Pandit, M. W. An additional ribosome-binding site on mRNA of highly expressed genes and A bifunctional site on the colicin fragment of 16S rRNA from *Escherichia coli*: Important determinants of the efficiency of translation-initiation. Nucl. Acids Res., *17*: 2973–2985, 1989.

74. De Smit, M. H., and Van Duin, J. Control of prokaryotic translation initiation by mRNA secondary structure. Prog. Nucl. Acids Res., *38*: 1–35, 1990.

75. Shine, J., and Dalgarno, L. The 3′-terminal sequence of *Escherichia coli* 16S ribosomal RNA: Complementarity to nonsense triplets and ribosomal binding sites. Proc. Natl. Acad. Sci. U.S.A., *71*: 1342–1346, 1974.

76. Stormo, G. D. Translation initiation. *In*: W. Reznikoff and L. Gold (eds.), Maximizing Gene Expression, pp. 195–224. Stoneham, Massachusetts: Butterworth Publishers. 1986.

77. McCarthy, J. E. G., and Gualerzi, C. Translational control of prokaryotic gene expression. Trends in Genetics, *6*: 78–85, 1990.

78. Petersen, G. B., Stockwell, P. A., and Hill, D. F. Messenger RNA recognition in *Escherichia coli*: A possible second site of interaction with 16S ribosomal RNA. EMBO J., *7*: 3957–3962, 1988.

79. Collins, M. D., Smida, J., and Stackebrandt, E. Phylogenetic evidence for the transfer of caseobacter polymorphus (Crombach) to the genus corynebacterium. Int. J. Syst. Bacteriol., *39*: 7–9, 1989.

80. Cianciotto, N., Serwold-Davis, T., Groman, N., Ratti, G., and Rappuoli, R. DNA sequence homology between *att*B-related sites of *Corynebacterium diph-*

theriae, Corynebacterium ulcerans, Corynebacterium glutamicum, and the *att*P site of γ-corynephage. FEMS Microbiol. Lett., *66*: 299–302, 1990.

81. Gill, D. M., and Dinius, L. L. Observations on the structure of diphtheria toxin. J. Biol. Chem., *246*: 1485–1491, 1971.

82. Raynaude, M., Turpin, A., Mangalo, R., Bizzini, B., and Pery, R. Croissance et toxinogenese. Ann. Inst. Pasteur, Paris, *87*: 599–616, 1954.

83. Kanei, C.-I., Uchida, T., and Yoneda, M. Isolation from *Corynebacterium diphtheriae* C7 (β) of bacterial mutants that produce toxin in medium with excess iron. Infect. and Immun., *18*: 203–209, 1977.

84. Hatano, M. Effect of iron concentration in the medium on phage and toxin production in a lysogenic, virulent *Corynebacterium diphtheriae*. J. Bacteriol., *71*: 121–122, 1956.

85. Yoneda, M., and Pappenheimer, A. M., Jr. Some effects of iron deficiency on the extracellular products released by toxigenic and nontoxigenic strains of *Corynebacterium diphtheriae*. J. Bacteriol., *74*: 256–264, 1957.

86. Miller, P. A., Pappenheimer, A. M., Jr., and Doolittle, W. F. Phage-host relationships in certain strains of *Corynebacterium diphtheriae*. Virology, *29*: 410–425, 1966.

87. Gill, D. M., Uchida, T., and Singer, R. A. Expression of diphtheria toxin genes carried by integrated and nonintegrated phage beta. Virology, *50*: 664–668, 1972.

88. Matsuda, M., and Barksdale, L. Phage directed synthesis of diphtherial toxin in non-toxigenic *Corynebacterium diphtheriae*. Nature, *210*: 911–913, 1966.

89. Matsuda, M., and Barksdale, L. System for the investigation of the bacteriophage-directed synthesis of diphtherial toxin. J. Bacteriol., *93*: 722–730, 1967.

90. Singer, R. A. Lysogeny and toxinogeny in *Corynebacterium diphtheriae*. *In*: A. W. Bernheimer (ed.), Mechanisms in Bacterial Toxinology, pp. 32–52. New York: John Wiley & Sons, 1976.

91. Uchida, T., and Yoneda, M. Estimation of the time required for the process of diphtheria toxin formation. Biken J., *10*: 121–128, 1967.

92. Pappenheimer, A. M., Jr. Diphtheria toxin. Ann. Rev. Biochem., *46*: 69–94, 1977.

93. Righelato, R. C., and van Hemert, P. A. Growth and toxin synthesis in batch and chemostat cultures of *Corynebacterium diphtheriae*. J. Gen. Microbiol., *58*: 403–410, 1969.

94. Hirai, T., Uchida, T., Shinmen, Y., and Yoneda, M. Toxin production by *Corynebacterium diphtheriae* under growth-limiting conditions. Biken J., *9*: 19–31, 1966.

95. Pappenheimer, A. M., Jr., Howland, J. L., and Miller, P. A. Electron transport systems in *Corynebacterium diphtheriae*. Biochim. Biophys. Acta, *64*: 229–242, 1962.

96. Uchida, T., and Pappenheimer, A. M., Jr. Mutation in β-phage genome affecting diphtheria toxin structural gene and toxin yield. *In*: A. Ohsaka, K.

Hayashi, and Y. Sawai (eds.), Anim., Plant Microb. Toxins, Proc. Int. Symp., 4th, pp. 353–362. New York: Plenum, 1976.

97. Matsuda, M., Kanei, C., and Yoneda, M. Degree of expression of the tox^+ gene introduced into corynebacteria by phage derived from diphtheria bacilli having different capacities to produce toxin. Biken J., *14*: 365–368, 1971.

98. Pappenheimer, A. M. Jr. Production and mode of action of diphtheria toxin. *In*: F. Patocka (ed.), Bacterial Toxins and Selected Topics in Virology, pp. 34–57. Prague: Charles University, 1967.

99. Locke, A., and Main, E. R. The relation of copper and iron to production of toxin and enzyme action. J. Infect. Dis., *48*: 419–435, 1931.

100. Pappenheimer, A. M., Jr., and Johnson, S. J. Studies in diphtheria toxin production. I: The effect of iron and copper. Br. J. of Exp. Pathol., *17*: 25–341, 1936.

101. Groman, N., and Judge, K. Effect of metal ions on diphtheria toxin production. Infect. Immun., *26*: 1065–1070, 1979.

102. Clarke, G. D. The effect of cobaltous ions on the formation of toxin and coproporphyrin by a strain of *Corynebacterium diphtheriae*. J. Gen. Microbiol., *18*: 708–719, 1958.

103. Gadre, S. V., and Rao, S. S. Iron and diphtheria toxin production. Prog. Drug Res., *19*: 283–287, 1975.

104. Edwards, D. C., and Seamer, P. A. The uptake of iron by *Corynebacterium diphtheriae* growing in submerged culture. J. Gen. Microbiol., *22*: 705–712, 1960.

105. Mueller, J. H. The influence of iron on the production of diphtheria toxin. J. Immunol., *42*: 343–351, 1941.

106. Cryz, S. J., Russel, L. M., and Holmes, R. K. Regulation of toxinogenesis in *Corynebacterium diphtheriae*: Mutations in the bacterial genome that alter the effects of iron on toxin production. J. Bacteriol., *154*: 245–252, 1983.

107. Kanei, C. I., Uchida, T., and Yoneda, M. Mutants of *Corynebacterium diphtheriae* PW8 that produce toxin in medium with excess iron. Appl. Environ. Microbiol., *42*: 1130–1131, 1981.

108. Murphy, J. R., Pappenheimer, A. M., Jr., and de Borms, S. T. Synthesis of diphtheria *tox*-gene products in *Escherichia coli* Ext. Proc. Natl. Acad. Sci. U.S.A. *71*: 11–15, 1974.

109. Murphy, J. R., Skiver, J., and McBride, G. Isolation and partial characterization of a corynebacteriophage β, *tox* operator constitutive-like mutant lysogen of *Corynebacterium diphtheriae*. J. Virol., *18*: 235–244, 1976.

110. Murphy, J. R., and Bacha, P. Regulation of diphtheria toxin production. *In*: D. Schlessinger, D. (eds.) Microbiology—1979, pp. 181–186. Washington, D.C.: American Society for Microbiology, 1979.

111. Welkos, S. L., and Holmes, R. K. Regulation of toxinogenesis in *Corynebacterium diphtheriae*. I. Mutations in bacteriophage β that alter the effects of iron on toxin production. J. Virol., *37*: 936–945, 1981.

112. Welkos, S. L., and Holmes, R. K. Regulation of toxinogenesis in *Corynebacterium diphtheriae*. II. Genetic mapping of a *tox* regulatory mutation in bacteriophage β. J. Virol., *37*: 946–954, 1981.

113. Fourel, G., Phalipon, A., and Kaczorek, M. Evidence for direct regulation of diphtheria toxin gene transcription by an Fe^{2+}-dependent DNA-binding repressor, DtoxR, in *Corynebacterium diphtheriae*. Infect. Immun., *75*: 3221–3225, 1989.

114. de Lorenzo, V., Wee, S., Herrero, M., and Neilands, J. B. Operator sequence of the aerobactin operon of plasmid ColV-K30 binding the ferric uptake regulator (*fur*) repressor. J. Bacteriol., *169*: 2624–2630, 1987.

115. Boyd, J., Oza, M. N., and Murphy, J. R. Molecular cloning and DNA sequence analysis of a diphtheria *tox* iron-dependent regulatory element (*dtxR*) from *Corynebacterium diphtheriae*. Proc. Natl. Acad. Sci. U.S.A. *87*: 5968–5972, 1990.

16
Diphtheria Toxin Expression in *Escherichia coli*

Lawrence Greenfield   *Roche Molecular Systems, Alameda, California*

I. INTRODUCTION

Escherichia coli was first used for high-level expression of *tox* because of the successful expression of many heterologous genes and the amount of knowledge concerning transcriptional and translational regulation in that organism. As noted in Chapter 15, the presumed promoter, ribosomal-binding site, and secretory leader have many features in common with those of *E. coli*, suggesting that *E. coli* should be able to express diphtheria toxin. In addition, a wide variety of *Corynebacterium* proteins have been expressed in *E. Coli*, including a thioredoxin from *C. nephridii* (1), the prephenate dehydratase gene (*phe*A) from *C. glutamicum* (2), and the *lys*A gene of *C. glutamicum* (3). One factor which might complicate expression is the finding that diphtheria toxin can inhibit amino acid incorporation in *E. coli* and *C. diphtheriae* cell-free systems (4,5).

II. EXPRESSION OF DIPHTHERIA TOXIN IN *E. COLI*

The expression of fragments of the *tox* structural gene in *E. coli* has given a variety of results (Table 1), in part due to the uniqueness of each construct: The position at which termination occurs within the *tox* fragment; the plasmid vehicle used, the *E. coli* host strain; the presence of additional amino acids fused to the carboxyl end; and the addition of *E. coli* promoters and ribosomal binding sites. In addition, the different methods quantifying the levels of *tox* peptides each have inherent problems. The ADP-ribosyltransferase activity differs depending on the length of the diphtheria toxin fragment. In contrast to intact toxin, which has little or no enzymatic activity (6,7), both CRM30 and CRM45 show considerable activity even without nicking (8). In addition, the presence of cold NAD in the bacterial

Table 1 Expression of Fragments of the *tox* Structural Gene in *E. coli*

Clone	Promoter	S/D	DT leader	Start	End
1 pTD44	DT	DT	Yes	<-300	>1920
2 pTD76	DT	DT	Yes	<-300	>1920
3 pDT101	DT, *lac*	DT, *lacZ*	Yes	-179	$+653$
4 pDT201	DT	DT	Yes	-179	$+653$
5 pTD134	DT, Tet?	DT	Yes	<-300	>1920
6 pSDTA10	DT	DT	Yes	-179	$+653$
7 pSDTA20	DT	DT	Yes	-179	$+653$
8 pSDTA30	DT	DT	Yes	-121	$+653$
9 pLSDTA40	DT, PL	DT	Yes	-121	$+653$
10 pTrpDTA100	*trp*	trpL	No	$+76$	$+653$
11 pLDTA200	P_L	gene N	No	$+76$	$+653$
12 pLOPDTA1000	P_L	gene N	No	$+76$	$+653$
13 pLOPMsp	P_L	gene N	No	$+76$	$+1220$
14 pRTF1A	DT	DT	Yes	-179	$+653$
15 pRTF1B	DT	DT	Yes	-179	$+653$
16 pRTF2A	DT	DT	Yes	-236	$+1219$
17 pRTF2B	DT, Tet?	DT	Yes	-236	$+1219$
18 pDT201	DT	DT	Yes	-179	$+653$
19 pABC313	DT	DT	Yes	-121	$+945$
20 pABM313	DT	DT	Yes	-121	$+945$
21 pABC402	DT	DT	Yes	-121	$+1206$
22 pABC508	DT	DT	Yes	-121	$+1528$
23 pABM508	DT	DT	Yes	-121	$+1528$
24 pDT1201	P_R	*cro*	Alt[13]	$+3$	$+653$
25 pABC1313	P_R	*cro*	Alt[13]	$+3$	$+945$
26 pABM1313	P_R	*cro*	Alt[13]	$+3$	$+945$
27 pABC1402	P_R	*cro*	Alt[13]	$+3$	$+1206$
28 pABM1402	P_R	*cro*	Alt[13]	$+3$	$+1206$
29 pABC1508	P_R	*cro*	Alt[13]	$+3$	$+1528$
30 pABM1508	P_R	*cro*	Alt[13]	$+3$	$+1528$
31 pABM508	DT	DT	Yes	-121	$+1528$
32 pABM1508	P_R	*cro*	Alt[13]	$+3$	$+1528$
33 pABM4508	P_R	*cro*	No	$+58$	$+1528$
34 pABM6508	P_{trc}	syn	No	$+76$	$+1528$
35 pβ197	DT	DT	Yes	-121	$+1755$
36 pDO1	ptac		Yes		
37 pDO2	ptac		Yes		
38 pKK-B6	ptrc	*lacZ*	No	$+658$	$+1755$
39 pCTE1	p_R	*cro*	No	$+58$	>1920
40 pCTF1	p_R	*cro*	No	$+58$	$+1363$
41 pBRDT-S148	DT	DT	Yes	-236	$+1755$

No. DT amino acids	Amino acids added to		Transc. termin.	Vector
	Amino	Carboxyl		
535	No	No	Pos[a]	pBR322
535	No	No	Pos2	pBR322
195	No	1022[c]	No	pUC8
194	No	29[d]	No	pUC8
535	No	No	Pos2	pBR322
194	No	37[e]	No	pBR322
198	No	No	No	pBR322
198	No	No	No	pBR322
198	No	No	No	pBR322
198	No[f]	No	No	pDG141
198	No[f]	No	No	pDG141
198	No[f]	No	No	pLOP
383	No[f]	No	No	pLOP
194	No	37[e]	No	pBR322
194	No	7[g]	No	pBR322
382	No	6[h]	No	pBR322
382	No	10[i]	No	pBR322
194	No	29[d]	No	pUC8
290	No	4[j]	No	pEMBL8
290	No	16[k]	No	pEMBL8
377	No	9[l]	No	pEMBL8
485	No	4[j]	No	pEMBL8
485	No	16[k]	No	pEMBL8
194	No	29[d]	No	pEBMLex3
290	No	4[j]	No	pEMBLex3
290	No	16[k]	No	pEMBLex3
377	No	9[l]	No	pEMBLex3
377	No	21[n]	No	pEMBLex3
485	No	4[j]	No	pEMBLex3
485	No	16[k]	No	pEMBLex3
485	No	16[k]	No	pEMBL8
485	No[b]	16[k]	No	pEMBLex3
485	7[o]	16[k]	No	pEMBLex3
485	No[b]	16[k]	Yes	pKK233-2
535	No	No	Pos[b]	pEMBL8
535	No	No	Pos[b]	pBR322
200	No	2[p]	No	pBR322
342	3[q]	No	Yes 2, 15	pKK233-2
535	9[r]	No	Pos[b]	pCQV2
429	9[r]			
535	No	No	Pos[b]	pBR322

Table 1 Continued

Plasmid	Expression level[s]		Secretion detected	Proteolysis detected	Ref.
	ELISA	Enzymatic			
1 pTD44	50–100μg/L				18
2 pTD76	50–100μg/L				18
3 pDT101					22
4 pDT201			Yes		22
5 pTD134	50–100μg/L			No	17
6 pSDTA10	0.03%	0.008 n/m/m	Yes		20
7 pSDTA20	0.04%	0.008 n/m/m			20
8 pSDTA30	0.03%	0.003 n/m/m			20
9 pLSDTA40	0.06%	0.021 n/m/m			20
10 pTrpDTA100	0.38	0.29 n/m/m			20
11 pLDTA200	2.59	1.73 n/m/m			20
12 pLOPDTA1000	7.06	5.45 n/m/m		No	20
13 pLOPMsp	5–10%				20
14 pRTF1A		5 μg/L[t]	Yes	Yes	21
15 pRTF1B		5 μg/L[t]	Yes	Yes	21
16 pRTF2A		5 μg/L[t]	Yes	Yes	21
17 pRTF2B		5 μg/L[t]	Yes	Yes	21
18 pDT201		1.4 mg/L		No	49
19 pABC313		5.5 mg/L		Yes	49
20 pABM313		6.0 mg/L		Yes	49
21 pABC402		3.7 mg/L		Yes	49
22 pABC508		5.0 mg/L		Yes	49
23 pABM508		4.2 mg/L		Yes	49
24 pDT1201		0.6 mg/L		No	49
25 pABC1313		1.3 mg/L		Yes	49
26 pABM1313		1.1 mg/L		Yes	49
27 pABC1402		1.7 mg/L		Yes	49
28 pABM1402		1.8 mg/L		Yes	49
29 pABC1508		1.2 mg/L		Yes	49
30 pABM1508		0.9 mg/L		Yes	49
31 pABM508	0.4 mg/L			Yes	29
32 pABM1508	0.7 mg/L			Yes	29
33 pABM4508	0.2 mg/L			Yes	29
34 pABM6508	10.0 mg/L			Yes	29
35 pβ197				No	29
36 pDO1		0.5 mg/L[u]			48
37 pDO3		0.5 mg/L[u]			48
38 pKK-B6	2.0 mg/L			No	76
39 pCTE1				Yes	69
40 pCTF1					69
41 pBRDT-S148			Yes		44

(Table 1 footnotes on p. 311)

^aSummary of expression levels obtained from several constructs. The columns are as follows: clone represents the plasmid designation; promoter indicates the origin of the promoter; S/ D indicates the Shine/Dalgarno sequence used; DT indicates whether the diphtheria toxin secretory leader was used in its native configuration (yes), altered (DT), or not used at all (no); start represents the first and end the last nucleotide derived from the β phage sequence (see Figure 3 for numbering); the number of amino acids of the mature toxin is indicated, followed by columns indicating the number and sequence of amino acids added to either the amino- or carboxy-terminal ends; transc. termin. indicates the use of a transcription terminator; vector represents the vector used in the construct; expression levels gives the levels as quoted in the publication.

^bThere is a possible termination sequence present downstream of the *tox* gene which is present in the β-phage genome.

^c8 amino acids from polylinker (Val of polylinker same as in DT), 1015 amino acids from β-galactosidase—ArgArgSerVal:AspLeuGlnProSerLeu-β-galactosidase (amino acids 7–1021).

^d29 amino acids from pUC8—ArgArgSer:ProGlyIleArgAsnHisGlyHisSerCysPheLeuCys-GluIleValIleArgSerGlnPheHisThrThrTyrGluProGluAla.

^e30 amino acids from pBR322—ArgArgSer:SerThrProAspSerSerTrpProAlaSerProAlaPro-GlnValArgLeuLeuAlaProIleSerProThrSerProMetGlyLysIleGlyLeuAlaThrSerGlySer.

^fPossibly Met, depending on if *E. coli* processes off the amino-terminal N-formyl Met.

^g7 amino acids from pBR322—ArgArgSer:ThrGlyArgValTrpSerPro.

^h6 amino acids from pBR322—SerPro:MetIleSerCysGlnThr.

ⁱ10 amino acids from pBR322—SerPro:IleSerPheAsnAlaValValTyrHisSer.

^jSynthetic 4 amino acids: AlaAlaAlaCys.

^kSynthetic, 3 alanines with MSH: AlaAlaAlaAlaSerTyrSerMetGluHisPheArgTrpGlyLys-ProVal.

^lSynthetic: ArgSerValAspArgProAlaAlaCys.

^mAlt:diphtheria toxin secretory leader altered by the addition of 2 amino acids at positions 2 and 3 such as the sequence MetSerArgLysLeuPheAlaSer . . . becomes MetAspProSer-ArgLysLeuPheAlaSer . . . .

ⁿ21 amino acids: synthetic 7 to α-MSH: AsnArg:ArgSerValAspArgProAlaAlaSerTyrSerMet-GluHisPheArgTrpGlyLysProVal.

^oFrom digestion of pAMB1508 with *Bam*HI and *Apa*I followed by blunt end formation with mung bean nuclease and ligation: MetProProSerAlaHisAla.

^pAdded LysSer.

^qAdded MetGlyGly, Met may be processed off.

^rAdded from fusion of *Bam*HI linker to the blunt-ended *Apa*I site in secretory leader to amino terminal end of processed protein: MetAspProProProSerAlaHisAla.

^s% indicates % of total cell protein that is immunologically cross-reactive with DTA; n/m/m indicates nmol/min of ADP-ribosyltransferase activity per milligram of total cell protein.

^t150 molecules enzymatically active peptide per cell in log phase growth: assuming Mwt fragment A of 21,145, translates to 5 μg/L.

^uUninduced.

cell alters the specific activity of the [^{14}C]NAD, making determination of specific activity more difficult. In the case of ELISA assays, *E. coli* proteins frequently cross-react with the antisera used. Furthermore, different parts of toxin molecule elicit varying degrees of immune response. All of these factors must to considered when trying to identify the role of each change in altering expression.

Recently, methods and vectors have been developed permitting the successful cloning in *Corynebacterium*. A. *C. glutamicum–Bacillus subtilis* chimeric vector has been successfully transformed into *Corynebacterium* with a frequency of 10^4 transformants per micrograms of plasmid DNA (9). This may permit further manipulation of *tox* in its original host.

A. Expression of Diphtheria Toxin In Vitro by *E. coli* Protein-Synthesizing Extracts

Diphtheria toxin can be synthesized by *E. Coli* S-30 protein synthesizing extract when programmed with purified DNA from corynebacteriophages $\beta_c^{tox^+}$ and β_c^{tox-45} (10), total RNA isolated from C7(β) and *C. diphtheriae* Park-Williams strain No. 8 (PW8) (11), or messenger RNA isolated from membrane-bound polysomes purified from *C. diphtheriae* strain PW8 grown under iron-limiting conditions (12,13). The synthesis is unaffected by high concentrations of iron, but can be inhibited by the addition of extracts from $C7_s(-)^{tox^-}$ (10). A precursor of molecular weight 66,000–68,000 ds is formed, which can be converted to the mature 62,000 d form by the addition of *E. coli* outer membrane (12,13). Addition of inverted *E. coli* inner membrane vesicles during the initial stages of in vitro translation results in intravesicular segregation of mature toxin (13). As has been found with many secreted *E. coli* proteins, vesicles cannot sequester the toxin if added after the toxin precursor has been completed (13). Vesicles formed from *E. coli* membranes in which their cytoplasmic surface has been first treated with pronase cannot segregate the toxin, suggesting that membrane-bound proteins are involved in the secretion process.

Therefore, in vitro, the *E. coli* synthetic machinery can transcribe β phage DNA, translate the resultant message into pro-diphtheria toxin, process the secretory leader, and secrete the mature form of toxin. This synthesis is unaffected by the presence of excess iron, but can be inhibited by addition of the *C. diphtheriae* repressor.

B. Expression of Diphtheria Toxin Peptides in *E. coli* from Its Native Regulatory Sequences

Despite the ability of *C. diphtheriae* to produce diphtheria toxin at 400–500 mg/L under optimal conditions (14–16), the initial expression in *E.*

coli was low. Cloning the entire structural gene for a mutant toxin, *tox*-228, as a 3.9 kb *Bam*HI fragment into plasmid pBR322 (pTD44, pTD76, pTD134) results in the synthesis of 50–100 ng/ml of the *tox*-228 product which is secreted into the periplasmic space of *E. coli* without additional amino acids fused at either end (see Table 1) (17,18). The product is expressed independent of the orientation of the inset, suggesting that the *tox* promoter and ribosomal binding sites are functional in *E. coli*.

Cloning of the 832-bp *Sau*3A fragment (positions −179 to +653) into pBR322 (pSDTA10, pSDTA20, pRTF1A, pRTF1B) (19–21) and pUC8 (pDT101, pDT201) (22,23) produces low levels of a fragment A–like peptide when expressed from the native DT promoter and ribosomal binding site (see Table 1). As a result of the location of the *Sau*3AI site within the structural gene, the expressed peptides are fused to plasmid-encoded amino acids until the first in-phase codon. However, directed termination within the diphtheria toxin sequence (pSDTA20) has no effect on the expression level. The orientation of the fragment within the vector has little effect on expression levels and the peptides were secreted into the periplasmic space (20–22). Placement of additional promoters in front of the *tox* promoter (pLSDTA40, pDT101), minimally increases the levels of expression (20,22) suggesting that inefficient transcription initiation is not the major factor limiting expression.

Similar results are obtained with a 1455-bp *Msp*I fragment (bases −236 to +1219; pRTF2A, pRTF2B) encoding all of fragment A and 189 amino acids of fragment B (19–21). Again, secretion and expression levels are independent of the orientation of the fragment within the vector and the number of plasmid-encoded amino acids fused to the carboxyl end of the peptide. In contrast to other expressed *tox* fragments, the 43,000-d product resulting from fusion of the *Msp*I encoded sequences to 10 plasmid-encoded amino acids (plasmid pRTF2B) appears to result in an enzymatically inactive peptide following purification from an SDS-polyacrylamide gel (21).

Therefore, despite the homology between the *tox* regulatory region and other *E. coli* promoters, expression of *tox*-related peptides is low in that organism. There are a variety of factors which may contribute to the low levels of expression. The short spacing of the "−10" and "−35" consensus sequences may result in weak transcription; the length and composition of this spacer is known to significantly effect promoter activity in *E. coli* (24–26). Alternatively, the high incidence of codons which are rarely used in *E. coli* may limit translation efficiency (reviewed in Ref 27); for example, removing rare codons from the sequence of tetanus toxin fragment C boosted the levels of expression of *E. coli* three- to fourfold (28). In addition, it has been pointed out that the mRNA is capable of forming a stem-loop structure with free energy of −7.1 kcal within the sequence from $+62$ to 82 (29). Finally, message stability (30,31) and proteolysis (see below) may also limit the overall product yield.

C. Mapping Sequences With the Diphtheria Toxin Operon that Function as a Promoter in *E. coli*

As previously discussed (Chapter 8), sequences extending between positions -114 to $+49$ (Figure 1B) probably are involved in promoter activity in *C. diphtheriae* (32). In addition, this region contains sequences highly homologous to the "-35" and "-10" regions associated with transcription initiation in *E. coli* (see Section II.B.2) (Figure 1A). Defining those sequences involved in promoter activity in *E. coli* should permit further manipulation of the region in order to elevate the expression levels of *tox*-related peptides.

Mapping of the regions of *tox* regulatory sequences acting as the promoter in *E. coli* was facilitated by fusing to monitorable proteins. A *Hae*III fragment (positions -311 to $+54$) can drive both transcription and translation of β-galactosidase fused in-phase to the first 18 amino acids of the DT secretory leader (17). Similarly, the sequence extending between positions -179 and $+54$ is able to drive transcription of galactokinase (*galK*) gene (23). The *tox* sequences express 10-fold higher levels of *galK* when the orientation of the fragment is toward the *galK* sequences. Sequences extending between -121 and $+54$ and -111 and -44 directed similar levels of *galK* expression (23). In vivo, removal of sequences upstream from the *Hind*III site (plasmid pSDTA30; see Table 1) does not alter the level of expression of DTA-like peptides (20). In addition, the *tox* sequences extending between -80 and -16 can drive low levels of expression of β-galactosidase in vivo (33).

The sequences homologous to the *E. coli* "-35" and "-10" consensus sequences both lie within the -80 to -16 stretch (Figure 1A). However, there are two possible "-10" sequences: TATAAT (-56 to -51) and TAGGAT (-50 to -45), with the latter having favorable spacing (see Section II.B.2). Mutational analysis combined with quantification and identification of the mRNA produced in *E. coli* harboring the modified sequences identified the TAGGAT as the major "-10" sequence involved in transcription initiation (39). Alteration of the TATAAT sequence to AATAAT did not reduce the major transcript initiated at positions $-39/$ -40, but eliminated the minor transcript starting at position -45 (Figure 1C1). In contrast, changing the TAGGAT sequence to AAGGAT dramatically reduced the $-39/-40$ transcript with a compensatory increase in the -45 transcript to a level 2.5 times that of the $-39/-40$ mRNA observed in *E. coli* with the wild-type sequence (Figure 1C2). Changing

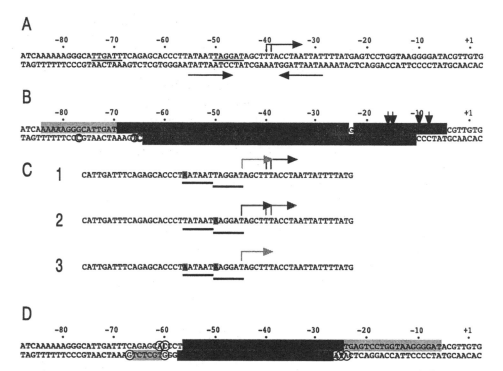

Figure 1 *tox* Promoter and operator region. The DNA sequence of the *tox* regulatory region (A) indicates the "−35" (TTGATT) and "−10" (TAGGAT) consensus sequences, the 5′ end of the mRNA (T at positions −40 and −39) and the diad symmetry believed to be involved in repressor binding (inward pointing arrows). The sequence protected from DNAse I digestion by binding of extracts from *C. diphtheriae* (B), referred to as "F.exp" (taken from Ref. 32) shows both strongly protected (dark box) and weakly protected (light box) regions. Methylation at the G residues at positions −16, −15, −10, and −8 (arrows) prevent the extract from protecting the region from DNAse I. The circles bases on the bottom strand (−65, −66, and −77) are positions of enhanced cleavage. Mutational analysis of the two possible "−10" consensus sequences (C) indicate that changing the sequence TATAAT to AATAAT (C1) decreased the amount of mRNA beginning at position −45, whereas substitution of the sequence AAGGAT for TAGGAT (C2) decreases the amount of the mRNAs initiated at positions −40 and −39. Modification of both sequences decreases all *tox* mRNA (taken from Ref. 39). The DNA sequence protected from DNAse I digestion by binding of an *C. diphtheriae* aporepressor in the presence of Fe^{2+} (F.rep.) (D) results in regions that are maximally protected (dark box) and variably protected (light box) as well as bases whose cleavage is enhanced (circled bases). (From Ref. 32.)

both sequences simultaneously decreased the level of all transcripts (Figure 1C3). Measurement of the ADP-ribosyltransferase activity produced in *E. coli* containing each of the modified sequences gives results consistent with those from mRNA quantification: The levels of activity are 60–90% of the wild-type activity for the $-56A$ mutation (Figure 1C1), 40–60% for the $-50A$ mutation (Figure 1C2), and 5–15% for the double mutation (Figure 1C3). Thus, although both "-10" homology regions function in *E. coli*, the sequence at position -50 to -45 predominates.

There is no obvious "-35" consensus with the correct spacing for the "-10" consensus at position -56 to -51. However, positively-regulated promoters frequently do not have a "-35" consensus sequence (34,35). It is interesting that the *gal* operon has two overlapping promoters, one of which is positively regulated (36).

Attempts have been made to quantitate the efficiency of the *tox* promoter in *E. coli* by measuring its ability to drive expression of other genes. The sequence extending between positions -179 and $+54$ expresses *galK* at levels intermediate between those of the *gal* and *lac* promoters in the same vectors when grown in medium containing 1% fructose (23). However, both the *lac* and *gal* promoters are subject to catabolite repression; in the presence of a number of carbohydrates, *E. coli* represses the synthesis of a number of operons involved with carbohydrate metabolism (36,37). Fructose and glucose are equally effective in catabolite repression (36). Therefore, the activities of the *lac* and *gal* promoters in a cya^+ and crp^+ cell grown in 1% fructose is likely to be significantly repressed, thereby underestimating their maximum efficiency. In another series, the DT sequences extending between -311 and $+54$ were substituted for the *lac* promoter in *lacZ* such that β-galactosidase was fused in frame to the first 18 amino acids of the DT secretory leader (17). In this construct, the diphtheria toxin promoter/ribosomal–binding site is estimated to be 10% as efficient as the *lac* promoter. Furthermore, it is suggested that the *tox* promoter is 40 times as active in *C. diphtheriae* than as in *E. coli* (17).

Attempts have been made to elevate the expression levels by the addition or substitution of other promoters. Addition of the strong lambda p_L promoter (pLSDTA40; see Table 1) only marginally increased the amount of DTA-like peptide produced (20). The finding that driving *tox* transcription by other promoters does not greatly increase the levels of expression supports the conclusion that transcription initiation is not the main factor limiting expression of *tox*-related peptides in *E. coli*.

Having located the region of *tox* responsible for transcription initiation in both *E. coli* and *C. diphtheriae*, it is of interest to define the 5′ end of the message. Mapping experiments with *E. coli*-produced mRNA by both S1 digestion of RNA/DNA hybrids and reverse transcriptase of the mRNA using specific DNA primers, located the same 5′ end to the T at position −39 (±1 base) (17,38,39). As in *C. diphtheriae*, a second minor transcript appears to be initiated at position −45 (39).

D. Secretion by the Native *tox* Secretory Leader in *E. coli*

Gram-negative bacteria contain an inner cytoplasmic membrane and an outer membrane both of which bound the periplasmic space. The process of secretion, transport of proteins from their site of synthesis to the periplasmic space, is quite complex. Most secreted proteins are made as precursors with an N-terminal secretory leader which is removed by signal peptidase (40). The ability of secretory leaders to insert into membranes correlates with their effectiveness at initiating secretion (40). In order to be properly secreted, proteins must have sequences that are compatible with the translocation process (40). The products of several genes are required for secretion across the inner membrane: *secA*, *secB*, *secD*, *secE*, *secF*, and *secY* (or *prlA*). Located in the peripheral membrane, SecA protein is stimulated by translocatable substrates and has ATPase activity (41). SecD and secF proteins are integral membrane proteins which may act at late steps in protein export (42. SecB protein, a cytoplasmic protein which is required in translocation of a subset of secreted proteins, maintains the precursor form in an unfolded state (40,43). SecY and SecE proteins are integral membrane proteins. Secretion of most proteins in *E. coli* requires the SecA, SecY and SecE proteins (40).

Constructs which contain the DT secretory leader (see Table 1) translocate the DT-like product into the periplasmic space (19–23),29,44,48): This includes whole toxin, DTA-like peptides, and CRM45-like peptides. Similarly, secretion in *C. diphtheriae* does not require the entire *tox* structural gene: both CRM30 and CRM45 are released in amounts comparable to that of intact toxin (8). Secretion of these peptides in *E. coli* appears efficient, in that 89% of the DTA-like peptide (amino acids 1–193) and 68% of the CRM45-like peptide (amino acids 1–383) are located in the periplasmic space; the remaining material is found in the cytoplasm and cytoplasmic membrane (21). The less efficient secretion of the CRM45-like peptide may results from interference by the hydrophobic portion of the B moiety contained within the fragment. Similarly, the apparent insolubility of the ABM408 protein whereupon most of the protein remained with the membrane fraction after French press lysis or the spheroplast fraction after lysozyme-EDTA treatment may also be due to inefficient secretion resulting in the protein remaining membrane-associated (29).

Secretion *E. coli* of a DTA-like peptide requires the *secA* locus (23). Despite the addition of 29 amino acids at the carboxyl end, *E. coli* efficiently secretes the DTA-like peptide encoded by clone DT201. The product is located in the periplasmic space and has a molecular weight consistent with the construct (i.e., 28,500 d). Expression in a temperature-sensitive *secA* strain under nonpermissive conditions results in the cytoplasmic accumulation of a precursor 31,000-d form (23).

In a series of constructs intended to express secreted product using the strong lambda p_R promoter, the *tox* secretory leader was altered in order to facilitate cloning: the amino-terminal sequence MetSerArg . . . was replaced by the sequence MetAspProSerArg (pDT1201, pABC1313, pABM1313, pABC1402, pABM1402, pABC1508, and pABM1508; see Table 1) (49). The expression of *tox*-related peptides from these clones is less than those from clones containing the *tox* promoter and normal secretory leader (pDT201, pABC313, pABM313, pABC402, pABC508, and pABM508). The positively charged amino-terminal end of signal peptides are know to play an important role in protein secretion. Alteration of the net charge at the amino terminal end of the secretory leader of prolipoprotein decreases the kinetics and alters the pathway of secretion (45,46).

E. Effect of Expressing Portions of the Diphtheria Toxin B Chain on *E. coli* Growth

Secretion of peptides encoded by the *Msp*I fragment (bases −236 to +1219; amino acids 1–382) affect *E. coli* growth (20). In contrast to bacteria secreting fragment A (amino acids 1–193), those expressing the first 189 amino acids of fragment B develop smaller colonies, grow to lower cell densities in broth, and form fragile protoplasts (20). A similar inhibition of growth may be seen following induction of the *Msp*I-encoded peptide by *E. coli* harboring plasmid pTKW1 (47) and plasmid ptacF2-E148S (48). Expression of from the P_R promoter of *tox* peptides terminating at amino acids 193, 290, 377, and 485 also results in growth cessation of the host upon induction of the protein by raising the temperature of the culture (49). Thus, it appears that the deleterious effect associated with expression of portions of the B fragment are unrelated to the methods of induction: pMSP (20) and pTKW1 (47) are expressed constitutively from the *tox* promoter; pLSMsp (20), and the P_R-derived constructs are induced by raising the temperature of the culture; and ptacF2–E148S is induced by addition of isopropyl-β-D-thiogalactopyranoside (IPTG) (48).

Several factors may contribute to expression-related lethality. The *Msp*I-encoded peptide includes three of the four hydrophobic domains (amino acids 269–289, 301–321, 338–358, 418–438) suggested to be involved with membrane insertion and translocation of fragment A across membranes (50). The *Msp*I-encoded peptide may interact with the *E. coli* membrane in that CRM45, which is only 4 amino acids longer (51), can insert into lipid bilayer membranes forming small channels which permit leakage of ions (52,53). Although fragment A can insert into membranes and form ion-conductance channels (54,55), its membrane lesions are smaller and unstable compared to that of whole toxin (55). It has been proposed that both parts of fragments A and B may work in concert to develop the membrane lesions necessary for the translocation of the catalytic portion across the lipid bilayer (55). Therefore, the membrane lesion may be larger and more permanent when portions of B chain are included in the constructs.

Interaction of the *Msp*I-encoded peptide with the membrane may be augmented when expression involves secretion. The *Msp*I-encoded peptide is secreted less efficiently than the *Sau*3AI-encoded peptide (21). This peptide terminates close to a hydrophobic stretch (50); it is possible that sequences within the remaining portion of fragment B are necessary for efficient secretion. With the P_R constructs (49), the efficiency of secretion may be further reduced owing to the alteration of the secretory leader by the insertion of Asp and Pro between amino acids 1 and 2. Thus, less efficient, altered secretion could explain why "overexpression" lethality is observed only when expression is under P_R control and not P_{tox} control, despite the higher levels of expression from the latter constructs (49). In addition, lethality may result from tying up of the secretory apparatus by inefficiently secreted protein, as has been observed with expression and secretion of high levels of *malE–lacZ* fusions (56).

Expression and secretion of full-length diphtheria toxin or fragments containing amino acids of the B chain sensitizes the host *E. coli* to low pH (48). Artificial acidification of the media results in inhibition of growth and cell death due to disturbances of inner membrane functions (e.g., membrane potential, active transport, and ion impermeability) without lysis. The viability of uninduced cultures producing low levels of full-length toxin is decreased 50-fold at pH 6 and 10^7-fold at pH 5. Following induction, this lethality is increased; at pH 6 viability of induced cultures is decreased by 10^5-fold. *Msp*I-constructs (amino acids 1 to 382) have similar inhibitions at pH 5 but not at the higher pH. The lethal effect was not observed at pH 7 nor in constructs expressing only fragment A of diphtheria toxin despite the ability of fragment A itself to cause pH-dependent lesions in lipid bilayers (54,55). Thus, the lethality is dependent on the pH, the amount of toxin peptide produced, and the amount of B chain encoded by the construct. It was proposed that these changes signify insertion of the diphtheria toxin peptide into the inner membrane, thereby altering the bacterial permeability barrier. The pH-dependence is related to the ability of whole toxin and truncated forms to undergo conformational changes when exposed to low pH (57,58) and to induce pores in lipid bilayers (52,59).

The results of the artificial acidification of the medium may be related to the "overexpression" lethality observed with expressing portions of the B chain in *E. coli*. Growing bacteria produce acetic acid (60) and may disproportionately accumulate cations over anions, both resulting in a progressive fall in the pH of the medium (61). Furthermore, it has been suggested that, owing to Donnan equilibrium the pH in the periplasm should be significantly below that in the medium (62). Thus, it is conceivable that under some growth conditions, the pH of the periplasm may achieve the pH range which facilitates membrane insertion of portions of the B chain when *tox*-related peptides are secreted.

F. Proteolysis of *tox*-Related Peptides in *E. coli*

The accumulation of an expressed protein in *E. coli* is dependent on its degradation rate as well as its rate of synthesis (63,64). Proteins which are rapidly degraded include incomplete polypeptides, complete proteins containing amino acid substitutions, free subunits of large multimeric complexes, posttranslationally damaged proteins, and certain polypeptides synthesized through recombinant DNA technology (64). In many cases, the abnormal folding of the protein contributes to its increased proteolysis rate. For example, disulfide bonds of proteins synthesized within the reducing environment of the *E. coli* cytosol (65,66) are not formed, contributing to their abnormal folding. Proteolytic degradation occurs in many steps. The first step involves an ATP-dependent endoprotease which cleaves larger proteins into smaller polypeptides; next, the smaller polypeptides are cleaved by endoproteases into small peptides which are then cleaved by soluble di- and tripeptidases into amino acids (64).

There are eight soluble endoproteases in *E. coli* (64). Five serine proteases are active against large proteins (proteases Do, Re, Mi, Fa, and So); three are cytosolic and two are periplasmic. Two are metalloendoproteases that cleave smaller polypeptides: One is cytosolic (protease Ci) and one is periplasmic (protease Pi). Finally, protease La, the product of the *lon* locus, is an ATP-dependent cytosolic protease active against large proteins. The initial, rate-limiting step in the proteolysis of most abnormal proteins appears to be cleavage by protease La (64). Not all abnormal proteins are substrates for protease La; *lon* mutations do not affect the degradation of many labile cloned polypeptides, especially shorter ones (64). Polypeptides with hydrophobic stretches of amino acids on their surface may be a preferred substrate of protease La.

Protease La is induced by large amounts of abnormal polypeptides as well as growth of the bacterial strain at elevated temperatures (e.g., 42°C, conditions frequently used to induce the P_R and P_L promoters). Transcription is regulated by the product of the *htpR* locus, the σ^{32} subunit of RNA polymerase (67). The σ^{32} subunit is involved in regulating 17 *E. coli* heat-shock proteins, some of which may be proteases other than protease La (63,64,68). Although *lon* mutants may reduce the degradation rate of many abnormal proteins and cloned products, *htpR lon* double mutants are preferred because they offer several advantages (e.g., lower levels of several other proteases, lack of over production of capsular polysaccharides, lack of ultraviolet light sensitivity) (63,64).

The expression of *tox*-related peptides in *E. coli* is an ideal substrate for proteolytic cleavage. It is a foreign protein of *E. coli*. Because of the NIH guidelines, only abnormal toxin can be expressed: either missense mutations (17,18,44,48,69) or termination fragments (19–23,29,49). The proteolytic sensitivity of diphtheria toxin fragments and missense mutations has been previously observed in *C. diphtheriae*. Fragment B45 (the B fragment of CRM45) is more sensitive to trypsin digestion than wild-type B fragment (8). CRM30 is more sensitive to proteolysis during purification, and results in both free fragment A and products of molecular weight intermediate between fragment A and 30,000 (8). Similarly, the missense mutations within *tox*-228 also increase the proteolytic sensitivity of the B fragment of CRM228 (8). When introduced into the erythrocyte cytoplasm, fragment A is stable with a half-life of greater than 24 hr; in contrast, fragment B and the A chains from CRM176 and CRM197 are rapidly degraded, with half-lives close to 2.5 hr (70). In addition, the reducing environment of the *E. coli* cytosol will prevent correct oxidation of the two disulfide bonds when the protein is expressed intracellularly (71). Finally, many of the expression vectors utilized promoters (P_R or P_L) that are induced by raising the temperature (20,29,49,69), conditions known to increase the level of protease expression in *E. coli* (68,72).

The degree of proteolysis of *tox* peptides expressed in *E. coli* is variable, and in some cases is responsible for lack of product detection. Factors contributing to the variability may include differing hosts, levels of expression, methods of analysis, site of termination of the *tox* sequences, and solubility of the product. Furthermore, the fusion of plasmid-encoded amino acids to the carboxyl end of the *tox*-like peptide can effect degradation rates, as has been described for murine retroviral reverse transcriptase (73). The site of accumulation of the synthetic proteins is also important in that the number of proteases differ for each compartment of *E. coli*: three periplasmic and five cytoplasmic proteases have been identified (64,74).

In several cases, *lon* or *htp*R *lon* mutants have been used to reduce the extent of proteolysis.

In most cases, minimal proteolysis is evident when whole toxin mutants are expressed from the *tox* promoter and secreted in protease-proficient strains of *E. coli*; this is true of *tox*-197 (29) and DT-Ser-148 (44). Similarly, CRM228 is detectable in analogous constructs (17). Intracellular expression of *tox*-228, modified at its amino terminal end by the addition of Met-AspProProProSerAlaHisAla and induced from the p_R promoter by temperature induction results in significant product degradation, even in a *lon* protease-deficient host (69). A protein composed of CRM228, in which a portion of the hepatitis B surface antigen is inserted between amino acids 293 and 294 and the carboxy-terminal 106 amino acids of fragment B are replaced by β-galactosidase, is also proteolytically unstable (75).

Expression of DTA-like peptides gives more conflicting results. Little proteolysis of fragment A is evident when expressed in plasmid pDT201, in which the *Sau*3A-2 fragment (-179 to $+653$) is inserted into pUC8 and secreted by the protease-proficient strain MC4100 (*E. coli* K12 F, *ara*D139, (*lac*U$_{169}$, *rel*A, *rps*L, *thi*), despite the fusion to 29 plasmid-encoded amino acids at the carboxyl-terminal (23,29,49). Expression and secretion of the same peptide under control of the P_R promoter (plasmid pDT1201) and induced by temperature elevation also results in little degradation (29,49). Little proteolysis is evident when a DTA-like fragment (amino acids 1–198, pLOPDTA1000) is expressed intracellularly at high levels by temperature induction in *E. coli* strain DG95 (*thi*-1, *end*A1, *hsd*R17, *sup*^{0}, (λN7N53Cl$_{857}$*sus*P80)) (20). In contrast, 90% of the secreted peptide is degraded within 1 hr by the strain GM48 (*dam*-3, *dcm*-6, *thr*-1, *leu*-6, *thi*-6, *lac*Y, *gal*K2, *gal*T22, *ara*-14, *ton*A3, *tsx*-78, *sup*E44, *hsd*R) containing constructs in which the same fragment is inserted into pBR322 (pRTF1); in this case 37 plasmid-encoded amino acids are fused to the carboxyl-terminal (21).

Addition of portions of the B fragment increases the proteolytic sensitivity of the product expressed in *E. coli*. Severe proteolytic degradation occurs when three *tox* fragments, terminating at amino acids 290 (pABC313), 377 (pABC402), 382 (pRTF2B), or 485 (pABC508), are constitutively expressed from the *tox* promoter and secreted in both protease-proficient and protease-deficient strains (21,44,49). Each of these constructs terminate within fragment B and have additional amino acids tailored onto the carboxyl end of the encoded protein: pABC313 terminates at amino acid 290 followed by AlaAlaAlaCys; pABC402 terminates at amino acid 377 followed by ArgSerValAspArgProAlaAlaCys; pRTF2B terminates at amino acid 382 and is fused to the sequence IleSerPheAsnAlaValValTyrHisSer; and construct pABC508 terminates at amino acid 485 followed by Ala-

AlaAlaCys. Substitution of the carboxyl cysteine by the sequence for α-melanocyte stimulating hormone results in a similar degree of proteolysis (49). The degree of proteolysis is related to the temperature of induction: proteolysis is greater when the cultures are grown at 42°C than when grown at 30°C (29). Because analogous constructs containing *tox*-228 showed little proteolysis, it was concluded that removal of 50 amino acids from the C-terminus of diphtheria toxin results in a conformation that is now susceptible to *E. coli* proteases (29,49). Additionally, the apparent proteolytic resistance of expressed fragment A suggests that addition of portions of fragment B makes the resultant peptide proteolytically susceptible (29,49). The degree of degradation is partially overcome by intracellular expression at higher levels from the P_R promoter in *htp*R *lon* protease-deficient host (49).

Intracellular expression a number of the constructs further decreases their proteolytic degradation despite the presence of more proteases in that compartment. High intracellular levels of a CRM45-like peptide (pLOPMsp) induced by temperature elevation can be obtained in a protease-proficient strain (20); in this case, the construct terminates at amino acid 383 without additional fusion. The construct pABM508, in which the sequence for α-MSH is fused to the carboxyl end of *tox* fragment at amino acid 485, is severely degraded when expressed constitutively from P_{tox} or by temperature induction form P_R and secreted into the periplasmic space by the *tox* secretory leader in both protease-proficient and protease-deficient strains (29). Intracytoplasmic expression by partial removal of the secretory leader (resulting in the addition of 6 or 7 amino acids to the amino-terminal end, pAMB4508) and expression from P_R increased the level of proteolysis in a protease-proficient strain (29). The increased proteolysis is partially overcome by expression in protease-deficient strains (*lon*, *htpR*, and *htpR lon*). Proteolysis is further reduced by completed removal of the secretory leader, replacement of the promoter by P_{trc} and expression in a *htpR lon* protease-deficient strain at 30°C (29). Finally, little protein degradation is observed when the diphtheria toxin B fragment is expressed intracytoplasmically in *E. coli* by IPTG induction of the *trc* promoter (76).

It appears that constructs in which the *tox* peptides are secreted by the *tox* secretory leader are still susceptible to the intracellular protease La (29,49). This is true with expression from P_{tox} and a native secretory leader as well as P_R and a modified secretory leader. The role of cytoplasmic proteases in degrading the secreted proteins can be explained if there is an uncoupling of translation and secretion as a result of inefficient recognition or processing of the *tox* secretory leader by the *E. coli* apparatus, which may become more apparent as the expression levels are increased.

G. Solubility of the *E. coli*–Produced Products

Most overproduced proteins are insoluble when expressed in *E. coli*, accumulating in the cytoplasm as "inclusion bodies" (77,78). Generally, proteins within inclusion bodies are bound by strong noncovalent interactions, and can be released only by strong chaotropic reagents, requiring subsequent refolding steps in order to obtain functional product (77,78). The process of inclusion body formation appears to occur from intermediates of the protein-folding pathway (78). Because the *cis-trans* isomerization of proline may be rate limiting in the folding, it has been speculated that proteins of low proline content may have increased solubility (77). Other factors which may contribute product solubility include the ability of the protein to bind metal ions, the presence of stretches of acidic amino acids, and the pH and ionic strength of the compartment in which the protein is synthesized (77,78). Additional protein factors may aid in proper folding of proteins: Prolyl *cis-trans*isomerase (PPI) catalyzes the slow (half-life of 1–7 min) isomerization of prolines in peptides; proteins exist which hold other proteins in more soluble unfolded conformations (e.g., SecB during secretion of some proteins, gene 31 protein, the GroEL and GroES); and there are proteins involved in isomerizing disulfide groups (e.g., thioredoxin and glutaredoxin). Finally, posttranslational modification (e.g., phosphorylation, hydroxylation, glycosylation, and proteolysis) may alter product solubility (77). The growth temperature can drastically influence the solubility of a number of *E. coli*–expressed proteins as well as the formation of inclusion bodies (78): P22 tailspike protein (79), ricin A chain (80), prosubtilisin (81), and fibroblastic growth factor (82) are insoluble when induced at elevated temperatures.

Although inclusion body formation is generally associated with intracellular expression, aggregation of secreted proteins has also been described (83). The composition of the secretory leader can play an important role in the product solubility: substitution of the native leader with the outer membrane protein A (OmpA) secretory leader results in extensive aggregation of the mature form despite the identical amino acid sequence of the product (83). It is suggested that the signal sequence may effect the formation of an intermediate that is prone to aggregation (83). It is interesting that the addition of nonmetabolizable sugars reduces product aggregation perhaps by directly effecting peptide folding (83).

A number of constructs produce insoluble *tox*-related peptides in *E. coli*. The secreted form of ABM508 (α-MSH fused to DT amino acid 485) is largely insoluble when expressed constitutively from the *tox* promoter (29). Similarly, predominantly insoluble proteins result by temperature induction of the p_R promoter expressing both nonsecreted (with a 7-amino acid N-terminal addition; see Table 1) and secreted (with a modified secretory leader; (see Table 1) products (29). The solubility of the product is effected by the incubation temperature and not the amount of B chain present; both fragment A and whole toxin (CRM197) remain partly insoluble when expressed at 42°C, but are predominantly soluble when expressed at 30°C (29). Because most assessments of product are performed on supernatants following centrifugation, different product solubility levels may partly explain discrepancies in the expression leaves found by various investigators. In addition, formation of inclusion bodies may protect proteins from proteolysis (77), thereby contributing to higher final product.

It is interesting that both fragment A alone and whole toxin can undergo thermal denaturation at temperatures ranging from 42 to 58°C (84), and that the effect of pH and temperature on inducing conformational changes in whole toxin are cooperative (85). In addition, CRM45 contains hydrophobic sites on its surface (86). Theses factors may contribute to the aggregation observed with induction by temperature elevation.

H. Iron Regulation of Cloned Diphtheria Toxin Fragments in *E. coli*

Many *E. coli* genes, including several elaborate iron uptake systems (e.g., the *ent*CEBA(P15), *iuc*ABCDA, *feb*BDGC, *fep*A*ent*D, and *fes* operons), the colicin I receptor (*cir*) (87), and the *E. coli* Shiga I–like toxin I (*stl*-I) (88) are regulated by the product of the chromosomal *fur* (ferric uptake regulation) repressor gene. As with the expression of *tox* on *C. diphtheriae*, when the Fur protein binds to Fe^{2+} or certain other divalent metal cofactors, it interacts with the operator regions of these *E. coli* iron-regulated genes, precluding RNA polymerase binding to its promoter, thereby negatively regulating transcription. Deoxyribonuclease protection experiments demonstrate that the Fur protein binding to its operator sequence protects 20 bp and contacts successive major groove turns on the DNA helix (89). Comparison of the regulatory region of several iron-controlled operons suggest a consensus palindromic sequence, referred to as the "iron box" (90) of 5'-GAT*AA*TGATAATC*A*T*T*ATC as the recognition sequence for active Fur, with the most conserved bases thought to be important italicized (87). In the aerobactin operon, two contiguous Fur-binding sites were identified by DNAse I protection studies: a primary binding site which overlaps the "−35" promoter consensus region, and a secondary binding site which overlaps the promoter "−10" consensus region (91); in the *cir* operon, the 19-bp consensus sequence starts at the beginning of the "−35" region of the promoter (89). The *tox* promoter region extending between base −56 and −29 contains 10 of the 19 "iron box" bases and conserves the AAT . . . ATT sequence. In addition, regions of lesser homology exist further upstream (47).

Measurement of the effect of iron on *tox*-related peptide expression in *E. coli* has given conflicting results. Synthesis of whole toxin by *E. coli* extracts is unaffected by the presence of iron, but is inhibited by the addition of extracts of *C. diphtheriae* (10). Similarly, expression in vivo of a construct consisting of fragment A in a pUC vector (pDT201) is not inhibited by 500 μM FeSO$_4$ in the medium (38). To further study the mechanism of iron-dependent regulation of the *tox* regulatory region, a construct in which the *tox* promoter/operator region replaced the *lacZ* promoter was inserted as a single copy into the chromosome of *E. coli* (33). β-Galactosidase expression was unaffected by the *fur* locus. When the gene for *fur* was placed on a multicopy plasmid together with the *tox* promoter/operator–*lacZ* fusion in a *fur*$^-$ *E. coli* host, the presence of iron had no effect on the β-galactosidase expression, although expression of iron-regulated *E. coli* outer membrane proteins was repressed (33).

In a different construct in which the *tox* regulatory region was transcriptionally fused to the *gal*K gene (in a pBR322 vector), thus driving galactokinase expression, *E. coli* iron-regulatory factors effected transcription of the *tox* promoter (47). Growth of such a fusion in Luria broth results in a constant level of galactokinase synthesis throughout exponential growth of the baterium. In contrast, upon changing to low-iron Luria broth, the level of enzyme increases fivefold after 3 hr. This increases occurs primarily at a time when bacterial growth slows (47), and is somewhat reminiscent of the increase seen during the terminal stages of growth of C7($\beta^{tox-201}$) and wild-type C7(β) (92). The increase occurs at the level of transcription and is prevented by the addition of iron to the medium. The iron-sensitivity of galactokinase expression when under control of the *tox* regulatory sequences is mediated through *E. coli fur* as indicated by the higher levels of expression in a *fur*-deficient host, and its partial repression by supplementation of *fur* on a compatible plasmid (47).

The role of iron in the regulation of *slt*-I genes is dependent on the vector (88). Toxin production by *E. coli* strains lysogenized with coliphage carrying the *slt*-I gene is regulated by the amount of iron in the medium. In contrast, expression of the *slt*-I gene when present on the high-copy plasmid pBR328 is not effected by the presence of iron. It is possible that the higher copy number of the pUC plasmid (93) could titrate the level of Fur protein, resulting in unregulated synthesis of the *tox* on the remaining plasmids. However, this cannot explain the lack of iron-regulation when placed as a single copy in the chromosome. It is possible that recognition of the *tox* operator by *E. coli* Fur protein depends on the degree of DNA winding.

I. High-Level Expression of *tox*-Encoded Peptides in *E. coli*

Thus, we have identified a number of factors which effect the overall levels of *tox*-related peptides in *E. coli*. These include proteolysis, inefficient secretion by the *E. coli* secretory apparatus when the peptides are expressed at high levels, and "overexpression" lethality associated with secretion of several diphtheria toxin derivatives.

Transcription initiation does not appear to be the factor limiting high-level expression of *E. coli*. The activity of the *tox* promoter in *E. coli* and the marginal increase in expression levels resulting from the addition of highly active promoters in front of the *tox* regulatory sequences (pDT101, pLSDTA40) supports this conclusion (20,22). Addition of the p_L promoter in front of either fragment A or a CRM45-like peptide only increases expression by two- to threefold (20). Expression of *tox* fragments terminating at amino acids 290 (pABC313), 377 (pABC402) and 485 (pABC508) by replacement of the *tox* promoter with the lambda p_R promoter appears unsuccessful: The level of ADP-ribosyltransferase activity is two to sixfold greater with the *tox* promoter, although it is postulated that the lower amount of product measured may be a consequence of synthesis of undetectable, insoluble, inactive aggregates at the higher temperature (49).

Replacement of the *tox* promoter and ribosomal binding site with regulatory regions which normally function efficiently in *E. coli* results in high-level expression of *tox* peptides. By circumventing secretion and replacement of the *tox* promoter and ribosomal binding site with the *trp*L Shine-Dalgarno sequence and *trp* promoter (pTrpDTA100; see Table 1) the expression of fragment A is increased 13- to 16-fold (20). Substitution by the P_L promoter and gene N Shine-Dalgarno sequence (pLDTA200) results in a 86- to 216-fold increase in expression over the unmodified vector (20). Finally, placement of the entire construct into a temperature-sensitive copy number mutant plasmid (pLOPDTA1000), such that there would be simultaneous plasmid amplification and induction of the promoter upon a temperature shift, results in a 240- to 680-fold increase: Fragment A accounts for 7% of the total cell protein synthesized (20). Similar results are obtained with the *Msp*I-encoded peptide (pLOPMsp) (20). Although the effects of temperature on the solubility of the products were not investigated, the amount measured presents only that which is soluble.

In contrast, intracellular expression of amino acids 1–485 fused at the carboxyl end to β-MSH (pABM4580; see Table 1) by removal of the secretory leader (actually modified such that it was nonfunctional but resulted in an N-terminal addition of fMetProProSerAlaHisAla to the mature protein), and replacement of P_{tox} and the *tox* secretory leader with the p_R promoter and *cro* ribosomal-binding site, expression remained low (29).

It is believed that the actual expression levels are higher, but owing to the insolubility of the protein at the high temperatures used for induction, the product may be undetected.

Expression in a *htp*R*lon* strain of a fusion peptide consisting of diphtheria toxin amino acids 1–485 linked to α-MSH is increased to 10 mg/L, or 7.7% of the total cell protein by removal of the *tox* secretory leader, replacement of the *tox* promoter with P_{trc}, and addition of two rho-independent transcription terminators at the end of the gene (pABM6508; see Table 1) (29). Despite induction at lower temperatures, 50% of the product remained insoluble; thus the true level of expression may be higher.

Therefore, optimum expression of diphtheria toxin fragments is obtained by several modifications. The inefficient secretion of several *tox*-containing fragments expressed at higher levels is circumvented by removal of the secretory leader and expressing intracellularly. Proteolysis is minimized by expression in *htp*R*lon* protease-deficient hosts. Finally, synthesis of insoluble product is avoided by using promoters which can be induced at lower temperatures (e.g., P_{trc} or P_{tac}).

J. Expression of Diphtheria Toxin Fragment B in *E. coli*

It is not surprising that *E. coli* can express products possessing ADP-ribosyltransferase activity since diphtheria toxin fragment A is stable and refolds into an enzymatically active form following denaturation (7,84). In contrast, fragment B is unstable and less soluble when separated from fragment A (7). Therefore, the ability of *E. coli* to express fragment B possessing cell-binding activity is expected to be a more difficult problem.

Expression of diphtheria toxin fragment B in *E. coli* was obtained by fusing the DNA sequence encoding amino acids 184–535 to the sequence ATG GGG GGA (encoding the amino acid sequence MetGlyGly) to provide an ATG start codon, the *trc* promoter, the *lacZ* ribosomal binding site, and two *rrnB* transcription terminator sequences (76). The result is the expression of 2 mg/L (2% of the total cell protein) of a soluble, non-secreted fragment B. No degradation is detectable. Purification is based on the insolubility of fragment B in 100 mM NaCl, thus necessitating denaturation (in 4 M guanidine hydrochloride) followed by renaturation (by dialysis) in order to obtain active protein. However, the level of resolubilization is low. The renatured recombinant fragment B binds to Vero cells as well as native *C. diphtheriae*–produced fragment B. Reconstitution of whole toxin from wild-type fragment A and recombinant fragment B in the presence of 2 M guanidine hydrochloride produces a toxin with similar in vitro toxicity against Vero cells as that of native whole toxin. Again, the reconstituted whole toxin with recombinant B chain has lower solubility than native toxin. Thus, although the biological activity and cell-binding activity is similar to that of native fragment B, the decreased solubility might suggest some differences in folding.

K. Expression of Whole Diphtheria Toxin in *E. coli*

To comply with the NIH guidelines, only mutant forms of whole diphtheria toxin have been cloned intact in *E coli*. CRM228 expressed from the P_{tox} promoter and secreted by the *tox* secretory leader expresses from 50–100 μg/L toxin in *E. coli* (17,18). When the *tox* secretory leader was removed (adding the sequence MetAspProProProSerAlaHisAla at the amino-terminal end of the mature protein) and the *tox* promoter and ribosomal-binding site replaced with the P_R promoter and *cro* ribosomal-binding site, expression remained poor (69). Following induction, the level of *tox* message is 10-fold higher than the wild-type construct expressed from the *tox* promoter and ribosomal-binding site, such that 0.01% of the total bacterial RNA contains *tox* sequences. Only low levels of an immunologically reactive protein of a molecular weight lower than expected for full-length product is expressed in a *lon* protease-deficient host. Substitution of the C-terminal 106 amino acids of fragment B with β-galactosidase (at the *Nru*I site present in CRM228 but not wild-type toxin) permits monitoring of toxin expression indirectly by assaying for β-galactosidase. Induction in a *lon* strain results in a 95-fold increase in β-galactosidase expression over the uninduced level. Little full-length product is expressed in the protease-deficient host. Even in the *lon*-host, significant proteolysis occurs. The proteolysis of both intact CRM228 and the modified forms (both the intracellular synthesized form with the N-terminal addition, and the β-galactosidase fusion) may relate to the CRM228 mutations. CRM228 produced in *C. diphtheriae* is more sensitive to proteolysis than native toxin (8). In comparison, expression of CRM197 in *E. coli* showed little proteolytic degradation (29).

A novel mutant of diphtheria toxin consisting of a wild-type B fragment and a fragment A with Glu-148 changed to Ser-148 has been created by site-directed mutagenesis and expressed in *E. coli* (44). Alteration of Glu-148 to Ser-148 decreases the ADP-ribosyltransferase specific activity below that of wild-type toxin by 3 to 4 orders of magnitude. When expressed from the *tox* promoter, ribosomal-binding site, and secretory leader, the expressed product is secreted into the periplasmic space with little evidence of proteolysis and nicking (44). The functionality of the B fragment within the enzymatically inactive mutant is assayed by nicking the mutant and mixing with a 7.5-fold molar excess of wild-type fragment A. The cytotoxicity of the mixture is increased 250-fold over the mutant itself, achieving a level of one-third that of native nicked DT. It is unclear whether the lower activity represents incomplete exchange with subsequent competition by residual DT-Ser-148, improper folding of a population of the fragment B molecules, or incomplete folding of all of the molecules.

III. EXPRESSION OF DIPHTHERIA TOXIN FRAGMENTS IN EUKARYOTIC CELLS

The sensitivity of the eukaryotic EF-2 to ADP-ribosylation by diphtheria toxin precludes significant expression of catalytically active fragments in eukaryotic cells. However, regulated expression can be utilized to ablate cell populations and can be used as a negative selection strategy for inactivation of the toxin gene. Thus, both wild-type and catalytically attenuated CRM176 have been used to construct plasmids for this purpose (94,95). Expression was achieved in a transient cotransfection assay using a truncated promoter region of the human metallothionein IIA gene in conjunction with downstream simian virus 40 splicing polyadenylation signals. Because a single molecule of fragment A is sufficient to kill a eukaryotic cell (96), the regulation of expression must be tight. Even low uninduced expression will be sufficient to kill the cell. To address this problem, the attenuated form CRM176, which has 8–10% of the activity of wild type (8,16), has been cloned in a eukaryotic expression system (95). It is hoped that the combination of a tightly regulated promoter with the attenuated enzymatic activity will permit the stable introduction of fragment A into eukaryotic cells.

IV. IN VITRO EXPRESSION OF DIPHTHERIA TOXIN

To ensure safety, cloning of *tox* has been limited to hypotoxic peptide fragments or inactive mutants. We have seen that expression of some proteins leads to lethality of the host, and that cellular proteases can be a formidable problem for both intracellular and secreted forms of *tox*. Finally, cloning and expression in *E. coli* can be labor intensive, and the desired product may represent only a minor fraction of the bacterial protein. In vitro techniques can circumvent a number of these problems and provide enough product to permit rapid screening of altered forms of toxin.

A series of diphtheria toxin deletions were generated and placed downstream of the SP6 and T7 promoters of the expression plasmids pGEM Blue (Promega) and Bluescript KS (Strategene), respectively (97). The expression vectors were constructed with an ATG start codon and eukaryotic translation initiation consensus sequence. The plasmids were linearized, transcribed in vitro, and the resultant message purified and translated in vitro using a micrococcal nuclease-treated rabbit reticulocyte lysate. This also permitted easy internal labeling of the protein. Although the yield of toxin fragments was low, on the order of 1.5–9.0 nM, there is sufficient amounts to perform the desired studies. In vitro synthesis of whole toxin results in a protein which binds to Vero cells 5- to 10-fold less than [^{125}I]labeled native toxin. The decrease in binding is likely due to inhibitors in the reticulocyte lysate since addition of lysate to natural toxin results in a similar inhibition. In addition, the binding to Vero cells of the in vitro synthesized whole toxin and toxin B chain fragments is blocked by prior saturation of the receptor with native toxin. Various DTB deletion mutants show a resistance to trypsinization similar to that of native toxin. Several of the constructs bound to Vero cell membranes and generate a 25 kd (P25) and 18 kd (P18) pronase E–resistant fragment following exposure to low pH similar to that of wild-type toxin. In all, these results suggest that the in vitro synthesized products fold in a manner similar to that found in toxin produced by *C. diphtheriae*.

The recent development of the polymerase chain reaction (PCR) (98–101) can circumvent many of these problems, thus permitting a method for rapid screening of mutants. The gene for intact toxin was reconstituted in vitro by ligating restriction fragments containing different portions of the toxin gene, and amplified by PCR using primers encompassing the entire sequence (102). The product was transcribed in vitro and the resultant message translated in a rabbit reticulocyte lysate system containing [^{33}S]methionine and depleted of NAD to prevent the synthesized toxin from inactivating the EF-2. The yield of toxin is low (100 ng/ml). Despite being synthesized in the absence of secretion, the resultant toxin had similar in vitro cytotoxic activity against Vero cells as native toxin synthesized by *C. diphtheriae*. As with native toxin, the in vitro product can translocate fragment A from the cell surface following acidification of the media. In addition, both toxins produce a Pronase-resistant 25-kd polypeptide when cells with surface-bound toxin are exposed to low pH. Again, as seen before, there is inhibition of binding of both the in vitro synthesized product and the native toxin by components within the translation mixture which could be reduced by dilution. Thus, the process of secretion is not required for correct folding of the toxin.

V. CONCLUSIONS

Both classic genetics and current molecular biological approaches have been applied to the study and manipulation of the gene for diphtheria toxin. The toxin has been cloned, sequenced, and expressed in *E. coli* to high levels and structure-function analysis has been undertaken. In this way, the binding region has been substituted with more specific targeting moieties (e.g., antibodies, hormones) and the extreme potency against eukaryotic cells has been controlled in order to generate therapeutically useful drugs. This chapter has focused on the genetic and molecular biological approaches which has generated high-level expression of toxin and toxin fragments. It is hoped that those interested in cloning other toxins may learn from the successes and failures which we have encountered with diphtheria toxin.

ACKNOWLEDGMENTS

I would like to thank Shing Chang and Jonathan Raymond for carefully reviewing this manuscript.

REFERENCES

1. Lim, C.-J., Fuchs, J. A., McFarlan, S. C., and Hogenkamp, H. P. C. Cloning, expression, and nucleotide sequence of a gene encoding a second thioredoxin from *Corynebacterium nephridii.* J. Biol. Chem. *262:* 12114–12119, 1987.
2. Follettie, M. T., and Sinskey, A. J. Molecular cloning and nucleotide sequence of the *corynebacterium glutamicum phe*A gene. J. Bacteriol., *167:* 695–702, 1986.
3. Yeh, P., Sicard, A. M., and Sinskey, A. J. Nucleotide sequence of the *lys*A gene of *Corynebacterium glutamicum* and possible mechanisms for modulation of its expression. Mol. Gen. Genet., *212:* 112–119, 1988.
4. Tsugawa, A., Ohsumi, Y., and Kato, I. Inhibitory effect of diphtheria toxin on amino acid incorporation in *Escherichia coli* cell-free system. J. Bacteriol., *104:* 152–157, 1970.
5. Goto, N., Kato, I., and Sato, H. The inhibitory effect of diphtheria toxin on amino acid incorporation by a bacterial cell-free system. Jpn. J. Exp. Med., *38:* 185–192, 1968.
6. Gill, D. M., and Pappenheimer, A. M., Jr., Structure-activity relationships in diphtheria toxin. J. Biol. Chem., *246:* 1492–1495, 1971.
7. Drazin, R., Kandel, J., and Collier, R. J. Structure and activity of diphtheria toxin: II. Attack by trypsin at a specific site within the intact toxin molecule. J. Biol. Chem., *246:* 1504–1510, 1971.
8. Uchida, T., Pappenheimer, A. M., Jr., and Greany, R. Diphtheria toxin and related proteins. I. Isolation and Properties of Mutant Proteins serologically related to diphtheria toxin. J. Biol. Chem., *248:* 3838–3844, 1973.
9. Yoshihama, M., Higashiro, K., Rao, E. A., Akedo, M., Shanabruch, W. G., Follettie, M. T., Walker, G. C., and Sinskey, A. J. Cloning vector system for *Corynebacterium glutamicum.* J. Bacteriol., *162:* 591–597, 1985.
10. Murphy, J. R., Pappenheimer, A. M., Jr., and de Borms, S. T. Synthesis of diphtheria *tox*-gene products in *Escherichia coli* extracts. Proc. Natl. Acad. Sci. U.S.A., *71:* 11–15, 1974.

11. Lightfoot, H. N., and Iglewski, B. H. Synthesis of diphtheria toxin in *E. coli* cell-free lysate. Biochem. Biophys. Res. Commun., *56*: 351–357, 1974.

12. Smith, W. P., Tai, P.-C., Murphy, J. R., and Davis, B. D. Precursor in cotranslational secretion of diphtheria toxin. J. Bacteriol., *141*: 184–189, 1980.

13. Smith, W. P. Cotranslational secretion of diphtheria toxin and alkaline phosphatase in vitro: Involvement of membrane protein(s). J. Bacteriol. *141*: 1142–1147, 1980.

14. Rappuoli, R., Michel, J. L., and Murphy, J. R. Restriction-endonuclease map of corynebacteriophage $\overline{\omega}_c^{tox+}$ isolated from the Park-Williams No. 8 strain of *Corynebacterium diphtheriae*. J. Virol. *45*: 524–530, 1983.

15. Righelato, R. C., and van Hemert, P. A. Growth and toxin synthesis in batch and chemostat cultures of *Corynebacterium diphtheriae*. J. Gen. Microbiol., *58*: 403–410, 1969.

16. Uchida, T., Pappenheimer, A. M., Jr. Mutation in β-phage genome affecting diphtheria toxin structural gene and toxin yield, *In*: A. Ohsaka, K. Hayashi, and Y. Sawai (Eds.), pp. 353–362, Anim., Plant Microb. Toxins, Proc. Int. Symp., 4th. New York: Plenum Press, 1976.

17. Kaczorek, M., Zettlmeissl, G., Delpeyroux, F., and Streeck, R. E. Diphtheria toxin promoter function in *Corynebacterium diphtheriae* and *Escherichia coli*. Nucl. Acids Res., *13*: 3147–3159, 1985.

18. Kaczorek, M., Delpeyroux, F., Chenciner, N., Streek, R. E., Murphy, J. R., Boquet, P., and Tiollais, P. Nucleotide sequence and expression of the diphtheria *tox*228 gene in *Escherichia coli*. Science, *221*: 855–858, 1983.

19. Greenfield, L., Bjorn, M. J., Horn, G., Fong, D., Buck, G. A., Collier, R. J., and Kaplan, D. A. Nucleotide sequence of the structural gene for diphtheria toxin carried by corynebacteriophage β. Proc. Natl. Acad. Sci. U.S.A., *80*: 6853–6857, 1983.

20. Greenfield, L., Dovey, H. F., Lawyer, F. C., and Gelfand, D. H. High-level expression of diphtheria toxin peptides in *Escherichia coli*. Bio/Technology, *4*: 1006–1011, 1986.

21. Tweten, R. K., and Collier, R. J. Molecular cloning and expression of gene fragments from Corynebacteriophage β encoding enzymatically active peptides of diphtheria toxin. J. Bacteriol., *156*: 680–685, 1983.

22. Leong, D., Coleman, K. D., and Murphy, J. R. Cloned fragment A of diphtheria toxin is expressed and secreted into the periplasmic space of *Escherichia coli* K12. Science, *220*: 515–517, 1983.

23. Leong, D., Coleman, K. D., and Murphy, J. R. Cloned diphtheria toxin fragment A is expressed from the *tox* promoter and exported to the periplasm by the *SecA* apparatus of *Escherichia coli* K12. J. Biol. Chem., *258*: 15016–15020, 1983.

24. Mulligan, M. E., Brosius, J., and McClure, W. R. Characterization in vitro of the effect of spacer length on the activity of *Escherichia coli* RNA polymerase at the *TAC* promoter. J. Biol. Chem., *260*: 3529–3538, 1985.

25. Brosius, J., Erfle, M., and Storella, J. Spacing of the −10 and −35 regions in the *tac* promoter: effect on its in vivo activity. J. Biol. Chem., *260*: 3539–3541, 1985.

26. Beutel, B. A., and Record, M. T., Jr. *E. Coli* promoter spacer regions contain nonrandom sequences which correlate to spacer length. Nucl. Acids Res., *18*: 3597–3603, 1990.

27. de Boer, H. A., and Kastelein, R. A. Biased codon usage: An exploration of its role in optimization of translation., *In*: W. Reznikoff and L. Gold (eds.), pp. 225–285. Maximizing Gene Expression. Stoneham, Massachusetts: Butterworths, 1986.

28. Makoff, A. J., Oxer, M. D., Romanos, M. A., Fairweather, N. F., and Ballantine, S. Expression of tetanus toxin fragment C in *E. coli*: High level expression by removing rare codons. Nucl. Acids Res., *17*: 10191–10292, 1989.

29. Bishai, W. R., Rappuoli, R., and Murphy, J. R. High-level expression of a proteolytically sensitive diphtheria toxin fragment in *Escherichia coli*. J. Bacteriol., *169*: 5140–5151, 1987.

30. Kennel, D. E. The instability of messenger RNA in bacteria, *In*: W. Reznikoff and L. Gold (eds.), pp. 101–142. Maximizing Gene Expression. Stoneham, Massachusetts: Butterworths, 1986.

31. Hargrove, J. L., and Schmidt, F. H. The role of mRNA and protein stability in gene expression. FASEB J., *3*: 2360–2370, 1989.

32. Fourel, G., Phalipon, A., and Kaczorek, M. Evidence for Direct regulation of diphtheria toxin gene transcription by an Fe^{2+}-dependent DNA-binding repressor, DtoxR, in *Corynebacterium diphtheriae*. Infect. Immun., *75*: 3221–3225, 1989.

33. Boyd, J., Oza, M. N., and Murphy, J. R. Molecular cloning and DNA sequence analysis of a diphtheria *tox* iron-dependent regulatory element (*dtxR*) from *Corynebacterium diphtheriae*. Proc. Natl. Acad. Sci. U.S.A., *87*: 5968–5972, 1990.

34. Horwitz, M. S. Z., and Loeb, L. Structure-function relationships in *Escherichia coli* promoter DNA. Prog. Nucl. Acid Res., *38*: 137–164, 1990.

35. Rosenberg, M., and Court, D., Regulatory sequences involved in the promotion and termination of RNA transcription. Annu. Rev. Genet., *13*: 319–353, 1979.

36. de Crombrugghe, B., and Pastan, I. Cyclic AMP, the cyclic AMP receptor protein, and their dual control of the galactose operon. *In*: J. H. Miller and W. S. Reznikoff (eds.), pp. 303–324. The Operon. Cold Spring Harbor, New York: Cold Spring Harbor Laboratory, 1980.

37. Gottschalk, G. Bacterial Metabolism, Springer-Verlag, New York, 1979.

38. Leong, D., and Murphy, J. R. Characterization of the diphtheria *tox* transcript in *Corynebacterium diphtheriae* and *Escherichia coli*. J. Bacteriol., *163*: 1114–1119, 1985.

39. Bord, J., and Murphy, J. R. Analysis of the diphtheria *tox* promoter by site-directed mutagenesis. J. Bacteriol., *170*: 5949–5952, 1988.

40. Model, P., and Russel, M. Prokaryotic secretion. Cell, *61*: 739–741, 1990.

41. Lill, R., Cunningham, K., Brundage, L. A., Ito, K., Oliver, D., and Wickner, W., SecA protein hydrolyzes ATP and is an essential component of the protein translocation ATPase of *Escherichia coli*. EMBO J., *8*: 961–966, 1989.

42. Gardel, C., Johnson, K., Jacq, A., and Beckwith, J. The *sec*D locus of *E. coli* codes for two membrane proteins required for protein export. EMBO J., *9*: 3209–3216, 1990.

43. Saier, M. H., Jr., Werner, P. K., and Müller, M. Insertion of proteins into bacterial membranes: Mechanism, characteristics, and comparisons with the eucaryotic process. Microbiol. Rev., *53*: 333–366, 1989.

44. Barbieri, J. T., and Collier, R. J. Expression of a mutant, full-length form of diphtheria toxin in *Escherichia coli*. Infect. Immun., *55*: 1647–1651, 1987.

45. Inouye, S., Soberon, X., Franceschini, T., Nakamura, K., Itakura, K., and Inouye, M. Role of positive charge on the amino-terminal region of the signal peptide in protein secretion across the membrane. Proc. Natl. Acad. Sci. U.S.A., *79*: 3438–3441, 1982.

46. Vlasuk, G. P., Inouye, S., Ito, H., Itakrua, K., and Inouye, M. Effects of the complete removal of basic amino acid residues from the signal peptide on secretion of lipoprotein in *Escherichia coli*. J. Biol. Chem. *258*: 7141–7148, 1983.

47. Tai, S-P., and Holmes, R. K. Iron regulation of the cloned diphtheria toxin promoter in *Escherichia coli*. Infect. Immun., *56*: 2430–2436, 1988.

48. O'Keefe, D. O., and Collier, R. J. Cloned diphtheria toxin within the periplasm of *Escherichia coli* causes lethal membrane damage at low pH. Proc. Natl. Acad. Sci. U.S.A., *86*: 343–346, 1989.

49. Bishai, W. R., Miyanohara, A., and Murphy, J. R. Cloning and expression in *Escherichia coli* of three fragments of diphtheria toxin truncated with fragment B. J. Bacteriol., *169*: 1554–1563, 1987.

50. Eisenberg, D., Schwarz, E., Komaromy, M., and Wall, R. Analysis of membrane and surface protein sequences with the hydrophobic moment plot. J. Mol. Biol., *179*: 125–142, 1984.

51. Giannini, G., Rappuoli, R., and Ratti, G. The amino-acid sequence of two non-toxic mutants of diphtheria toxin: CRM45 and CRM197. Nucl. Acids Res., *12*: 4063–4069, 1984.

52. Kagan, B. L., Finkelstein, A., and Colombini, M. Diphtheria toxin fragment forms large pores in phospholipid bilayer membranes. Proc. Natl. Acad. Sci. U.S.A., *78*: 4950–4954, 1981.

53. Misler, S. Gating of ion channels made by a diphtheria toxin fragment in phospholipid bilayer membranes. Proc. Natl. Acad. Sci. U.S.A., *80*: 4320–4324, 1983.

54. Montecucco, C., Schiavo, G., and Tomasi, M. pH-Dependence of the phospholipid interaction of diphtheria-toxin fragments. Biochem. J., *231*: 123–128, 1985.

55. Jiang, G.-S., Solow, R., and Hu, V. W. Fragment A of diphtheria toxin cause pH-dependent lesions in model membranes. J. Biol. Chem., *265*: 17170–17173, 1989.

56. Benson, S. A., Hall, M. N., and Silhavy, T. J. Genetic analysis of protein export in *Escherichia coli* K12. Ann. Rev. Biochem., *54*: 101–134, 1985.

57. Blewitt, M. G., Chung, L. A., and London, E. Effect of pH on the conformation of diphtheria toxin and its implications for membrane penetration. Biochemistry, *24*: 5458–5464, 1985.

58. Cabiaux, V., Brasseur, R., Wattiez, R., Falmagne, P., Ruysschaert, J.-M., and Goormaghtigh, E. Secondary structure of diphtheria toxin and its fragments interacting with acidic liposomes studied by polarization infrared spectroscopy. J. Biol. Chem., *264*: 4928–4938.

59. Brasseur, R., Cabiaux, V., Falmagne, P., and Ruysschaert, J.-M. pH Dependent Insertion of a diphtheria toxin B fragment peptide into the lipid membrane: A conformational analysis. Biochem. Biophys. Res. Commun., *136*: 160–168, 1986.

60. Anderson, K. W., Grulke, E., and Gerhardt, P. Microfiltration culture process for enhanced production of rDNA receptor cells of *Escherichia coli*. Bio/Technology, *2*: 891–896, 1984.

61. Mandelstram, J., McQuillen, K., Dawes, I. Growth: Cells and population. *In*: J. Mandelstam, K. McQuillen, and I. Dawes (eds.). Biochemistry of Bacterial Growth. New York, pp. 99–123, John Wiley, 1982.

62. Stock, J. B., Rauch, B., and Roseman, S. Periplasmic space in *Salmonella typhimurium* and *Escherichia coli*. J. Biol. Chem., *252*: 7850–7861, 1977.

63. Gottesman, S. Minimizing proteolysis in *Escherichia coli*: Genetic solutions. Methods Enzymol., *185*: 119–129, 1990.

64. Goldberg, A. L., Goff, S. A. The selective degradation of abnormal proteins in bacteria. *In*: W. Reznikoff and L. Gold (eds.), pp. 287–314. Maximizing Gene Expression. Soneham, Massachusetts: Butterworth, 1986.

65. Creighton, T. E. Proteins in solution. *In*: Proteins: Structures and Molecular Properties. New York: W. H. Freeman, pp. 311–312, 1983.

66. Fahey, R. C., Hunt, J. S., Windham, G. C. On the cysteine and cystine content of proteins. Differences between intracellular and extracellular proteins. J. Mol. Evol., *10*: 155–160, 1977.

67. Grossman, A. D., Erickson, J. W., and Gross, C. A. The *htp*R gene product of *E. coli* is a sigma factor for heat-shock promoters. Cell, *38*: 383–390, 1984.

68. Neidhardt, F. C., VanBogelen, R. A., and Vaughn, V. The genetics and regulation of heat-shock proteins. Ann. Rev. Genet., *18*: 295–329, 1984.

69. Zettlmeissl, G., Kaczorek, M., Moya, M., and Streek, R. E. Expression of immunogenically reactive diphtheria toxin fusion proteins under the control of the P_R promoter of bacteriophage lambda. Gene, *41*: 103–111, 1986.

70. Yamaizumi, M., Uchida, T., Takamutsu, K., and Okada, Y. Intracellular stability of diphtheria toxin fragment A in the presence and absence of anti-fragment A. antibody. Proc. Natl. Acad. Sci. U.S.A., *79*: 461–465, 1982.

71. Stader, J. A., Silhavy, T. J. Engineering *Escherichia coli* to secrete heterologous gene products. Methods Enzymol., *185*: 166–187, 1990.

72. Goff, S. A., Casson, L. P., and Goldberg, A. L. Heat shock regulatory gene *htp*R influences rates of protein degradation and expression of the *lon* gene in *Escherichia coli*. Proc. Natl. Acad. Sci. U.S.A., *81*: 6647–6651, 1984.

73. Roth, M. J., Tanese, N., Goff, S. P. Purification and characterization of murine retroviral reverse transcriptase expressed in *Escherichia coli*. J. Biol. Chem., *260*: 9326–9355, 1985.

74. Swamy, K. H. S., and Goldberg, A. L. Subcellular distribution of various proteases, in *Escherichia coli*. J. Bacteriol., *149*: 1027–1033, 1982.

75. Phalipon, A., and Kaczorek, M. Genetically engineered diphtheria toxin fusion proteins carrying the hepatitis B surface antigen. Gene, *55*: 255–263, 1987.

76. Cabiaux, V., Phalipon, A., Wattiez, R., Falmagne, P., Ruysschaert, J. M., and Kaczorek, M. Expression in a biologically active diphtheria toxin fragment B in *Escherichia coli*. Mol. Microbiol., *2*: 339–346, 1988.

77. Schein, C. H. Production of soluble recombinant proteins in bacteria. Bio/Technology, *7*: 1141–1149, 1989.

78. Mitraki, A., and King, J. Protein folding intermediates and inclusion body formation. Bio/Technology, *7*: 690–697, 1989.

79. Haase-Pettingell, C. A., and King, J. Formation of aggregates from a thermolabile in vivo folding intermediate in P22 tailspike maturation. A model for inclusion body formation. J. Biol. Chem., *263*: 4977–4983, 1988.

80. Piatak, M., Lane, J. A., Laird, W., Bjorn, M. J., Wang, A., and Williams, M. Expression of soluble and fully functional ricin A chain in *Escherichia coli* is temperature-sensitive. J. Biol. Chem., *263*: 4837–4843, 1988.

81. Takagi, H., Morinaga, Y., Tsuchiya, M., Ikemura, H., and Inouye, M. Control of folding of proteins secreted by a high expression secretion vector, pIN-III-*omp*A: 16-fold increase in production of active subtilisin E in *Escherichia coli*. Bio/Technology, *6*: 948–950, 1988.

82. Squires, C. H., Childs, J., Eisenberg, S. P. Polverini, P. J., and Sommer, A. Production and characterization of human basic fibroblast growth factor from *Escherichia coli*. J. Biol. Chem., *263*: 4837–4843, 1988.

83. Bowden, G. A. and Georgiou, G. Folding and aggregation of β-lactamase in the periplasmic space of *Escherichia coli*. J. Biol. Chem, *265*: 16760–16766, 1990.

84. Kyger, E., and Wright, T. Thermal stability of different forms of diphtheria toxin. Arch. Biochem. Biophys., *228*: 569–576, 1984.

85. Zhao, J.-M. and London, E. Similarity of the conformation of diphtheria toxin at high temperature to that in the membrane-penetrating low-pH state. Proc. Natl. Acad. Sci. U.S.A., *83*: 2002–2006, 1986.

86. Boquet, P., Silverman, M. S., Pappenheimer, A. M., Jr., and Vernon, W. B. Binding of triton X-100 to diphtheria toxin, crossreacting material 45, and their fragments. Proc. Natl. Acad. Sci. U.S.A., *73*:4449–4453, 1976.

87. Crosa, J. H. Genetics and molecular biology of siderophore-mediated iron transport in bacteria. Microbiol. Rev., *53*: 517–530, 1989.

88. Weinstein, D. L., Holmes, R. K., and O'Brien, A. D. Effects of iron and temperature on shiga-like toxin I production by *Escherichia coli*. Infect. Immun., *56*: 106–111, 1988.

89. Griggs, D. W., and Konisky, J. Mechanism for Iron-regulated transcription of the *Escherichia coli* gene: Metal-dependent binding of Fur protein to the promoters. J. Bacteriol., *171*: 1048–1054, 1989.

90. Bagg, A., and Neilands, J. B. Molecular mechanism of regulation of siderophore-mediated iron assimilation. Microbiol. Rev., *51*: 509–518.

91. de Lorenzo, V., Wee, S., Herrero, M., and Neilands, J. B. Operator sequence of the aerobactin operon of plasmid ColV-K30 binding the ferric uptake regulator (*fur*) repressor. J. Bacteriol., *169*: 2624–2630, 1987.

92. Welkos, S. L., and Holmes, R. K. Regulation of toxinogenesis in *Corynebacterium diphtheriae*. I. Mutations in Bacteriophage β that alter the effects of iron on toxin production. J. Virol., *37*: 936–945, 1981.

93. Yanisch-Perron, C., Vierira, J., and Messing, J. Improved M13 phage cloning vectors and host strains: Nucleotide sequences of the M13mp18 and pUC19 vectors. Gene, *33*: 103–119, 1985.

94. Maxwell, A. F., Maxwell, F., and Glode, L. M. Regulated expression of a diphtheria toxin A-chain gene transfected into human cells: Possible strategy for inducing cancer cell suicide. Cancer Res., *46*: 4660–4664, 1986.

95. Maxwell, F., Maxwell, I. H., Glode, L. M. Cloning, sequence determination, and expression in transfected cells of the coding sequence for the *tox* 176 attenuated diphtheria toxin A chain. Mol. Cell. Biol., *7*: 1576–1579, 1987.

96. Yamaizumi, M., Mekada, E., Uchida, T., and Okada, Y. One molecule of diphtheria toxin fragment A introduced in to A cell can kill the cell. Cell, *15*: 245–250, 1978.

97. McGill, S., Stenmark, H., Sandvig, K., and Olsnes, S. Membrane interactions of diphtheria toxin analyzed using in vitro synthesized mutants, EMBO J., *8*: 2843–2848, 1989.

98. Mullis, K. B., and Faloona, F. Specific synthesis of DNA in vitro via a polymerase catalyzed chain reaction. Methods Enzymol., *155*: 335–350, 1987.

99. Erlich, H. A., Gibbs, R., and Kazazian, H. H., Jr. (eds.). Polymerase Chain Reaction. Cold Spring Harbor, New York: Cold Spring Harbor Laboratory Press, 1989.

100. Erlich, H. A. (ed.) PCR Technology: Principles and Applications for DNA Amplification. New York: Stockton Press, 1989.

101. Innis, M. A., Gelfand, D. H., Sninsky, J. J., and White T. J. (eds.). PCR Protocols: A Guide to Methods and Applications. San Diego: Academic Press, 1990.

102. Olsnes, S., Stenmark, H., McGill, S., Hovig, E., Collier, R. J., and Sandvig, K. Formation of active diphtheria toxin in vitro based on ligated fragments of cloned mutant genes. J. Biol. Chem. *264*: 12749–12751, 1989.

17

The Structure of Diphtheria Toxin as a Guide to Rational Design

Peter J. Nicholls and Richard J. Youle   *National Institute of Neurological Disorders and Stroke, National Institutes of Health, Bethesda, Maryland*

I. BASIC STRUCTURE AND MECHANISM OF ACTION

Diphtheria toxin (DT; 535 amino acids, M_r 58,342) is an acidic single-chain globular protein synthesized and secreted by the gram-positive bacterium *Corynebacterium diphtheriae* (1). The structural gene encoding the protein, *tox*, is contained within the genome of certain β and ω bacteriophages (2). Only strains of the bacterium which are lysogenized by phage produce toxin (3), and then only under host transcriptional control (4) in conditions of limited iron availability (5). Secretion involves the removal of a 25-residue leader peptide; once the mature protein is free from the bacterial cell, it is readily cleaved ("nicked") by proteases at an arginine-rich region located in a disulfide loop, yielding an A fragment (DTA, 193 amino acids, M_r 21,167) and a B fragment (DTB, 342 amino acids, M_r 37,195), linked by a disulfide bond between Cys-186 and Cys-201. Nicking is an essential prerequisite to biological activity because intact DT is not enzymatically active (5). Reduction of the interfragment disulfide bond yields free DTA (which is stable at 100°C, and within the pH range 2–12) and DTB (which is relatively unstable, and precipitates even at neutral pH). The main structural features of DT are summarized in Table 1.

The A fragment is located at the amino-terminal of the toxin molecule, and binds NAD^+ at a single site with high affinity (K_d 8.3 μM). DTA is responsible for the enzymatic activity of the toxin; it catalyzes the transfer of ADP-ribose from NAD^+ to a posttranslationally modified histidine residue (diphthamide) on elongation factor 2 (EF-2), thus inactivating it

Table 1 Important Structural Features of Diphtheria Toxin

Feature	Residues	Postulated function
DT (intact)	1–535	Toxic to many eukaryotic cells, inhibits protein synthesis.
DTA chain	1–193	Catalytic domain of DT. Enters cell cytosol and inactivates EF-2. Stable to extremes of pH and temperature. Nontoxic unless associated with DTB.
DTB chain	194–535	Involved in receptor binding and translocation of DTA to cytosol. Unstable (precipitates at neutral pH unless associated with DTA). Nontoxic.
S-S bonds	186–201	Bond between Cys-186 and Cys-201 links DTA and DTB; must be in oxidized state for binding/translocation. Reduced on exposure to cytosol, releasing active DTA.
	461–471	Bond between Cys-461 and Cys-471 may play a role in translocation, depending on receptor being used (66,85)
Proteolytically sensitive site	190–193	Arg residues at positions 190, 192, 193. Cleavage ("nicking") essential for toxicity; thought to be facilitated by proteases on cell surface (66,85).
Receptor binding domain	482–535	
Hydrophobic domains 1st	269–289	Involved in interaction of the toxin with the cell membrane (binding and/or translocation) (9).
2nd	301–321	
3rd	338–358	
4th	418–439	
Surface lipid-associating domain	210–223 237–248	SLAD (10,15)
Transverse lipid-associating domain	326–344 346–371	TLAD (10,15)
Membrane penetrating segment	418–439	MPS (10,15)
Cationic (or P) site	455–474	Cationic polyphosphate-binding site (11)

(5,6). Elongation factor 2 is required for the transfer of peptidyl-tRNA from the aminoacyl site to the peptidyl site of the 80S ribosome, so the catalytic inactivation of EF-2 ultimately leads to a degree of inhibition of protein synthesis sufficient to cause cell death. Although a single molecule of DTA in the cytosol can be fatal to a cell (7), fragment A alone applied extracellularly is not highly toxic because it has no efficient mechanism for binding to, or crossing, the cell membrane (5). In addition to its ADP-ribosyltransferase activity, DTA also manifests weak NAD^+-glycohydrolase activity, the physiological significance of which is unknown (5).

The DTB fragment is located at the carboxyl-terminus of the molecule, and contains a second disulfide bond between cysteine residues 461 and 471. This fragment is responsible for binding to a receptor (as yet unidentified, but present in the membrane of the majority of eukaryotic cells), and also for translocating DTA into the cytosol (5). The DT receptor-binding domain has been localized to the 6-kd carboxyl-terminus of the B fragment (8), and regions in the other 31 kd of DTB have been proposed as candidate structural elements, possibly involved in binding and/or translocation; these include four hydrophobic domains (9), surface and transverse lipid-associating domains, a membrane-penetrating segment (10), and a cationic (or P) site (11). These elements are discussed in more detail in the section on primary structure of the toxin.

Intoxication of cells by DT is a multistage process, involving (1) DTB-mediated binding to specific receptors (12), (2) endocytosis of the toxin-receptor complex into endosomal vesicles (13), (3) a conformational change in DT structure induced by acidification of the endosome (14), which exposes previously buried hydrophobic regions of the protein (15), (4) DTB-mediated delivery of DTA to the cytosol, together with reduction of the interfragment disulfide bond, and (5) inhibition of protein synthesis catalyzed by free DTA (5). There is currently considerable scientific interest in these processes, partly because a more thorough understanding of the mechanism of action of DT may aid the design of potentially therapeutic drugs; if the cytotoxic properties of DT could be redirected to specifically target an undesirable cell population, sparing non−target cells, a potent reagent may result. Such molecules were initially produced by chemically linking DT to cell type-specific monoclonal antibodies, and hence became known as immunotoxins (ITs) (16). More recently, alternative targeting moieties have been utilized, and attempts have been made to improve the potency of the reagents by genetic engineering (17–21). Although initial results of trails involving the in vivo use of chemically produced ITs have been somewhat disappointing (22), a fuller understanding of how the structure of DT relates to its function may facilitate the development of more successful therapeutic agents. It is the structure-function aspect of DT which is the focus of this chapter.

II. DETAILED STRUCTURE OF DT

A. Primary Structure

Despite the fact that some information regarding the primary structure of DT had been ascertained by sequencing DTA (23,24)–and DTB (10)– derived peptides, the definitive primary structure was obtained by cloning and sequencing the *tox* gene. After initial genetic and physical mapping studies had localized *tox* to a specific region of the phage genome (25– 27), the mutant gene encoding the nontoxic CRM228 was cloned and sequenced (28). Subsequently, the sequences of wild-type DT genes from both β (29) and ω (30) corynephages were determined, and the order of amino acids deduced. Although the ability to clone and sequence *tox* from cross-reacting material (CRM)–producing bacterial strains facilitated the identification of amino acids important in the structure-function of DT, to our knowledge, only eight mutants have been fully sequenced so far (see Table 2).

Detailed analysis of primary structure has identified regions of putative functional significance in DT. There are several segments within the B fragment which are likely to interact with the eukaryotic cell membrane (9,15), and possibly play a role in the process by which fragment A is translocated to the cytosol: (1) Four hydrophobic domains (residues 269–289, 301–321, 338–358, and 418–439) have been located in DTB, and are thought to be involved in the interaction with the cell membrane (9); (2) Dirkx and colleagues have identified two distinct lipid-associating domains in DTB (10,15); a hydrophilic surface lipid-associating domain (SLAD, composed of two hydrophilic alpha-helices, and located toward the amino-terminal [residues 210–223 and 237–248]), and a transverse lipid-associating domain (TLAD, composed of two adjacent hydrophobic alpha-helices, located in the central region of DTB [residues 326–344 and 346–371]); (3) a strongly hydrophobic region (corresponding exactly to the position of the fourth hydrophobic domain, residues 418–493), referred to as the membrane-penetrating segment (MPS); (4) the cationic polyphosphate-binding site (P site, residues 455–474), part of the DT nucleotide-binding site (but distinct from the receptor-binding site), is located within the carboxyl-terminal 8.5-kd region of the B fragment. Ligand binding at the P site inhibits the interaction with the cell surface DT receptor, and binding of NAD^+ site

Table 2 Properties of the CRM Mutants

CRM	M_r[a]	Mutation(s)[b]	Enz[c]	Bind[d]	Trans[e]	Tox[f]	Ref.
		CRMs with Altered Enzymatic Activity					
CRM228	58	Gly-79 to Asp	0	10–15	ND	0	28
		Glu-162 to Lys					
		Ser-197 to Gly					
		Pro-378 to Ser					
		Gly-431 to Ser					
CRM197	58	Gly-52 to Glu	0	>100	ND	0	32, 54
CRM176	58	Gly-128 to Asp	10	ND	ND	0.2–0.4	51
		CRMs with Altered Binding and/or Translocation Activity					
CRM45	45	Δ Thr-386	100	0	ND	0	32
CRM107	58	Ser-525 to Phe	100	0.0125	100	0.01	33
CRM103	58	Ser-508 to Phe	100	1	100	0.1	33, 34
CRM102	58	Pro-308 to Ser	100	1	10	0.1	33, 34
		Ser-508 to Phe					
CRM228	58	Gly-79 to Asp	0	10–15	ND	0	28
		Glu-162 to Lys					
		Ser-197 to Gly					
		Pro-378 to Ser					
		Gly-431 to Ser					
CRM1001	58	Cys-471 to Tyr	100	100	0	0	52

[a]Approximate M_r, in kilodaltons; wild-type DT M_r is 58 kd.
[b]Mutations within the CRM molecule; Δ means deletion of the rest of molecule following residue listed.
[c,d,e]Enzymatic (nicked form), binding and translocation activities, respectively, wild-type DT being 100% in each case.
[f]Toxicity on DT-sensitive cell line, wild-type DT being 100%. Values less than 0.0001% are considered as zero.

and P site ligands is competitive, in the sense that occupancy of either site blocks ligand binding to the other (11). Collier and colleagues have proposed a structural model in which the DTA NAD$^+$-binding site lies immediately adjacent to the DTB P site (11; discussed in Ref. 31).

The DT receptor-binding domain has been localized to a 6 kd region at the carboxyl-terminal of the B chain (residues 482–535); a peptide isolated from this region by hydroxylamine cleavage is capable of preventing whole toxin from binding to its receptor on Vero cells (8). Further evidence implicating this region in binding is provided by mutant, binding-deficient DT molecules (32–34) (see section below on CRMs with altered binding and/or translocation function), and also by truncated DT mutants produced by genetic engineering (20,21).

There is a considerable amount of literature regarding the interaction of intact DT, DTA, and DTB with artificial lipid bilayers, under varying conditions of pH and temperature (reviewed in Ref. 31). However, the membranes of living cells are considerably more complex than artificial ones, and such studies are not covered in this chapter.

B. Secondary Structure

Unlike *Pseudomonas* exotoxin A (ETA) (35), the crystal structure of DT has not yet been fully solved, although studies have been in progress for several years (36–39). Both nucleotide-bound and nucleotide-free forms of DT (36,37), and free DTA (38), crystallize from a number of concentrated salt solutions. Unfortunately, the crystals produced are of variable morphology, and are not ideal for x-ray structure analysis. The electron density map of a DT dimer has been determined to a moderate resolution (3 Å) by single-crystal x-ray diffraction (39). The dimer has dimensions of approximately $65 \times 100 \times 60$ Å, and portions of the polypeptide backbone have been identified in an electron density map. As of 1987, 126 (out of 535) residues had been assigned to nine alpha-helices and three beta-strands in each monomer, and at least two domains are prominently evident. The first domain contains the three beta-strands, and bears some resemblance to portions of the active domain of ETA (35). The other domain is predominantly helical, with one helix in particular being prominently exposed. Although attempts to unambiguously determine the three-dimensional structure of DT are ongoing, at the time of this writing, the definitive answer is not available. Nevertheless, Rappuoli and colleagues have used molecular modeling methods to generate a preliminary structure for the NAD^+-binding site of DT, based on its sequence homology with ETA (40). The catalytic domains of the two proteins share similar overall folding and organization; they are superimposable at the level of the hydrophobic groove that constitutes the NAD^+-binding site, whereas their differences are generally localized to external surface residues.

Since the definitive three-dimensional structure of the DT molecule has not yet emerged, a considerable amount of effort has been expended on the identification of individual residues or regions which play a role in the function of DT. These studies are discussed at length in the following section.

III. STRUCTURE/FUNCTION ANALYSIS OF DT BY MUTAGENESIS OF THE *tox* GENE

Studies aimed at inducing mutations in the *tox* gene with a view to obtaining a clearer picture of how the structure of DT relates to its function can be divided into two types: (1) those which are nonspecific, in that mutations (either deletions or substitutions) are introduced into the protein randomly (2,41); and (2) those which are specific, in the sense that predetermined residues are targeted for substitution (42–44) or deletion (45). Both types of studies require the expression of protein from a mutagenized gene, and an appropriate assay to determine functional changes in the new molecule when compared to wild-type DT (46).

A. Nonspecific Mutagenesis

The most commonly used method to introduce mutations at random into the DT gene is to expose the corynephage-carrying *tox* to N-methyl-N'-nitro-N-nitrosoguanidine (NG) sometimes after induction of lysogeny by exposure to ultraviolet (UV) light (2,41,47–49). An alkylating agent, NG is a potent mutagen in bacterial systems, inducing substitutions, transversions and large deletions (but rarely frameshift mutations) preferentially at the point of replication of the DNA strand (50). Although all possible substitutional events are observed, several reports have suggested that mutations tend to cluster in a limited area of the chromosome (50). In the case of corynephages, NG is most effective when applied to phages infecting *C. diphtheriae* cells during the exponential phase of growth. After mutagenesis, the protein secreted by the host is characterized, and interesting candidates (for instance, those which have drastically reduced toxicity to DT-sensitive cells) are screened for reactivity with a polyclonal anti-DT antibody. Isolates positive by this test have known as CRM mutants (41). It is important to bear in mind that although the procedure used to produce the CRMs involves nonspecific mutagenesis, the selection stage is specific in the sense that only DT mutants with altered function are isolated. Amino acid substitutions at unimportant residues would not result in impaired function, and therefore would go undetected.

The CRMs generally fall into one of two groups: (1) those with altered enzymatic activity (28,32,51), and (2) those which are deficient in binding and/or translocation function. Group 2 can be subdivided further into (a) mutants which are of lower M_r than wild-type DT (2,32,47), and (b) those which have approximately the same M_r as wild-type Dt (28,33,34,52). Mutations in members of groups (1) and (2b) are likely to represent substitution events leading to the replacement of residue(s) important for toxin function with other, inappropriate amino acids; members of group (2a) have most often suffered non-sense mutations, leading to premature termination of the growing polypeptide chain (48). Clearly, any molecule which is enzymatically inactive is, by definition, nontoxic, because it is not capable of shutting down protein synthesis even if it gets into a cell. A mutant which is enzymatically active but nontoxic to a DT-sensitive cell can be presumed to be binding and/or translocation deficient.

Although many CRMs have been generated (49), only those of known primary structure will be discussed in this chapter. The ability to correlate the precise nature of a mutation with its effect on protein function can give us a significant insight into understanding how the structure of DT relates to its activity.

CRMs with Altered Enzymatic Activity.   CRMs with no enzymatic activity (28,41), or with reduced activity when compared on an equimolar basis with wild-type DT (7,41,51) have been isolated; in general, the ability of a mutant to ADP-ribosylate EF-2 is related to its toxicity to sensitive cells (assuming, of course, that the binding and translocation functions of the B chain are intact). For example, CRM176 A chain (51) retains 8–10% enzymatic activity when compared to wild-type DTA, and is approximately 250- 500-fold less toxic (41). CRM228 (28) and CRM197 (32), on the other hand, are devoid of enzymatic activity, and are consequently nontoxic (41).

CRM176 differs from wild-type DT in that a single residue, Gly-128, is replaced with Asp (51). This mutation results in the reduction in enzymatic activity and toxicity (7,41), and is interesting in the light of the fact that glycine residues have been shown to be strongly conserved in "concensus" sequences thought to be general features of adenosine nucleotide–binding folds in proteins of diverse origin (32). The replacement of glycine by an acidic amino acid occurs in CRMs 176 (Gly-128 to Asp), 197 (Gly-52 to Glu), and 228 (Gly-79 to Asp), suggesting that the distribution of charged amino acids may be of particular important in relation to DTA function (51). Since it is known that Glu-148 forms part of the DTA NAD^+-binding site (44,53), it is possible that the additional negatively charged amino acid at position 128 in CRM176 reduces enzymatic activity by disrupting NAD^+ binding (51).

CRM197 contains a single amino acid substitution, the glycine residue at position 52 in the A chain being replaced with glutamic acid. This results in the loss of NAD^+ and ATP-binding activity, which is probably the primary cause of the lack of enzymatic activity (and therefore toxicity). A single amino acid replacement is sufficient to cause a significant structural difference between CRM197 and wild-type DT, as suggested by the mutant's increased trypsin sensitivity, different circular dichroic spectrum, and altered A chain mobility on SDS-PAGE (54,55). Hu and Holmes have postulated that the natural conformation of CRM197 A chain resembles that of acidified wild-type DTA, and that a charge interaction between Glu-52 and the cationic P site may facilitate the insertion of the mutant A chain into the membrane (56). Fluorescence quenching studies (57) and 2-hydroxy-5-nitrobenzyl bromide modification (58) have demonstrated the likely formation of a charge transfer complex between the nicotinamide moiety of NAD^+ and one of the two tryptophan residues in DTA, most likely Trp-50. Photolabeling experiments have recently provided strong evidence that the tyrosine residue at position 65 in DTA is also at the NAD^+-binding site (59), so the presence of a negatively charged amino acid at position 50 in CRM197 may disrupt charge interactions significantly enough to abrogate NAD^+ binding.

Although it is widely accepted that the enzymatic and translocation/binding functions of DT can be unequivocally assigned to the A chain and B chain, respectively, evidence exists that suggests a role for DTA in binding. Mekada and Uchida have demonstrated that nicked CRM197 has a significantly higher affinity for the eurkaryotic cell surface than wild-type DT (54). The binding properties of a hybrid protein formed between wild-type DT A chain and CRM197 B chain were indistinguishable from those of wild-type DT, suggesting that a conformation change in A, but not B, is responsible for the increase in CRM197 binding. Montecucco and colleagues (55) and Hu and Holmes (56) have demonstrated that CRM197 A chain interacts more readily than wild-type DTA with lipid membranes, probably because of increased affinity for the polar head groups of negatively charged phospholipids. The nonspecific interaction between CRM197 A chain and the cell membrane could result in a localized increase in concentration, so that the overall binding constant is the product of A chain affinity for lipids and B chain affinity for the DT receptor.

The gene encoding CRM228 was produced by NG mutagenesis at the same time as those encoding CRMs 45 and 197 (41); it contains 16 nucleotide changes within the coding region relative to wild-type *tox*, 5 of which result in amino acid substitutions (28) (see Table 2). Two of these substitutions fall within the A chain: the Gly residue at position 79 is replaced with Asp, and Glu-162 is replaced with Lys. Although it is tempting to speculate that one (or both) of these A chain substitutions result in CRM228 being devoid of enzymatic activity, the data need to be interpreted with caution because it is conceivable that the other (B chain) mutations may also be involved, possibly by inducing a gross conformational change in the protein which masks the active site.

CRMs with Altered Binding and/or Translocation Function.
Lower M_r than wild-type DT. As discussed earlier, CRMs which are of lower M_r than wild-type DT are most often the result of mutagen-induced non-sense mutations in the *tox* gene, which lead to the premature termi-

nation of the growing polypeptide chain (41). For example, the mutation in the gene which encodes CRM45 (2) introduces a C to T substitution, thus generating a translation termination codon immediately after residue Thr-386 (32). The deletion of the carboxyl-terminal 149 amino acids (16,530 d) results in 41,826-d protein which is fully enzymatically active, but which is not capable of binding to the DT receptor, thus localizing the binding domain (or a region essential for the binding domain to assume the correct configuration) to this deleted segment. It should be noted that CRM45 also lacks the fourth hydrophobic domain, one of four such regions proposed by Eisenberg and colleagues to be involved in delivery of DTA to the cytosol (9).

Reichlin and colleagues have compared a DTA–like molecule (CRM-26, approximate M_r 26,000) and CRM45 in conjugates with thyrotropin-releasing hormone (TRH), and found that the CRM26-TRH conjugates are 200- to 500-fold less toxic to cells expressing the TRH receptor than conjugates containing CRM45 (19). Since both of these hybrid molecules are enzymatically active, and both possess a functional binding domain, there must be structural elements present in the amino-terminal 20 kd of DTB which play a role in delivery of DTA to the cytosol. It is precisely this region of DTB which has been shown to contain the first three hydrophobic domains. It would be interesting to compare the toxicity of CRM45-TRH with that of a similar conjugate made with a CRM possessing all four hydrophobic domains; increased toxicity in the latter would suggest an active role for the fourth hydrophobic domain in the translocation process.

Pappenheimer and colleagues generated a series of hybrid molecules with full biological activity by combining the A chain from CRM45 with the B chains from wild-type DT, CRM197 or CRM176 (60,61). A similar hybrid incorporating the B chain from CRM228 was considerably less active (Table 3). These results are in total agreement with the data which have since become available regarding the mutations present in CRMs 45, 197, 176, and 228; the A chain of CRM45 and B chains of CRMs 197 and 176 are mutation free, so hybrid toxins formed between these components should be identical to wild-type DT. The B chain of CRM228 contains three amino acid substitutions which reduce its binding activity to 10–15% of that of wild-type DT (28); replacing the enzymatically inactive A chain with a fully active one generates a molecule which is approximately seven-fold less toxic in animals than wild-type DT, suggesting that CRM228 B chain is fully capable of translocating A chain through the cell membrane. *Same M_r as wild-type DT.* Since the demonstration that the enzymatic and binding/translocation activities of DT could be assigned to the A and B fragments, respectively, there has been considerable interest in attempt-

Table 3 Properties of "Hybrid" CRM Mutants

CRM	$M_r{}^a$	Mutation(s)[b]	Enz[c]	Bind[d]	Trans[e]	Tox[f]	Ref.
CRM45 A + wt B	58	None	100	100	100	100	60, 61
CRM45 A + CRM197 B	58	None	100	100	100	100	60, 61
CRM45 A + CRM176 B	58	None	100	100	100	100	60, 61
CRM45 A + CRM228 B	58	Ser-197 to Gly Pro-378 to Ser Gly-431 to Ser	100	ND	ND	15	60, 61

[a]Approximate M_r, in kilodaltons; wild-type DT M_r is 58 kd.
[b]Mutations within the hybrid CRM molecule.
[c,d,e]Enzymatic (nicked form), binding and translocation activities, respectively, wild-type DT being 100% in each case.
[f]Toxicity measured by minimum lethal dose in test animal, standardized for wild-type DT toxicity being 100%.

ing to isolate binding and translocation functions to distinct segments of DTB. Leppla and Laird demonstrated that three 58,000-d DT mutants, CRMs 107, 103, and 102, retained full enzymatic activity with respect to wild-type DT, but were defective in receptor binding; subsequent sequencing of the genes encoding these CRMs confirmed B chain substitutions in all three (CRM102, Pro-308 to Ser and Ser-508 to Phe; CRM103, Ser-508 to Phe, CRM107, Ser-525 to Phe) (33,34), and provided further evidence that a domain important in DT binding can be localized to the carboxyl-terminal 17 kd portion of the molecule. It should be noted that sequencing studies in our laboratory have demonstrated that the Leu-390 to Phe substitution originally reported for CRM107 does not, in fact, exist, and that this mutant contains a single change, Ser-525 to the Phe (31). We do not know whether or not the serine residues at positions 525 and 508 are directly involved in the interaction of the toxin with the cell surface receptor; it is possible that mutation of either of these residues results in a conformational change in the protein that impairs the efficiency of a spatially disparate binding domain.

CRM102 and CRM103 have 1% of the binding activity, and 0.1% of the toxicity of wild-type DT; the respective figures for CRM107 and 0.0125 and 0.01%. Chemically conjugating a monoclonal antibody specific for human T cells to either CRM103 or CRM 107 generates an immunotoxin with full target cell toxicity—indistinguishable from that of wild-type DT

linked to the same antibody—without restoring non–target cell toxicity
(33). This indicates that when provided with a new, functional binding
domain, CRMs 103 and 107 are as efficient as wild-type DT at translocating
fragment A into the cell cytosol; clearly, the single amino acid substitution
present in the B chain of these two mutants significantly reduces their
ability to bind the DT receptor, apparently without compromising the
translocation process. This was the first demonstration that the receptor
binding and translocation functions of DTB could be separated.

CRM102 contains the same binding domain amino acid substitution as
CRM103 (Ser-508 to Phe), and in addition contains a second mutation
(Pro-308 to Ser) which falls within the second of four hydrophobic domains
proposed by Ruysschaert and colleagues to play a role in the translocation
of the enzymatically active A fragment across the endosomal membrane
and into the cytosol (10). Providing CRM102 with a new, functional binding
domain (by chemical conjugation with antibodies against the T-cell or
transferrin receptors, or with transferrin itself) generates immunotoxins
which are 10-fold less toxic than wild-type DT or CRM103 linked to the
same binding moieties, indicating that the Pro-308 mutation inhibits translocation by approximately 90% (34). Interestingly, although CRM102 conjugates exhibit only 10% of the translocation activity of the control conjugates when targeting occurs via the new binding domain, they are equally
as active in this respect when binding is to the DT receptor. It has been
proposed that buried hydrophobic regions within DTB play a role in the
translocation process (62); several such hydrophobic regions have been
identified, four of which bear similarities to the transmembrane regions of
integral membrane proteins (9). Proline residues (which have a net preference for aqueous environments) have been shown to be selectively excluded from transmembrane segments of single-spanning nontransport proteins, whereas they constitute a significant proportion of the residues in
similar regions of multiple-spanning transport proteins (63). Brandl and
Deber recognized that proline is conserved at specific locations in such
regions, its presence suggesting it has functional significance (63). Similarly
located proline residues are conserved in the second, third, and fourth
hydrophobic domains of DTB (34); Ruysschaert and colleagues have proposed that cis-trans isomerization of the bond between specific prolines
and adjacent residues is responsible for the acid-dependent conformational
change required for the DT translocation process (64). This may be the
explanation for the translocation deficiency found when CRM102 is provided with a new binding domain (34). The mutation at position 308 removes a proline residue very close to the conserved region in the second
hydrophobic domain, possibly disrupting the cis-trans isomerization process
enough to significantly inhibit translocation. This inhibition is not apparent

when the DT receptor-mediated toxicity of CRM102 is measured, possibly reflecting a difference in the efficiency or mechanism of translocation with the different receptor.

CRM1001 is an interesting mutant which contains a single amino acid substitution, Cys-471 being replaced with Tyr. This disrupts the formation of the DTB intrafragment disulfide bond found in wild-type DT between Cys residues 461 and 471. CRM1001 is 5000-fold less toxic to Vero cells than wild-type DT, and the reduction in toxicity cannot be explained by reduced binding to the cell surface receptor, or by a loss of enzymatic activity (65). Tridente and colleagues have shown that the mutant protein does not interact with the lipid bilayer in the way that wild-type DT does, and postulate that the mutation disrupts the low pH−induced conformational change in the protein that initiates membrane penetration (65).

Having discussed DT mutants produced by nonspecific methods, we will now move on to those in which specific residues are targeted for modification.

Specific mutagenesis.   Despite the fact that interesting variants of the *tox* gene have been produced by nonspecific mutagenesis, the lack of a convenient genetic transfer system in *C. diphtheriae* has generated significant interest in developing *E. coli*–based expression systems (43). Site-directed mutagenesis—followed by protein expression in *E. coli*—is probably the most commonly used method for determining the contribution individual amino acids make to the function of a protein; the great power of this technique is that predefined residues can easily be replaced with any other amino acid of choice. Nevertheless, in the case of DT, such studies have been limited, almost certainly because of the restrictions imposed by the Recombinant DNA Advisory Committees (RAC) on the cloning of genes for such highly toxic proteins.

A Chain Mutants.

Enzymatic and substrate binding mutants.   The only site-directed mutagenesis study on DTA reported to date involves replacement of Glu-148 with aspartic acid (42,44), glutamine (44), or serine (43,44). On the other hand, experiments utilizing photoaffinity labeling (53,59,67), amino acid modification (58,68,69), and protection from proteolysis (70) have implicated His-21, Lys-39, Tyr-65, Glu-148, and Trp-153 in the catalytic function of DTA (see Table 4). These latter studies may point genetic engineers in the right direction regarding which amino acids to target for replacement (in attempts to develop new vaccines against diphtheria, for example), and are therefore considered relevant for discussion in this section.

As early as 1973, it had been established that tyrosine and lysine play roles in the activity of DTA chain (71). Four years later, Michel and Dirkx modified the two tryptophan residues of DTA by treatment with 2-hydroxy-

Table 4 Residues in DT A Implicated in Its Catalytic Function

Those Identified by Nonspecific Mutagenesis

Residue	Method	Ref.	Comment
Gly-52	Nitrosoguanidine mutagenesis	32	CRM197
Gly-128	Nitrosoguanidine mutagenesis	51	CRM176

Those Identified by Specific Mutagenesis or Protein Modification

Residue	Method	Ref.	Comment
His-21	DEPC-modification	68, 69	Reversal by treatment with hydroxylamine
Lys-39	Protection from proteolysis	70	
Tyr-65	Photoaffinity-modification	59	
Glu-148	Photoaffinity-modification	53, 67	
	Site-directed mutagenesis	42, 43, 44	Glu to Asp (42), Glu to Ser (43) Glu to Asp, Gln, Ser (44)
	Deletion mutagenesis	45	Deleted Glu-148 only
Trp-153	2-Hydroxy-5-nitrobenzyl bromide modification	58	

Note: Evidence from studies on CRM228 suggest that Gly-79 and/or Glu-162 may also be involved in the catalytic function of DT A. Since it has not been determined which residue(s) is important, neither has been included in this table.

5-nitrobenzyl bromide, and demonstrated that the protein loses ADP-ribosyltransferase activity as a function of the number of residues affect (58). Modification of Trp-153 had by far the greatest effect; spectroscopic and immunological data indicated that no major conformational change in DTA had occurred, suggesting that Trp-153 may be involved in substrate binding or catalysis. Around the same time, similar studies performed with tetranitromethane as the modifying agent corroborated this hypothesis (72).

In 1984, Carroll and Collier (67) showed that exposing mixtures of (carbonyl-^{14}C)NAD$^+$ and DTA to UV radiation induced the formation of covalently linked protein-ligand photoproducts, and that the label was associated with a single residue in DTA, glutamic acid at position 148

(53). The structure of the photoproduct was found to be amino-gamma-(6-nicotinamidyl)butyric acid, generated by the formation of a new carbon-carbon bond between the gamma methylene group of Glu-148 and C-6 of the nicotinamide ring, decarboxylation of the Glu-148 side chain, and disruption of the linkage between the nicotinamide ring and ADP-ribose. Fragment A modified in this manner was found to be enzymatically inactive, suggesting that this amino acid is also at or near the catalytic site of the molecule. Further evidence to support this hypothesis came from mutagenesis studies, in which Glu-148 was replaced with aspartic acid (42,44), glutamine (44), or serine (43,44). After expression in *E. coli*, the mutant proteins were found to exhibit 100-, 250-, and 300-fold reductions in ADP-ribosyltransferase activity, respectively, when compared to wild-type DTA, but showed little or no reduction in affinity for NAD^+ or EF-2, and no increase in susceptibility to trypsin. Since the side chains of Glu and Asp differ only by a single methylene group, this mutation is a particularly conservative one, indicating that the precise spatial orientation of the carboxylate moiety at position 148 may be crucial for catalytic activity. Deletion of Glu-148 from DTA (45) essentially abolishes enzymatic activity, and when associated with wild-type DTB, the mutant protein is nontoxic to DT-sensitive cell lines. Functionally similar, and presumably homologous, residues have been identified in ETA (Glu-553) (73) and pertussis toxin (Glu-129) (74), and the role that they may play in the mechanism of catalysis has been discussed at length (44).

Zhao and London have demonstrated that a specific trypsin cleavage within whole DT occurs at Lys-39, suggesting that this residue is accessible, and therefore at the surface of the protein (70). Prebinding the DTA substrate NAD^+, or a competing ligand, adenylyl(3'-5')uridine 3'-phosphate (ApUP), resulted in protection from proteolysis. This may be due to steric hindrance, or to a conformational change in the DT molecule, either of which could hinder the approach of the protease. It should be noted that, at present, there is no evidence that Lys-39 plays any role in catalysis, or that it is in direct contact with bound NAD^+.

Sequence comparisons between DT and ETA indicate that the motif Tyr-His-Gly-Thr is conserved between DTA (residues 20–23) and ETA (residues 439–442), and lies in a putative NAD^+-binding cleft evident in the crystal structure of the latter; residue 21 in DTA (the only His in this fragment) corresponds to the second residue in this motif, and was therefore postulated to be important for enzymatic activity. Recently, analogous conserved histidine residues have been identified in cholera toxin, pertussis toxin, and others (44). Montecucco and colleagues have demonstrated that modification of DTA His-21 with diethylpyrocarbonate results in the loss of NAD^+ binding, and NAD^+-glycohydrolase and ADP-ribosyltransferase activities (68,69), and that this effect can be reversed by treatment with hydroxylamine. These results strongly suggest that His-21 is at, or near, the DT NAD$^+$-binding site. To date, no attempts to study the function of His-21 by mutagenesis have been reported.

Most recently, Montecucco and colleagues have demonstrated Tyr-65 of DTA is photolabeled by 8-azido derivatives of adenine and adenosine. This effect is blocked by NAD$^+$, suggesting that Tyr-65 is at the NAD$^+$-binding site. Once again, although no mutagenesis studies have been reported, comparison with ETA has supported this hypothesis.

Significance of protease-sensitive site.   Two groups have recently studied the significance of the trypsin-sensitive region between DTA and DTB (75,76). Murphy and colleagues used site-directed mutagenesis to examine the role played by Arg-190, Arg-192, and Arg-193 in the intoxication of high-affinity interleukin-2 (IL-2) receptor–bearing cells by a fusion protein between truncated DT and IL-2 (75). These three arginine residues are positioned in the proteolytically sensitive 14–amino acid loop subtended by the disulfide bond between Cys-186 and Cys-201 in wild-type DT, and it has been shown that nicking by trypsin can occur after any one of the three; DTA is thus a heterogeneous mixture of three peptides, each with a different carboxyl-terminus (23). Substitution of Arg-193 with glycine results in a 1000-fold reduction in toxicity of the DT–IL-2 fusion protein, whereas replacement of either Arg-190 or Arg-192 had little effect on toxicity. On the other hand, Olsnes and colleagues recovered DTA from the cytosol of intoxicated cells, and demonstrated that only enzymatic fragments with a single carboxyl-terminus Arg residue (that is, those cleaved immediately after Arg-190) are translocation competent, even though all three A fragment types were represented at the cell surface (76). This is interesting in the light of the fact that only a minority of nicked toxin molecules bound to receptors can be induced to translocate their A fragment to the cytosol on exposure to low pH. The position and number of arginine residues in a polypeptide has been shown to be crucial for its ability to interact in the correct manner with lipid membranes; it is possible that joint membrane insertion of the carboxyl-terminal end of DTA and the amino-terminal end of DTB initiates the translocation process, but only if the A fragment contains a single C-terminus arginine residue (76).

B-Chain Mutants.

Mutants deficient in binding and/or translocation.   As in the case of CRMs deficient in binding and/or translocation function, B chain mutants have been produced by truncating the *tox* gene in the region encoding DTB, and also by altering the nucleotide order at specified locations, allowing the production of mutants with the desired amino acid substitution.

Some of the deletion studies have been primarily aimed at producing potentially therapeutic drugs by substituting the DT-binding domain with an alternative, cell type specific, targeting moiety, and are therefore limited in the amount of useful information they provide with respect to the function of DT. Nevertheless, some important facts regarding the DT translocation process have recently come to light, and merit discussion in this section.

Pastan and colleagues have fused the genes encoding a truncated form of DT, DT388 (residues 1–388) with an element encoding the antigen-binding domain of a monoclonal antibody (178,18); once expressed and purified, the chimeric protein exhibits cell type-specific toxicity, but is considerably less toxic than a similar fusion protein between a truncated form of ETA (PE40) and the same antigen-binding domain (18). This difference may reflect the fact that truncation of DT after residue 388 results in the loss of the fourth hydrophobic domain, membrane-penetrating segment, and P site, and the mutant may therefore have significantly reduced translocation activity with respect to wild-type DT. Colombatti and colleagues have demonstrated that the most potent DT-based immunotoxins are produced by including as much as possible of DTB (80). The authors compared chemically produced ITs between DTA, MspSA (a truncated form of DT lacking the C-terminal 17 kd) or full-length DT and an antibody specific for the human CD3 antigen. The conjugate containing full-length DT was 100-fold more toxic than the conjugate containing MspSA, and this in turn was 100-fold more toxic than the one containing DTA. MspSA includes the first three of the four hydrophobic domains, and is essentially equivalent to DT388.

Murphy and colleagues have genetically replaced the diphtheria toxin–binding domain with genes encoding the lymphocytokine IL-2 (20), and alpha-melanocyte–stimulating hormone (MSH) (21). Both chimeric toxins have been shown to be selectively cytotoxic to cells carrying the appropriate cell surface receptor (52). The choice of the precise location of the truncation within the DTB-encoding region should be influenced by the following considerations: (1) the DT-binding domain (localized to the carboxyl-terminal 6 kd of the protein) should be removed to abolish nonspecific toxicity; (2) the hydrophobic domains in DTB postulated to play a role in translocation should be preserved because chemical conjugates between CRM truncation mutants and thyrotropin-releasing hormone (TRH) were only active providing the hydrophobic domains were present (19). In the case of both DT–IL-2 and DT-MSH fusion proteins, the DTB sequence was truncated after the alanine residue at position 485 in mature DT, and neither protein exhibited any toxic effects to cells expressing the DT receptor. However, this still does not unequivocally answer the question of whether the binding domain lies within the deleted region, or whether its presence is required for the correct folding of a distal receptor-binding domain. Intoxication of target cells bearing IL-2 or MSH receptors was found to be blocked by lysosomotrophic agents (which act by reversing the acidification of endocytic vesicles), suggesting that the fusion proteins, like DT itself, must pass through an acidic cellular compartment to exert their effect. In an attempt to identify more precisely residues required for the delivery of fragment A to the cytosol of target cells, Murphy and colleagues have used cassette and deletion mutagenesis to study the DT–IL2 fusion protein (77). Deletion of the 97 amino acids between Thr-386 and His-484 (the region containing the B chain disulfide bond, fourth hydrophobic domain, MPS, and P site) increases the potency and apparent association constant for IL-2 receptor–bearing target cells, in comparison to a fusion protein including residues 386–484; in contrast, deletion of the regions between Asp-290 and Gly-482 (191 residues, containing the second, third, and fourth hydrophobic domains, TLAD, MPS, and P site), or between Asn-203 and Ile-289 (85 residues, containing the first hydrophobic domain and SLAD) results in a 1000-fold reduction in potency (binding being unaffected in the case of the former deletion, and reduced fourfold in the case of the latter). This study raises several interesting points which are worthy of discussion. First, neither DTB sequences distal to Thr-386 nor an intact disulfide bond between Cys-641 and Cys-471 is required for cytotoxicity of DT–IL-2, whereas the presence of both of these is essential for wild-type DT acting via the DT receptor. Since we know from the Thr-386 to His-484 deletion that removal of the fourth hydrophobic domain, MPS, and P site does not reduce the toxicity of the fusion protein, the remaining deletion mutants implicate the first, second, and third hydrophobic domains, SLAD, and TLAD in the intoxication process. It is interesting that the only mutant lacking the SLAD also shows reduced binding efficiency; this fits the model proposing a role for the SLAD in interacting with cell surface lipids, thus producing a localized increase in toxin concentration prior to binding to the DT receptor. It should be noted that deletion of such large regions of the protein may have significant effects on the conformation of distal structures, and the results need to be evaluated with caution.

As early as 1973, DT had been synthesized in vitro by adding corynephage DNA to cell-free lysates of *E. coli* (78,79). Recently, Sandvig and colleagues have used this strategy to design DT mutants by programming a rabbit reticulocyte lysate with RNA encoding toxin (81); the RNA was produced by transcription from a polymerase chain reaction (PCR)–generated DNA template containing a promoter for T3 RNA polymerase.

These are two features inherent in this approach which make it an attractive method for analyzing toxin structure-function relationships. First, it obviates the need for gene cloning, thus bypassing the strict regulations governing manipulation of toxin genes in vivo. Second, PCR is an extremely powerful and rapid method for introducing mutations (substitutions, insertions, or deletions) at any point in a gene. Olsnes, Sandvig, and colleagues have used the in vitro transcription/translation method to investigate the role of DTB in the intoxication process (82–84). Their results can be summarized as follows: (1) Protease protection studies on DT mutants demonstrate that the region of DTB extending from approximately residue 300 to the C-terminus is inserted into the cell membrane at low pH (82); (2) N-terminal deletion mutants of DTB exhibited significantly impaired receptor binding, and no autonomous binding domain could be identified in the fragment. This suggests that regions at *both* ends of DTB are involved in receptor recognition (83); (3) Mutants lacking the N-terminus 3–10 kd of DTB are able to insert into the cell membrane and form ion channels, but the channels are nonselectively permeable (84).

Significance of the DTB intrafragment disulfide bond.   Several investigations have been aimed at evaluating the role of the DTB intrafragment disulfide bond in toxin function. The observation that chemical modification of Cys-461 and Cys-471 abolished receptor binding suggested a role for the these residues in the binding process (85). In contrast, Madshus and colleagues used the in vitro transcription/translation method to produce a full-length mutant in which Cys residues 461 and 471 were replaced with Ser, and found that the mutant was functionally indistinguishable from wild-type DT with respect to trypsin sensitivity, receptor binding, translocation of DTA to the cytosol and membrane channel formation (66). However, a molecule in which Cys-471 was replaced with Tyr (the mutation found in CRM1001) was found to be deficient in membrane interactions at low pH. The authors speculate that the bulky inflexible side chain of tyrosine may prevent clustering of the charged residues which constitute the P site (amino acids 455–474), whereas the smaller, more flexible side chain of serine may be accommodated into the structure without distorting it significantly.

IV. CONCLUSIONS

The main gaps in our understanding of the mode of action of DT concern the mechanism of attachment to the cell surface, and penetration of the catalytic domain into the cell cytosol. The enzymatic mechanism of DTA is, by comparison, relatively well understood. Recently developed techniques for rapidly producing and analyzing mutants should facilitate the analysis of the structural elements in DT postulated to play a role in the intoxication process. These data, together with the solution of the DT crystal structure, should provide us with sufficient information to develop a fuller understanding of how the structure of this complex protein relates to its function.

REFERENCES

1. Barksdale, L. *Corynebacterium diphtheriae* and its relatives. Bact. Rev., *34*: 378–422, 1970.
2. Uchida, T., Gill, D. M., and Pappenheimer, A. M., Jr. Mutation in the structural gene for diphtheria toxin carried by temperate phage β. Nature (New Biol.), *233*: 8–11, 1971.
3. Freeman, V. J., Studies on the virulence of bacteriophage-infected strains of *Corynebacterium diphtheriae*. J. Bacteriol., *61*: 675–688, 1951.
4. Kaczorek, M., Zettlmeissl, G., Delpeyroux, F., and Streeck, R. E. Diphtheria toxin promoter function in *Corynebacterium diphtheriae* and *Escherichia coli*. Nucl. Acids Res., *13*: 3147–3159, 1985.
5. Collier, R. J., Diphtheria toxin: Mode of action and structure. Bact. Rev. 39: 54–85, 1975.
6. Honjo, J., Nishizuka, Y., Hataishi, O., and Kato, I. Diphtheria toxin-dependent adenosine diphosphate ribosylation of aminoacyl transferase II and inhibition of protein synthesis. J. Biol. Chem., *243*: 3553–3555, 1968.
7. Yamaizumi, M., Mekada, E., Uchida, T., and Okada, Y. One molecule of diphtheria toxin fragment A introduced into a cell can kill the cell. Cell, *15*: 245–250, 1978.
8. Rolf, J. M., Gaudin, H. M., and Eidels, L. Localization of the diphtheria toxin receptor-binding domain to the carboxyl-terminal Mr ~6000 region of the toxin. J. Biol. Chem., *265*: 7331–7337, 1990.
9. Eisenberg, D., Schwarz, E., Komaromy, M., and Wall, R. Analysis of membrane and surface protein sequences with the hydrophobic moment plot. J. Mol. Biol., *179*: 125–142, 1984.
10. Falmagne, P., Capiau, C., Lambotte, P., Zanen, J., Cabiaux, V., and Ruysschaert, J-M. The complete amino acid sequence of diphtheria toxin fragment B. Correlation with its lipid-binding properties. Biochim. Biophys. Acta, *827*: 45–50, 1985.
11. Lory, S., Carroll, S. F., and Collier, R. J. Ligand interactions of diphtheria toxin II. Relationships between the NAD site and the P site. J. Biol. Chem., *255*: 12016–12019, 1980.
12. Middlebrook, J. L., Dorland, R. B., and Leppla, S. H. Association of diphtheria toxin with Vero cells. Demonstration of a receptor. J. Biol. Chem., *253*: 7325–7330, 1978.
13. Moya, M. A., Dautry-Versat, A., Goud, B., Louvard, D., and Boquet, P. Inhibition of coated pit formation in Hep2 cells blocks the cytotoxicity of diphtheria toxin but not that of ricin toxin. J. Cell Biol., *101*: 548–553. 1985.

14. Sandvig, K., Tonnessen, T. I., Sand, O., and Olsnes, S. Requirement of a transmembrane pH gradient for the entry of diphtheria toxin into cells at low pH. J. Biol. Chem., *261*: 11639–11645, 1986.

15. Lambotte, P., Falmagne, P., Capiau, C., Zanen, J., Ruysschaert, J-M., and Dirkx, J. Primary structure of diphtheria toxin fragment B: Structural similarities with lipid-binding domains. J. Cell. Biol., *87*: 837–840, 1980.

16. Arthur E. Frankel (ed.). Immunotoxins. New York: Kluwer Academic Publishers, 1988.

17. Chaudhary, V. K., Gallo, M. G., FitzGerald, D. J., and Pastan, I. A recombinant single-chain immunotoxin composed of anti-Tac variable regions and a truncated diphtheria toxin. Proc. Natl. Acad. Sci. U.S.A., *87*: 9491–9494, 1990.

18. Batra, J. K., FitzGerald, D. J., Chaudhary, V. K., and Pastan, I. Single-chain immunotoxins directed at the human transferrin receptor containing Pseudomonas exotoxin A or diphtheria toxin: Anti-TFR(Fv)-PE40 and DT388-Anti-TFR(Fv). Mol. Cell. Biol., *11*: 2200–2205, 1991.

19. Bacha, P., Murphy, J. R., and Reichin, S. Thyrotropin-releasing hormone-diphtheria toxin-related polypeptide conjugates. Potential role of the hydrophobic domain in toxin entry. J. Biol. Chem., *258*: 1565–1570, 1983.

20. Williams, D. P., Parker, K., Bacha, P., Bishai, W., Borowski, M., Genbauffe, F., Strom, T. B., and Murphy, J. R. Diphtheria toxin receptor binding domain substitution with interleukin-2: Genetic construction and properties of a diphtheria toxin-related interleukin-2 fusion protein. Protein Eng., *1*: 493–498, 1987.

21. Murphy, J. R., Bishai, W., Borowski, M., Miyanohara, A., Boyd, J., and Neagle, S. Genetic construction, expression, and melanoma-selective cytotoxicity of a diphtheria toxin-related alpha-melanocyte-stimulating hormone fusion protein. Proc. Natl. Acad. Sci. U.S.A., *83*: 8258–8262, 1986.

22. Rybak, S., and Youle, R. J. Clinical use of immunotoxins: Monoclonal antibodies conjugated to protein toxins. Herbert F. Oettgen (ed.). Immunol. Allergy Clin. North Am. Hum. Cancer Immunol., *2*: 359–380, 1991.

23. DeLange, R. J., Drazin, R. E., and Collier, R. J., Amino-acid sequence of fragment A, an enzymatically active fragment from diphtheria toxin. Proc. Natl. Acad. Sci. U.S.A., *73*: 69–72, 1976.

24. Michel, A., Zanen, J., Monier, C., Crispeels, C., and Dirkx, J. Partial characterization of diphtheria toxin and its subunits. Biochim. Biophys. Acta, *257*: 249–256, 1972.

25. Costa, J. J., Michel, J. L., Rappuoli, R., and Murphy, J. R. Restriction map of corynephages β_c and β_{vir} and physical localization of the diphtheria *tox* operon. J. Bacteriol., *148*: 124–130, 1981.

26. Buck, G. A., and Groman, N. B. Physical mapping of β-converting and gammma-nonconverting corynebacteriophage genomes. J. Bacteriol., *148*: 131–142, 1981.

27. Buck, G. A., and Groman, N. B. Identification of deoxyribonucleic acid restriction fragments of β-converting corynebacteriophages that carry the gene for diphtheria toxin. J. Bacteriol., *148*: 153–162, 1981.

28. Kaczorek, m., Delpeyroux, F., Chenciner, N., Streeck, R. E., Murphy, J. R., Boquet, P., and Tiollais, P. Nucleotide sequence and expression of the *tox*228 gene in *Escherichia coli*. Science, *221*: 855–858, 1983.

29. Greenfield, L., Bjorn, M. J., Horn, G., Fong, D., Buck, G. A., Collier, R. J., and Kaplan, D. A. Nucleotide sequence of the structural gene for diphtheria toxin carried by corynebacteriophage β. Proc. Natl. Acad. Sci. U.S.A., *80*: 6853–6857, 1983.

30. Ratti, G., Rappuoli, R., and Giannini, G. The complete nucleotide sequence of the gene coding for diphtheria toxin in the corynephage omega (tox⁺) genome. Nucl. Acids Res., *11*: 6589–6595, 1983.

31. Johnson, V. G., and Youle, R. J., Intracellular routing and membrane translocation of diphtheria toxin and ricin. *In*: C. J. Steer and J. A. Hanover (eds.), Intracellular Trafficking of Proteins. Cambridge University Press, Cambridge, England, pp. 183–225, 1991.

32. Giannini, G., Rappuoli, R., and Ratti, G. The amino-acid sequence of two nontoxic mutants of diphtheria toxin: CRM45 and CRM197. Nucl. Acids Res., *12*: 4063–4069, 1984.

33. Greenfield, L., Johnson, V. G., and Youle, R. J. Mutations in diphtheria toxin separate binding from entry and amplify immunotoxin selectivity. Science, *238*: 536–539, 1987.

34. Johnson, V. G., and Youle, R. J. A point mutation of proline 308 in diphtheria toxin B chain inhibits membrane translocation of toxin conjugates. J. Biol. Chem., *264*: 17739–17744, 1989.

35. Allured, V. S., Collier, R. J., Carroll, S. F., and McKay, D. B. Structure of exotoxin A of *Pseudomonas aeruginosa* at 3.0-Å resolution. Proc. Natl. Sci. U.S.A., *83*: 1320–1324, 1986.

36. Collier, R. J., Westbrook, E. M., McKay, D. B., and Eisenberg, D. X-ray grade crystals of diphtheria toxin. J. Biol. Chem., *257*: 5283–5285, 1982.

37. McKeever, B., and Sarma, R. Preliminary crystallographic investigation of the protein toxin from *Corynebacterium diphtheriae*. J. Biol. Chem., *257*: 6923–6925, 1982.

38. Kantardjieff, K., Collier, R. J., and Eisenberg, D. X-ray grade crystals of the enzymatic fragment of diphtheria toxin. J. Biol. Chem., *264*: 10402–10404, 1989.

39. Kantardjieff, K., Dijkstra, B., Westbrook, E. M., Barbieri, J. T., Carroll, S. F., Collier, R. J., and Eisenberg, D. Structural studies on diphtheria toxin. UCLA Symp. Mol. Cell. Biol. (New Ser.), *69*: 187–200, 1987.

40. Domenighini, M., Montecucco, C., Ripka, W. C., and Rappuoli, R. Computer modelling of the NAD binding site of ADP-ribosylating toxins: active-site structure and mechanism of NAD binding. Mol. Microbiol., *5*: 23–31, 1991.

41. Uchida. T., Pappenheimer, A. M., Jr., and Greany, R. Diphtheria toxin and related proteins. I. Isolation and properties of mutant proteins serologically related to diphtheria toxin. J. Biol. Chem., *248*, 3838–3844, 1973.

42. Tweten, R. K., Barbieri, J. T., and Collier, R. J. Diphtheria Toxin: Effect of substituting aspartic acid for glutamic acid 148 on ADP-ribosyltransferase activity. J. Biol. Chem., *260*: 10392–10394, 1985.

43. Barbieri, J. T., and Collier, R. J. Expression of a mutant, full-length form of diphtheria toxin in *Escherichia coli*. Infect. Immun., *55*: 1647–1651, 1987.

44. Wilson, B. A., Reich, K. A., Weinstein, B. R., and Collier, R. J. Active-site mutations of diphtheria toxin: Effects of replacing glutamic acid-148 with aspartic acid, glutamine or serine. Biochemistry, *29*: 8643–8651, 1990.

45. Emerick, A., Greenfield, L., and Gates, C. Enzymatically inactive diphtheria A fragment: Expression in *E. coli* and toxicity characterization. DNA, *4*: 78, 1985.

46. Carroll, S. F., and Collier, R. J. Diphtheria toxin: quantification and assay. Methods Enzymol., *165*: 218–225, 1988.

47. Matsuda, M., Kanei, C., and Yoneda, M. A phage-mutant directed synthesis of a fragment of diphtheria toxin protein. Biochem. Biophys. Res. Commun., *46*: 43–49, 1972.

48. Holmes, R. K. Characterization and genetic mapping of nontoxinogenic (*tox*) mutants of corynebacteriophage β. J. Virol., *19* 195–207, 1976.

49. Laird, W., and Groman, N. Isolation and characterization of *tox* mutants of corynebacteriophage β J. Virol., *19*: 220–227, 1976.

50. Drake, J. W. The molecular basis of mutation. San Francisco. Holden-Day, 1970.

51. Maxwell, F., Maxwell, I. H., and Glode, L. M. Cloning, sequence determination, and expression in transfected cells of the coding sequence for the *tox* 176 attenuated diphtheria toxin A chain. Mol. Cell. Biol., 7: 1576–1579, 1987.

52. Murphy, J. R. Diphtheria-related peptide hormone gene fusions: A molecular genetic approach to chimeric toxin development. *In*: Arthur E. Frankel (ed.), Immunotoxins. New York: Kluwer Academic Publishers, 1988.

53. Carroll, S. F., McCloskey, J. A., Crain, P. F., Oppenheimer, N. J., Marschner, T. M., and Collier, R. J. Photoaffinity labeling of diphtheria toxin fragment A with NAD: Structure of the photoproduct at position 148. Proc. Natl. Acad. Sci. U.S.A., *82*: 7237–7241, 1985.

54. Mekada, E., and Uchida, T. Binding properties of diphtheria toxin to cells are altered by mutation in the fragment A domain. J. Biol. Chem., *260*: 12148–12153, 1985.

55. Papini, E., Colonna, R., Schiavo, G., Cusinato, F., Tomasi, M., Rappuoli, R., and Montecucco, C. Diphtheria toxin and its mutant CRM 197 differ in their interaction with lipids. FEBS Lett., *215*: 73–78, 1987.

56. Hu, V. W., and Holmes, R. K. Single mutation in the A domain of diphtheria toxin results in a protein with altered membrane insertion behavior, Biochim. Biophys. Acta, *902*: 24–30, 1987.

57. Kandel, J., Collier, R. J., and Chung, D. W. Interaction of fragment A from diphtheria toxin with nicotinamide adenine dinucleotide. J. Biol. Chem., *249*: 2088–2098, 1974.

58. Michel, A., and Dirkx, J. Occurrence of tryptophan in the enzymatically active site of diphtheria toxin fragment A. Biochim. Biophys. Acta, *491*: 286–295, 1977.

59. Papini, E., Santucci, A., Schiavo, G., Domenighini, M., Neri, P., Rappuoli, R., and Montecucco, C. Tyrosine 65 is photolabeled by 8-azidoadenine and

8-azidoadenosine at the NAD binding site of diphtheria toxin. J. Biol. Chem., *266*: 2494–2498, 1991.

60. Uchida, T. Pappenheimer, A. M., Jr., and Harper, A. A. Reconstitution of diphtheria toxin from two nontoxic cross-reacting mutant proteins. Science, *175*: 901–903, 1972.

61. Uchida, T., Pappenheimer, A. M., Jr., and Harper, A. A. Diphtheria toxin and related proteins. III. Reconstitution of hybrid "diphtheria toxin" from nontoxic mutant proteins. J. Biol. Chem., *248*: 3851–3854, 1973.

62. Boquet, P., Silverman, M. S., Pappenheimer, A. M., Jr., and Vernon, W. B. Binding of triton X-100 to diphtheria toxin, crossreacting material 45, and their fragments. Proc. Natl. Acad. Sci. U.S.A., *73*: 4449–4453, 1976.

63. Brandl, C., and Deber, C. Hypothesis about the function of membrane-buried proline residues in transport proteins. Proc. Natl. Acad. Sci. U.S.A., *83*: 917–921, 1986.

64. Deleers, M., Beugnier, N., Falmagne, P., Cabiaux, V., and Ruysschaert, J. M. Localization in diphtheria toxin fragment B of a region that induces pore formation in planar lipid bilayers at low pH. FEBS Lett., *160*: 82–86, 1983.

65. Dell'Arciprete, L., Colombatti, M., Rappuoli, R., and Tridente, G. A C terminus cysteine of diphtheria toxin B chain involved in immunotoxin cell penetration and cytotoxicity. J. Immunol., *140*: 2466–2471, 1988.

66. Stenmark, H., Olsnes, S., and Madshus, I. H. Elimination of the disulphide bridge in fragment B of diphtheria toxin: Effect on membrane insertion, channel formation, and ATP binding. Mol. Microbiol., *5*: 595–606, 1991.

67. Carroll, S. F., and Collier, R. J. NAD binding site of diphtheria toxin: Identification of a residue within the nicotinamide subsite by photochemical modification with NAD. Proc. Natl. Acad. Sci. U.S.A., *81*: 3307–3311, 1984.

68. Papini, E., Schiavo, G., Sandona, D., Rappuoli, R., and Montecucco, C. Histidine 21 is at the NAD^+ binding site of diphtheria toxin. J. Biol. Chem, *264*: 12385–12388, 1989.

69. Papini, E., Schiavo, G., Rappuoli, R., and Montecucco, C. Histidine-21 is involved in diphtheria toxin NAD^+ binding. Toxicon, *28*: 631–635, 1990.

70. Zhao, J-M., and London, E. Localization of the active site of diphtheria toxin. Biochemistry, *27*: 3398–3403, 1988.

71. Beugnier, N., and Zanen, J. Mise en evidence d'une tyrosine dans le site enzymatique de la toxine diphtherique. Arch. Int. Physiol. Biochem., *81*: 581, 1973.

72. Beugnier, N., and Zanen, J. Diphtheria toxin: The effect of nitration and reductive methylation on enzymatic activity and toxicity. Biochim. Biophys. Acta, *490*: 225–234, 1977.

73. Caroll, S. F., and Collier, R. J. Active site of *Pseudomonas aeruginosa* exotoxin A. Glutamic acid 553 is photolabeled by NAD and shows functional homology with glutamic acid 148 of diphtheria toxin. J. Biol. Chem., *262*: 8707–8711, 1987.

74. Barbieri, J. T., and Collier, R. J. Expression of a mutant, full-length form of diphtheria toxin in *Escherichia coli*. Infect. Immun., *55*: 1647–1651, 1987.

75. Williams, D. P., Wen, Z., Watson, R. S., Boyd, J., Strom, T. B., and Murphy, J. R. Cellular processing of the interleukin-2 fusion toxin DAB$_{486}$-IL-2 and efficient delivery of diphtheria fragment A to the cytosol of target cells requires Arg 194. J. Biol. Chem., *265*: 20673–20677, 1990.

76. Moskaug, J. O., Sletten, K., Sandvig, K., and Olsnes, S. Translocation of diphtheria toxin A-fragment to the cytosol. Role of the site of interfragment cleavage. J. Biol. Chem., *264*: 15709–15713, 1989.

77. Williams, D. P., Snider, C. E., Strom, T. B., and Murphy, J. R. Structure/function analysis of interleukin-2-toxin (DAB$_{486}$-IL-2). Fragment B sequences required for the delivery of fragment A to the cytosol of target cells. J. Biol. Chem., *265*: 11885–11889, 1990.

78. Murphy, J. R., Pappenheimer, A. M., Jr., and Tayart De Borms, S. Synthesis of diphtheria *tox*-gene products of *Escherichia coli* extracts. Proc. Natl. Acad. Sci. U.S.A., *71*: 11–15, 1974.

79. Lightfoot, H. N., and Iglewski, B. H. Synthesis of diphtheria toxin in *E. coli* cell-free lysate. Biochem. Biophys. Res. Commun., *56*: 351–357, 1974.

80. Colombatti, M., Greenfield, L., and Youle, R. J. Cloned fragment of diphtheria toxin linked to T cell-specific antibody identifies regions of B chain active in cell entry. J. Biol. Chem., *261*: 3030–3035, 1986.

81. Olsnes, S., Stenmark, H., McGill, S., Hovig, E., Collier, R. J., and Sandvig, S. Formation of active diphtheria toxin in vitro based on ligated fragments of cloned mutant genes. J. Biol. Chem., *264*: 12749–12751, 1989.

82. Moskaug, J. O., Stenmark, H., and Olsnes, S. Insertion of diphtheria toxin B-fragment into the plasma membrane at low pH. Characterization of the topology of inserted regions. J. Biol. Chem., *266*: 2652–2659, 1991.

83. Stenmark, H., McGill, S., Olsnes, S., and Sandvig, K. Permeabilization of the plasma membrane by deletion mutants of diphtheria toxin. EMBO J., *8*: 2849–2853, 1989.

84. McGill, S., Stenmark, H., Sandvig, K., and Olsnes, S. Membrane interactions of diphtheria toxin analyzed using in vitro synthesized mutants. EMBO J., *8*: 2843–2848, 1989.

85. Wright, H. Y., Marston, A. W., and Goldstein, D. J. A functional role for cysteine disulfides in the transmembrane transport of diphtheria toxin. J. Biol. Chem., *259*: 1649–1654, 1984.

18
Protein Engineering of Diphtheria Toxin
Development of Receptor-Specific Cytotoxic Agents for the Treatment of Human Disease

John R. Murphy, Fadi G. Lakkis, Johanna C. vanderSpek, and Paige Anderson   *The University Hospital, Boston, Massachusetts*

Terry B. Strom   *Beth Israel Hospital and Harvard Medical School, Boston, Massachusetts*

I. INTRODUCTION

Diphtheria toxin is produced and secreted by strains of *Corynebacterium diphtheriae* that are lysogenic for one of a number of toxigenic coryne-bacteriophages (1). Uchida et al. (2) demonstrated that corynebacterio-phages β carried the structural gene for diphtheria toxin by the isolation and characterization of mutants that directed the synthesis of nontoxic serologically related cross-reacting material (CRM). In mature form, native diphtheria toxin is a 58,348-d protein which may be cleaved into two poly-peptides following exposure to serine proteases (e.g., trypsin) (3–4). The N-terminal 21,167 fragment of toxin, fragment A, is the catalytically active toxophore responsible for the adenosine diphosphorylribosylation (ADPR) of elongation factor 2 (EF-2) within the cytosol of intoxicated cells. Frag-ment B, the C-terminal 37,199-d fragment of diphtheria toxin, has been shown to carry the membrane-associating regions which facilitate the de-livery of fragment A to the eukaryotic cell cytosol, and the receptor-binding domain (2).

It is now well accepted that the diphtheria intoxication process involves at least the following steps: (A) the binding of toxin to its cell surface receptor (5); (B) internalization of bound toxin by receptor-mediated endocytosis (6); (C) cleavage of the arginine-rich proteolytically sensitive loop between fragments A and B; (D) upon acidification of the endocytic vesicle (7), the partial denaturation of the hydrophobic membrane associating domains of fragment B (8–9), which leads to the formation of a pore or channel (10–11); and through which (E) fragment A–associated ADP-ribosyltransferase is delivered to the cytosol of the cell. Yamizumi et al. (12) have clearly demonstrated that the introduction of a single molecule of fragment A to the cytosol of the cell is sufficient to cause the death of that cell.

The use of diphtheria toxin and the related CRMs as model systems for the development of targeted toxins stems from the early observations of Uchida et al. (3). These investigators found that a fully active diphtheria toxin could be reconstituted by combining the functional A fragment from the A^+B^- mutant CRM45, with a functional B fragment from the A^-B^+ mutant CRM197. Importantly, the reconstituted "toxin" was as potent as native diphtheria toxin. Since this study clearly demonstrated that purified fragments of diphtherial toxin could be used to reassemble a functional toxin, it was attractive to attempt to combine nontoxic fragments of heterologous protein toxins in an attempt to form "hybrid" toxins that would retain the functional properties of each of their respective components. Toward this end, Olsnes et al. (14) demonstrated that a biologically active toxin could be assembled from ricin A chain coupled through disulfide linage to the B chain of abrin. These studies demonstrated that the chain, or fragment-specific function of these toxins, could be retained in the assembly of a biologically active hybrid, or chimeric, toxin molecule. Indeed, the foundations of the immunotoxin technology finds its roots in these early studies. As is well known, the immunotoxins are composed of monoclonal antibodies that are directed to a target cell surface antigen coupled by disulfide linkage to the toxophore fragments of either plant or microbial toxins.

Early studies with the immunotoxins demonstrated that the choice of toxophore was important. For example, with a given monoclonal antibody, a potent immunotoxin could be assembled through disulfide linkage with the A chain of ricin; however, if the fragment A of diphtheria toxin was chosen as the toxophore, the immunotoxin was almost without exception found to be either of low potency or biologically inactive. These early observations suggested that the delivery of diphtheria toxin fragment A across the eukaryotic cell membrane and into the cytosol of target cells required additional structures. Boquet et al. (15) demonstrated that fragment B isolated from the CRM45 mutant of diphtheria toxin was able to bind [^{3}H]Triton X-100, and as such had properties which were similar to integral membrane proteins. In addition, both Donovan et al. (10) and Kagen et al. (11) reported that the truncated B fragment purified from CRM45 was able to spontaneously insert into artificial membranes and form conductive pores or channels. Importantly, these channels were only formed under conditions of differential pH across the membrane, and the conditions that were necessary mimicked those of an acidified endocytotic vesicle. These studies suggested that functional domains within fragment B of diphtheria toxin were likely to be essential for the delivery of fragment A across the eukaryotic cell membrane and into the cytosol.

In order to test the hypothesis that functional domains within fragment B were necessary for the delivery of diphtheria toxin fragment A to the cytosol of target cells, Bacha et al. (16) assembled two conjugate toxins: CRM26–TRH and CRM45–TRH. Thyrotropin-releasing hormone (TRH) is a tripeptide composed of pyro-glutamate-histidine-proline-amide, and both the pyro-Glu and Pro-amide moieties are required for TRH-receptor binding. The conjugate toxins were assembled by modifying the imidazole ring of histidine such that a disulfide bond could be formed with N-suc-cinimidyl-3-(2-pyridyldithio)proprionate (SPDP)–modified CRM26 or CRM45. The results of these experiments were absolutely clear: while both CRM26–TRH and CRM45–TRH were able to bind to the TRH receptor on the target cell surface, only the CRM45–TRH was specifically cytotoxic. These experiments strongly suggested that only CRM45–TRH contained sufficient structural information to facilitate the delivery of fragment A across the membrane. Moreover, these experiments suggested that highly potent diphtheria toxin–related peptide hormone conjugate molecules could be assembled. Perhaps more importantly, these experiments suggested that the native receptor binding domain of diphtheria toxin could be replaced with peptide ligands that recognized eukaryotic cell surface receptors. As long as the targeted receptors were internalized by receptor-mediated endocytosis, the diphtheria toxin–based conjugate proteins containing the fragment B hydrophobic membrane-associating domains should be biologically active.

II. GENETIC CONSTRUCTION OF DIPHTHERIA TOXIN–BASED FUSION TOXINS

As described above, the early experiments with diphtheria toxin and diphtheria toxin–based conjugate toxins demonstrated that in order to be biologically active a toxic protein must contain at least the following minimal structure: (1) a receptor-binding domain, (2) an ADP-ribosyltransferase, and (3) the structural domains necessary to facilitate the delivery of the ribosyltransferase across the membrane and into the cytosol. Based upon these apparent requirements, we reasoned that the portion of fragment B coding for the native toxin receptor-binding domain could be sequentially replaced with DNA coding for polypeptide hormones or growth factors. Such a fusion protein would include the diphtheria toxin fragment A–associated ADP-ribosyltransferase, the membrane translocation domains of fragment B, and either a polypeptide hormone or growth factor that was fused to the truncated toxin through a peptide bond. Since the ligand component of the fusion toxin should be the only factor to determine cell specificity, only those cells expressing the appropriate cell surface receptors should be targeted. The recombinant toxin should then be internalized and intoxicate *only* target cells. In addition to the minimal structural features necessary for assembly of these new receptor directed toxins, it was also clear that these recombinant fusion proteins had to be expressed in *Escherichia coli* in a protease-resistant form. Once expressed, the individual functional domains of these new recombinant toxins had to fold into a biologically active conformation.

Despite these caveats, the genetic construction of cell surface receptor-directed fusion toxins offered great potential. This potential was based on the precision and definition intrinsic to genetic engineering methodologies. For example, the assembly of traditional immunotoxins, conjugate proteins assembled by chemical cross-linking of monoclonal antibodies (mAbs) with fragments of microbial or plant toxins, leads to the formation of hybrid proteins in which the linkage between the mAbs and the toxophore is dependent upon the sites which react with the cross-linking reagent (e.g., SPDP). As a result, the immunotoxins are comprised of a racemic mixture in which the disulfide linkage between the mAbs and the toxophore is both relatively unstable and its relative position(s) is undefined. In marked contrast, the assembly of fusion toxins at the level of the gene ensures that the protein product expressed from that gene is a monomeric species and the fusion junction between the toxophore and the ligand components is precisely defined. Perhaps more importantly, once a given fusion toxin is genetically constructed, expressed, and shown to be selectively cytotoxic for receptor-bearing target cells, it would then be possible to employ site-directed point and deletion mutagenesis in order to study structure-function relationship of the recombinant toxins, as well as to optimize their respective cytotoxic potency for target cells. Simply stated, if it were possible to genetically construct biologically active fusion toxins, then it would be possible to systematically take the recombinant toxin genes apart and/or modify them in order to study how the fusion toxins work.

At present, a number of diphtheria toxin–based fusion proteins have been genetically constructed, expressed, and purified, and shown to selectively intoxicate only those eukaryotic cells which express the targeted surface receptor (17–21). While all of these fusion toxins have been shown to be selectively cytotoxic for their respective target cell populations, our understanding of interleukin-2 (IL-2) receptor–targeted cytotoxin DAB_{486}–IL-2 is the most highly developed, and therefore will be considered in detail.

Diphtheria Toxin–Related IL-2 Fusion Toxin: DAB_{486}–IL-2

We have recently described the genetic construction and properties of a fusion toxin that was assembled from a truncated form of diphtheria toxin and human interleukin-2 (IL-2) (18). In this construct, the 3′ end of the *tox* structural gene encoding the C-terminal 50 amino acids of diphtheria toxin was removed and replaced with a synthetic gene encoding amino acids 2–133 of human IL-2 in correct translational reading-frame. Since the native diphtheria toxin receptor-binding domain was replaced with IL-2 sequences, the resulting chimeric toxin, DAB_{486}–IL-2, expressed from this gene fusion is directed toward cells bearing the IL-2 receptor. Importantly, cells lacking the IL-2 receptor were found to be universally resistant to the inhibitory action of DAB_{486}–IL-2. Bacha et al. (22) have confirmed and extended this study and have shown that the cytotoxic action of this fusion toxin is mediated through the IL-2 receptor and can be blocked by excess free recombinant IL-2, as well as by antibodies that bind to the p55 subunit (Tac antigen) of the high-affinity form of the IL-2 receptor. Moreover, since lysosomotrophic agents (e.g., chloroquine) block the cytotoxic action of DAB_{486}–IL-2, it is apparent that the fusion toxin must pass through an acidic compartment in order to deliver its ADP-ribosyltransferase to the cytosol of target cells. Bacha et al. (22) also demonstrated that inhibition of protein synthesis in target cells was due to the specific ADP-ribosylation of elongation factor 2. Thus, the cytotoxic action of DAB_{486}–IL-2 is (1) mediated through the IL-2 receptor, (2) requires passage through an acidic compartment in a manner analogous to native diphtheria toxin, and (3) catalyzes the ADP-ribosylation of elongation factor 2.

Recently, Walz et al. (23) have demonstrated that the sequential events following the binding of DAB_{486}–IL-2 to the IL-2 receptor on phytohemaglutinin (PHA)–activated T cells reflects *both* the IL-2 and the ADP-ribosyltransferase components of the fusion toxin. In a manner identical to native IL-2, the fusion toxin was found to stimulate the expression of mRNA for c-*myc*, interferon gamma, IL-2 receptor, and IL-2 for the first 7 hr of exposure. However, after 7 hr exposure, the action of DAB_{486}–IL-2 is analogous to that of cycloheximide. By this time, the effects of inhibition of protein synthesis by the ADP-ribosylation of elongation factor 2 predominate and the steady state levels of c-*myc* and IL-2 receptor mRNA are decreased. Importantly, Walz et al. (23) have demonstrated that an ADP-ribosyltransferase defective mutant $DA(197)B_{486}$–IL-2 does not inhibit protein synthesis, and is capable of signal transduction in PHA-activated T cells which results in cell proliferation. This study clearly demonstrated that the functional domains of both the diphtheria toxin–related and IL-2 components of DAB_{486}–IL-2 are retained: (A) interaction of the fusion toxin with the IL-2 receptor results in increased levels of specific mRNAs, and (B) the delivery of the ADP-ribosyltransferase to the cytosol results in an irreversible inhibition of protein synthesis that leads to cell death.

It is well known that the high-affinity form of the IL-2 receptor is composed of at least two subunits: a low-affinity 55-kd glycoprotein (p55, Tac antigen) and an intermediate-affinity 75-kd glycoprotein (p75) (24–29). Moreover, it is known that both the high-affinity (p55 + p75) and intermediate-affinity receptor, but *not* the low-affinity receptor, undergo accelerated internalization after binding native IL-2 (28–31). Based on these observations, Waters et al. (32) have examined the receptor-binding requirements of DAB_{486}–IL-2 for the efficient intoxication of target cells. Dose-response analysis of high-, intermediate-, and low-affinity IL-2 receptor–bearing cells demonstrates that *only* cells lines which bear the high-affinity form of the IL-2 receptor are sensitive to the cytotoxic action of DAB_{486}–IL-2 ($IC_{50} = 1 \times 10^{-10}$ M). In marked contrast, cell lines which bear either isolated p55 chains or p75 chains are resistant to the action of the fusion toxin and require approximately 1000-fold higher concentrations of DAB_{486}–IL-2 (10^{-7} M) to achieve an IC_{50} (i.e., concentration of fusion toxin required to achieve a 50% inhibition of protein synthesis as measured by [^{14}C]leucine incorporation).

Since peripheral blood mononuclear cells (PBMCs) with natural killer (NK) activity have been reported to bear only the p75 subunit of the IL-2 receptor on the cell surface and are responsive to IL-2 and appear to be precursors of lymphokine-activated killer (LAK) cell activity, Waters et al. (32) also examined the effect of DAB_{486}–IL-2 on NK cell activity. In these experiments, PBMCs from healthy donors were cultured in the presence or absence of IL-2 and DAB_{486}–IL-2 and the subsequent NK cell activity was measured using K-562 target cells in a 4-hr ^{51}Cr-release assay. Anti-CD3–induced T-cell cytotoxicity was measured in the same assay using an anti-CD3 mAb-producing target cell line. As shown in Table 1, concentrations of DAB_{486}–IL-2 greater than 1×10^{-7} M are required to inhibit NK cell activity. Thus, human peripheral blood monocytes with NK activity are as resistant to the action of DAB_{486}–IL-2 as are continuous cell lines which only express the p75 subunit of the IL-2 receptor.

Weissman et al. (30) have shown that the p55 subunit of the IL-2 receptor does not mediate efficient internalization of bound IL-2. By comparison, native IL-2 bound to the p75 subunit of the receptor is known to be internalized as rapidly as by the high-affinity receptor [$t_{1/2}$ = 15 min] (33). Since cell lines bearing only the p75 subunit were resistant to the action of DAB_{486}–IL-2, we reasoned that this resistance was due to altered binding of the fusion toxin to this subunit of the receptor. Waters et al. (32) have determined the receptor binding properties of DAB_{486}–IL-2 by competitive displacement experiments using ^{125}I-labeled IL-2. As shown in Table 2, approximately 200-fold higher concentrations of DAB_{486}–IL-2 are required to displace radiolabeled IL-2 from the high-affinity (p55 + p75) receptor. It is of interest to note that the concentration of DAB_{486}–IL-2 needed to displace radiolabeled ligand from the p55 subunit are only 18-fold higher than that of native IL-2, whereas 120-fold higher concentrations are required for the p75 subunit.

The receptor-binding experiments described above strongly suggest that the relative resistance of cells bearing the p75 subunit to the cytotoxic action of DAB_{486}–IL-2 is due to altered binding to the p75 subunit of the receptor. Although both the p75 and the high-affinity heterodimer share the common property of rapidly internalizing bound ligand, the binding of

Table 1 Effect of DAB_{486}–IL-2 on Human NK Cell Activity

	NK cell activity (% specific lysis) effector/target ratio		
Culture conditions	40	20	10
Medium	26	16	9
rIL-2	43	31	27
DAB_{486}–IL-2 (10^{-7} M)	12	9	7
DAB_{486}–IL-2 (10^{-7} M) + rIL-2	20	14	10
DAB_{486}–IL-2 (10^{-8} M)	27	18	14
DAB_{486}–IL-2 (10^{-8} M) + rIL-2	42	31	19
DAB_{486}–IL-2 (10^{-9} M)	24	20	14
DAB_{486}–IL-2 (10^{-9} M) + rIL-2	38	32	22

Adapted from Ref. 32.

Table 2 Relative Ability of rIL-2 and DAB$_{486}$–IL-2 to Displace ^{125}I–rIL-2 from the High-, Intermediate-, and Low-Affinity IL-2 Receptor

Cell line	IL-2 receptor	50% Displacement	
		DAB$_{486}$–IL-2	rIL-2
HUT 102/6TG	p55, p75	8.1×10^{-9} M	3.8×10^{-11} M
YT2C2	p75	5.3×10^{-7} M	4.4×10^{-9} M
MT-1	p55	4.0×10^{-7} M	2.2×10^{-8} M

Adapted from Ref. 32.

IL-2 to the p75 subunit is characterized by slow kinetics of association/dissociation. In contrast, the high-affinity receptor displays the fast "on" rate of p55 and the slow "off" rate of p75 (34–35). Thus, an alteration in DAB$_{486}$–IL-2 binding to the p75 subunit should dramatically influence the kinetics of this fusion toxin's binding to the intermediate vis-à-vis high-affinity receptor. Clearly, the results of the competitive displacement studies described by Waters et al. (32) are consistent with this interpretation. As determined by the concentration of fusion toxin required to inhibit ^{125}I–IL-2 binding, it is evident that DAB$_{486}$–IL-2 displays altered binding to *both* subunits of the IL-2 receptor; however, binding to the p75 subunit is more significantly affected than binding to the p55 subunit.

Collins et al. (36) and Ju et al. (37) have reported that the N-terminal sequences of native IL-2, particularly Asp-20, are essential for binding to the p75 subunit of the IL-2 receptor. These studies raise the possibility that the altered binding of DAB$_{486}$–IL-2 to the p75 subunit results from steric constraints imposed on the fusion toxin:p75 interaction. In the case of the diphtheria toxin–related IL-2 fusion protein, human IL-2 sequences as fused to the C-terminal end of a truncated form of the toxin. As a result, the fusion junction between the diphtheria toxin–related and IL-2 sequences are likely to place "Asp-20" (i.e., Asp-505 in DAB$_{486}$–IL-2) in an internal or less favorable position for binding to the p75 subunit. This interpretation is consistent with the results of Lorberboum-Galski et al. (38) who have described the cytotoxic action of a *Pseudomonas* exotoxin A–based IL-2 fusion toxin, IL-2–PE40. In this instance, cells expressing the high-affinity form of the IL-2 receptor have been found to be only 8–20 times more sensitive to the cytotoxic action of IL-2–PE40 than cell lines expressing either isolated p75 or p55 chains of the receptor. Moreover, only six-fold differences in p75 binding were reported for IL-2–PE40 relative to native IL-2. In the case of IL-2–PE40, the fusion junction of the two proteins is between the C-terminal end of IL-2 and the N-terminal of PE40. Since the N-terminal of this fusion protein consists of IL-2 sequences, Asp-20 of the exotoxin A–based fusion toxin also corresponds to Asp-20 of native IL-2, and should therefore be available for binding to the p75 subunit of the receptor. The observed cytotoxicity of this fusion toxin for p75 only–bearing cells supports this hypothesis.

Structure-Function Analysis of DAB_{486}–IL-2

Site-Directed Mutational Analysis of the Protease-Sensitive Loop Between Fragments A and B. As described above, native diphtheria toxin may be separated into an A and a B fragment following mild proteolytic digestion with trypsin. Trypsin has been shown to cleave the peptide backbone of toxin after Arg-190, Arg-193, and Arg-194. Following digestion and reduction of the disulfide bond between Cys-186 and Cys-201, the corresponding species of fragment A and fragment B can be resolved. Since proteolytic "nicking" or processing of this region by the eukaryotic cell in order to release fragment A was anticipated, Williams et al. (39) used site-directed mutational analysis to examine the role played by these three arginine residues in the intoxication of high-affinity IL-2 receptor–bearing cells by DAB_{486}–IL-2. As shown in Table 3, the arginine residues (i.e., Arg-191, Arg-193, and Arg-194) (it should be noted that since the *tox* signal sequence has been deleted from this construct, the fusion toxin accumulates in the cytosol of recombinant *E. coli* and contains an N-terminal methionine, the amino acid numbering of DAB–IL-2 fusion toxins is $+1$ out of phase with that of native diphtheria toxin) were systematically changed to glycine by site-directed mutagenesis. Following DNA sequencing to ensure that the desired appropriate mutations were introduced, DAB_{486}–IL-2 and the mutant fusion toxins were evaluated for their capacity to inhibit ^{14}C-leucine incorporation. While single amino acid substitutions at Arg-191 and Arg-193 were accompanied by a slight loss of cytotoxic potency, the substitution of Arg-194 with Gly, $DAB(RVRG)_{486}$–IL-2, resulted in a greater than 3-log loss of cytotoxic potency. Importantly, pretreatment of $DAB(RVRG)_{486}$–IL-2 with trypsin in order to introduce nicks in the peptide backbone at Arg-191 and Arg-193 resulted in a restoration of cytotoxic activity. Since pretreatment of the nontoxic triple mutant $DAB(GVGG)_{486}$–IL-2 with trypsin did not restore biological activity, we conclude that Arg-194 is the site of cellular processing.

Both Amphipathic and Membrane-Spanning Helices of DAB_{486}–IL-2 Are Required for the Intoxication of Target Cells. In order to begin to analyze the structural features of diphtheria toxin fragment B–related sequences

Table 3 Cytotoxic Activity of DAB_{486}–IL-2 and Related Fragment A/B Loop Mutant Fusion Toxins on High-Affinity IL-2 Receptor–Bearing HUT 102/6TG Cells

| Fusion toxin | Partial sequence of A/B loop | | | | Tryspin-nicked | |
	Residue #191	192	193	194	IC_{50}	IC_{50}
DAB_{486}–IL-2	R	V	R	R	2×10^{-10}	2×10^{-10}
$DAB(GVRR)_{486}$–IL-2	G	V	R	R	4×10^{-10}	ND
$DAB(RVGR)_{486}$–IL-2	R	V	G	R	4×10^{-10}	ND
$DAB(RVRG)_{486}$–IL-2	R	V	R	G	1×10^{-7}	1×10^{-9}
$DAB(GVGG)_{486}$–IL-2	G	V	G	G	1×10^{-7}	1×10^{-7}

ND = no data.
Adapted from Ref. 39.

of DAB_{486}–IL-2 required for the delivery of fragment A across the endocytic vesicle membrane and into the cytosol, we have used cassette and deletion mutagenesis. In the case of DAB_{486}–IL-2, the diphtheria toxin fragment B–related sequences included in the fusion protein are from Ser-195 to Ala-486, and include the disulfide bond between Cys-462 and Cys-472. Williams et al. (40) have recently demonstrated that the in-frame deletion of 97 amino acids from Thr-387 to His-485 of DAB_{486}–IL-2 increases the potency ($IC_{50} = 2$–5×10^{-11} M) and lowers the apparent dissociation constant (K_d) of the resulting DAB_{389}–IL-2 for high-affinity IL-2 receptor–bearing T cells. In marked contrast, the deletion of 191 amino acids between Asp-291 and Gly-483 results in a ≥ 1000-fold loss of cytotoxic potency in the fusion toxin DAB_{295}–IL-2. It is of particular interest to note that the latter deletion does not affect the apparent K_d for the high-affinity IL-2 receptor. Since the structural domains between Asp-291 and Gly-483 include two hydrophobic helical regions of fragment B that have been postulated to facilitate the delivery of fragment A across the endocytic vesicle membrane by forming a pore or channel (10–11,15), we conclude that these regions of fragment B are essential in facilitating the delivery of fragment A across the membrane in the intoxication process.

In addition, Williams et al. (40) have shown that the amphipathic membrane surface-binding region of fragment B between Asn-204 and Ile-290 is also essential for the cytotoxicity of the DAB–IL-2 fusion toxins. It is important to note that the genetic deletion of this region of fragment B in both $DAB(205$–$289)_{486}$–IL-2 and $DAB(205$–$289)_{389}$–IL-2 also decreases biological activity by greater than 1000-fold. Most interestingly, these in-frame internal deletion mutations also affect the apparent K_d of the fusion toxin for the high-affinity IL-2 receptor. Since the region that has been deleted carries an amphipathic domain(s) (41), it is reasonable to postulate that this region of fragment B associates with the T-cell membrane surface and stabilizes the binding of the fusion toxins with the target cell surface. The latter study has identified two important features of DAB_{486}–IL-2: (A) the fusion of human IL-2 sequences to Ala-486 of a truncated form of diphtheria toxin does *not* yield a fusion toxin with maximal cytotoxic potency, and (B) the K_d of the fusion toxin for the IL-2 receptor appears to directly correlate with the level of cytotoxic potency. Most importantly, this study demonstrates that protein engineering of the fusion junction between diphtheria toxin–related and IL-2 sequences should yield fusion toxins with increased biological potency.

Protein Engineering of DAB_{389}–IL-2 to Increase Cytotoxic Potency. We have analyzed the primary amino acid sequence of DAB_{486}–IL-2 and DAB_{389}–IL-2 fusion toxins for predicted secondary structure using the P/C GENE software (release 6.01, intelligenetics, Mountain View, CA) (42). In particular, the FLEXPRO program was used to predict the flexibility of the DAB–IL-2 fusion toxins at each point of their sequence. In this program, the polypeptide chain flexibility at an amino acid is calculated from the average value of the atomic temperature factor (B value) of the alpha-carbon atom, as affected by the adjacent amino acids. Flexible locations have high B values because their displacements can be large. The "neighbor-correlated" average, or normalized B values, for each amino was found from a set of proteins whose three-dimensional structural was known. The predicted flexibility at an amino acid is the weighted sum of the normalized B values (taking account of neighbors) of the seven amino acids closest to that point in the sequences.

Analysis of both DAB_{486}–IL-2 and DAB_{389}–IL-2 toxins revealed a common "most" flexible region—amino acids 2–8 of their respective human IL-2 component. Since this region was also found to be unordered in the 3-Å crystal structure of IL-2 (43), we reasoned that the apparent flexibility of amino acids 2–8 of IL-2 in the fusion toxins might allow IL-2 some degree of mobility with respect to the diphtheria toxin–related sequences. In order to test this hypothesis, we have genetically constructed insertion mutants of DAB_{389}–IL-2 in which amino acids 2–8 of IL-2 sequence were duplicated at the fusion junction. These constructions were made by cloning oligonucleotide linkers encoding for duplications of the first eight amino acids of IL-2 into the unique *SphI* site, which defines the fusion junction between diphtheria toxin–related and IL-2 sequences. Codon usage in the oligonucleotide linkers was varied in order to ensure genetic stability of the inserts.

Dose-response curves of the DAB_{389}–IL-2 duplication mutants on high-affinity IL-2 receptor bearing HUT 102 6/TG cells demonstrated that the IC_{50} of the DAB_{389}–$(1-10)_2$-IL-2 form of fusion toxin is consistently 40- to 60-fold lower than that of DAB_{486}–IL-2 ($\sim 5 \times 10^{-12}$ M)! Kiyokawa et al. (42) have shown by displacement studies using ^{125}I–rIL-2 that receptor-binding affinity is directly related to the cytotoxic potency of a given fusion toxin variant. These studies demonstrate that modification of the region of the fusion junction between diphtheria toxin–related and IL-2 sequences can dramatically effect the cytotoxic activity of the DAB–IL-2 fusion toxins. The apparent increase in the flexibility of this region by the insertion of unordered sequence has been shown to result in a corresponding increase in receptor-binding affinity and cytotoxic potency. Moreover, the results of these studies also suggest that the insertion of a flexible hinge region between the toxophore and the receptor-binding domain of other fusion toxins may be important in maximizing their respective potency.

Shiga-like Toxins: Structure-Function and Mode of Action

The Shiga and Shiga-like toxins have been the subject of a recent excellent review by O'Brien and Holmes (44). Conradi (45) first described Shiga toxin in the autolysates of *Shigella dysenteriae* 1. While early studies focused on the apparent paralytic-lethal effects of the crude toxin, it was not until the 1960 that the toxin was shown to be cytotoxic for selected mammalian cells (46). It is now known that Shiga holotoxin is composed of a single A subunit of approximately 32,000 molecular weight, and a pentameric B subunit in which the monomeric peptide chain has a molecular weight of 7700 (47). It is of interest to note, that the A subunit of Shiga toxin must be proteolytically nicked and reduced in order for the enzymatically active A1 subunit to be released (48). Recently, Endo and Tsurgi (49) have shown that the enzymatic activity of the Shiga and Shiga-like A1 fragment is identical to that of the plant toxin ricin, and catalyzes the cleavage of the N-glycosidic bond at adenine 4324 in the 28S ribosome. Thus, ricin A chain and the Shiga toxin A1 subunits inhibit protein synthesis in target cells by an identical mechanism. The structural gene for Shiga-like toxin has recently been cloned and sequenced (50–51), and as a result the restriction endonuclease digestion map of the gene and the deduced amino acid sequence for this toxin is known in fine detail. Indeed, the availability of the Shiga-like toxin structural gene has allowed for the genetic construction of the first ribosome-inactivating protein-based tripartite fusion toxin: Shiga-like–DT"B"–IL-2 described below.

Triparite Fusion Toxins: Shiga-A1–DT"B"–IL-2

It is important to note that the basic underlying mechanisms of inhibition of protein synthesis employed by diphtheria toxin fragment A and Shiga-like A subunit (i.e., ADP-ribosylation of EF-2 and RNase activity, respectively) have important theoretical implications for the cytotoxic action of the fusion toxins. In the case of the diphtheria toxin–related fusion proteins, the apparent sensitivity of a given cell to the toxin is dependent upon the basal rate of protein synthesis (52). Since *only* EF-2, which cycles off the ribosome, is sensitive to the fragment A–catalyzed ADP-ribosylation, the *higher* the basal rate of protein synthesis in a given cell, the *higher* its apparent sensitivity to the toxin. In the case of the ribosome-inactivating toxins (i.e., Shiga-like A, ricin A chain, abrin A chain, gelonin), the basal rate of protein synthesis is *not* a factor in the determination of sensitivity of a given cell line. Because of this difference, it is of interest to genetically construct a family of fusion toxins in which the Shiga-like A subunit is used to replace diphtheria toxin fragment A as the catalytic subunit, and then to compare the relative potency of both the Shiga-like A and diphtheria toxin fragment A fusion toxins on receptor-specific target cells.

We selected the A1 chain of Shiga-like toxin to replace diphtheria toxin fragment A in the genetic construction of the first tripartite toxin for the following reasons: (1) both Shiga-like A and diphtheria fragment A are approximately the same molecular mass, (2) the introduction of a single molecule of Shiga-like A chain to the cytosol of a target cell will result in the death of the cell, and (3) the modification of the gene for Shiga-like A chain required to construct the tripartite fusion toxin gene is straightforward and was readily accomplished.

As described above, the gene for Shiga-like toxin has recently been cloned and sequenced (50–51). Shiga-like toxin is composed of an A subunit (32,225 d) that is noncovalently bound to a pentameric B subunit (7961 d). The A subunit of Shiga-like toxin contains a single disulfide bridge that subtends a protease-sensitive loop, and upon trypsin nicking the A1 fragment is released. The A1 fragment of Shiga-like toxin has been shown to be an enzyme which specifically cleaves the N-glycosidic bond at adenine 4324 in the 28S ribosomal RNA. It should be noted that the isolated A and B subunits of Shiga-like toxin are *not* toxic for intact eukaryotic cells.

As discussed above, biochemical/genetic analysis of DAB_{486}–IL-2 mutants has recently shown (39) that fragment A *must* be released from the fusion toxin in order to intoxicate target lymphocytes. We have found that the substitution Arg-194 with Gly results in an approximate 3-log loss of cytotoxic potency of the fusion toxin. Since prenicking the mutant fusion toxin with trypsin restores full biological activity, these observations strongly suggest that DAB_{486}–IL-2 binds to the IL-2 receptor, is internalized by receptor-mediated endocytosis, and is then *processed* by a cellular protease at Arg-194 in order to release fragment A, which is then delivered to the cytosol. While these observations suggest that an intact disulfide bridge between Cys-187 and Cys-202 in DAB_{486}–IL-2 is essential for full biological activity, this hypothesis needs to be tested experimentally.

As a result of the above observations, we have developed a vector for the genetic construction of the tripartite fusion toxin which retains the Cys-187 to Cys-202 disulfide bond. We have taken advantage of a unique *NsiI* restriction endonuclease site that is positioned at Cys-202 and the *NcoI* site containing the ATG of the translational initiation signal at the beginning of fragment A in plasmid pABI6508 (18). Following digestion of plasmid pABI6508 with *NsiI* and *NcoI*, we cloned an *NsiI-ApaI-NcoI* linker that restores the genetic information encoding the Cys-187 to Cys-202 disulfide loop. The modified plasmid has been designated pPA101. Importantly, this linker also introduces a unique *ApaI* site immediately upstream of Cys-187. As a result, any gene that can be modified at its 5′ end by the introduction of an *NcoI* site, and at its 3′ end by the introduction of an *ApaI* site can be inserted into pPA101, giving rise to a tripartite fusion gene.

We have modified the structural gene for Shiga-like toxin A chain by the introduction of a 5′ *NcoI–TaqI* linker, and a 3′ *XmmI–ApaI* linker. The modified gene for Shiga-like A chain was then introduced into the *NcoI* and *ApaI* sites of pPA101 to form plasmid pPASA101. Following growth of recombinant *E. coli* and expression of the tripartite toxin gene, the Shiga-like A–DT"B"–IL-2 fusion toxin was purified by immunoaffinity chromatography on an anti–IL-2 matrix. Preliminary experiments show that the tripartite toxin can be readily purified to apparent homogeneity by immunoaffinity chromatography followed by HPLC chromatography. Furthermore, the tripartite fusion toxin has been shown to carry determinants that are reactive on immunoblots probed with anti–Shiga-like A, anti–IL-2, and anti–diphtheria toxin sera. Thus, the three domains of the tripartite toxin fold in such a way as to present their own unique and characteristic immunodominant epitopes. Most exciting, however, is the observation that the tripartite toxin is biologically active against high-affinity IL-2 receptor–bearing T lymphocytes. It should be noted that the biological activity of the tripartite toxin can be specifically blocked with excess free IL-2. These results demonstrate that the action of this fusion toxin is mediated through the IL-2 receptor. While these results are preliminary, they suggest that a wide variety of fusion toxins can be genetically assembled.

REFERENCES

1. Buck, G. A., Gross, R. E., Wong, P., Lore, T., and Groman, N. DNA relationships among some *tox*-bearing corynebacteriophage that carry the gene for diphtheria toxin. J. Bacteriol., *148*: 153–162, 1985.
2. Uchida, T., Gill, D. M., and Pappenheimer, A. M., Jr. Mutation in the structural gene for diphtheria toxin carried by temperate phage β. Nature, *233*: 8–11, 1971.
3. Collier, R. J., and Kandel, J. Structure activity of diphtheria toxin. I. Thiol-dependent disassociation of a fragment of toxin into enzymatically active and inactive fragments. J. Biol. Chem., *246*: 1496–1503, 1971.
4. Gill, D. M., and Pappenheimer, A. M., Jr. Structure activity relationships in diphtheria toxin. J. Biol. Chem., *264*: 1492–1496, 1971.
5. Middelbrook, J. L., Dorland, R. B., Leppla, S. H. Association of diphtheria toxin with Vero cells: Demonstration of a receptor. J. Biol. Chem., *253*: 7325–7330, 1978.
6. Moya, M., Dautry-Versat, A., Goud, B., Louvard, D., and Boquet, P. Inhibition of coated pit formation in Hep$_2$ cells blocks the cyotoxicity of diphtheria toxin but not that of ricin toxin. J. Cell. Biol., *101*: 548–559, 1985.
7. Sandvig, K., Tonnessen, I., Sand, O., and Olsnes, S. Requirement of a transmembrane pH gradient for the entry of diphtheria toxin into cells at low pH. J. Biol. Chem., *261*: 11639–11645, 1986.
8. Pappenheimer A. M., Jr. Diptheria: Studies on the biology of an infectious disease. Harvey Lect., *76*: 45–73, 1982.
9. Kieleczawa, J., Zhao, J.-M., Luongo, C. L., Dong, L-Y. D., and London, E. The effect of high pH upon diphtheria toxin conformation and model membrane association: Role of partial unfolding. Arch. Biochem. Biophys., *282*: 214–220, 1990.
10. Donovan, J. J., Simon, M. I., Draper, R. K., and Montal, M. Diphtheria toxin forms transmembrane channels in planar lipid bilayers. Proc. Natl. Acad. Sci. U.S.A., *78*: 172–176, 1981.
11. Kagan, B. L., Finkelstein, A., and Colombini, M. Diphtheria toxin fragment forms large pores in phospholipid bilayer membranes. Proc. Natl. Acad. Sci. U.S.A., *78*: 4950–4954, 1981.
12. Yamaizumi, K., Mekada, E., Uchida, T., and Okada, Y. One molecule of diphtheria toxin fragment A introduced into a cell can kill the cell. Cell, *15*: 245–250, 1978.
13. Uchida, T., Pappenheimer, A. J., Jr., and Harper, A. Diptheria toxin and related proteins: III. Reconstitution of hybrid "diphtheria toxin" from nontoxic mutant proteins. J. Biol. Chem., *248*: 3851–3854, 1972.
14. Olsnes, S., Pappenheimer, A. M, Jr., and Meren, R. Lectins from *Abrus precatorius* and *Ricinus communis*. 2. Hybrid toxins and their interaction with chain-specific antibodies. J. Immunol., *113*: 842–847, 1974.
15. Boquet, P., Silverman, M. S., Pappenheimer, A. M., Jr., and Vernon, B. J. Binding of triton X-100 to diphtheria toxin, cross reacting material 45, and their fragments. Proc. Natl. Acad. Sci. U.S.A., *73*: 4449–4453, 1976.

16. Bacha, P., Murphy, J. R., and Reichlin, S. Thyrotropin–releasing hormone–diphtheria toxin-related polypeptide conjugates: Potential role of the hydrophobic domain in toxin entry. J. Biol. Chem., *258*: 1565–1570, 1983.

17. Murphy, J. R., Bishai, W., Borowski, M., Miyanohara, A., Boyd, J., and Nagle, S. Genetic construction, expression, and melanoma-selective cytotoxicity of a diphtheria toxin-related α-melanocyte stimulating hormone fusion protein. Proc. Natl. Acad. Sci. U.S.A., *83*: 8258–8261, 1986.

18. Williams, D. P., Parker, K., Bacha, P., Bishai, W., Borowski, M., Genbauffe, F., Strom, T. B. and Murphy, J. R. Diphtheria toxin receptor binding domain substitution with interleukin-2: Genetic construction and properties of a diphtheria toxin-related interleukin-2 fusion protein. Protein Eng., *1:* 492–498.

19. Lakkis, F., Steele, A., Pacheco-Silva, A., Rubin-Kelley, V., Strom, T. B., and Murphy, J. R. Interleukin-4 receptor targeted cytoxicity: Genetic construction and *in vivo* immunosuppressive activity of a diphtheria toxin-related murine interleukin-4 fusion protein. Eur. J. Immunol., 1991.

20. Shaw, J. P., Akiyoshi, D. E., Rhoad, A. E., Sullivan, B., Thomas, J., Genbauffe, F., Bacha, P., and Nichols, J. C. Cytotoxic properties of DAB_{486}-EGF and DAB_{389}-EGF, EGF receptor targeted fusion toxins. J. Biol. Chem., 1991.

21. Jean L-F. L., and Murphy, J. R. Interleukin-6 receptor targeted cytotoxicity: Genetic construction and properties of a diphtheria toxin-related interleukin-6 fusion protein. Protein Eng. 1991.

22. Bacha, P., Williams, D. P., Waters, C., Williams, J. M., Murphy, J. R., and Strom, T. B. Interleukin-2 receptor targeted cytotoxicity: Interleukin-2 receptor mediated action of a diphtheria toxin-related interleukin-2 fusion protein. J. Exp. Med., *167*: 612–622, 1988.

23. Walz, G., Zanker, B., Brand, K., Waters, C., Genbauffe, F., Zeldis, J. B., Murphy, J. R., and Strom, T. B. Sequential effects of interleukin-2 diphtheria toxin fusion protein on T-cell activation. Proc. Natl. Acad. Sci. U.S.A., *85*: 9485–9488, 1989.

24. Sharon, M., Klausner, R. D., Cellen, B. R., Chizzonite, R., and Leonard, W. L. Novel interleukin-2 receptor subunit detected by cross linking high affinity conditions. Science, *234*: 859–863, 1986.

25. Tsudo, M. R., Kozak, W., Goldman, C. K., and Waldmann, T. A. Demonstration of a non-Tac peptide that binds interleukin-2: A potential participant in a multichain interleukin-2 that binds interleukin-2. A potential participant in a multichain interleukin-2 receptor complex. Proc. Natl. Acad. Sci. U.S.A., *83*: 9694–9698, 1986.

26. Teshigawara, K., Wang, H-M., Kato, K., and Smith, K. A. Interleukin-2 high affinity receptor expression requires two distinct binding proteins. J. Exp. Med., *165*: 223–238, 1987.

27. Dukovich, M., Wano, Y., Bich-Thuy, L. T., Katz, P., Cullen, D., Kehrl, J. H., and Greene, W. C. A second human interleukin-2 binding protein that may

be a component of high affinity interleukin-2 receptors. Nature, *327*: 518–522, 1987.

28. Robb, R. J., Rusk, C. M., Yodoi, J., and Greene, W. C. Interleukin 2 binding molecule distinct from the Tac protein: Analysis of its role in the formation of high affinity receptors. Proc. Natl. Acad. Sci. U.S.A., *84*: 2002–2006, 1987.

29. Tanaka, T., Saiki, O., Doi, S., Negoro, S., and Kishimoto, S. Novel receptor-mediated internalization of interleukin-2. J. Immunol., *140*: 866–870, 1988.

30. Weissman, A. M., Harford, J. B., Svetlik, P. B., Leonard, W. L., Depper, J. M., Waldmann, T. A., Greene, W. C., and Klausner, R. D. Only high-affinity receptors for interleukin-2 mediate internalization of ligand. Proc. Natl. Acad. Sci. U.S.A. *83*: 1463–1466, 1986.

31. Fugii, M., Sugamura, K., Sano, K., Nakai, M., Sugita, K., and Kinuma, Y. High affinity receptor-mediated internalization and degradation of interleukin-2 in human T-cells. J. Exp. Med., *163*: 550–562, 1987.

32. Waters, C. A., Schimke, P. A., Snider, C. E., Itoh, K., Smith, K. A., Nichols, J. C., Strom, T. B., and Murphy, J. R. Interleukin-2 binding requirements for entry of a diphtheria toxin related interleukin-2 fusion protein into cells. Eur. J. Immunol., *20*: 785–791, 1990.

33. Robb, R. J., and Greene, W. C. Internalization of interleukin-2 binding molecule distinct from the Tac protein: Analysis of its role in formation of high-affinity receptors. Proc. Natl. Acad. Sci. U.S.A., *84*: 2002–2006, 1987.

34. Wang, H-M., and Smith, K. A. The interleukin-2 receptor. Functional consequences of its biomolecular structure. J. Exp. Med., *166*: 1055–1069, 1987.

35. Lowenthal, J. W., and Greene, W. C. Contrasting interleukin-2 binding properties of the alpha (p55) and beta (p70) protein subunits of the human high affinity interleukin-2 receptor. J. Exp. Med., *166*: 1156–1161, 1987.

36. Collins, L., Tsien, W-H., Seals, C., Hakimi, J., Webber, D., Bailon, P., Hoskings, J., Greene, W. C., Toome, V., and Ju, G. Identification of specific residues of human interleukin-2 that affect binding to the 70-kDa subunit (p70) of the interleukin-2 receptor. Proc. Natl. Acad. Sci. U.S.A., *85*: 7709–7713, 1988.

37. Ju, G., Collins, G., Kaffka, K. L., Tsien, W.-H., Chizzonite, R., Crow, R., Bhatt, R., and Kilian, P. L. Structure function analysis of human interleukin-2. Identification of amino acid residues required for biological activity. J. Biol. Chem., *262*: 5723–5731, 1987.

38. Lorberboum-Galski, H., Kozak, R. W., Waldmann, T. A., Bailon, D., FitzGerald, D., and Pastan, I. Interleukin-2 (IL-2) PE40 is cytotoxic to cells displaying either the p55 or p70 subunit of the IL-2 receptor. J. Biol. Chem., *263*: 18650–18656, 1988.

39. Williams, D. P., Wen, Z., Watson, S., Boyd, J., and Murphy, J. R. Cellular processing of the interleukin-2 fusion toxin DAB_{486}–IL-2 and efficient delivery of diphtheria toxin fragment A to the cytosol of target cells requires Arg[194]. J. Biol. Chem., *265*: 20673–20677, 1990.

40. Williams, D. P., Snider, C. E., Strom, T. B. and Murphy, J. R. Structure/function analysis of IL-2 toxin (DAB486–IL-2): Fragment B sequences

required for the delivery of fragment A to the cytosol of target cells. J. Biol. Chem., *265*: 11885–11889, 1990.

41. Lambotte, P., Falmagne, P., Capiau, C., Zanen, J., Ruysschaert, J. M., and Dirx, J. Primary structure of diphtheria toxin fragment B: structural similarities with lipid binding domains. J. Cell. Biol., *87*: 837–840, 1980.

42. Kiyokawa, T., Williams, D. P., Snider, C. E., Strom, T. B., and Murphy, J. R. Protein engineering of IL-2 toxin (DAB–IL-2): Increased flexibility of the diphtheria toxin interleukin-2 fusion junction increases cytotoxic potency. Protein Eng., *4*: 463–468, 1991.

43. Brandhuber, B. J., Boone, T., Kenney, W. C., and McKay, D. B. Three dimensional structure of interleukin-2. Science, *238*: 1707–1710, 1987.

44. O'Brien, A., and Holmes, R. K. Shiga and Shiga-like toxins. Microbiol. Rev., *51*: 206–220, 1987.

45. Conradi, H. Ueber loslishe, durch aseptische Autolyse, erhaltene Giftstoffe von Ruhr-und Typhus bazillen. Dtsch. Med. Wochenschr., *29*: 26–28, 1903.

46. Vicari, G., Olitzki, A. L., and Olitzki, Z. The action of the thermolabile toxin of *Shigella dysentariae* on cells cultivated in vitro. Br. J. Exp. Pathol., *41*: 179–189, 1960.

47. Donohue-Rolfe, A., Keusch, G. T., Edson, C., Thorley-Lawson, J., and Jacewicz, M. Pathogenesis of *Shigella* diarrhea. IX. Simplified high yield purification of *Shigella* toxin and characterization of subunit composition and function by the use of subunit-specific monoclonal and polyclonal antibodies. J. Exp. Med., *160*: 1767–1781.

48. Olsnes, S., Reisbig, R., and Eiklid, K. Subunit structure of *Shigella* cytotoxin. J. Biol. Chem., *256*: 8732–8738, 1981.

49. Endo, Y., and Tsurugi, K. RNA-glycosidase activity or ricin A-chain. Mechanism of action of the toxic lectin on eukaryotic ribosomes. J. Biol. Chem., *262*: 8128–8130, 1987.

50. Caulderwood, S. B., Auclair, F., Donohue-Rolfe, A., Keusch, G. T., and Mekalanos, J. J. Nucleotide sequence of the Shiga-like toxin genes of *Escherichia coli*. Proc. Natl. Acad. Sci. U.S.A., *84*: 4364–4368, 1987.

51. Strockbine, N. A., Jackson, M. P., Sung, L. M., Holmes, R. K., and O'Brien, A. D. Cloning and sequencing of the genes for Shiga toxin from *Shigella dysenteriae* type 1. J. Bacteriol., *170*: 1116–1122, 1988.

52. Pappenheimer, A. M., Jr. Diphtheria. Annu. Rev. Biochem., *46*: 69–94, 1977.

19

In Vivo Studies with Chimeric Toxins
Interleukin-2 Fusion Toxins as Immunosuppressive Agents

Michael E. Shapiro,*‡ Robert L. Kirkman,†‡ Vicki Rubin Kelley,‡ Patricia Bacha,° Jean C. Nichols,° and Terry B. Strom*‡
*Beth Israel Hospital, †Brigham and Women's Hospital, and ‡Harvard Medical School, Boston, Massachusetts; °Seragen, Inc., Hopkinton, Massachusetts

I. INTRODUCTION

The ability to direct targeted immunosuppression to those immune cells responsible for allograft rejection or for the development of autoimmune disease without producing a nonspecifically immunoincompetent host has been an elusive goal. Current therapy with cyclosporine, azathioprine, steroids, or some of the newer chemical agents produces broad, nonselective immunosuppression resulting in opportunistic infections and the potential for malignancy without preventing graft loss. The development of antilymphocyte globulin preparations and monoclonal anti–pan-T cell antibodies, such as OKT3, has provided powerful tools to treat allograft rejection, but these agents target the entire T-cell population and, as a result, produce serious adverse effects.

The de novo acquisition of high-affinity membrane receptors for interleukin-2 (IL-2) is a critical event in the course of T-cell activation (1,2), and these receptors are thus of particular interest for immunosuppressive therapy. Interaction of IL-2 with its receptor is required for the clonal expansion and continued viability of activated T cells (3). Interleukin-2 receptors are present on the cell surface as high-, intermediate-, and low-affinity binding sites. The high-affinity site is a complex of a 55-kd alpha chain (Tac antigen) and a 75-kd beta chain. The isolated 75-kd chain possesses intermediate affinity, is characterized by slow kinetics of association, and is responsible for rapidly internalizing bound ligand. The 55-kd chain is characterized by rapid binding of ligand, but low affinity when expressed alone (4). A number of monoclonal antibodies with specificity for both rodent and human interleukin-2 receptors have been reported, primarily directed against the 55-kd chain (5,6). These antibodies block the binding of IL-2 to the high-affinity receptors, and have been extensively studied in mice (7,8), rats (9), nonhuman primates (10,11), and humans (12,13) as immunosuppressive agents for the prevention of allograft rejection and autoimmune diseases (14). Excellent results in mice, leading to indefinite survival of cardiac allografts, have been followed by more modest effect in nonhuman and human primates. Disadvantages of the antibody therapy have included the lack of cell killing in primate and human models and the development of antibodies to the infused monoclonals, limiting their effectiveness. The design of interleukin-2 fusion toxins is thus of great interest.

The construction of IL-2 fusion toxins, using IL-2 sequences to replace the receptor-binding domains of diphtheria toxin (DAB_{486}–IL-2) or *Pseudomonas* exotoxin A (IL-2–PE40) is the subject of other chapters in this book. Thus, the structure and in vitro activity of these fusion toxins will not be discussed in detail here. DAB_{486}–IL-2 is a 68-kd protein which requires interaction with the IL-2 receptor to gain entry to the cell. Following acidification of endocytic vesicles, the NAD^+ ADP-ribosyltransferase component of the fusion protein is released, which interacts with elongation factor 2 (EF-2) in the target cell cytosol, leading to cell death (15,16). IL-2–PE40 is a 54-kd protein which acts in much the same manner as DAB_{486}–IL-2 through ADP-ribosylation of EF-2. In vitro, both fusion proteins intoxicate cells bearing high-affinity IL-2 receptors, an action which can be blocked by molar excess of recombinant IL-2 or monoclonal anti-p55 IL-2 receptor antibodies. However, intoxication with DAB_{486}–IL-2 is selective for cells which bear the high-affinity receptor (16), whereas IL-2–PE40 intoxicates cells which bear all forms of the IL-2 receptor (17). The IL-2 fusion toxins have been studied in a wide range of animal models of both transplantation and autoimmune disease and will be presented in this chapter. Studies of antitumor behavior of these agents in animals and preliminary human studies with DAB_{486}–IL-2 will not be discussed.

II. IN VIVO STUDIES OF DELAYED-TYPE HYPERSENSITIVITY

Initial in vivo studies of DAB_{486}-IL-2 were designed to demonstrate selective and specific actions against T-cell–mediated immune responses. The delayed-type hypersensitivity (DTH) to simple haptens is an elegant, reproducible model of such behavior. Mice are typically immunized by either subcutaneous injection or skin painting with hapten, and the immune response measured following rechallenge 1 week later.

Following subcutaneous immunization with trinitrobenzenesulfonic acid (TNBS), mice are rechallenged 6 days later by hind footpad injection. Footpad thickness is measured 24 hr later, and compared to the uninjected control. Mice were treated with either DAB_{486}-IL-2 (5 mg/day), CRM45, a 45-kd nontoxic mutant form of diphtheria toxin lacking the receptor-binding domain, or M7-20 (5 mg/day), a rat immunoglobulin M (IgM) anti-p55 IL-2 receptor monoclonal antibody from the day of immunization until the day of rechallenge (day 6) intraperitoneally. Untreated mice developed footpad swelling of 40–50 DTH units. CRM45 had no effect on the DTH response. DAB_{486}-IL-2 produced profound depression of the DTH response (18–20 U, $p < 0.001$), and this effect was significantly more pronounced than that of the anti–IL-2R antibody (35 U, $p < 0.005$) [18]. DAB_{486}-IL-2 was shown to be effective at doses as low as 50 ng/day (IP). In addition two-color flow cytometric analysis of lymphocytes showed selective depletion of IL-2R$^+$CD4$^+$ and IL2R$^+$CD8$^+$ cells from draining popliteal lymph nodes. Significantly, the effect of DAB_{486}-IL-2 was not compromised by preimmunization of mice with diphtheria toxoid.

Using a more complex, two-hapten DTH model, Bastos et al. [19] have shown that the effect of DAB_{486}-IL2 is antigen-specific and results in specific deletion of the antigen-activated T-cell clones. Mice were immunized and challenged with TNBS during week 1 (first phase) during which they were treated with daily subcutaneous injections of DAB_{486}-IL-2, DA(197)B$_{486}$-IL-2 (a nontoxic mutant form of DAB_{486}-IL-2 which binds to the IL-2 receptor) or anti-CD3 monoclonal antibody. After 1–4 weeks of rest, animals were reimmunized with TNBS and with dinitro-flourobenzene (DNFB) at a different site without further immunosuppression, and rechallenged 6 days later (second phase). Animals treated with DAB_{486}-IL-2 had profound depression of their DTH response, which persisted with a 1-week rest and rechallenge (Table 1). This response was specific to the hapten presented during the course of immunotherapy (TNBS) and immune responsiveness to a second hapten (DFNB) was not affected. In contrast, anti-CD3 monoclonal antibody therapy produced long-lasting suppression of DTH with suppression of the DNFB response as well.

Table 1 DAB_{486}–IL-2 Induces Antigen-Specific DTH Unresponsiveness

Treatment	TNBS 1st Phase	2nd Phase	DNFB
Control	39 ± 6	56 ± 7	9 ± 9
DAB_{486}–IL-2 (10 mg/day)	$5 \pm 3^{a,b}$	$14 \pm 4^{a,b}$	3 ± 6
DAB_{486}–IL-2 (50 mg/day)	$9 \pm 2^{a,b}$	$9 \pm 2^{a,b}$	9 ± 3
Anti-CD3	$4 \pm 1^{a,b}$	$3 \pm 2^{a,b}$	3 ± 4^c
$DA(197)B_{486}$–IL-2 (50 mg/day)	21 ± 5^a	29 ± 8^a	39 ± 9

[a] $p < 0.001$ vs control.
[b] $p < 0.05$ vs $DA(197)B_{486}$–IL-2.
[c] $p < 0.05$ vs control.
Source: From Ref. 19; used with permission.

$DA(197)B_{486}$–IL-2 also produced moderate immunosuppression, possibly by competitively inhibiting IL-2 binding to its receptor.

With a 4-week delay between exposure to TNBS, the mice escaped from immunosuppression by DAB_{486}–IL-2, but not anti-CD3. This escape could be abrogated by thymectomy of the animal prior to the initial exposure to TNBS. These data suggest that DAB_{486}–IL-2 acts by deletion of the clone(s) of cells responding to a specific antigen and expressing high-affinity IL-2 receptors at that time. Escape from this "clonal deletion" may occur by education of new T-cell clones in the thymus and can be blocked by thymectomy.

III. IN VIVO TRANSPLANTATION MODELS

A. Murine Cardiac Transplantation

Both IL-2–PE40 and DAB_{486}–IL-2 have been studied in a vascularized heterotopic heart allograft model in mice (20,21). Heart allografts were performed according to the method of Corry et al. (22).

Mice were treated with DAB_{486}–IL-2 as a single daily dose for 10 days intraperitoneally. B10.BR (H-2k) hearts were transplanted into C57B1/10(H-2b) mice. No toxicity was clinically evident in any of the transplanted mice, and simultaneous toxicity studies showed only mild renal tubular damage at doses 40-fold higher than those used in the transplanted animals. Some prolongation of graft survival was seen in two mice at 0.5 mg/day, and four of five grafts had indefinite survival at 1.0 mg/day (Table 2). The CRM45 diphtheria toxin fragment had no effect on graft survival in this model (20).

Table 2 IL-2 Toxins Prolong Murine Cardiac Allograft Survival

Treatment	Total daily dose (mg)	Allograft survival (days)
Saline[a,b]		9, 10, 10, 10, 11, 14, 15, 17, 18, 18, 18, 19, 20, >60
DAB$_{486}$–IL-2[a]	0.5	12, 16, 19, 41, >50
	1.0	27, >50, >50, >50, >50
CRM45[a]	0.66	11, 12, 12, 13, 13
IL-2–PE40[b]	0.2	10, 10, 10, 13, 13
	2.0	10, 20, 21, 27, >100
	5.0	11, 14, >100, >100, >100
	10.0	>100, >100, >100, >100, >100
IL-2–PE40(Asp-553)[b]	10.0	9, 11, 14, 16, 17

[a]Data from Ref. 20.
[b]Data from Ref. 21.

Similarly, animals receiving IL-2–PE40 at 5 or 10 mg/day had indefinite allograft survival in three of five and five of five animals, respectively (see Table 2). Pharmacokinetic studies in these mice showed that IL-2–PE40 levels dropped to approximately 100 ng/ml by 6 hr after dosing, and therefore IL-2–PE40 was given in two divided doses intraperitoneally daily (21). Thus, both IL-2–PE40 and DAB$_{486}$–IL-2 were able to produce indefinite graft survival in this model.

As antibody formation is a limiting factor in the use of monoclonal anti–IL-2 receptor antibodies, the antibody response to both IL-2–PE40 and DAB$_{486}$–IL-2 was studied. In both cases, low titers of antibody could be measured 2 weeks following a 10- to 14-day course of treatment at therapeutic doses of IL-2 fusion toxin. The clinical significance of these levels of antibody is not known.

B. Murine Pancreatic Islet Transplantation

The transplantation of pancreatic islets to reverse streptozotocin-induced diabetes in mice has also been used as a model to examine the immunosuppressive action of DAB$_{486}$–IL-2 (23). As purified islets, free of contaminating acinar tissue, survive indefinitely, a crude preparation containing immunogenic material was used. DAB-2 (H-2d) islets were prepared and transplanted into B6AF1 (H-2b × H-2a) mice, which received DAB$_{486}$–IL-2 or CRM45 (5 mg/day IP) for 20 days beginning on the day of transplant. Control animals rejected their islet grafts, as measured by reappearance of hyperglycemia with a mean survival of 13.5 days. CRM45 animals rejected all grafts at a mean of 19 days (p = n.s.). DAB_{486}–IL-2–treated mice, however, had prolonged graft survivals, with only one rejection during the course of the IL-2 toxin therapy and five of seven grafts surviving in excess of 30 days. One animal had indefinite (>100 days) graft survival. In an effort to increase the number of indefinitely surviving grafts, six animals received DAB_{486}–IL-2 in divided doses for 10 days. Two of six demonstrated prolonged survival, with the other four rejecting islet grafts between the thirteenth and nineteenth days.

Thus, in this strongly immunogenic islet transplant model, DAB_{486}–IL-2 was able to prolong graft survival. Higher doses were required than in the cardiac allograft model, and indefinite survival was more difficult to demonstrate.

C. Primate Renal Transplant

We have extensively employed a non–human primate model of renal transplantation in preclinical trials of anti–IL-2 receptor–directed immunotherapy (10,11). Cynomolgus monkeys (*Macaca fasicularis*), weighing 4–7 kg, are anesthetized with ketamine/halothane anesthesia. A mismatched kidney is transplanted to the recipient aorta and vena cava and a ureteroneocystostomy performed. Bilateral native nephrectomies are then performed. Following treatment with anti-Tac (a mouse-antihuman, IgG_{2a} anti–p55–IL-2 receptor antibody) at 2 mg/kg on alternate days, survival is extended to a mean of 19 days compared with 12 days for control animals. No synergy with cyclosporine was noted (11).

Based on the encouraging results obtained with DAB_{486}–IL-2 and IL-2–PE in the murine cardiac allograft model, we embarked on a trial of these agents in the primate renal model. Eighteen monkeys were treated with DAB_{486}–IL-2 following renal transplantation. Initially, monkeys received DAB_{486}–IL-2 at doses from 0.05 mg/kg IV to 0.10 mg/kg IV with or without cyclosporine. With the exception of one monkey who survived 42 days, no graft prolongation over control was seen. In a few monkeys, the rejection process appeared to be accelerated. The addition of cyclosporine at 2–10 mg/kg/day did not appear to affect the outcome.

Since the serum half-life of DAB_{486}–IL-2 in monkeys was five min, and the time required for intoxication of lymphocytes was 30–60 min, three monkeys were treated with bolus DAB_{486}–IL-2 followed by a 60-min infusion; this schedule was designed to keep the levels of fusion toxin above 100 ng/ml for an extended period. Animals also received 5–10 mg/kg of cyclosporine. Although only one of these monkeys had a prolonged survival (23 days), only mild elevations of creatinine were observed at the time of euthanasia, urine outputs were normal, and no pathological evidence of cellular graft rejection was observed. Only one of these three animals (the longest survivor) had a rise in soluble IL-2 receptors in the serum, indicative of mild rejection, as opposed to earlier monkeys receiving bolus therapy. While infusion DAB_{486}–IL-2 therapy in combination with cyclosporine appeared to prevent allograft rejection in these three animals, only one had prolonged survival. All displayed evidence of significant hepatotoxicity as manifested by elevated AST and ALT levels and pathological changes on liver histology (M. E. Shapiro, P. Bacha, unpublished).

As both DAB_{486}–IL-2 and cyclosporine alone have associated hepatotoxicity, an additional two monkeys were transplanted using lower (2 mg/kg/day) cyclosporine doses and a balanced anesthesia omitting halothane. Once again, a positive effect on rejection was noted, but survival was not prolonged.

Rejection without graft prolongation was obtained using IL-2–PE-40 in the identical model (R. L. Kirkman, unpublished). Graft survival appeared shortened compared with controls.

Although DAB_{486}–IL-2 appears to have an effect on graft rejection in the primate model, this effect seems to occur at a toxic dose of the fusion toxin. It is possible that the failure of IL-2 fusion toxin to perform better is the result of high local IL-2 levels in the allograft, which could displace the fusion toxin from its receptor because of its greater affinity for the IL-2 receptor. Thus, effective immunosuppression in this model may require a modification of the DAB molecule to increase its affinity for the receptor. Preliminary studies with a modified form of DAB–IL-2, DAB_{389}–IL-2 (24) suggest that this may be a promising molecule for use in transplantation.

IV. IN VIVO MODELS OF AUTOIMMUNE DISEASE

A. Adjuvant Arthritis

Chronic adjuvant arthritis is an autoimmune disease which can be induced in genetically susceptible rat strains by immunization with mycobacterial adjuvant (25). The disease is characterized by subacute polyarthritis involving the distal extremities which is pathologically and clinically similar to human rheumatoid arthritis. Activated T-lymphocytes have been shown to be important in the development of the disease (26). Since elimination of activated T lymphocytes may be useful in the treatment of rheumatoid arthritis, the effects of both IL-2 PE40 (27) and DAB_{486}–IL-2 (28) treatment on rat adjuvant arthritis have been investigated. Using similar protocols, rats received IL-2–toxin at 500 μg/kg daily for 10 days at the induction of immunization with adjuvant (days 1–10 in the IL-2–PE40 study,

1–9 in the DAB_{486}–IL-2 study). Peak arthritis index (a scoring system for measuring inflammation) was reduced approximately two-thirds in the treated groups compared to buffer or inactive mutant toxin controls. In the DAB_{486}–IL-2 study, lymph node cells were shown to have specifically depressed proliferative responses to mycobacterial antigens. Preimmunization with diphtheria toxoid had no effect on the ability of DAB_{486}–IL-2 to ameliorate the arthritis symptoms. Based on these encouraging results, trials of DAB_{486}–IL-2 in humans with severe rheumatoid arthritis are underway.

B. Experimental Allergic Encephalomyelitis

Experimental allergic encephalomyelitis (EAE) is an autoimmune disease that can be experimentally induced in several animal species by immunization with a crude spinal cord homogenate or purified myelin basic protein. Involvement of T lymphocytes in EAE has been well documented and is supported by the observations that EAE can be transferred by freshly isolated lymphocytes or T-cell clones (29). Subhuman primates develop a hyperacute form of the disease characterized by the rapid onset of paralysis which is universally fatal within a few days if untreated (30).

Lewis rats were immunized with guinea pig myelin basic protein in CFA. DAB_{486}–IL-2, 0.75 mg/kg SC was administered daily on days 1–14. Control animals developed severe ascending paralysis after day 10. DAB_{486}–IL-2–treated rats developed very mild symptoms which resolved in a few days.

DAB_{486}–IL-2 has also been studied in a cynomolgus monkey model of EAE. The disease was induced by injection of emulsified monkey spinal cord intradermally. DAB_{486}–IL-2 treatment was started at the first signs of disease and continued for 7 days, if the animal survived. Animals received 0.075 mg/kg/day as an intravenous bolus or a 0.235 mg/kg/day bolus followed by 0.12 mg/kg/day as a 1-hr infusion. Clinical responses were seen in one of six bolus-treated animals and two of three infusion-treated monkeys. One of the infusion-treated monkeys survived until the experiment terminated; the other two responders eventually relapsed and were euthanized. Thus, in this difficult model of EAE, preliminary results are encouraging.

V. CONCLUSIONS

Preclinical studies have demonstrated that both IL-2–PE40 and DAB_{486}–IL-2 have beneficial effects in rodent models of arthritis and cardiac allografts. Immunosuppressive activity of DAB_{486}–IL-2 has been further shown in murine models of delayed-type hypersensitivity and islet transplantation as well as in rodent and primate models of EAE. Based on efficacy studies in preclinical animal models, clinical trials are underway with DAB_{486}–IL-2 in severe rheumatoid arthritis, recent-onset autoimmune diabetes mellitus, and IL-2 receptor–expressing malignancies. These two IL-2 fusion toxins are clearly cytotoxic for activated T lymphocytes and have efficacy in animal models of T-cell–mediated disease, including rodent transplantation, suggesting the validity of this approach. Effective transplantation immunosuppression should be obtainable using second-generation constructs with greater affinity for the IL-2 receptor, which can better compete with the levels of IL-2 normally associated with allografts.

REFERENCES

1. Leonard, W. J., Depper, J. M., Uchiyama, T., Smith, K. A., Waldmann, T. A., and Greene, W. C. A monoclonal antibody that appears to recognize the receptor for human T-cell growth factor; partial characterization of the receptor. Nature, *300*: 267, 1982.
2. Cantrell, P. A., and Smith, K. A. The interleukin 2 T-cell system: A new cell-growth model. Science, *224*: 1312, 1984.
3. Robb, R. J., Munck, A., and Smith, K. A. T-cell growth-factor receptors. Quantitation, specificity and biological relevance. J. Exp. Med., *154*: 1455, 1981.
4. Tsudo, M., Kozak, R. W., Goldman, C. K., and Waldmann, T. A. Demonstration of a new peptide (non-Tac) that binds IL-2: A potential participant in a multichain IL-2 receptor complex. Proc. Natl. Acad. Sci. U.S.A., *83*: 9694, 1986.
5. Uchiyama, T., Broder, S., and Waldmann. T. A. A monoclonal antibody (anti-Tac) reactive with activated functionally mature human T cells. J. Immunol., *126*: 1393, 1981.
6. Gaulton, G. N., Bangs, J., Maddock, S., Springer, T., Eardley, D. D., and Strom, T. B. Characterization of a monoclonal rat anti-mouse interleukin 2 (IL-2) receptor antibody and its use in the biochemical characterization of the murine IL-2 receptor. Clin. Immunol. Immunopathol., *94*: 283, 1985.
7. Kirkman, R. L., Barrett, L. V., Gaulton, G. N., Kelley, V. E., Ythier, A., and Strom, T. B. Administration of an anti-interleukin 2 receptor monoclonal antibody prolongs cardiac allograft survival in mice. J. Exp. Med., *162*: 358–362, 1985.
8. Kirkman, R. L., Barrett, L. V., Gaulton, G. N., Kelley, V. E., Kolton, W. A., Schoen, F. J., Ythier, A., and Strom, T. B. The effect of anti-interleukin 2 receptor monoclonal antibody on allograft rejection. Transplantation, *40*: 719, 1985.
9. Kupiec-Weglinski, J. W., Diamantstein, T., Tilney, N. L., and Strom, T. B. Anti-interleukin 2 receptor monoclonal antibody spares T suppressor cells and prevents acute allograft rejection. Proc. Natl. Acad. Sci. U.S.A., *83*: 2624, 1986.
10. Shapiro, M. E., Kirkman, R. L., Reed, M. H., Puskas, J. D., Mazoujian, G., Letvin, N. L., Carpenter, C. B., Milford, E. L., Waldman, T. A., Strom,

T. B., and Schlossman, S. F. Monoclonal anti-IL-2 receptor antibody in primate renal transplantation. Transplant. Proc., *19*: 594–598, 1987.

11. Reed, M. H., Shapiro, M. E., Strom, T. B., Milford, E. L., Carpenter, C. B., Weinberg, D. S., Reimann, K. A., Letvin, N. L., Waldmann, T. A., and Kirkman, R. L. Prolongation of primate renal allograft survival by anti-Tac, on anti-human IL-2 receptor monoclonal antibody. Transplantation, *47*: 55–59, 1989.

12. Soulillou, J. P., Peyronnet, P., LeMauff, B., Housmant, M., Olive, D., Mowas, C., Delaage, M., Kirn, M., and Jacques, Y. Prevention of rejection of kidney transplants by monoclonal antibody directed against interleukin 2 receptor. Lancet, *1*: 1139–1342, 1987.

13. Kirkman, R. L., Shapiro, M. E., Carpenter, C. B., McKay, D. B., Milford, E. L., Ramos, E. L., Tilney, N. L., Waldmann, T. A., Zimmerman, C. E., and Strom, T. B. A randomized prospective trial of anti-Tac monoclonal antibody in human renal transplantation. Transplantation, *51*: 107, 1991.

14. Kelley, V. E., Gaulton, G. N., Hattori, M., Ikegami, H., Eisenbarth, G., and Strom, T. B. Anti-interleukin 2 receptor antibody suppresses murine diabetic insulitis and lupus nephritis. J. Immunol., *140*: 59, 1988.

15. Williams, D., Parker, K., Bacha, P., Bishai, W., Borowski, M., Genbauffe, F., Strom, T. B., and Murphy, J. R. Diphtheria toxin receptor binding domain substitution with interleukin 2: Genetic construction and properties of a diphtheria toxin-related interleukin 2 fusion protein. Protein Eng., *1*: 493, 1987.

16. Waters, C. A., Schimke, P. A., Snider, C. E., Itoh, K., Smith, K. A., Nichols, J. C., Strom, T. B., and Murphy, J. R. Interleukin 2 receptor-targeted cytotoxicity. Receptors binding requirements for entry of a diphtheria toxin-related interleukin 2 fusion protein into cells. Eur. J. Immunol., *20*: 785–791, 1990.

17. Lorberboum-Galski, H., Kozak, R. W., Waldmann, T. A., Bailon, P., Fitzgerald, D. J., and Pastan, I. Interleukin 2 (IL2) PE40 is cytotoxic to cells displaying either the p55 or p70 subunit of the IL-2 receptor. J. Biol. Chem., *263*: 18650, 1988.

18. Kelley, V. E., Bacha, P., Pankewycz, O., Nichols, J. C., Murphy, J. R., and Strom, T. B. Interleukin 2–diphtheria toxin fusion protein can abolish cell-mediated immunity in vivo. Proc. Natl. Acad. Sci. U.S.A., *85*: 3980–3984, 1988.

19. Bastos, M. G., Pankewycz, O., Rubin-Kelley, V. E., Murphy, J. R., and Strom, T. B. Concomitant administration of hapten and IL-2–toxin (DAB486-IL2) results in specific deletion of antigen activated T cell clones. J. Immunol., *145*: 3535–3539, 1990.

20. Kirkman, R. L., Bacha, P., Barrett, L. V., Forte, S., Murphy, J. R., and Strom, T. B. Prolongation of cardiac allograft survival in murine recipients treated with a diphtheria toxin-related interleukin-2 fusion protein. Transplantation, *47*: 327–330, 1989.

21. Lorberboum-Galski, H., Barrett, L. V., Kirkman, R. L., Ogata, M., Willingham, M. C., Fitzgerald, D. J., and Pastan, I. Cardiac allograft survival in mice treated with IL-2-PE40, Proc. Natl. Acad. Sci. U.S.A., *86*: 1008–1012, 1989.

22. Corry, R. J., Winn, H. J., and Russell, P. S. Primarily vascularized allografts of hearts in mice. Transplantation, *16*: 343, 1973.
23. Pankewycz, O., Mackie, J., Hassarjian, R., Murphy, J. R., Strom, T. B., and Kelley, V. E. Interleukin-2–diphtheria toxin fusion protein prolongs murine islet cell engraftment. Transplantion, *47*: 318–322, 1989.
24. Williams, D. P., Snider, C. E., Strom, T. B., and Murphy, J. R. Structure/ function analysis of interleukin-2–toxin (DAB486IL-2): Fragment B sequences required for the delivery of fragment A to the cytosol of target cells. J. Biol. Chem., *265*: 11885–11889, 1990.
25. Pearson, C. M., and Wood, F. D. Studies of polyarthritis and other lesions induced in rats by injection of mycobacterial adjuvant. I. General clinical and pathological characteristics and some modifying factors. Arthritis Rheum., *2*: 440, 1959.
26. Prud'homme, G. J., and N. A. Parfry. Biology of disease: Role of T helper lymphocytes in autoimmune diseases. Lab. Invest., *59*: 158, 1988.
27. Case, J. P., Lorberboum-Galski, H., Lafyatis, R., Fitzgerald, D., Wilder, R. L., and Pastan, I. Chimeric cytotoxin IL2-PE40 delays and mitigates adjuvant-induced arthritis in rats. Proc. Natl. Acad. Sci. U.S.A., *86*: 287–291, 1989.
28. Forte, S. E., Perper, S. J., Trentham, D. E., Nichols, J. C., and Bacha, P. Anti-arthritic effects demonstrated by an interleukin-2 receptor targeted cytotoxin (DAB486IL-2) in rat adjuvant arthritis. J. Immunol., submitted.
29. Zamil, S. S., and Steinman, L. The T lymphocyte in experimental allergic encephalomyelitis. Ann. Rev. Immunol., *8*: 579–621, 1990.
30. Rose, L. M., Alvord, E. C., Hruby, S., Jackevicius, S., Petersen, R., Warner, N., and Clark, E. A. In vivo administration of anti-CD4 monoclonal antibody prolongs survival in longtailed macaques with experimental allergic encephalomyelitis. Clin. Immunol. Immunopathol., *45*: 405–423, 1987.

20
Initial Clinical Experiences with an Interleukin-2 Fusion Toxin (DAB$_{486}$–IL-2)

Carole M. Meneghetti and C. F. LeMaistre   *University of Texas Health Science Center at San Antonio, San Antonio, Texas*

I. INTRODUCTION

Within the past decade, increasing interest has been generated in the development of targeted therapies for the treatment of malignancies. The introduction of these novel agents has allowed investigators to exploit tumor-associated receptors or "targets" with the potential to deliver therapy selectively to the tumor while sparing normal, healthy cells. DAB$_{486}$–IL-2, the first genetically engineered fusion toxin to come to clinical trials, is one such agent selective for cells expressing the high-affinity interleukin-2 receptor (IL-2R) (1–3). The interleukin-2 receptor is a multiunit complex found on the cell surface of activated T lymphocytes and monocytes as well as in certain malignancies (4–13). In order to understand why the interleukin-2 receptor is an appropriate target, it is helpful to review the characterization of interleukin-2 (IL-2) and the IL-2R in the human immune response (5).

II. IL-2 AND IL-2R IN THE HUMAN IMMUNE RESPONSE

The human immune system has an unlimited capacity to recognize and generate a response to antigen; a component of this response involves the activation of T lymphocytes (5). T cells are responsible for mediating regulator functions such as help or suppression as well as effector functions such as the cytotoxic destruction of antigen and the production of lymphokines. For an immune response to be effective, T cells must move from a resting state to an activated state. This involves two sets of signals within the cell. First, appropriately processed and presented antigen binds with antigen receptors on the T-cell surface, triggering the T cell to enter the activated state and to synthesize and secrete IL-2. Interleukin-2 is a lymphokine which exerts its effect by binding to the high-affinity IL-2R. This interaction with antigen causes both the expression of IL-2 and a 200-fold increase in its receptor on the cell surface. The binding of IL-2 to its receptor provides the second signal which causes T-cell proliferation and the generation of specific regulatory and effector cells.

The high-affinity IL-2R has two identified components, the p55 and the p75 subunits (4,5). Either of these subunits may be expressed independently on the cell surface, as the low-affinity and intermediate-affinity receptor, respectively, or together as the high-affinity IL-2R. These two subunits bind different regions of the IL-2 molecule; kinetic binding studies demonstrate how the p55 and the p75 subunits cooperate to interact with IL-2. The p55 subunit exhibits a rapid "on-off" binding time with the IL-2 molecule of <5–10 sec whereas "on-off" binding time for the p75 subunit is approximately 5–20 min. It follows that the association rate of IL-2 with the IL-2R is dependent upon the p55 subunit, and that the dissociation rate is dependent on the p75 subunit of the high-affinity IL-2R.

In addition to being present on activated T cells, IL-2Rs have been detected on the cell surfaces of certain B- and T-cell leukemias and lymphomas, in certain autoimmune diseases, and in activated T lymphocytes present during allograft rejection (5–14). In addition, a soluble form of the IL-2R (a 45-kd subfragment of p55) is detectable in the serum of individuals with various hematological malignancies, including adult T-cell leukemia (ATL), chronic lymphocytic leukemia (CLL), and hairy cell leukemia (4,7–14). Individual levels of secretory IL-2R (sIL-2R) are variable but may be an indicator of the extent of tumor in certain malignancies. Strategies for targeting the IL-2R have included monoclonal antibodies, antibody-toxin conjugates, and genetically engineered ligand toxins (15–20).

The IL-2R is therefore an attractive "target" for designing novel cancer therapies because of the restricted expression of the IL-2R on normal cells and the constituitive expression of the IL-2R by a number of hematological malignancies. Two groups have reported the construction of genetically engineered fusion toxins designed to be selectively cytotoxic for cells expressing the receptor for IL-2 (18,19). One of these conjugates, DAB_{486}–IL-2 is currently in phase I–II clinical trials in the treatment of IL-2R expressing hematological malignancies.

III. THE CHIMERIC FUSION PROTEIN DAB_{486}–IL-2

DAB_{486}–IL-2 is a chimeric fusion protein created through the replacement of the binding domain of the diphtheria toxin with sequences of the gene for human IL-2 (18). DAB$_{486}$–IL-2 selectively inhibits protein synthesis in vitro in T cells and malignant cells which express the high-affinity IL-2R and spares cells which do not express this receptor, including those bearing the low or intermediate IL2-R subunits (18,20). The binding of the high-affinity IL-2R by DAB$_{486}$–IL-2 triggers the internalization of the receptor complex into an acidic endosome (18). The diphtheria toxin component then translocates into the cytoplasm where it inhibits protein synthesis. Only one molecule of diphtheria toxin need enter a cell in order to effect cell death, underscoring the potency of this agent. Additionally, this mechanism of action is attractive not only for its selectivity, but because no other anticancer agents work via protein synthesis inhibition.

In a phase I clinical trial of DAB$_{486}$–IL-2, 18 patients with IL-2R (p55 positive) expressing hematological malignancies were enrolled (1,22). Patients were treated in a cohort dose-escalation design with single and multiple doses. DAB$_{486}$–IL-2, was administered as a 1–5 min intravenous bolus infusion at doses ranging from 0.0007 to 0.1 mg/kg/day. In the initial cohort, patients received an initial dose, then 48 hr later, three daily doses, then 1 week later, 7 seven daily doses. In each of the remaining cohorts, patients received three, then 1 week later, 7 daily doses for a total of 10 daily doses. Patients received retreatment a minimum of 4 weeks after their first day of therapy. Responding patients were placed on a maintenance schedule consisting of seven daily infusions every 3–4 weeks after their initial dose.

There were 14 men and 4 women with a median age of 59 years (range 22–81) who received at least one course of DAB$_{486}$–IL-2 (Table 1) (1). Eligible patients had hematological malignancies unresponsive to standard therapy; the patients entered onto this trial had a median of four prior treatment regimens (range two to eight). Nine of these individuals had B-cell non-Hodgkin's lymphomas; four had B-cell chronic lymphocytic leukemias; one had T-cell chronic lymphocytic leukemia; three had Hodgkin's lymphomas, and one had a cutaneous T-cell lymphoma. Tumor samples from each of the patients were tested for expression of the low-affinity IL-2R by antibody staining with a median of 43% (range 14–94%) of tumor cells expressing p55. All patients had elevated soluble IL-2R with a median pretreatment value of 8016 U/ml (range 1036–11,346 U/ml).

DAB$_{486}$–IL-2 was well tolerated at all dose levels and treatment schedules (Figure 1) (1). Five patients experienced mild fevers, typically within 4–6 hr following the end of the DAB$_{486}$–IL-2 infusion, which were controlled with the administration of acetaminophen. Three patients developed chest tightness during drug administration, which responded to increasing the infusion length to between 10 and 90 min. Two patients who had preexisting renal compromise from prior cis-platinum therapy developed a transient increase in serum creatinine. The dose-limiting toxicity

Table 1 Patient Demographics

Patient	Age/Sex	Diagnosis
001	67/M	HD
002	55/M	NHL
003	63/M	NHL
004	65/M	NHL
005	81/M	CTCL
006	53/F	NHL
007	52/M	NHL
008	54/F	NHL
009	58/M	CLL
010	72/M	NHL
011	53/M	NHL
012	59/M	CLL
013	69/F	CLL
014	58/M	CLL
015	44/M	CLL
016	32/M	HD
017	60/F	NHL
018	22/M	HD

HD = Hodgkin's disease; NHL = non-Hodgkin's lymphoma; CTCL = cutaneous T-cell lymphoma; and CLL = chronic lymphocytic leukemia.

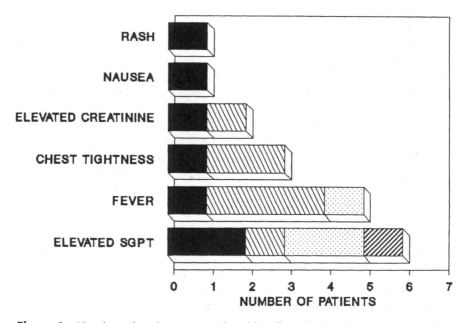

Figure 1 Number of patients expressing side effects during their treatment with DAB_{486}–IL-2. Toxicity grading is according to standard NCI criteria. Grade I ■; Grade II ▨; Grade III ▦; Grade IV ▨.

for this trial was a grade III–IV transient, asymptomatic, reversible elevation in SGOT and SGPT without associated elevations in other liver function tests. These transaminase elevations were not cumulative and did not preclude retreatment with DAB$_{486}$–IL-2.

There were three significant responses documented in this initial clinical trial (1). One patient with a follicular large cell lymphoma progressive on multiple chemotherapies had a complete remission, which has continued for greater than 18 months. Two additional intensively pretreated patients, one with chronic lymphocytic leukemia and one with a follicular small cell lymphoma, each had a partial remission of 5 and 12 months, respectively. There does not, as yet, appear to be a predictable relationship between tumor histology, the presence of IL-2R, the level of soluble IL-2R, the presence of antibodies to DAB$_{486}$–IL-2, and antitumor response.

Based on the initial Phase I trial results, a Phase Ib trial was initiated to explore a longer infusion schedule and the importance of IL-2 receptor levels in response. Twenty-one patients have been treated with a 90 minute infusion daily for five days. The trial is not complete, but two important observations have already been made (23). First, prolonged blood levels can be maintained with a long infusion. Second, the expression of the IL-2 receptor is critical to response since antitumor activity was seen in only one patient whose cancer did not express p55. However, 13 out of 14 patients whose tumors were p55 positive demonstrated antitumor activity.

IV. CONCLUSIONS

Additional phase I and phase II trials are planned or are underway examining means of improving the efficacy of this IL-2 fusion toxin in autoimmune disease, as well as in cancer therapy (1–3). In the development of these types of agents, issues associated with potency, pharmacokinetics, and dosing schedule may be more important than with traditional antineoplastic agents. For example, the relatively rapid clearance of DAB$_{486}$–IL-2 might preclude the contact time necessary to effect binding by the maximal number of tumor cells such that a long infusion may be much more effective. Further, a more in-depth understanding of the characteristics of the IL-2R in malignant cells is critical. Despite evidence of constitutive expression, little is known about the functionality of these IL-2Rs. Abnormalities in the affinity, modulation, or processing of these receptors may have profound implications upon the ability of DAB$_{486}$–IL-2 to achieve an antitumor response. These differences will also dictate dosing schedules as the optimal treatment plans will be best developed with an understanding of the intervals between treatment and receptor expression rather than considerations of the length of time needed for patients to recover from systemic toxicity. At present, the importance of these agents goes beyond the benefit experienced by a few patients in early testing. These tumor responses underscore the power of recombinant DNA technologies in the rational design of therapies based upon our understanding of cancer biology.

REFERENCES

1. LeMaistre, F., Rosenblum, M., Reuben, J., Parkinson, D., Meneghetti, C., Parker, K., Shaw, J., Deisseroth, A., and Woodworth, T. Phase I study of genetically engineered DAB_{486}IL-2 receptor expressing malignancies. Blood, 76: 360a, 1990.

2. LeMaistre, C. F., Von Hoff, D., Meneghetti, C., Adkins, D., Rosenblum, M., Reuben, J., Deisseroth, A., Parker, K., Shaw, J., and Woodworth, T. DAB_{486}IL-2 is effective therapy for some patients with IL-2 receptor expressing malignancies. Prog. ASCO, 10: 280a, 1991.

3. Sewell, K. L., and Trentham, D. E. Rapid improvement in refractory rheumatoid arthritis by an interleukin-2 receptor targeted immunotherapy. Proc. ACR 55th Annual Scientific Meeting., Abst. 141a, 1991.

4. Waldmann, T. A. The structure, function, and expression of interleukin-2 receptors on normal and malignant lymphocytes. Science, 232: 727–732, 1986.

5. Waldmann, T. A. The multi-subunit interleukin-2 receptor. Annu. Rev. Biochem., 58: 875–911, 1989.

6. Waldmann, T. A., Goldman, C. K., Robb, R. J., Depper, J. M., Leonard, W. J., Sharrow, S. O., Bongiovanni, K. F., Korsmeyer, S. J., and Greene, W. C. Expression of interleukin 2 receptors on activated human B cells. J. Exp. Med., 160: 1450–1466, 1984.

7. Casey, T. T., Olson, S. J., Cousar, J. B., and Collins, R. D. Immunophenotypes of Reed-Sternberg cells: A study of 19 cases of Hodgkin's disease in plastic-embedded sections. Blood, 74: 2624–2628, 1989.

8. Steis, R. G., Marcon, L., Clark, J., Urba, W., Longo, D. L., Nelson, D. L., and Maluish, A. E. Serum soluble IL-2 receptor as a tumor marker in patients with hairy cell leukemia. Blood, 71: 1304–1309, 1988.

9. Kay, N. E., Burton, J., Wagner, D., and Nelson, D. L. The malignant B-cells from B-chronic lymphocytic leukemia patients release TAC-soluble interleukin-2 receptors. Blood, 72: 447–450, 1988.

10. Yogura, H., Tamaki, T., Furitsu, T., Tomiyama, Y., Nishiura, T., Tominaga, N., Katagiri, S., Yonezawa, T., and Tarui, S. Demonstration of high-affinity interleukin-2 receptors on B-chronic lymphocytic leukemia cells: Functional and structural characterization. Blut, 60: 181–186, 1990.

11. Semenzato, G., Foa, R., Agostini, C., Zambello, R., Trentin, L., Vinante, F., Benedetti, F., Chilosi, M., and Pizzolo, G. High serum levels of soluble

interleukin-2 receptor in patients with B chronic lymphocytic leukemia. Blood, *70*: 396–400, 1987.

12. Strauchen, J. A., and Breakstone, B. A. IL-2 receptor expression in human lymphoid lesions. Am. J. Pathol., *126*: 506–512, 1987.

13. Rosolen, A., Masayuki, N., Poplack, D. G., Cole, D., Quinones, R., Reaman, G., Trepel, J. B., Cotelingam, J. D., Sausville, E. A., Marti, G. E., Jaffe, E. S., Neckers, L. M., and Colamonici, O. R. Expression of interleukin-2 receptor B subunit in hematopoietic malignancies. Blood, *73*: 1968–1972, 1989.

14. Finberg, R. W., Wahl, S. M., Allen, J. B., Soman, G., Strom, T. B., Murphy, J. R., and Nichols, J. C. Selective elimination of HIV infected cells with an interleukin-2 receptor-specific cytotoxin. Science, *252*: 1703–1705, 1991.

15. Ferrara, J. L. M., Marion, A., McIntyre, J. F., Murphy, G. F., and Burakoff, S. J. Amelioration of acute graft-versus-host disease due to minor histocompatibility antigens by *in vitro* administration of anti-interleukin 2 receptor antibody. J. Immunol., *137*: 1874–1877, 1986.

16. Cavazzona-Calvo, M., Fromont, C., Le Deist, F., Lusardi, M., Coulombel, L., Derocq, J. M., Gerota, I., Griscelli, C., Fisher, A. Specific elimination of alloreactive T cells by an anti-interleukin-2 receptor B chain-specific immunotoxin. Transplantation, *50*: 1–7, 1990.

17. Kondo, T., FitzGerald, D., Chaudhary, VK, Adhya, S., and Paston, I. Activity of immunotoxins constructed with modified Pseudomonas exotoxin A lacking the cell recognition domain. J. Biol. Chem., *263*: 9470–9475, 1988.

18. Williams, D. P., Parker, K., Bucha, P., Bishai, W., Borowshi, M., Genbauffe, F., Strom, T. B., Murphy, J. R. Diphtheria toxin receptor binding domain substitution with interleukin-2: Genetic construction and properties of a diphtheria toxin-related interleukin-2 fusion protein. Protein Eng. *1*(6): 493–498, 1987.

19. Lorberboum-Galski, H., Kozak, R. W., Waldmann, T. A., Bailon, P., Fitzgerald, D. J. P., Pastan, I. Interleukin 2 (IL2) PE40 is cytotoxic to cells displaying either the p55 or p70 subunit of the IL2 receptor. J. Biol. Chem., *263*: 18650–18656, 1988.

20. Bacha, P., Williams, D. P., Waters, C., Williams, J. M., Murphy, J. R., and Strom, T. B. Interleukin 2 receptor-targeted cytotoxicity. J. Exp. Med., *167*: 612–622, 1988.

21. Rosenblum, M. G., LeMaistre, C. F., Reuben, J. M., Meneghetti, C. M., Parker, K., Shaw, J. P., and Woodworth, T. Clinical pharmacology studies of DAB$_{486}$IL2. Blood, *76*: 3142a, 1990.

22. LeMaistre, C. F., Meneghetti, C., Rosenblum, M., Reuben, J., Parker, K., Shaw, J., Deisseroth, A., Woodworth, T., and Parkinson, D. R. Phase I trial of an IL-2 fusion toxin (DAB$_{486}$IL-2) in hematologic malignancies expressing the IL-2 receptor. Blood, *79*: 51–58, 1992.

23. LeMaistre, C. F., Craig, F., Meneghetti, C., McMullin, B., Banks, P., Reuben, J., Rosenblum, M., Parker, K., Woodworth, T., and VonHoff, D. D. Phase I-II trial of an IL-2 fusion toxin (DAB$_{486}$IL-2) in IL-2 receptor positive and negative malignancies. Blood, *78*: 493a, 1991.

VI
Pseudomonas Exotoxins

21
Pseudomonas aeruginosa Exotoxin A

G. Jiliani Chaudry, Priscilla L. Holmans,* Royston C. Clowes,† and Rockford K. Draper   *University of Texas at Dallas, Richardson, Texas*

I. INTRODUCTION

Pseudomonas aeruginosa is a common, opportunistic, nosocomial pathogen infecting patients with weakened immunity, such as those with burns. The major proteins secreted by *P. aeruginosa* are exotoxin A, exoenzyme S, phospholipase C, elastase, and an alkaline protease. Exotoxin A (M_r = 66,583) is the major protein secreted by most strains and appears to be a virulence factor in many *Pseudomonas* infections (1–3). The amount of toxin secreted depends on the *Pseudomonas* strain (4). Strain PA103 (5) secretes the most toxin and is preferred for purifying large quantities of the toxin (6). Strain PA01 has been used in most genetic studies (7), but produces less toxin than PA103.

Exotoxin A synthesis by *P. aeruginosa* is not constitutive; production begins when the cells enter stationary phase and requires low iron concentration in the growth medium, 0.01–0.05 μg/ml being the optimal range. Outside this range, toxin production drops sharply, and very little is synthesized if iron exceeds 0.1 μg/ml (8).

The LD_{50} of exotoxin A is about 100 ng in 20-g mice and about 20μg in 350-g rats (6). The sensitivity of cultured mammalian cells to the toxin varies; whereas some mouse cells, such as NIH 3T3 and L cells, are the most sensitive, others, such as RAG cells, are about 1500 times less sensitive than mouse L cells (9). Exotoxin A kills cells by catalyzing the covalent attachment of the ADP-ribosyl moiety of NAD^+ to elongation

---

†Deceased.

Present affiliation: University of Texas Southwestern Medical Center at Dallas, Dallas, Texas

factor 2 (EF-2), which inactivates EF-2 and arrests cytoplasmic protein synthesis (10–12). The acceptor for ADP-ribose in EF-2 is a single modified histidine, called diphthamide, unique to eukaryotic EF-2 (13,14). The catalytic activity of exotoxin A appears identical to that of diphtheria toxin. Despite their similar enzymic activities, exotoxin A and diphtheria toxin are antigenically dissimilar; antibodies against one do not cross-react with the other (6,10). There is no major sequence homology between the two toxins, either at the nucleotide or the amino acid level (15). Recently, however, some homology that may be functionally significant was observed when glutamic acid 553 of exotoxin A (16–19) and glutamic acid 148 of diphtheria toxin (20), the residues essential for the catalytic activities of the respective toxins, were aligned.

To inactivate EF-2, the catalytic center of exotoxin A must penetrate a membrane of the target cell and enter the cell cytosol. The entry process requires receptor-mediated endocytosis (21) and involves three steps: (A) the toxin binds a cell surface receptor of unknown nature; (B) the toxin-receptor complex is internalized by endocytosis, and (C) at least the enzymically active part of the toxin escapes from a vesicle to enter the cytosol, where it inactivates EF-2. The mechanism of escape is unknown, but may involve insertion of the toxin into the lipid bilayer induced by the low pH within endocytic vesicles.

For maximal cytotoxic activity, exotoxin A must be in the proenzyme form, the form secreted by *Pseudomonas*. The proenzyme has very little ADP ribosyl transferase activity in vitro unless activated by simultaneous treatment with a denaturant, such as urea, and a reducing agent, such as dithiothreitol (22). However, activation strongly reduces the cytotoxicity of the toxin for intact cells (11), suggesting that the proenzyme is required for efficient receptor-mediated endocytosis, and that the toxin is activated after entry into the target cells.

The three-dimensional structure of the proenzyme form of exotoxin A was solved to 3-Å resolution by Allured et al. (23). The mature, secreted toxin has three structural domains: domain Ia (residues 1–252), domain Ib (residues 365–404), domain II (residues 253–364), and domain III (residues 405–613) (Figure 1). Although it was clear that the catalytic site of the toxin is in domain III (15,24), empirical data suggesting functions for domains I and II were lacking at the time the structure was solved. Domain I is primarily an antiparallel beta structure resembling the receptor-binding moieties of certain proteins, such as influenza hemagglutinin, and domain II comprises six alpha-helices long enough to span the lipid bilayer. Based on these features, Allured et al. (23) suggested that domain I may function in receptor binding and domain II in membrane penetration. In 1987, evidence supporting this prediction became available; when all or part of

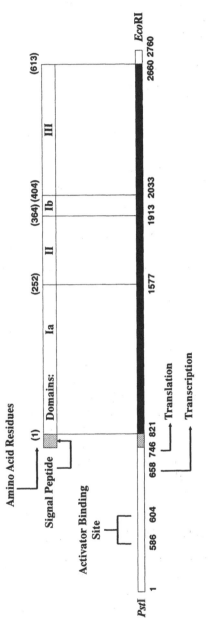

Figure 1 Features of the ETA gene encoded on the cloned *PstI–EcoRI* fragment. Nucleotides are numbered from the *PstI* site (left, nucleotide 1) to the *EcoRI* site (right, nucleotide 2760). The upstream activator-binding site is between nucleotides 586–604, the transcriptional start site is at 658, the translational start is at 746, and the end of the structural gene is at 2660. Features of the ETA protein are shown above the *PstI–EcoRI* fragment. Numbers in parentheses refer to amino acid residues, beginning with mature protein lacking the signal peptide. Boundaries of the domains are indicated by amino acid numbers and are matched to nucleotide numbers of the *PstI–EcoRI* fragment.

domain I is deleted, the resulting variants fail to compete with wild-type toxin for binding to cell surface receptors (25,26). Variants lacking domain II, on the other hand, retain normal receptor binding and enzymic activity, but have reduced cytotoxicity, suggesting that they have impaired translocation after internalization, thus implicating domain II in membrane penetration (25).

Recent mutational analysis of domain II supports the notion that it functions in membrane penetration. Toxin variants with each of the cysteines at positions 265 and 287 substituted with serines, eliminating the only disulfide bond in domain II, are significantly less cytotoxic than the wild-type toxin, but retain enzymic activity and receptor binding (27). Deleting parts of domain II also impairs cytotoxicity, but not binding or enzymic activity (28). Further, certain independent substitutions for each of the arginines at 276, 279, 330, and 337 in domain II greatly reduce cytotoxicity without impairing enzymic activity or receptor binding (29). The data further suggest that at least the arginine at 276 undergoes a specific, saturable interaction after internalization of the toxin (29).

Although a comprehensive functional map of the active site is still unavailable, several workers nevertheless have identified specific amino acids in domain III that are essential for the enzymic activity of the toxin. Photolabeling studies have shown that glutamic acid at 553 functions in NAD^+ binding (16). Substituting it with aspartic acid drastically reduces the enzymic activity of the toxin (18), and deleting it abolishes the activity (19). Selective iodination of tyrosine at 481 (30), or deletion of histidine at 426 (31) also abolishes the enzymic activity of the toxin, and tryptophan at 558 has been implicated in the enzymic activity (16,17). However, ADP-ribosylation may not be the only function of domain III; recently; Chaudhary et al. (32) reported that certain mutations in the carboxyl-terminal sequence Arg-Glu-Asp-Leu-Lys generate variants of exotoxin A that retain enzymic activity and receptor binding, but fail to intoxicate mouse cells. Further, the authors have suggested this pentapeptide may be the recognition sequence for an intracellular component that functions in effecting the translocation of the toxin (32).

II. CLONING THE EXOTOXIN A GENE

The exotoxin A gene (*toxA*) maps to 85 min on the chromosome of strain PA01 (33). The entire gene from the chromosome of strain PA103 was first cloned in *Escherichia coli* and sequenced by Gray et al. (15), who synthesized oligonucleotide probes corresponding to the amino acid sequence of the toxin to identify the cloned DNA fragments encoding the toxin, and then by recombinant techniques reconstructed the entire *toxA* gene from these fragments. The gene was part of a 2.76-kbp *PstI–Eco*RI segment of PA103 chromosomal DNA, the nucleotides of which were numbered 1–2760, beginning at the *Pst*I site. (Figure 2). The translational start (ATG) is at position 746 and the translational stop at position 2660. The cloned gene, however, did not express from its postulated natural promoter (that is, upstream of position 746) in *E. coli*, but did express in *E. coli* under control of the *trp* promoter. This is consistent with the fact that no consensus sequence is present in the -10 and -35 promoter regions upstream of the translational start for binding the σ^{70} RNA polymerase holoenzyme of *E. coli* (15) (see Chapter 15).

Initial attempts in our laboratory to clone *toxA* from PA103 in the broad host-range plasmid pRO1614 (34) resulted in cloning a segment that encoded a truncated, but enzymically active, exotoxin A polypeptide (24). The plasmid containing this fragment is pRC345 (Table 1). Experiments with pRC345 in *E. coli* HB101 (35) or a nontoxigenic *P. aeruginosa* host, MAM2 (see Section III.A), indicated that cell lysates had ADP-ribosyl transferase activity and contained a 50-kda protein and several smaller polypeptides reactive with antibodies to exotoxin A, but none of the toxin-related polypeptides were secreted. The absence of a polypeptide the size of the native toxin suggested that only a segment of *toxA* had been cloned, evidently sufficient to encode an enzymically active and antigenically reactive part of the toxin. Subcloning and further analysis by restriction enzyme mapping suggested the fragment was expressed from the Tet^r gene promoter P_2, resulting in the synthesis of a fusion polypeptide, part tetracycline resistance protein and part exotoxin A.

Using the information of Gray et al. (15), we cloned the 2.76-kbp *PstI–Eco*RI fragment from both strains PA103 and PA0286 (36). Employing a nick-translated 520-bp *toxA* fragment from pRC345 as a probe to detect the exotoxin A structural gene, we cloned in pUC9 (37) the 2.76-kbp *PstI–Eco*RI fragment carrying the toxin structural gene from PA0286, an auxotrophic derivative of PA01 (38). For these cloning experiments, done in *E. coli* HB101, we used pUC9 and not pRO1614 because cloning in pRO1614 resulted in spontaneous loss of a part of the cloned 2.76-kbp *PstI–Eco*RI fragment, as well as a part of the Tet^r gene.

Cloning the toxin gene from strain PA103 was not straightforward; the chromosomal DNA of our PA103 did not cleave with *Pst*I to yield the 2.76-kbp *PstI–Eco*RI fragment reported by Gray et al. (15). But we succeeded in cloning the 2.76-kbp fragment in pUC9 initially as part of a 10-kbp *Eco*RI fragment from PA103. Plasmid DNA containing the 10-kbp fragment isolated from HB101 was then cleaved with *Pst*I to yield the 2.76-kbp *PstI–Eco*RI fragment carrying the entire toxin gene. We then subcloned this fragment in pUC9 to derive pRC360, and in pRC357 to derive

```
        10        20        30        40        50        60        70        80        90       100       110       120
CTGCAGCTGGTCAGGCCGTTTCCGCAACGCTTGAAGTCCTGGCCGATATACCGGCAGGCCAGCCATCGTTCGACGAATAAAGCCACCTCAGCCATGATGCCCTTTCCATCCCCAGCGGA

       130       140       150       160       170       180       190       200       210       220       230       240
ACCCCGACATGGACGCCCAAAGCCCTGCTCCTGGCAGCCTGCCTGCCGCCCATTCGGCCGACGCGGCCGACGCTCGACAATGCTCTCTCCGCCTGCCTCGCCGCCGGCTCGGTGCAC

       250       260       270       280       290       300       310       320       330       340       350       360
CGCCACGCCGGAGGGCCAGTTGCACCTGCCCACTCACCCTTGAGGCCCGGCGCTCCACCGGGCAATGCGGCTCCGGCGCTGGTGCGATATCGGCTGCTGGCCAGGGGGCCCAGCG

       370       380       390       400       410       420       430       440       450       460       470       480
CCGACAGCCTTCGTGCTTCAAGAGGGCTGCTCCAGGACAGCCGGCGAGCGGCGCCGCCGCTTGGCGAGCGGCGACGGACTTGGTCGTGGTCTGCACCCTGGGTTG

       490       500       510       520       530       540       550       560       570       580       590       600
TCAGGCGCCTGACTGACGAGGCCGGGCTGCCACCACGGCCGAGATGGACGGCCTGCATGTATCCTCCGACTCGGCAAGCCTCCCGTTCGCACATTCCAACCACTCTGCAATCCAGTTCATAA

       610       620       630       640       650       660       670       680       690       700       710       720
ATCCATAAAGCCCTCTTCGCTCCCGCCGGCATCCCCGGCAATCCCCGACACCCTAGGCCCCGCGCTCTCCGCGGCTGCCCGACAAGAAAACCAACCGCTCGATCAGCCTCATCC

                     730       740       750       760       770       780       790       800       810       820       830       840
TTCACCCATCACAGGAGCCATCGACTGATCATGACCATTGGATCCCCGTTCCCGTTCCAGGCTGCCCGGGCGTCGTCCGGGCCGGCTCCGCCGCCGAGGAAGCCTTCGACCT
             MetHisLeuIleProHisTrpIleProLeuValAlaSerLeuGlyLeuLeuAlaGlyLyrSerSerAlaAlaAlaGluGluAlaPheAspLe

        850       860       870       880       890       900       910       920       930       940       950       960
CTGGAACGAATGCGCCAAAGCCTGCGTGCTCGACCTCAAGACGGGCTGCGTTCCAGCCGACGTGCCGTTCCAGCCGCCATGGCGTCGACCCGCCATCGCCGACACCAACCGGCAACCGCAGGCCTTGCTGCACTACTCCAT
uTrpAsnGluCysAlaLysAlaCysValLeuAspLeuLysThrGlyCysValProAlaIleAlaAspThrAsnArgSerSerAlaArgGlyValAlaGlyIleAlaGly

                970       980       990      1000      1010      1020      1030      1040      1050      1060      1070      1080
GGTCCTGGAGGGCGGCAACGACGGCCTCAAGCTGCCATCGACACAACGCCCTCAGCATCCAGCGACGGCCGTGACCATCGCCGCCTCGAAGGCCGGCGTGGAGCGCAACAAGCCGGTGCGCTA
tValLeuGluGlyGlyAsnAspAlaLeuLysLeuAlaIleAspAsnAlaLeuSerIleThrSerAspGlyLeuThrIleArgLeuGluGlyLeuGluProAsnLysProValArgTy

               1090      1100      1110      1120      1130      1140      1150      1160      1170      1180      1190      1200
CAGCTACACGCCGCCAGGCGCGCCGAGTTGGTCGCTGAACTGGCTGTACCGATGGGGCCCACGAGAAGCCCTCGAACATCAAGGTGTTCATCCACGAACTGAACGCGCAACCAGCTCAG
rSerTyrThrArgGlnAlaArgGlySerTrpSerLeuAsnTrpLeuValProIleGlyHisGluLysProSerAsnIleLysValPheIleHisGluLeuAsnAlaGlyAsnGlnLeuSe

           1210      1220      1230      1240      1250      1260      1270      1280      1290      1300      1310      1320
CCACATGTCGCCGATCTACACCATCGAGATGGGCGACGAGTTGCTGGCGAAGCTGCCGCCGATGCCACCTTCTTCGTCCAGGGCGCACGAGAGCAACGAGATGCAGCCGCAGCGCCGCCAT
rHisMetSerProIleTyrThrIleGluMetGlyAspGluLeuLeuAlaLysLeuAlaLysLeuAlaArgAspAlaThrPhePheValArgAlaHisGluSerAsnGluMetGlnProThrLeuAlaIl

       1330      1340      1350      1360      1370      1380      1390      1400      1410      1420      1430      1440
CAGCCATGCCGGGGTCAGCGTGGTCATGGCCCAGACCGCGCCGGGAAAAGCGCTGGAGGCGAATGGGCCCAGCGCCAAGGTGTTGTGCCTCGGACCCGGTTGGCTCACAA
eSerHisAlaGlyValSerValValMetAlaGlnThrGlnProAlaArgGlutArgGluLysAsrGluTrpAlaSerGlyLysValLeuCysLeuAspProLeuAspGlyValTyrAs

   1450      1460      1470      1480      1490      1500      1510      1520      1530      1540      1550      1560
CTACCTCGCCCAGCAACGCTGCAACCTGGGAAGGCAAGATCTACCGGGTGCTCGCCGGCAACCCGGCGAAGCATGAAGCCAGGTCATCAGTCATCG
nTyrLeuAlaGlnGlnArgCysAsnLeuAspAspThrTrppGluGlyLysIleTyrArgValLeuAlaGlyAsnProAlaLysHisAspLeuGlyAsnProLeuAspProLeuTyrArg
```

```
     1570      1580      1590      1600      1610      1620      1630      1640      1650      1660      1670      1680
CCTGCACTTTCCCGAGGCGGCAGCCTGGCCGCGCGACCGGCCACCAGGCTTGCCACCTGCCGCTGGAGACTTTCACCCGTCATCGCCAGCCGCGGCGGCTGGAACAACTGGAGCAGTG
gLeuHisPheProGluGlyGlySerLeuAlaAlaLeuThrAlaHisGlnAlaCysHisLeuProLeuGluThrPheThrArgHisArgGlnProArgGlyTrpGlnLeuGluGlnCy

     1690      1700      1710      1720      1730      1740      1750      1760      1770      1780      1790      1800
CGGCTATCCGGTGCAGGCGTGGTCGCCCTCTACCTGCCGCGGCGGCTGTCGTGGAACCAGGTCGACCAGGTTATCGCCAACGCCCTGGCCCAGCCGGCGGCGCCGACCTGGGCCGA
sGlyTyrProValGlnAlaTrpLeuAlaLeuTyrLeuArgLeuAlaAlaArgLeuSerTrpAsnGlnValIleArgAsnAlaLeuAlaSerProGlyGlyAspLeuGlyGl

     1810      1820      1830      1840      1850      1860      1870      1880      1890      1900      1910      1920
AGCGATCCGGAGCAGGCCGGACAGGCCCGTCTGGCCCTGACCCTGGCCGCGGCGGAGAGCGAGCGGTTCGTCCGGCAGGGCACCGGCAACGACGAGGCCGGCGGCCAAACCGCGACGT
uAlaIleArgGluGlnProGluGlnAlaGlyGlnAlaArgLeuAlaLeuThrLeuAlaAlaAlaGluSerGluArgGlyArgGlyGlyAsnAspGluAlaGlyAlaAlaAsnAlaAspVa

     1930      1940      1950      1960      1970      1980      1990      2000      2010      2020      2030      2040
GGTGAGCCTGACCTGCCCGGTGAATGCGGCCCGGCGGACAGCGGCGACGCCCTGCTGGAGCGACCAGCTATCCCACTGGCCGCAGTTCCTCGGCGACGGGCCGGCAGCGT
lValSerLeuThrCysProValAlaAlaGlyGluCysAlaGlyGlyProLaAspSerGlyAspAlaLeuLeuGluArgAsnTyrProThrGlyAlaGluPheLeuGlyAspGlyGlyAspVa

     2050      2060      2070      2080      2090      2100      2110      2120      2130      2140      2150      2160
CAGCTTCAGCACCCGCGGCCAGCAGACGGTGGAGCCGGCTGCCAGGGCACCGCCAACTGGAGGAGCGCGGCTATGTGTTCGTGGTACCAGGCACCTTCCTCGAAGCGGC
lSerPheSerThrArgGlyThrGlnAsnTrpThrValGlnLeuGluGlnAlaHisArgGlnLeuGluGlyArgGlyTyrValPheValGlyTyrHisGlyThrPheLeuGluAlaAlaAl

     2170      2180      2190      2200      2210      2220      2230      2240      2250      2260      2270      2280
GCAAAGCATCGTCTTCGGCGGCGGGTGCCGCGCCAGGACCTCGACGCGCGGTTCTATATCGCCGGCGGCGATCCGGCGCTGCCCTACGGCCTACGGCCAGGACCAGGAACC
aGlnSerIleValPheGlyGlyGlyValAlaArgAlaArgSerGlnAsnAspLeuAspAlaIleThrArgGlyPheTyrIleAlaGlyGlyAspProAlaLeuAlaTyrGlyTyrAlaGlnAspGlnGluPr

     2290      2300      2310      2320      2330      2340      2350      2360      2370      2380      2390      2400
CGACCCACGCGCCGGATCCGCCAACGGTGCCCTGCTGCCGGTCTATGTGCCGCGCTGCGCCGACCAGCCTGCCGAGCCTGCCGAGCGGCGGCGGGCGGGCCAGGT
oAspProAlaArgGlyArgIleArgAsnGlyAlaLeuLeuArgValTyrValProArgSerCysProArgGlyTyrPheThrSerLeuThrLeuAlaAlaAlaProGlyLeuAlaLaGlyGluVa

     2410      2420      2430      2440      2450      2460      2470      2480      2490      2500      2510      2520
CGAACGGCTGATCGGCCATCCCGTCGCCCTGCGCGTGGACGCCATCACCGGCCATCGCTGGTGGCCCGGCGAGGAGGAAGGCGGGCGCGCCTGGAGACCATTCTCCGGCGCTGGCCGTGGCCGAGCGCACCGTGGTGAT
lGluArgLeuIleGlyHisProValAlaLeuArgValAspAlaIleThrGlyHisArgTrpTrpProGlyGluGluGluGlyGlyArgLeuGluThrIleLeuGlyTrpProLeuAlaGluArgThrValValIl

     2530      2540      2550      2560      2570      2580      2590      2600      2610      2620      2630      2640
TCCCTCGGCGATCCCCACCGACCCCGGCAACGTCGGCGGCGACCTCGACCCCCGTTCGAGCCTCCCGACCAAGGAACAGGCGATACCCGCCCTGCCGGACTACGCCCAGCCCGGCCAAACC
eProSerAlaIleProThrAspProGlyAsnValGlyGlyAspLeuAspProArgSerIleProAspLysGlyGlnAlaIleSerAlaLeuProAspTyrAlaSerGlnProGlyLysPr

     2650      2660      2670      2680      2690      2700      2710      2720      2730      2740      2750      2760
GCCGGCGAGGACCTGAAGTAACTGCCCGACCGGCCGGCTCCCTTCGCAGGAGCCGGCCTTCTCGGGGGCCTGGCCATACATCAGGTTTTCCTGATGCCAGCCCAATCGAATATGAATTC
oProArgGluAspLeuLysEnd
```

Figure 2 Nucleotide sequence of the *PstI–Eco*RI fragment containing the structural gene for exotoxin A. The sequence is from Gray et al. (15). The amino acid sequence of the toxin is indicated underneath the nucleotide sequence.

Table 1   Directory of Plasmids

Plasmid	Description
pUC9	Derived from pBR322 by Vieira and Messing (37).
pRO1614	Contains a 1.85-kbp *Pst*I fragment from the plasmid RP1 in the *Pst*I site of pBR322. The 1.85-kbp fragment allows replication of the plasmid in *Pseudomonas*. Amp^{r}Tet^r (34).
pRC345	Constructed by cloning a 17.7-kbp fragment of PA103 chromosomal DNA into the unique *Bam*HI site of pRO1614. The 17.7-kbp fragment carries part of *tox*A, from nucleotides 1753–2760 of the 2.76-kbp *Pst*I–*Eco*RI fragment in Figure 1. Amp^{r}Tet^r (24).
pRC354	Constructed by cloning the 1.85-kbp *Pst*I fragment of pRO1614 into the *Pst*I site of pUC9. Amp^r (36).
pRC357	Derived from pRC354 by deleting the *Pst*I site distal to the unique *Eco*RI site of pUC9. Amp^r (36).
pRC350	Constructed by cloning into pUC9 the 2.76-kbp *Pst*I–*Eco*RI fragment from PA0286 chromosomal DNA. Amp^r (36).
pRC351	Constructed by cloning the 1.85-kbp *Pst*I fragment of pRO1614 into the *Pst*I site of pRC350. Amp^r (36).
pRC352	Derived from pRC351 by deleting the *Pst*I site distal to the unique *Eco*RI site of pUC9. Amp^r (36).
pRC361	Constructed by cloning a 10-kbp *Eco*RI fragment from the chromosome of PA103J into the unique *Eco*RI site of pUC9. The 10-kbp fragment includes the 2.76-kbp *Pst*I–*Eco*RI fragment carrying *tox*A. Amp^r (36).
pRC360	Constructed by cloning the 2.76-kbp *Pst*I–*Eco*RI fragment of pRC361 into pUC9. Amp^r (36).
pRC362	Constructed by cloning the 2.76-kbp *Pst*I–*Eco*RI fragment of pRC361 into pRC357. Amp^r (36).

pRC362 (see Table 1). Since *Pst*I did not cleave the chromosomal DNA from our PA103, and since it did cleave the cloned 10-kbp *Eco*RI fragment from the same strain after replication in *E. coli*, we concluded that DNA from the PA103 strain we used as the source of the toxin structural gene was altered, probably by a restriction site modification system, so that it did not cleave with *Pst*I. Thus, our strain of PA103 appears different than that of Gray et al. (15).

Gray et al. (15) compared the deduced sequence of exotoxin A with the partial, chemically determined sequence of the secreted toxin. This analysis revealed that the protein is synthesized as a single polypeptide of 638 amino acids, of which the first 25 constitute the signal sequence. The mature, secreted toxin lacks the signal sequence, apparently cleaved during secretion. The cloned fragments encoding the C-terminal region of the toxin expressed enzymic activity (15,24), suggesting that the catalytic center is located in the C-terminal region of the protein. This is unlike diphtheria toxin, in which the catalytically active region is located in the N-terminal fragment A.

We sequenced two segments of the cloned 2.76-kbp fragment carrying *toxA* from strains PA103 or PA0286, one from nucleotides 553–961 and the other from nucleotides 2020–2560 (36). The sequence of both fragments from PA103 was identical to that reported by Gray et al. (15) for PA103. The sequence of the two fragments from PA0286, however, differed from that of PA103 at five sites in the 553–961 region and four in the 2020–2560 region. Of these, three differences are in the upstream region and six in the structural gene. Four differences in the structural gene lead to amino acid substitutions; isoleucine 3 of the signal sequence in PA103 for threonine in PA0286, serine 16 of the signal sequence in PA103 for phenylalanine in PA0286, valine 407 of mature toxin in PA103 for isoleucine in PA0286, and serine 515 of the mature toxin in PA103 for glycine in PA0286. Thus, the amino acid sequence of the toxin from the two strains of *Pseudomonas aeruginosa* is very similar, but not identical. Further, these amino acid differences affect neither the biological activity of the toxin nor its secretion from the native strains (36).

III. EXPRESSION SYSTEMS

We routinely isolate exotoxin A from nontoxigenic *Pseudomonas* hosts carrying the cloned *toxA* gene. This has several advantages: (1) large amounts of the toxin are synthesized, presumably because the cloned gene is on a multicopy plasmid; (2) exotoxin A is well over 50% of the total protein secreted, which facilitates purification; (3) the secretion of toxin variants with dipeptide insertions or single amino acid deletions remains unimpaired, although the secretion of variants with larger deletions is impaired; and (4) culture supernatants can be tested directly to screen for variants with reduced cytotoxicity (39). We have developed a variety of hosts and vectors useful in expressing the exotoxin A gene.

A. *Pseudomonas* Hosts

We have made nontoxigenic *Pseudomonas* hosts from both strains PA01 and PA103. MAM2 was obtained after nitrosoguanidine mutagenesis of PA0286, an auxotrophic derivative of the wild-type strain PA01. The failure to secrete exotoxin A was detected by the Elek test (4), an immunodiffusion technique, and later confirmed by a radioimmune assay and Western blot analysis of MAM2 culture supernatants and cell lysates. The mutation responsible for the phenotype of MAM2 mapped to 85 min, near the toxin structural gene, but the precise nature of the lesion in MAM2 remains undetermined (24). MAM4, a Rec$^-$ derivative of MAM2 (36), is the host we have used extensively to express the cloned wild-type toxin gene as well as its mutant forms.

We recently derived a nontoxigenic host from PA103 by replacing the chromosomal *toxA* gene with the Tet^r gene. To do this, the plasmid pRC360Δ*toxA*1 was constructed from pRC360 (see Table 1) by substituting a major part of *toxA*, as well as the upstream control sequence on the 2.76-kbp fragment, with Tet^r gene from pBR322. pRC360Δ*toxA*1 also carries the Amp^r gene of pUC9, but lacks the 1.85-kbp *Pst*I fragment required for replication in *Pseudomonas*. The rationale for making the nontoxigenic host by this method is that Tet^{r}Amp^s progeny will result only if the Tet^r replaces a segment of the chromosomal *toxA* by homologous recombination. Since pRC360Δ*toxA*1 cannot replicate in PA103, it would impart antibiotic resistance only by integrating into the host chromosome. We hybridized the chromosomal DNA of several Tet^r-Amp^s clones with probes specific for the *toxA* gene and the Tet^r gene. Whereas the chromosomal DNA of the parental strain, PA103, hybridized with the *toxA* probe but not the Tet^r probe, the chromosomal DNA of several Tet^r-Amp^s clones hybridized with the Tet^r probe but not the *toxA* probe. The final strain, PA103Δ*toxA*1, secreted no toxin, whereas the parental strain grown under identical conditions secreted large amounts of toxin. We concluded that PA103Δ*toxA*1 was nontoxigenic because a major segment of its chromosomal *toxA* structural gene, along with the upstream regulatory sequence, had been substituted with the Tet^r gene.

B. Vectors

We have constructed a variety of shuttle plasmids suitable for expressing *toxA* in both *E. coli* and *P. aeruginosa* (24,36). Most plasmids are based on the *E. coli* cloning vector pUC9 and contain the 1.85-kbp *Pst*I fragment from pRO1614 for replication in *P. aeruginosa*. A description of these plasmids is provided in Table 1.

IV. TRANSCRIPTION AND REGULATION OF *toxA*

A. Transcription Regulation in Procaryotes

Since the discovery in 1966 that the lactose repressor protein is responsible for the selective induction of β-galactosidase by *E. coli* in the presence of lactose, extensive studies have been made both on the lactose repressor, the bacteriophage lambda repressor, and other bacterial gene-regulating proteins. Sequence-specific DNA-binding proteins either inhibit or stimulate initiation of RNA synthesis by binding next to the promoter. The repressor binds to the operator sequence preventing RNA polymerase from starting RNA synthesis at the promoter, thereby blocking transcription of the adjacent region of DNA. In contrast, an activator binds next to a promoter and improves RNA polymerase binding to the promoter and facilitates transcription. Examples of prokaryotic activators include the catabolite activator protein and the ntrC protein of *E. coli*. The binding sites for bacterial regulatory proteins can be remote from the promoter. This is accomplished by DNA loops to bring the regulatory protein near the promoter. The *Pseudomonas* exotoxin gene apparently has an upstream sequence for binding by an activator as described below.

Transcription

The transcriptional start site for the exotoxin A gene has been investigated by S1 nuclease mapping. Grant and Vasil (40) reported start sites at positions 657 and 684 on the cloned 2.76-kbp *Pst*I–*Eco*RI fragment of strain PA103 (Figure 1). Work from our laboratory (36) using a 1488-bp *Pst*I–*Bgl*II fragment from pRC352 (see Table 1) as a probe initially indicated that the transcriptional start site was between positions 615 and 675 on the *Pst*I–*Eco*RI *toxA* fragment. We then hybridized a smaller probe, the 357-bp *Pvu*I–*Sal*I fragment 5′ end labeled at the *Sal*I end, with mRNA isolated from MAM4(pRC352). Digestion with S1 nuclease suggested the initiation of transcription occurred between positions 645 and 665. This probe also yielded identical results with mRNA from both PA103 and PA0286, although the bands were much less intense than with MAM4(pRC352), consistent with the fact that the two wild-type strains produce much less toxin than does MAM4 carrying *toxA* on a multicopy plasmid. We finally used an even smaller probe to precisely determine the position of the transcriptional start site. The 214-bp *Pvu*I–*Bam*HI fragment from pRC360, 5′ end labeled at the *Bam*HI end, was hybridized with mRNA from MAM4 carrying pRC362, differing from pRC360 only in that it has the 1.85-kbp *Pst*I fragment from pRO1614 for replication in *Pseudomonas* (see Table 1). S1 mapping indicated that the initiation site for exotoxin A mRNA was at position 658, 88 bp upstream of the translational start site. This site is within 1 bp of the longer transcript reported by Grant and Vasil (40). The role of the smaller transcript reported by Grant and Vasil (40) is unclear at present.

Regulation

Exotoxin A biosynthesis is under the control of iron. Evidence from several groups has suggested that at least part of the regulation is at the transcriptional level (36,40,41), although other factors, such as mRNA turnover, may also be involved (41). The negative control of *toxA* expression by iron in MAM4 carrying pRC352 or pRC362 (see Table 1) is retained (36), suggesting the target sequence for this regulation, directly or indirectly, is between nucleotides 1 and 745 of the cloned 2.76-kbp *Pst*I–*Eco*RI fragment (see Figure 1). To identify upstream regions involved in regulating *toxA* expression, we deleted DNA in this region by Bal31 digestion (42). Plasmids harboring the deletions were expressed in MAM4 under low iron conditions and toxin in culture supernatants was quantitated by a radioimmune assay. Deletions between nucleotides 1 and 585 of the *Pst*I–*Eco*RI fragment had no effect on toxin production, while some deletions between 585 and the transcriptional start site at 658 abolished expression, indicating that the 72-bp region between 585 and 658 contained *toxA* transcription regulatory sequences (42).

Critical sequences within the 72-bp region were identified by linker scanning, a mutagenesis technique in which the original DNA sequence is changed by substituting it with synthetic oligonucleotides (43). Using the *Eco*RI linker pGGAATTCC, a series of mutants was constructed such that in any given mutant the linker replaced the original sequence at a particular site within the 72-bp region. Plasmids containing the linkers were expressed in MAM4 and the toxin produced was assayed by a radioimmune assay. Regions -72 to -54 and -9 to -2 within the 72-bp upstream sequence were absolutely critical for the toxin gene expression; linkers substituted in these regions reduced expression to only 3–5% that of wild type. Substitutions in the -53 to -14 region reduced expression to no more than 50%, although simultaneous substitution and addition of 4 nucleotides in the -53 to -39 region reduced expression to 5% that of the wild type (42).

One model to explain these findings is that a *Pseudomonas* RNA polymerase binds sequences centered at -41 and -6 and a transcription regulatory factor binds the -72 to -54 region. That the putative regulator-binding sequence is required for expression suggests the regulator is an activator, not a repressor, of transcription. The unknown activator may bind first and facilitate the binding of RNA polymerase and the subsequent initiation of transcription (42).

Recent studies have indicated that a positive activator, the *toxR* product, is required for efficient expression of *toxA* (44,45). *ToxR* is a polypeptide of about 25 kd, which may be the positive regulator of *toxA* (45).

But so far, the putative protein remains uncharacterized. Recently, Hamood et al. (46) expressed the cloned *toxR* (also called *regA*) gene under a T7 promoter in *E. coli* and identified a 29-kd polypeptide product. The putative *toxR* product, however, appeared to be membrane associated and failed to bind the *toxA* upstream sequence (46). Thus, the *toxR* product itself may not act by directly binding to regulatory regions of *toxA*.

V. MUTATIONAL ANALYSIS

Our interest in studying exotoxin A by mutational analysis is threefold: (1) to generate and characterize structural variants of the toxin to elucidate function; (2) to produce variants that have significantly reduced cytotoxicity because of impaired receptor binding; and (3) to assess the possibility that variants impaired in receptor binding may be useful in preparing hybrid toxins. Our initial approach was to generate deletion mutants. We used maintenance of secretion by the *Pseudomonas* host MAM4 as a biological quality-control test to assess whether the mutated proteins adopted a conformation related to the native toxin; if secretion were impaired, the altered toxins may not have retained sufficient resemblance to the native toxin for meaningful tests of function. Most of the deletion variants were not secreted (39), leading us to use other techniques for generating structural variants of the toxin. Note, however, that extensive studies have been done with deletion mutants extracted and renatured from the cytosol of *E. coli* (25,26) that have supported the suggestion that domain I is involved in receptor recognition and domain II in membrane penetration (23).

A. TAB Linker Mutants

Failing to obtain deletion mutants that were secreted by the *Pseudomonas* host MAM4, we turned to the dipeptide insertional technique of Barany called TAB (two amino acid Barany) linker mutagenesis (47). In this technique, a hexanucleotide encoding a dipeptide is inserted at an existing restriction enzyme site, either retaining or destroying this site, but always generating a new one. The dipeptide the hexanucleotide encodes depends on whether it is inserted after the first, the second, or the third base of the existing codon. The original reading frame of the gene, however, is retained in all three cases.

When choosing a restriction enzyme and a hexanucleotide for TAB linker mutagenesis, three considerations are important: (1) restriction sites for the enzyme should be numerous and widely distributed in the DNA segment to be mutagenized so that a variety of mutants can be generated; (2) the hexanucleotide chosen should insert a dipeptide predicted to perturb function; and (3) the hexanucleotide chosen should insert a restriction site, ideally unique, that can be easily mapped. For the TAB linker mutagenesis of exotoxin A structural gene, we chose the restriction enzyme *Taq*I. This enzyme has 10 sites in domain I, 1 in domain II, and 5 in domain III. Since there are 10 sites in domain I, *Taq*I is a good choice for generating domain I structural variants. TAB linker mutagenesis of the exotoxin A structural gene was done as follows (39):

1. pRC362 was linearized by partial cleavage with *Taq*I.
2. Either the hexanucleotide pCGAGCT, which generates an *Sst*I site, or pCGAATT, which generates an *Eco*RI site, was ligated with the linearized pRC362.
3. The resulting plasmid was then cleaved with either *Sst*I or *Eco*RI to remove excess linkers that may have been ligated as multimers.
4. A Km^r gene cassette, excised from pUC4-KISS (Pharmacia, Piscataway, N.J.) to give ends compatible with the newly created site, was inserted into the cleaved plasmid.
5. The plasmid was transformed in *E. coli* HB101 and the mutants selected on Km plates. Next, the plasmid DNA from Km^r clones was isolated and the site of the hexanucleotide insertion in each determined by restriction enzyme mapping.
6. The Km^r cassette was excised from the plasmids of interest and the resulting plasmids were transformed into the nontoxigenic recipient MAM4 for further characterization.

B. Characterization of Variants

Potentially interesting variants were initially screened for secretion by MAM4 using the immunological Elek test (4). The secreted toxins were then quantitated more accurately by a radioimmune assay with culture supernatants. All the dipeptide insertion derivatives of exotoxin A examined have been secreted normally by MAM4. The cytotoxicity of variants secreted into culture supernatants was measured by the effect on protein synthesis with mouse LMTK⁻ cells to identify variants that had reduced cytotoxicity. We identified several such variants. One of these, with Glu-Leu inserted between the residues 60 and 61, designated ETA-60EL61, had a 100-fold reduction in cytotoxicity. We then inserted a second Glu-Leu at the same site, hoping it would further reduce cytotoxicity, but the variant ETA-60ELEL61 proved no less cytotoxic than ETA-60EL61. Next, we considered the possibility that insertion at this site of an amino acid with a larger side chain, such as phenylalanine, might impair cytotoxicity more than 100-fold. We chose the hexanucleotide pCGAATT because it would insert a Glu-Phe between residues 60 and 61. This rationale proved valid; the resulting variant, ETA-60EF61, was about 500-fold less cytotoxic than normal toxin after purification from culture supernatants.

To investigate whether the lesion in ETA-60EF61 impaired interaction with receptors, mouse LMTK$^-$ cells were incubated at 4°C for 2 hr with either the variant or the wild-type toxin to allow receptor binding. Enough of each toxin was used to reduce protein synthesis by 60–80% if the cells were immediately shifted to 37°C after removing the unbound toxin to allow toxin action. The cells were washed to remove unbound toxin molecules and incubation at 4°C was continued for different lengths of time. During this time, the toxins could dissociate from the cells and a more rapid dissociation of the variant toxin would indicate a defect in receptor interaction. At the end of each time period the cells were washed, incubated at 37°C for 4 hr, and protein synthesis was assessed. The results indicated that protein synthesis in cells treated with wild-type toxin was inhibited even after prolonged incubation at 4°C, suggesting that normal toxin dissociated slowly from receptors. However, cells treated with ETA-60EF61 rapidly recovered the ability to synthesize protein, suggesting that the variant toxin dissociated rapidly from the cell surface (39). We concluded that the dipeptide insertion in domain I between residues 60 and 61 adversely affected the affinity of the toxin for cell surface receptors.

Jinno et al. (48) reported changing each of the 11 lysine residues in domain I to glutamic acid residues and found that only substitution of lysine 57 with glutamic acid reduced cytotoxicity for mouse cells, and their evidence suggested that this substitution impaired the receptor binding function of the toxin. The proximity of this change to the insertion site in ETA-60EF61 emphasizes the importance of this region of domain I in receptor interaction. But how the mutations in this region perturb the structure to reduce the toxin's affinity for the cell surface receptors is unclear. One of the major structural features of domain I, however, is a cavity between the loopout projections formed by residues 20/21 and 81/82. The surface of this cavity is lined by three antiparallel beta-strands formed by residues 55–61, 63–69, and 72–80. Thus, one way a mutation in this region of domain I could impair receptor binding is by perturbing the antiparallel arrangement of these beta-strands. The total or partial loss of this arrangement might impair receptor binding by displacing an amino acid that interacts with the receptor, or by perturbing the overall structure to affect the contact points between the receptor and the toxin.

DEDICATION

We dedicate this chapter to the memory of Royston C. Clowes.

REFERENCES

1. Liu, P. V. Extracellular toxins of *Pseudomonas aeruginosa*. J. Infect. Dis., *130*: 594–599, 1974.
2. Cross, A. S., Sadoff, J. C., Iglewski, B. H., Sokol, P. A. Evidence for the role of toxin A in the pathogenesis of infection with *Pseudomonas aeruginosa* in humans. J. Infect. Dis., *142*: 538–546, 1980.
3. Young, L. S. The role of exotoxins in the pathogenesis of *Pseudomonas* infections. J. Infect. Dis., *142*: 626–630, 1980.
4. Bjorn, M. J., Vasil, M. L., Sadoff, J. C., and Iglewski, B. H. Incidence of exotoxin A production by *Pseudomonas* species. Infect. Immun., *16*: 362–365, 1977.
5. Liu, P. V. The role of various fractions of *Pseudomonas aeruginosa* in its pathogenesis. III. Identity of the lethal toxins produced *in vitro* and *in vivo*. J. Infect. Dis., *116*: 481–489, 1966.
6. Leppla, S. H. Large scale purification and characterization of the exotoxin of *Pseudomonas aeruginosa*. Infect. Immun., *14*: 1077–1086, 1976.
7. Holloway, B. W., Krishnapillai, V., and Morgan, A. F. Genetics of *Pseudomonas*. Microbiol. Rev., *43*: 73–102, 1979.
8. Iglewski, B. H., and Sadoff, J. C. Toxin inhibitors of protein synthesis: production, purification, and assay of *Pseudomonas aeruginosa* toxin A. Methods Enzymol., *68*: 780–793, 1979.
9. Middlebrook, J. L., and Dorland, R. B. Response of cultured mammalian cells to the exotoxins of *Pseudomonas aeruginosa* and *Corynebacterium diphtheriae*: Differential cytotoxicity. Can. J. Microbiol., *23*: 183–189, 1977.
10. Iglewski, B. H., and Kabat, D. NAD-dependent inhibition of protein synthesis by *Pseudomonas aeruginosa* toxin. Proc. Natl. Acad. Sci. U.S.A., *72*: 2284–2288, 1975.
11. Vasil, M. L., Kabat, D., and Iglewski, B. H. Structure-activity relationships of an exotoxin of *Pseudomonas aeruginosa*. Infect. Immun., *16*: 353–361, 1977.
12. Iglewski, B. H., Liu, P. V., and Kabat, D. Mechanism of action of *Pseudomonas aeruginosa* exotoxin A: Adenosine diphosphate-ribosylation of mammalian elongation factor 2 *in vitro* and *in vivo*. Infect. Immun., *15*: 138–144, 1977.
13. Van Ness, B. G., Howard, J. B., and Bodley, J. W. ADP-ribosylation of elongation factor 2 by diphtheria toxin. J. Biol. Chem., *255*: 10717–10720, 1980.
14. Dunlop, P. C., and Bodley, J. W. Biosynthetic labeling of diphthamide in *Saccharomyces cerevisiae*. J. Biol. Chem., *258*: 4754–4758, 1983.
15. Gray, G. L., Smith, D. H., Baldridge, J. S., Harkins, R. N., Vasil, M. L., Chen, E. Y., and Heneyker, H. L. Cloning, nucleotide sequence, and expression in *Escherichia coli* of the exotoxin A structural gene of *Pseudomonas aeruginosa*. Proc. Natl. Acad. Sci. U.S.A., *81*: 2645–2649, 1984.
16. Carroll, S. F., and Collier, R. J. Active site of *Pseudomonas aeruginosa* exotoxin A. Glutamic acid 553 is photolabeled by NAD and shows functional

homology with glutamic acid 148 of diphtheria toxin. J. Biol. Chem., *262*: 8707–8711, 1987.

17. Carroll, S. F., and Collier, R. J. Amino acid sequence homology between the enzymic domains of diphtheria toxin and *Pseudomonas aeruginosa* exotoxin. A. Mol. Microbiol., *2*: 293–296, 1988.

18. Douglas, C. M., and Collier, R. J. Exotoxin A of *Pseudomonas aeruginosa*: Substitution of glutamic acid 553 with aspartic acid drastically reduces toxicity and enzymatic activity. J. Bacteriol., *169*: 4967–4971, 1987.

19. Lukac, M., Pier, G. B., Collier, R. J. Toxoid of *Pseudomonas aeruginosa* exotoxin A generated by deletion of an active-site residue. Infect. Immun., *56*: 3095–3098, 1988.

20. Carroll, S. F., and Collier, R. J. NAD binding site of diphtheria toxin: Identification of a residue within the nicotinamide subsite by photochemical modification with NAD. Proc. Natl. Acad. Sci. U.S.A., *81*: 3307–3311, 1984.

21. FitzGerald, D., Morris, R. E., and Saelinger, C. B. Receptor-mediated internalization of *Pseudomonas* toxin by mouse fibroblasts. Cell, *21*: 867–873, 1980.

22. Leppla, S. H., Martin, O. C., and Muehl, L. A. The exotoxin of *P. aeruginosa*: A proenzyme having an unusual mode of activation. Biochem. Biophys. Res. Commun., *81*: 532–538, 1978.

23. Allured, V. S., Collier, R. J., Carroll, S. F., McKay, D. B. Structure of exotoxin A of *Pseudomonas aeruginosa* at 3.0-angstrom resolution. Proc. Natl. Acad. Sci. U.S.A., *83*: 1320–1324, 1986.

24. Mozola, M. A., Wilson, R. B., Jordan, E. M., Draper, R. K., and Clowes, R. C. Cloning and expression of a gene segment encoding the enzymatic moiety of *Pseudomonas aeruginosa* exotoxin A. J. Bacteriol., *159*: 683–687, 1984.

25. Hwang, J., FitzGerald, D. J., Adhya, S., and Pastan, I. Functional domains of *Pseudomonas* exotoxin identified by deletion analysis of the gene expressed in *E. coli.* Cell, *48*: 129–136, 1987.

26. Guidi-Rontani, C., and Collier, R. J. Exotoxin A of *Pseudomonas aeruginosa*: Evidence that domain I functions in receptor binding. Mol. Microbiol., *1*: 67–72, 1987.

27. Madshus, I. H., and Collier, R. J. Effects of eliminating a disulfide bridge within domain II of *Pseudomonas aeruginosa* exotoxin A. Infect. Immun., *57*: 1873–1878, 1989.

28. Siegall, C. B., Chaudhary, V. K., FitzGerald, D. J., and Pastan, I. Functional analysis of domains II, Ib, and III of *Pseudomonas* exotoxin. J. Biol. Chem., *264*: 14256–14261, 1989.

29. Jinno, Y., Ogata, M., Chaudhary, V. K., Willingham, M. C., Adhya, S., FitzGerald, D., and Pastan, I. Domain II mutants of *Pseudomonas* exotoxin deficient in translocation. J. Biol. Chem., *264*: 15953–15959, 1989.

30. Brandhuber, B. J., Allured, V. S., Falbel, T. G., and McKay, D. B. Mapping the enzymatic active site of *Pseudomonas aeruginosa* exotoxin A. Proteins, *3*: 146–154, 1988.

31. Wozniak, D. T., Hsu, L.-Y., and Galloway, D. R. His-426 of the *Pseudomonas aeruginosa* exotoxin A is required for ADP-ribosylation of elongation factor II. Proc. Natl. Acad. Sci. U.S.A., *85*: 8880–8884, 1988.

32. Chaudhary, V. K., Jinno, Y., FitzGerald, D., and Pastan, I. *Pseudomonas* exotoxin contains a specific sequence at the carboxyl terminus that is required for cytotoxicity. Proc. Natl. Acad. Sci. U.S.A., *87*: 308–312, 1990.

33. Hanne, L. F., Howe, T. R., and Iglewski, B. H. Locus of the *Pseudomonas aeruginosa* toxin A gene. J. Bacteriol., *154*: 383–386, 1983.

34. Olsen, R. H., DeBusscher, G., and McCombie, W. R. Development of broad-host-range vectors and gene banks: Self-cloning of the *Pseudomonas aeruginosa* PAO chromosome. J. Bacteriol., *150*: 60–69, 1982.

35. Boyer, H. W., and Roulland-Dussoix, D. A complementation analysis of the restriction and modification of DNA in *Escherichia coli*. J. Mol. Biol., *41*: 459–472, 1969.

36. Chen, S. T., Jordan, E. M., Wilson, R. B., Draper, R. K., and Clowes, R. C. Transcription and expression of the exotoxin A gene of *Pseudomonas aeruginosa*. J. Gen. Microbiol., *133*: 3081–3091, 1987.

37. Vieira, J., and Messing, J. The pUC plasmids, an M13mp7-derived system for insertion mutagenesis and sequencing with synthetic universal primers. Gene, *19*: 259–268, 1982.

38. Stanisich, V. A., and Holloway, B. W. Conjugation in *Pseudomonas aeruginosa*. Genetics, *61*: 327–339, 1969.

39. Chaudry, G. J., Wilson, R. B., Draper, R. K., and Clowes, R. C. A dipeptide insertion in domain I of exotoxin A that impairs receptor binding. J. Biol. Chem., *264*: 15151–15156, 1989.

40. Grant, C. C. R., and Vasil, M. L. Analysis of Transcription of the exotoxin A gene of *Pseudomonas aeruginosa* PAO. J. Bacteriol., *168*: 1112–1119, 1986.

41. Lory, S. Effect of iron on accumulation of exotoxin A-specific mRNA in *Pseudomonas aeruginosa*. J. Bacteriol., *168*: 1451–1456, 1986.

42. Meei-Ling, T., and Clowes, R. C. Localization of the control region for expression of exotoxin A in *Pseudomonas aeruginosa*. J. Bacteriol., *171*: 2599–2604, 1989.

43. McKnight, S. L., and Kingsbury, R. Transcription control signals of a eukaryotic protein-coding gene. Science, *217*: 316–324, 1982.

44. Hedstrom, R. C., Funk, C. R., Kaper, J. B., Pavlovskis, G. R., and Galloway, D. R. Cloning of gene involved in regulation of exotoxin A expression in *Pseudomonas aeruginosa*. Infect. Immun., *51*: 37–42, 1986.

45. Wozniak, D. J., Cram, D. C., Daniels, C. J., and Galloway, D. R. Nucleotide sequence and characterization of *toxR*: A gene involved in exotoxin A regulation in *Pseudomonas aeruginosa*. Nucl. Acids Res., *15*: 2123–2135, 1987.

46. Hamood, A. N., and Iglewski, B. H. Expression of the *Pseudomonas aeruginosa toxA* positive regulatory gene (regA) in *Escherichia coli*. J. Bacteriol., *172*: 589–594, 1990.

47. Barany, F. Single-stranded hexameric linkers: A system for in-phase insertion mutagenesis and protein engineering. Gene, *37*: 111–123, 1985.

48. Jinno, Y., Chaudhary, V. K., Kondo, T., Adhya, S., FitzGerald, D. J., and Pastan, I. Mutational analysis of domain I of *Pseudomonas exotoxin. J. Biol. Chem., 263*: 13203–13207, 1988.

22
Expression of Growth Factor–Toxin Fusion Proteins

Gwynneth M. Edwards, Deborah Defeo-Jones, Steven M. Stirdivant, David C. Heimbrook, and Allen Oliff *Merck Sharp & Dohme Research Laboratories, West Point, Pennsylvania*

I. INTRODUCTION

A variety of hybrid fusion proteins linking specific growth factors with segments of bacterial toxins are under development as novel medical therapeutics (1–4). These agents have the potential to selectively eliminate or suppress clinically important cell populations in vivo that express specific cell surface receptors. For example, hybrid fusion proteins consisting of melanocyte-stimulating hormone (MSH) fused to a segment of diphtheria toxin could be used to kill melanoma cells that express MSH receptors. Alternatively, fusion proteins consisting of transforming growth factor α (TGFα) fused to a segment of the *Pseudomonas* exotoxin could be used to kill breast cancer cells that express epidermal growth factor (EGF) receptors. Growth factor–toxin fusion proteins may also be used to regulate immune responses by eliminating subsets of lymphocytes or mononuclear cells that express specific cytokine receptors (2,3,5). For example, interleukin-2 fused to segments of either the diphtheria or *Pseudomonas* exotoxin may prove useful as anti-inflammatory agents (3,6). However, before the therepeutic potential of these new agents can be realized, a series of practical problems involving the design, synthesis, and isolation of "nonnatural" recombinant proteins from microbial hosts must be solved.

A variety of genetic constructions employing several types of microbial-expression plasmids have been used successfully to produce growth factor–toxin fusion proteins (1,3,7). No single construction has demonstrated superiority for all hybrid proteins. Rather each recombinant expression system has its own advantages and disadvantages, and must be evaluated independently for each protein candidate. In particular, the optimal transcriptional promoters, ribosome-binding sites, and distances between the promoter and translational start site for each genetic construct are often different for different growth factor–toxin fusion proteins. A series of less well-defined parameters influence the isolation of biologically active recombinant proteins from bacteria. In the case of growth factors, the secondary and tertiary structure of these proteins clearly affects their ability to recognize and bind to their cognate receptors (8,9). Therefore, it is imperative that proper intramolecular disulfide bonds and protein folding occur during synthesis and/or purification of these molecules. A variety of biochemical techniques, including denaturation/renaturation and reduction/oxidation protocols, can be used during protein purification to modify the conformation of the final recombinant protein product. Alternatively, changing the microbial host vector used to synthesize recombinant proteins may alter the conformation of some fusion proteins. In the case of ADP-ribosylating toxins like diphtheria or *Pseudomonas* exotoxin, these proteins are generally toxic to eukaryotic organisms, and therefore must be produced in bacterial hosts (10,11). However, yeast host strains and expression vectors have recently been developed that allow production of recombinant hormone–toxin fusion proteins (12). These new yeast expression systems may foster proper disulfide bond formation without having to perform complex postsynthetic oxidation/reduction reactions.

The following chapter draws upon our experience with TGFα and *Pseudomonas* exotoxin to highlight the common problems associated with converting growth factor–toxin fusion proteins into practical therapeutics.

II. EXPRESSION SYSTEMS

A primary consideration for the expression of TGFα–PE40 fusion proteins with proper biological activity is the cysteine content of such proteins. Transforming growth factor α contains six cysteine residues which must associate in the correct disulfide pairs for biological activity. PE40 contains two disulfide pairs. Expression of TGFα–PE40 fusion protein clearly had the potential for incorrect disulfide pairings both within and between its two domains. Therefore, it seemed prudent to attempt expression of this fusion protein in a system that would fold the protein correctly. Yeasts are capable of expressing correctly folded, biologically active heterologous proteins when the proteins are directed into the yeast's secretory pathway (17). Unfortunately, expression of fusion proteins containing ADP-ribosylating toxins, like *Pseudomonas* exotoxin, is generally not feasible in eukaryotic cells owing to the toxin's enzymatic inhibition of cellular protein synthesis. All attempts at expressing PE40–TGFα in normal yeast strains failed, presumably as a result of inhibition of yeast cell protein synthesis.

As noted earlier, it may now be possible to circumvent this problem by using newly derived yeast strains that are resistant to the ADP-ribosylating activity of diphtheria toxin.

Biologically active, heterologous proteins can also be secreted from bacterial host vectors (18). Bacterial signal sequences, when fused to the N-terminus of a foreign protein, direct its secretion into the periplasmic space. This is accompanied by the processing away of the signal sequence and refolding of the protein. An OmpA signal sequence was added to the N-terminus of PE40–TGFα (OmpA–PE40–TGFα) and the fusion protein expressed in *E. coli* under the T7 polymerase expression system (19). Periplasmic preparations of the induced cells revealed that the vast majority of the fusion protein was not transported into this space but remained within the cytoplasm as insoluble inclusion bodies. Further analysis showed that the protein still contained the OmpA signal sequence. When the expression of the reverse construct (OmpA–TGFα–PE40) was investigated, a distribution of induced protein was seen between insoluble inclusion bodies and the soluble fraction of the lysed cells. Whether soluble or insoluble, the induced protein contained two major species thought to represent OmpA and non-OmpA containing protein. Only a very small portion of the induced fusion protein (<1%) was found in the periplasmic space, and this was largely degraded as judged by Western analysis. Since secretion of useful quantities of biologically active TGFα–PE40 fusion proteins in yeast or bacteria appeared infeasible, expression of cytoplasmic fusion protein devoid of extraneous signal sequence was attempted in *E. coli*.

No single expression system works well for all proteins. This fact is well illustrated by our early attempts to express fusion proteins containing both TGFα and PE40 domains. We utilized three different systems to express PE40–TGFα: a T7 RNA polymerase promoter system described by Studier et al. (13), a P_L temperature-sensitive promoter system (14), and a Trp-Lac (Tac) expression system (15). All three systems gave poor yields of recombinant protein. Minimal production of recombinant proteins in bacterial expression systems employing strong promoters can be due to several factors. For example, translational inefficiency can occur if secondary structures in the mRNA specified by the novel fusion gene bury the RNA's initiating ATG codon. This possibility was particularly germane to the expression of PE40–TGFα because of a theoretical hairpin structure found at the extreme 5' end of this fusion gene. Three ways of altering the PE40–TGFα chimeric gene to expose the initiating ATG codon were investigated. First, extra amino acid codons were added to the 5' end of the PE40 gene, thereby moving the initiating ATG upstream. Next, several codons were deleted from the 5' end of the PE40 gene, thus moving the initiating ATG downstream. Last, third-base "wobble" codons were inserted into the existing 5′ end of the PE40 gene. All three of these changes caused significant increases in the expression of PE40–TGFα protein.

Interestingly, none of the difficulties associated with the expression of PE40–TGFα were encountered when PE40–TGFα genetic construct was inverted and expressed as TGFα–PE40. Placing the TGFα sequences at the 5′ end of the chimeric fusion gene eliminated the translational problem associated with the PE40–TGFα genetic construction. High-level expression of the TGFα–PE40 hybrid fusion protein occurred in several expression vectors, although the best protein production was ultimately achieved using a modified Tac expression system (16). This novel Tac system employs a small "first" cistron before the protein sequence intended for expression. A Tac vector modified in this manner has been shown to increase the level of expression of acidic fibroblast growth factor 10-fold over the production from the parent vector. We found this to be the case for expression of TGFα–PE40 as well.

High-level expression of target proteins can also be influenced by the bacterial strain used as the host vector. In our expression studies on TGFα–PE40, we examined recombinant protein levels in several bacterial strains harboring the same recombinant plasmids. Two of these strains, JM109 and DH5α, were representative of most bacterial species. JM109 cells contain the Lac Iq gene which represses the Tac promoter. DH5α cells do not contain the Lac Iq gene. Therefore, DH5α cells permit constitutive expression from the Tac promoter. TGFα–PE40 expression levels were consistently better in JM109 than DH5α cells. At least part of the explanation for this result may be due to a toxic effect of the TGFα–PE40 protein on the bacterial cells. Constitutive expression of TGFα–PE40 protein in DH5α cells may prevent the bacterial cell mass from reaching the density needed for optimal recombinant protein production.

III. ISOLATION OF GROWTH FACTOR–TOXIN FUSION PROTEINS

Both TGFα–PE40 and PE40–TGFα fusion proteins expressed well in JM109 cells using the modified Tac promoter system. The induced proteins were completely soluble and accounted for >1% of the total cell protein. The PE40–TGFα used in these studies contained an additional five amino acids at the N-terminus of the fusion protein to promote translation of the PE40–TGFα mRNA in bacterial host cells (see above). The biological activity of the TGFα–PE40 and PE40–TGFα fusion proteins was investigated in crude *Escherichia coli* lysates with equivalent ADP-ribosylating activity. In cell-killing assays using EGF-receptor bearing A431 cells, TGFα–

PE40 appeared to be seven times more toxic to these cells (IC_{50} = 220 pM) than PE40–TGFα (IC_{50} = 1.5 nM). Control lysates showed no cell-killing activity. However, nonreducing gel analysis revealed that both protein preparations contained a substantial proportion of inappropriately disulfide cross-linked material. Therefore, the apparent differences in cytotoxicity between these fusion proteins may have reflected their degree of improperly refolded protein. Both proteins would have to be purified, and an attempt made to refold them correctly before a meaningful comparison of their biological activities could be made.

Since the TGFα–PE40 fusion protein was soluble, initial purification of this protein was attempted using nondenaturing techniques. TGFα–PE40 bound well to both DEAE-Sepharose and phenyl-Sepharose, making anionic exchange and hydrophobic chromatography attractive for the purification of this protein. However, despite the apparent stability of the TGFα–PE40 fusion protein in the lysate, extensive degradation of TGFα–PE40 occurred during manipulation of the lysate on both DEAE-sepharose and phenyl-sepharose columns. The inclusion of a battery of different classes of protease inhibitors to column buffers failed to prevent this breakdown. Only when the lysis supernatant was first denatured by the addition of solid urea and subsequent columns run in the presence of 6 M urea could the degradation during chromatography be prevented. In anticipation of the need to generate purified TGFα–PE40 fusion protein with the correct disulfide arrangement, the initial denaturing step was coupled with a chemical reaction that converted all cysteine residues to S-sulfonate derivatives. This technique has the advantage of "deactivating" cysteines during manipulations and adds negative charge to the protein, which can be exploited during purification. The reaction can easily be reversed and has been successfully employed to purify and refold biologically active proinsulin from bacteria (20). Thus, a scheme to purify TGFα–PE40 fusion proteins evolved that entailed lysing cells in 8 M guanidine hydrochloride, converting the cysteines to their S-sulfonate derivatives, and isolating the resulting protein by anion exchange and gel filtration chromatography. Many protocols to renature and refold the purified S-sulfonated fusion proteins were investigated and the results compared in terms of ADP-ribosylating activity, EGF receptor–binding activity, and the ability to specifically kill EGF receptor–bearing cells. The best results were obtained by dialysing the S-sulfonated fusion protein in the presence of β-mercaptoethanol at dilute (0.1 mg/ml) protein concentrations. After purification and renaturation, the biological activities TGFα–PE40 and PE40–TGFα were compared. TGFα–PE40 was now found to be very much more toxic to the A431 cells than PE40–TGFα (IC_{50}'s = 21 and 620 pM, respectively). Neither protein in its S-sulfonated form was toxic to EGF receptor–bearing cells, and the toxicity of both molecules was improved after purification and refolding based on total ADP-ribosylating equivalents from whole bacterial cell lysates. Because of its greater potency, the fusion protein with TGFα at the N-terminus and PE40 in the C-terminal position was chosen for further studies.

Despite the toxicity of TGFα–PE40 to EGF receptor–bearing cells, the fusion protein showed little activity in the EGF receptor–binding assay compared to native TGFα (IC_{40}'s = 540 nM and 5.5 nM, respectively). One of the factors that influences the receptor-binding activity of TGFα, and thus TGFα–PE40, is the formation of intrachain disulfide bonds. In an attempt to improve the receptor-binding capabilities of a TGFα–PE40 fusion protein, a series of mutants were investigated that manipulated the cysteine residues in the PE40 domain (7). Two deletion mutants were made which fused the TGFα domain to either a 34,000- or 25,000-d segment from the C-terminus of the original PE40 molecule (TGFα–PE34 and TGFα–PE25, respectively). TGFα–PE34 lacks the first 59 amino acids of the PE40 translocation domain, including two cysteine residues, and TGFα–PE25 lacks 130 amino acids and all four cysteine residues of the PE40 polypeptide. After purification and refolding, TGFα–PE34 and TGFα–PE25 were compared to TGFα–PE40 in the receptor-binding assay. Indeed, as the cysteine content of the fusion protein decreased, the EGF receptor–binding activity increased (IC_{50}'s = 540, 340, and 180 nM for TGFα–PE40, TGFα–PE34, and TGFα–PE25, respectively), but the cell-killing activity decreased dramatically in the deletion mutants (IC_{50}'s = 21, 4,200, and 860 pM for TGFα–PE40, TGFα–PE34, and TGFα–PE25, respectively). This result suggested that removing cysteines in the PE40 portion may facilitate refolding of the TGFα domain, but deletion of large segments of the translocation domain of PE40 was deleterious to the cytotoxic activity of the fusion protein.

In an attempt to improve the receptor-binding activity of the TGFα–PE40 fusion protein while preserving the cell killing activity, three additional mutants were prepared by site-directed mutagenesis. The cysteines in PE40 were changed to alanines in pairs that normally form disulfide bonds. For convenience, the cysteine residues at amino acid residues 265 and 287 were designated locus A and those at residues 372 and 379 were designated locus B. When alanines were introduced at either location, lower case letters are used. Using this nomenclature, the three mutants were named TGFα–PE40 Ab, TGFα–PE40 aB, and TGFα–PE ab, and the parent molecule was now called TGFα–PE40 AB. After purification and refolding, the three new mutant proteins were compared for EGF receptor binding and cell-killing activity. Both TGFα–PE40 aB and TGFα–PE40 ab were approximately 15 times more potent in the receptor-binding assay than either TGFα–PE40 AB or TGFα–PE40 Ab. This result suggested that the cysteine pair at locus A adversely affected the receptor-binding activity of TGFα–PE40 hybrid molecules. Since TGFα–PE40 Ab did not bind more efficiently than TGFα–PE40 AB, the cysteine residues in locus B did not appear to interfere with receptor binding. It should be noted that even when all the PE40 cysteines are replaced by alanines, as in TGFα–PE40 ab, the receptor-binding activity is still approximately sevenfold less than that of native TGFα. Therefore, other factors besides the cysteine residues in PE40 must contribute to the reduced binding efficiency of TGFα–PE40. One of these factors might be the presence of the relatively large PE40 domain attached to the C-terminal of TGFα which could sterically hinder its ability to bind to the EGF receptor. The cysteines in the A locus also seem important for cell-killing activity. TGFα–PE40 aB and TGFα–PE40 ab exhibited reduced toxicity toward A431 cells compared to TGF–PE40 AB or TGFα–PE40 Ab despite the fact that they bound to the EGF receptor with greater affinity. Presumably, the disulfide bond that forms between cysteine residues at locus A contributes to the conformation of TGFα–PE40 that is optimal for intoxication of mammalian cells. Again, the cysteines at the B locus appeared unimportant to the cytotoxic properties of TGFα–PE40.

In order to better understand the poor receptor binding of the purified, refolded TGFα–PE40 AB as compared to the TGFα–PE40 ab fusion protein, its disulfide arrangement was examined more closely. A unique methionine residue had been engineered between the TGFα and PE40 domains of the fusion proteins. By cleaving the purified and refolded TGFα–PE40 protein with cyanogen bromide (CNBr) and examining the products by HPLC, an estimate of the percentage of correctly refolded protein could be made. If the TGFα domain contained incorrect intradomain disulfides, then the TGFα isolated following CNBr cleavage would be inactive in the EGF receptor–binding assay. If incorrect interdomain disulfides had formed between the TGFα and PE40 domains, then the CNBr cleavage products would only be separable in the presence of reducing reagent (Figure 1). On cleavage of refolded TGFα–PE40 AB fusion protein with cyanogen bromide, most of the TGFα was released and shown to have near-normal receptor-binding activity (IC_{50} = 12.0). However, approximately 25% of the TGFα could only be released from a complex with PE40 by treatment with DTT, indicating a significant amount of interdomain disulfide cross-linking in TGFα–PE40 AB. When the same experiment was repeated with TGFα–PE40 ab, all of the TGFα was recovered from the cyanogen bromide reaction. These data indicated that the majority of the disulfide bonds in the TGFα–PE40 AB had formed correctly. The diminished ability of TGFα–PE40 AB to bind to the EGF receptor must, therefore, be due in

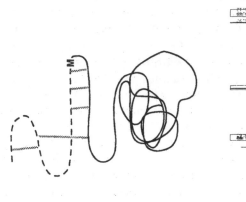

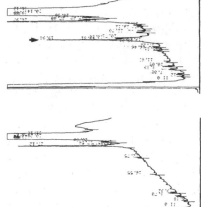

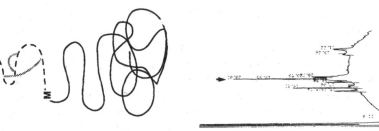

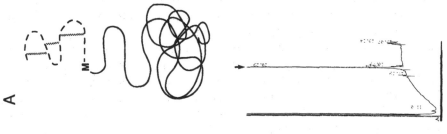

A210

Figure 1 Schematic representation of TGFα—PE40 species containing three different types of cysteine pairings for intrachain disulfide bonds. Dashed lines represent TGFα domains, and solid lines represent PE40 domains of TGFα—PE40 proteins. Panel A shows properly paired disulfide bonds within the TGFα domain. HPLC chromatogram corresponding to this species of TGFα—PE40 exhibits a single peak of TGFα isolated by cyanogen bromide cleavage of TGFα—PE40 ab after optimal oxidation/reduction procedures to reform disulfide bonds. Arrow indicates the elution time of properly folded TGFα species. Panel B shows theoretically improperly paired disulfide bonds within the TGFα domain. HPLC chromatogram of TGFα species isolated from TGFα—PE40 ab following cyanogen bromide cleavage but without optimal oxidation/reduction procedure. Note multiple TGFα species, one of which exhibits proper TGFα retention time, whereas others exhibit aberrant chromatographic properties. Panel C shows theoretically improperly paired disulfide bonds between the TGFα and PE40 domains. HPLC chromatogram of TGFα species obtained by cyanogen bromide cleavage of TGFα—PE40 AB followed by treatment with dithiothreitol (DTT)—right panel, or no DTT—left panel. Note that no TGFα is seen after cyanogen bromide treatment unless the interdomain disulfide bonds are reduced with DTT prior to examination by HPLC (7).

part to the disulfide pair at the A locus generating a conformation which interferes with the ability of the TGFα domain to bind to the EGF receptor. Although TGFα–PE40 ab was slightly less potent in the cell-killing assay than TGFα–PE40 AB, its superior receptor-binding characteristics prompted its selection for further in vivo testing.

IV. SCALE-UP AND REFOLDING

One of the most critical requirements for successful large-scale isolation of a recombinant fusion protein from bacteria is proper formation of disulfide bonds. An appropriate refolding step is generally essential for recovery of biologically active foreign proteins from bacteria. While this step is cumbersome on a laboratory scale, it can become nearly impossible on the production scale, since refolding is most efficient when performed on very dilute protein solutions (e.g., 0.1 mg/ml) (21). Optimal refolding conditions vary widely for different proteins, and must be determined empirically. With TGFα–PE40, we have found that sulfitolysis of the disulfides, followed by reduction with β-mercaptoethanol and air oxidation provides good recovery of properly folded material. However, limitations of scale still hinder this process owing to the inordinate volumes required to refold gram quantities of recombinant protein.

As described above, there are many difficulties associated with attempting to obtain a recombinant fusion protein from bacteria in clinically useful quantities. Many of these difficulties, such as protein toxicity to the host vector cells, refolding, and the development of a practical large-scale isolation procedure can usually be overcome given sufficient efforts. Ultimately, however, bacteria are not eukaryotic cells, and may not always be capable of producing biologically active mammalian cell proteins. This point is especially relevant for proteins which require posttranslational modification.

Besides the problem of obtaining properly folded proteins, most bacteria also lack appropriate enzymes to phosphorylate, glycosylate, amidate, or acylate eukaryotic proteins. While disulfide bond refolding can be circumvented, albeit inefficiently, via chemical means, other modifications (e.g., C-terminal amidation) are more difficult to mimic. For example, gastrin-releasing peptide (GRP) is a C–terminally amidated peptide hormone which elicits a variety of biological effects in mammalian cells. Data from our laboratory and others (22) have demonstrated that the C-terminal Met-amide residue in this hormone is important for biological activity. Because of a possible link between GRP and cancer (23), we generated a series of PE40–GRP fusion proteins in bacteria. Since the host bacteria were unable to generate an appropriate C-terminal amide residue, these fusion proteins were unable to bind to GRP receptors. In an attempt to solve this problem, we added an extra glycine at the C-terminus of the fusion protein. Treatment of this extended fusion protein with a commercial preparation of peptide α-amidating monooxygenase converted the C-terminal -Met-Gly-OH to -Met-NH$_2$ with greater than 95% efficiency. This enzymatically modified protein was then able to specifically bind to GRP receptors on the surface of mammalian cells in culture. Such a reaction might be feasible on a process scale through the postpurification use of immobilized enzymes. Nonetheless, this example illustrates the difficulties which can arise owing to the lack of appropriate posttranslational protein processing in bacteria.

V. CONCLUSIONS

The major obstacle to bringing several growth factor–toxin fusion proteins to clinical trials is the production and purification of sufficient quantities of biologically active protein. Many factors can influence the feasibility of obtaining highly purified recombinant proteins in large quantities. In the case of TGFα and *Pseudomonas* exotoxin fusion proteins, four factors proved to be critical to the isolation of large quantities of a biologically active final product: (1) the relative positions of the growth factor and toxin domains, (2) selection of the microbial host vector, (3) attention to both intra- and interchain disulfide bonds, and (4) postsynthetic refolding to obtain the optimal biologically active conformation. These factors and many other details of molecular biology and biochemistry require careful consideration during the design and production of all growth factor–toxin fusion proteins.

REFERENCES

1. Chaudhary, V. K., FitzGerald, D. J., Adhya, S., and Pastan, I. Activity of a recombinant fusion protein between transforming growth factor type alpha and *Pseudomonas* toxin. Proc. Natl. Acad. Sci. U.S.A., *84:* 4538–4542, 1987.
2. Siegall, C. B., Chaudhary, V. K., FitzGerald, D. J., and Pastan, I. Cytotoxic activity of an interleukin 6-*Pseudomonas* exotoxin fusion protein on human myeloma cells. Proc. Natl. Acad. Sci. U.S.A., *85:* 9738–9742, 1988.
3. Kelley, V. E., Bacha, P., Pankewycz, O., Nichols, J. D., Murphy, J. R., and Strom, T. B. Interleukin 2-diphtheria toxin fusion protein can abolish cell-mediated immunity in vivo. Proc. Natl. Acad. Sci. U.S.A., *85:* 3980–3984, 1988.
4. Murphy J. R., Bishai, W., Borowski, M., Miyanohara, A., Boyd, J., and Nagle, S. Genetic construction, expression, and melanoma-selective cytotox-

icity of a diphtheria toxin-related α-melanocyte-stimulating hormone fusion protein. Proc. Natl. Acad. Sci. U.S.A., *83*: 8258–8262, 1986.

5. Ogata, M., Chaudhary, V. K., FitzGerald, D. J., and Pastan, I. Cytotoxic activity of a recombinant fusion protein between interleukin 4 and *Pseudomonas* exotoxin. Proc. Natl. Acad. Sci. U.S.A., *86*: 4215–4219, 1989.

6. Lorberboum-Galski, H., FitzGerald, D., Chaudhary, V., Adhya, S., and Pastan, I. Cytotoxic activity of an interleukin 2-*Pseudomonas* exotoxin chimeric protein produced in *Escherichia coli*. Proc. Natl. Acad. Sci. U.S.A., *85*: 1922–1926, 1988.

7. Edwards, G. M., Defeo-Jones, D., Tai, J. Y., Vuocolo, G. A., Patrick, D. R., Heimbrook, D. C., and Oliff, A. Epidermal growth factor receptor binding is affected by structural determinants in the toxin domain of transforming growth factor-alpha-*Pseudomonas* exotoxin fusion proteins. Mol. Cell. Biol., *9*: 2860–2867, 1989.

8. Defeo-Jones, D., Tai, J. Y., Wegrzyn, R. J., Vuocolo, G. A., Baker, A. E., Payne, L. S., Garsky, V. M., Oliff, A., and Riemen, M. W. Structure-function analysis of synthetic and recombinant derivatives of transforming growth factor alpha. Mol. Cell. Biol., *8*: 2999–3007, 1988.

9. Lazar, E., Vicenzi, E., Van Obberghen-Schilling, E., Wolff, B., Dalton, S., Watanabe, S., and Sporn, M. B. Transforming growth factor alpha: an aromatic side chain at position 38 is essential for biological activity. Mol. Cell. Biol., *9*: 860–864, 1989.

10. Iglewski, B. H., and Kabat, D. NAD-dependent inhibition of protein synthesis by *Pseudomonas aeruginosa* toxin. Proc. Natl. Acad. Sci. U.S.A., *72*: 2284–2288, 1975.

11. Pappenheimer, A. M., Jr. Diphtheria toxin. Annu. Rev. Biochem., *46*: 69–94, 1977.

12. Perentesis, J. P., Genbauffe, F. S., Veldman, S. A., Galeotti, C. L., Livingston, D. M., Bodley, J. W., and Murphy, J. R. Expression of diphtheria toxin fragment A and hormone-toxin fusion proteins in toxin-resistant yeast mutants. Proc. Natl. Acad. Sci. U.S.A., *85*: 8386–8390, 1988.

13. Rosenberg, A. H., Lade, B. N., Chui, D. S., Lin, S. W., Dunn, J. J., and Studier, F. W. Vectors for selective expression of cloned DNAs by T7 RNA polymerase. Gene, *56*: 125–135, 1987.

14. Bernard, H. U., Remaut, E., Hershfield, M. V., Das, H. K., Helinski, D. R., Yanofsky, C., and Franklin, N. Construction of plasmid cloning vehicles that promote gene expression from the bacteriophage lambda pL promoter, Gene, *5*: 59–76, 1979.

15. deBoer, H. A., Comstock, L. J., and Vasser, M. The tac promoter: A functional hybrid derived from the trp and lac promoters. Proc. Natl. Acad. Sci. U.S.A, *80*: 21–25, 1983.

16. Linemeyer, D. L., Kelly, L. J., Minke, J. G., Gimenez-Gallego, G., DiSalvo, J., and Thomas, K. A. Expression in *Escherichia coli* of a chemically synthesized gene for biologically active bovine acidic fibroblast growth factor. Bio/Technology *5*: 960–965, 1987.

17. Ratner, M. Protein expression in yeast. Bio/Technology 7: 1129–1131, 1989.

18. Schein, C. H., Production of soluble recombinant proteins in bacteria. Bio/ Technology 7: 1141–1149, 1989.

19. Chaudhary, V. K., Yong-Hua, X., FitzGerald, D., Adhya, S., and Pastan I. Role of domain II of *Pseudomonas* exotoxin in the secretion of proteins into the periplasm and medium by *Escherichia coli*. Proc. Natl. Acad. Sci. U.S.A., *85*: 2939–2943, 1988.

20. Wetzel, R., Kleid, D. G., Crea, R., Heyneker, H. L., Yansura, D. G., Hivose, T., Kraszewski, A., Riggs, A. D., Itakura, K., and Goeddel, D. V. Expression in *Escherichia coli* of a chemically synthesized gene for a "mini-c" analog of human proinsulin. Gene, *16*: 63–71, 1981.

21. Light, A. Protein stability, protein modifications, and protein folding. Biotechniques, *3*: 298–306, 1985.

22. Heimbrook, D. C., Saari, W. S., Balishin, N. L., Friedman, A., Moore, K. S., Riemen, M. W., Kiefer, D. M., Rotberg, N. S., Wallen, J. W., and Oliff, A. Carboxyl-terminal modification of a gastrin releasing peptide derivative generates potent antagonists. J. Biol. Chem., *264*: 11258–11262, 1989.

23. Cuttitta, F., Carney, D. N., Mulshine, J., Moody, T. W., Fedorko, Fischler, A., Minna, J. D. Bombesin-like peptides can function as autocrine growth factors in human small-cell cancer. Nature, *316*: 823–826, 1985.

23
The Structure of *Pseudomonas* Exotoxin A as a Guide to Rational Design

Peter J. Nicholls and Richard J. Youle   *National Institute of Neurological Disorders and Stroke, National Institutes of Health, Bethesda, Maryland*

I. BASIC STRUCTURE AND MECHANISM OF ACTION

Pseudomonas exotoxin A (ETA) has interesting similarities to and differences with diphtheria toxin in its mechanism of killing cells. Exotoxin A binds to a cell surface receptor of unknown identity, is endocytosed into the cell, and crosses the membrane to reach the cytosol. Inside the cytosol a fragment of the toxin ADP-ribosylates elongation factor 2 (EF-2), inhibiting protein synthesis and killing the cell.

Cloning and sequencing of the toxin gene (1) revealed a 638-amino acid preprotein with a 25-amino acid leader sequence yielding a mature protein of 66,583 ds. The crystal structure of ETA reveals three distinct domains labeled from N- to C-terminal Ia, Ib, II, and III (2). The properties of amino acid sequences within these four regions are summarized in Table 1. Domain III functions in ADP-ribosylation of EF-2. Domain I appears to bind cell surface receptors and domain II is proposed to function in the transport of domain III into the cell cytosol.

II. DETAILED STRUCTURE OF ETA

A. Domain I

The N-terminal domain of ETA appears to include the cell surface–binding domain, although direct demonstration of this activity in a specific domain is lacking because of the difficulty in measuring ETA binding to cells.

Table 1   Important Structural Features of *Pseudomonas* Exotoxin A

Domain	Amino Acids	Function
Ia	1–252	Cell surface binding
Ib	365–404	Unknown
II	253–364	Cell entry (proposed)
III	405–613	ADP-ribosylation of EF-2

Deletion studies of ETA mutants that lack ADP-ribosylation activity show that those mutant toxins that contain most of domain I can compete for wild-type ETA toxicity, whereas fragments lacking domain I do not compete (3,4). When ETA was cross-linked to monoclonal antibodies using the amino-modifying cross-linking agent iminothiolane, it was noted that the modified ETA lost toxicity via the ETA receptor but not via new binding domains attached to the toxin (5). Proposing that a modified amino group in domain I accounted for a lack of binding of ETA to its native receptor, Pastan and colleagues mutagenized all 12 lysine residues in domain I, individually and in combinations, to glutamic acid (88). Mutation of Lys-57 to Glu reduced the toxicity of ETA 50-fold, whereas no other lysine in domain I appeared essential for toxicity. The mutation at Lys-57 also inactivated the capacity of an ADP-ribosylation–incompetent ETA mutant to compete for native ETA toxicity. Together these results indicate that domain I and Lys-57 are involved in the binding of ETA to cells.

Extensive studies of ETA fragments fused with new binding moieties also indicate that the C-terminal, domain I, contains the binding region of the toxin (6). A truncated ETA called PE40, lacking the first 252 amino acids of ETA, has been fused to many new binding domains, such as transforming growth factor (TGF), and is greatly diminished in toxicity to mouse cells, yet it acquires potent toxicity to cells with TGF receptors. These studies indicate that the translocation function of PE40 fusion proteins is preserved to some extent and that the native binding activity is deleted. This would localize the binding domain to the N-terminal 252 amino acids in domain Ia. If only 245 amino acids are deleted from the amino-terminus much of the native toxicity is retained, pointing to the region between amino acids 245 and 252 as potentially playing a role in binding to cells (7). Several additional point mutants have been constructed within this region 245–252 to reduce non–target cell toxicity, yet maintain potent target cell toxicity of immunotoxins made from the mutant (8). When the three basic amino acids, His-246, Arg-247, and Arg-249, were changed to Glu, the resulting mutant was more than 2000-fold inactivated in vitro and more than 100-times less toxic to animals. These mutations did not appear to affect the translocation activity to a great degree based upon the maintenance of potency of immunotoxins made which included the three Glu mutations (9). These results are consistent with a model that the region of domain Ia, between amino acids 245 and 252, contributes to the binding of the toxin to cells.

Domain Ib can be deleted from amino acid 365–380 without any loss of toxicity of TGFα–PE40 fusion proteins (10), and elimination of the disulfide bond in domain Ib, by mutating Cys-372 and -379 to alanines had no effect on the activity of TGFα–PE40 fusion proteins (11).

B. Domain II

Domain II has been proposed to play a role in the entry of domain III into the cytosol based upon the loss of toxicity of domain II mutants without loss in cell surface–binding activity or loss of ADP-ribosylation activity. In TGF–PE40 fusion proteins with amino acids 253–380 deleted, toxicity to target cells was decreased more than 180 times (10). Eliminating about half of domain II, amino acids 253–308, decreased toxicity of the fusion protein 100-fold. Elimination of domain Ib did not inhibit the toxicity of TGF–PE40 fusions. How these various deletions would affect native ETA remains unknown, however. Elimination of the disulfide bond within domain II of ETA by mutation of Cys-265 and -287 to serines resulted in a mutant 80-fold less toxic than the wild type (12). Interestingly, ammonium chloride blocked the toxicity of the domain II mutant to a lesser degree than the wild-type ETA, exactly as was seen in the diphtheria toxin (DT) translocation mutant, cross-reacting material 102 (CRM102) (13). Elimination of this same disulfide in domain II of a TGF–PE40 immunotoxin by mutation of the Cys-265 and -287 residues to serines (10) or alanines (11) resulted in a 10- to 150-fold reduction in toxicity. Together these results indicate that domain II plays an important role in ETA toxicity distinct from cell binding and ADP-ribosylation.

Point mutation of Arg-276 in domain II inactivates ETA more than 2000-fold and mutation of Arg-279 inactivates toxicity 440-fold. Pastan and colleagues present evidence for a saturable intracellular step in ETA toxicity that involves Arg-276 of domain II (88). However, subsequent work indicates that the mutations at Arg-276 and Arg-279 may inhibit intracellular proteolysis, possibly an essential activation step for the toxin (14).

Chaudhary and coworkers suggest that the C-terminal of ETA, in domain III, plays a role in intracellular routing of ETA to the endoplasmic reticulum (15). Munro and Pelham (17) have shown that proteins with the carboxyl-terminal sequence Lys-Asp-Glu-Leu (KDEL in the single-letter code) are retained in the endoplasmic reticulum. As these proteins move from the endoplasmic reticulum (ER) to the Golgi apparatus, they get recycled back to the ER, presumably via a KDEL-specific receptor. The C-terminal sequence of ETA is Arg-Glu-Asp-Leu-Lys (REDLK). If the last two amino acids are deleted the toxin ADP-ribosylation activity is maintained, yet toxicity is reduced more than 1000 times. Full toxicity requires a basic residue at amino acid 609, acidic amino acids at positions 610 and 611, and a leucine at amino acid 612. Substitution of the last five amino acids with KDEL results in a fully active toxin. If the toxin does interact with the KDEL receptor, the KDEL receptor may redirect the intracellular routing of the toxin to facilitate the passage to the cytosol. No similar structure is apparent in DT; however, an intracellular receptor interaction by ricin does appear to be necessary for this toxin to enter the cytoplasm (17,18). Similar to ricin B chain, ETA domain II or domain III may bind an intracellular receptor that alters intracellular routing, and thereby increases the transport of the toxin to the cytosol. Definition of the exact transport activity of ETA and the role of domains II and III in this activity awaits further study. However, it does appear that ETA can carry other polypeptides into cells (19).

C. Domain III

Expression of C-terminal fragments of the toxin that correspond with domain III of the crystal structure demonstrates that the last 306 amino acids contain the ADP-ribosylation activity (1). Thorough analysis of the N-terminal boundary of the ADP-ribosylation activity shows that the C-terminal 213 amino acids retain full activity, whereas the C-terminal 208 amino acids have lost over 50% of the ADP-ribosylation activity (10). Deletion of the C-terminal 36 amino acids or more blocked ADP-ribosylation activity more than 90% (20). The C-terminal location of this activity in ETA contrasts with the N-terminal location of the ADP-ribosylation function in DT. In light of the similar mechanism of action of ETA and DT, it was surprising initially to find no detectable homology between ETA and DT (1). However, careful comparison of the DT and ETA sequences by eye reveals a minor but significant homology between the ADP-ribosylation domains of the two toxins (2, 21–24). Figure 1 shows the alignment of 50 amino acids in these two regions of DT and ETA. Several of the conserved amino acids have been identified as active site residues involved in the ADP-ribosylation of EF-2 in both DT and ETA. This region shows distant homology not only with diphtheria toxin, but with other ADP-ribosylating toxins (22), and possibly even shows homology with elongation factor 2 itself (25).

```
        430       440       450       460       470       480       490       550
PE   EERGYVFVGYHGTFLEAAQSIVFGGVRARS---QDLDAIWRGFYIAGDPALAYGYAQDQEPDARGRIRNGALLRVYVP----
DT   SFVMENFSSYHGTKPGYVDSIQKGIQKPKSGTQGNYDDDWKGFYSTDNKYDAAGYSVDNENPLSG--KAGGVVKVTYPGLTK
        20        30        40        50        60        70        80        90

        510       520       530       540       550       560       570
PE   ----RSSLPGFYRTSLTLAAPEA-AGEV--ERLIGH--PLPLRLDAITGPEEGGRLETILGWPLAERTVVIPSAIPTDPRN
DT   VLALKVDNAETIKKELGLSLTEPLMEQVGTEEFIKRFGDGASRVVLSLPFAEGSSSVEYINNWE-QAKALSVELEINFETRG
        100       110       120       130       140       150       160       170
```

Figure 1   Amino acid homology between DT and ETA.

Elegant studies have shown that ETA glutamic acid 553 is involved in ADP-ribosylation of EF-2. Point mutagenesis of Glu to Asp inactivated the enzymatic activity 3200-fold and reduced toxicity to mouse cells 400,000-fold (26,27). Mutagenesis of Glu-553 to cysteine inactivated the enzymatic activity more than 10,000 times. Chemical modification of this mutant Cys-553 with iodoacetamide to reintroduce a carboxyl group slightly longer than that of the original Glu enhanced the enzymatic activity 2500-fold (28,29). This Glu-553 of ETA corresponds to Glu-148 of DT, also shown by photolabeling to be involved in ADP-ribosylation activity of ETA (30). Tyrosine-470 and -481 have been mutated to examine their role in toxin activity (28). Both of these residues lie adjacent to the active site of ETA (23), and are in conserved positions of DT and ETA (see Figure 1). Changing Tyr-470 to Phe did not affect the ADP-ribosylation activity nor the cytotoxicity of ETA. Mutation of Tyr-481 to Phe inactivated enzymatic activity and cytotoxicity 10-fold but did not affect the NAD^+-glycohydrolase activity of ETA. Tyr-481 is, therefore, proposed to play a role in the interaction of EF-2 with the toxin.

An inactive mutant ETA, CRM66 was isolated and found to lack ADP-ribosylation activity (31). Sequencing of this mutant revealed a His to Tyr mutation at codon 426 (32). This histidine, 126 amino acids before the active site Glu-553, is also conserved in DT 126 amino acids in front of Glu-148 and also in other ADP-ribosylating toxins (32). The crystal structure shows that His-426 is not in the NAD^+-binding pocket, but lies along a cleft in domain III. The authors propose that this residue is involved in the interaction of ETA with EF-2.

III. FUTURE PROSPECTS

The mechanisms of action of DT and ETA have fascinating similarities and differences. We still do not know the identity of the receptor for either toxin, although we know they do not bind to the same receptor. The cell entry mechanism also appears to differ between DT and ETA, although the intracellular mechanism of inhibition of protein synthesis is the same.

The long history of research in DT mechanism relative to the recent burst of interest in ETA is reflected in the approaches taken to probe structure-function relationships. In the 1970s, several groups used random mutagenesis and screening of mutants to generate CRM mutants of DT. This led to a strong understanding of DT even in the absence of the crystal structure. Many of the most informative mutations in DT CRMs would not have been engineered by rational design. Research in ETA, coming largely after the revolution in molecular biology, has relied heavily upon point mutagenesis to probe structure-function. Although the work has moved rapidly with knowledge of the crystal structure to aid design, many questions remain regarding the area of receptor binding and the mechanism of entry into cells. The power of point mutagenesis is to test specific residues and models, whereas the traditional approach of random mutagenesis allows the discovery of unanticipated residues, regions, and mechanisms. Hopefully, the two approaches will yield a thorough understanding of the mechanism of toxin action that will undoubtedly further the use of these toxins in rational drug design.

REFERENCES

1. Gray, G. L., Smith, D. H., Baldridge, J. S., Harkins, R. N., Vasil, M. L., Chen, E. Y., and Heyneker, H. L. Cloning, nucleotide sequence, and expression in *Escherichia coli* of the exotoxin A structural gene of *Pseudomonas aeruginosa*. Proc. Natl. Acad. Sci. U.S.A., *81*: 2645–2649, 1984.

2. Allured, V. S., Collier, R. J., Carroll, S. F., and McKay, D. B. Structure of exotoxin A of *Pseudomonas aeruginosa* at 3.0-Angstrom resolution. Proc. Natl. Acad. Sci. U.S.A., *83*: 1320–1324, 1986.

3. Guidi-Rontani, C., and Collier, R. J. Exotoxin A of *Pseudomonas aeruginosa*: evidence that domain 1 functions in receptor binding. Mol. Microbiol., *1*: 67–72, 1987.

4. Jinno, Y., Ogata, M., Chaudhary, V. K., Willingham, M. C., Adhya, S., FitzGerald, D., and Pastan, I. Domain II mutants of *Pseudomonas* exotoxin deficient in translocation. J. Biol. Chem., *264*: 15953–15959, 1989.

5. Pirker, R., FitzGerald, D. J., Hamilton, T. C., Ozols, R. F., Laird, W., Frankel, A. E., Willingham, M. C., and Pastan, I. Characterization of immunotoxins active against ovarian cancer cell lines. J. Clin. Invest., *76*: 1261–1267, 1985.

6. FitzGerald, D., and Pastan, I. Targeted toxin therapy for the treatment of cancer. J. Natl. Canc. Inst., *81*: 1455–1463, 1989.

7. Jinno, Y., Chaudhary, V. K., Kondo, T., Adhya, S., FitzGerald, D. J., and Pastan, I. Mutational analysis of domain I of *Pseudomonas* exotoxin. Mutations

in domain I of *Pseudomonas* exotoxin which reduce cell binding and animal toxicity. J. Biol. Chem., *263*: 13203–13207, 1988.

8. Chaudhary, V. K., Yosihiro, J., Gallo, M. G., Fitzgerald, D., and Pastan, I., Mutagenesis of *Pseudomonas* exotoxin in identification of sequences responsible for the animal toxicity. J. Biol. Chem., *265*: 16303–16310, 1990.

9. Lorberboum-Galski, H., Garsia, R. J., Gately, M., Brown, P. S., Clark, R. E., Waldmann, T. A., Chaudhary, V. K., Fitzgerald, D. J., and Pastan, I. IL2-PE664Glu, a new chimeric protein cytotoxic to human-activated T lymphocytes. J. Biol. Chem., *265*: 16311–16317, 1990.

10. Siegall, C. B., Chaudhary, V. K., FitzGerald, D. J., and Pastan, I. Functional analysis of domains II, Ib, and III of *Pseudomonas* exotoxin. J. Biol. Chem., *264*: 14256–14261, 1989.

11. Edwards, G. M., DeFeo-Jones, D., Tai, J. Y., Vuocolo, G. A., Patrick, D. R., Heimbrook, D. C., and Oliff, A. Epidermal growth factor receptor binding is affected by structural determinants in the toxin domain of transforming growth factor–alpha–*Pseudomonas* exotoxin fusion proteins. Mol. Cell. Biol., *9*: 2860–2867, 1989.

12. Madshus, I. H., and Collier, R. J. Effects of eliminating a disulfide bridge within domain II of *Pseudomonas aeruginosa* exotoxin A. Infect. Immun., *57*: 1873–1878, 1989.

13. Johnson, V. G., and Youle, R. J. A point mutation of proline 308 in diphtheria toxin B chain inhibits membrane translocation of toxin conjugates. J. Biol. Chem., *264*: 17739–17744, 1989.

14. Ogata, M., Chaudhary, V. K., Pastan, I., and FitzGerald, D. J. Processing of *Pseudomonas* exotoxin by a cellular protease results in the generation of a 37,000-Da toxin fragment that is translocated to the cytosol. J. Biol. Chem., *265*: 20678–20685, 1990.

15. Chaudhary, V. K., Jinno, Y., FitzGerald, D., and Pastan I. *Pseudomonas* exotoxin contains a specific sequence at the carboxyl terminus that is required for cytotoxicity. Proc. Natl. Acad. Sci. U.S.A., *87*: 308–312, 1990.

16. Munro, S., and Pelham, H. R. A C-terminal signal prevents secretion of luminal ER proteins. Cell, *48*: 899–907, 1987.

17. Youle, R. J., Murray, G. J., and Neville, D. M. Jr. Studies on the galactose binding site of ricin and the hybrid toxin man 6P-ricin Cell, *23*: 551–559, 1981.

18. Youle, R. J., and Colombatti, M. Hybridoma cells containing intracellular anti-ricin antibodies show ricin meets secretory antibody before entering the cytosol. J. Biol. Chem., *262*: 4676–4682, 1987.

19. Prior, T. I., FitzGerald, D. J., and Pastan, I. Barnase toxin: A new chimeric toxin composed of *Pseudomonas* exotoxin A and barnase. Cell, *64*: 1017–1023, 1991.

20. Chow, J. T., Chen, M. S., Wu, H. C. P., and Hwang, J. Identification of the carboxyl-terminal amino acids important for the ADP-ribosylation activity of *Pseudomonas* exotoxin A. J. Biol. Chem., *264*: 18818–18823, 1989.

21. Zhao, J.-M., and London, E. Localization of the active site of diphtheria toxin. Biochemistry, *27*: 3398–3403, 1988.

22. Gill, D. M., Sequence homologies among the enzymically active portions of ADP-ribosylating toxins. *In*: F. J. Fehrenbach et al. (eds.), Bacterial Protein Toxins, pp. 315–323. New York: Gustav Fischer, 1988.

23. Carroll, S., and Collier, R. J. Amino acid sequence homology between the enzymic domains of diphtheria toxin and *Pseudomonas aeruginosa* exotoxin A. Mol. Microbiol., *2*: 293–296, 1988.

24. Brandhuber, B. J., Allured, V. S., Fabel, T. G., and McKay, D. B. Mapping the enzymatic active site of *Pseudomonas aeruginosa* exotoxin. A. Proteins, *3*: 146–154, 1988.

25. Iglewski, W. J., and Fendrick, J. L. ADP ribosylation of elongation factor II in animal cells. *In*: J. Moss, and M. Vaughan (eds.), ADP-Ribosylating Toxins and G Proteins, pp. 511–524. American Society for Microbiology, Washington, D.C., 1990.

26. Douglas, C. M., and Collier, R. J., Exotoxin A of *Pseudomonas aeruginosa*: Substitution of glutamic acid 553 with aspartic acid drastically reduces toxicity and enzymatic activity. J. Bactiol., *169*: 4967–4971, 1987.

27. Douglas, C. M., and Collier, R. J., *Pseudomonas aeruginosa* exotoxin A: Alterations of biological and biochemical properties resulting from mutation of glutamic acid 553 to aspartic acid. Biochemistry, *29*: 5043–5049, 1990.

28. Lukac, M., and Collier, R. J. *Pseudomonas aeruginosa* exotoxin A: Effects of mutating tyrosine-470 and tyrosine-481 to phenylalanine. Biochemistry, *27*: 7629–7632, 1988.

29. Lukac, M., and Collier, R. J. Restoration of enzyme activity and cytotoxicity of mutant, E553C, *Pseudomonas aeruginosa* exotoxin A by reaction with iodoacetic Acid. J. Biol. Chem, *263*: 6146–6149, 1988.

30. Carroll, S., and Collier, R. J. Active site of *Pseudomonas aeruginosa* exotoxin. A. J. Biol. Chem., *262*: 8707–8711, 1987.

31. Cryz, S. J. Jr., Friedman, R. L., and Iglewski, B. H. Isolation and characterization of a *Pseudomonas aeruginosa* mutant producing a nontoxic, immunologically crossreactive toxin A protein. Proc. Natl. Acad. Sci. U.S.A., *77*: 7199–7203, 1980.

32. Wozniak, D. J., Hsu, L. Y., and Galloway, D. R. His-426 of the *Pseudomonas aeruginosa* exotoxin A is required for ADP-ribosylation of elongation factor II. Proc. Natl. Acad. Sci. U.S.A., *85*: 8880–8884, 1988.

24
Generation of Chimeric Toxins

David FitzGerald, Vijay K. Chaudhary, Robert J. Kreitman, Clay B. Siegall,* and Ira Pastan *Laboratory of Molecular Biology, National Cancer Institute, National Institutes of Health, Bethesda, Maryland*

I. INTRODUCTION

Chimeric toxins are generated by removing or crippling the receptor-binding domains of bacterial or plant toxins and substituting in their place proteins or peptides that bind mammalian cells. The cell-binding protein will then dictate which cell type the chimeric toxin will bind and kill. After binding to the cell surface, the chimera is internalized and a toxin fragment is delivered to the cell cytosol where it catalyzes the cell's demise.

Over the past 10–15 years, a variety of strategies have been used to make chimeric toxins. One approach is to use the tools of molecular biology to fuse portions of toxin genes with genes encoding cell-binding proteins. Gene fusions are then introduced into an appropriate expression system and recombinant chimeric toxins produced. This approach is relatively new and may have certain advantages over traditional methods of making immunotoxins. Recombinant chimeric toxins have been made with diphtheria toxin (DT) and *Pseudomonas* exotoxin (PE) in combination with a variety of hormones, antibodies, and other cell-binding proteins. Reasons for choosing the recombinant approach include ease of production, low cost, production of a homogeneous protein, and the flexibility to custom design the final product by introducing desired mutations.

Until recently, toxin conjugates such as immunotoxins were made exclusively by the use of chemical cross-linking reagents (17,19,22,35,42,54). Toxins or their A chains were cross-linked to cell-binding proteins. This approach necessitates the purification of large amounts of both the toxin and the cell-binding protein and then the use of heterobifunctional cross-linking reagents to join the two together. The efficiency of making con-

---

**Present affiliation*: Bristol-Myers Squibb Company, Evansville, Indiana

jugates in this manner is often quite low and usually leads to the generation of heterogeneous products. The traditional manner of making immuno-toxins has been a useful way to obtain research-grade material. However, as a way to manufacture drugs for clinical use, this approach has limitations. Here we will explore the development of recombinant chimeric toxins. The benefits of this approach, such as the production of a homogeneous product and the ability to custom design the product, will be discussed. Some of the possible negative aspects will also be mentioned.

II. HISTORY

The first two recombinant chimeric toxins were made using a truncated segment of the structural gene of diphtheria toxin (see Chapter 18). Be-cause there were fears that cloned diphtheria toxin could pose a health hazard, molecular constructions had to be performed under P4 contain-ment. This undoubtedly slowed that rate at which progress was made. Nevertheless, recombinant chimeric toxins were made by combining trun-cated DT with the genes for alpha-melanocyte–stimulating hormone (αMSH) and interleukin-2 (IL-2) (36,55). In both instances, sequences known to mediate DT binding were removed and replaced with sequences encoding the binding ligand. The replacement sequences were placed at the 3' end of the fusion. This is the same location occupied by sequences coding for DT binding (Figure 1).

Soon after the construction of the DT chimeras, a series of PE-derived chimeras was constructed (41). As with the DT chimeras, the sequences responsible for PE binding were removed and replaced by genes for binding ligands (see Figure 1). The binding ligands that have been fused to PE include: transforming growth factor α (TGFα), IL-2, IL-4, IL-6, insulin-like growth factor I, CD4, and the Fv fragment of monoclonal antibodies (9–13,16,32,38,44,45,50). Gene fusions with plant toxins have not been reported until very recently. Apparently, one of the problems was the inability of cellular proteases to cleave the A chain from the binding ligand. However, with the introduction of an arginine-rich linker sequence, it is now seems that the production of active chimeras, similar to those described for DT and PE, will be possible with ricin A chain and other plant toxins (37). The arginines provide proteolytic cleavage sites that can be used by cellular proteases to separate the cell-binding ligand from the A chain.

The sequences of DT that code for cell binding are at the 3' end of its structural gene, and therefore the C-terminal end of the DT protein (14). The situation is inverted for PE, with its binding domain at the 5' end of the gene and the N-terminal end of the protein (23) (see Figure 1). When making chimeric toxins, the significance of where the binding seg-ment is located in the native toxin has not been fully explored, but some

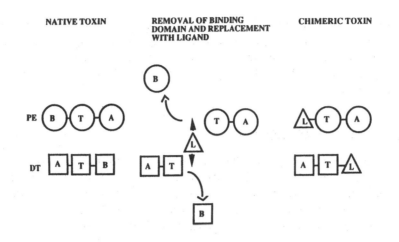

NATIVE TOXIN   REMOVAL OF BINDING DOMAIN AND REPLACEMENT WITH LIGAND   CHIMERIC TOXIN

Figure 1 Generation of chimeric toxins. The binding domains of either PE or DT can be removed and replaced with a binding ligand of choice. The ligand is often a growth factor, a cytokine, or a single-chain antibody. DNA is deleted from the 5' end of the PE structural gene (approximately 750 bp—250 amino acids— are removed) or the 3' end of the DT structural gene (either 150 bp or 450 bp— 50 and 150 amino acids—are removed; see Chapter 18 for details). B = binding; T = translocation; A = activity; and L = replacement ligand that is introduced to target the toxin to specific cells.

practical rules have emerged. Chimeras appear to be most active when the replacement binding ligand occupies the same location as the binding domain from the native toxin. Alpha-melanocyte stimulating hormone and IL-2 were both placed at the C-terminal end of the DT protein. No reports describe the placement of the binding ligands at the N-terminal of DT. Likewise chimeras with the PE gene are routinely made by placing binding ligands at the N-terminus of the chimera. There is one exception to this rule: TGFα can be placed at either end of the truncated form of PE, termed PE40 (10,51). However, when TGFα is placed at the N-terminus of PE40, the toxin is 30-fold more active than when TGFα is placed at the C-terminus. In all other instances, placing a binding ligand at the C-terminus of PE leads to the production of an inactive chimera.

III. GENE CONSTRUCTION

Fusion genes are constructed using routine molecular biology techniques. Signal sequences for secretion can be included if desired; however, many recombinant proteins can be purified easily from insoluble inclusion bodies, so often signal sequences are removed rather than added. As well as the structural chimeric gene, usually composed of a cDNA sequence joined to a gene fragment of the toxin, there are control sequences for promoting and terminating RNA transcription and sequences for ribosome binding. In most cases, the chimeric gene is housed on a multicopy plasmid which can be transformed into the appropriate host cell for either propagation or expression.

IV. GENE EXPRESSION

Various expression systems have been used to produce chimeric toxins. Most of these have been in *Escherichia coli*, although production in *Bacillus* species may also be possible. Expression in yeast is likely to be difficult since toxins are active against the eukaryotic protein synthetic machinery. Production in cell-free systems is possible and may be quite useful in certain instances (39).

Theoretically, proteins expressed in *E. coli* can be recovered from one of four locations: (1) the growth medium, (2) the periplasm, (3) the soluble cytoplasm, and (4) the insoluble fraction–termed inclusion bodies. Even though native toxins are often secreted into the growth medium, it has not been possible to secrete chimeras in a similar fashion. Some secretion to the periplasm has been reported, but proteins secreted there can be subject to proteolysis. In fact, most chimeras produced at high levels are recovered from inclusion bodies and have to be denatured and renatured before they can be converted into "usable" protein.

A. Purification of Chimerics

If proteins are purified from inclusion bodies, they must first be solubilized using denaturants. For PE chimeras, this is done with 7 M guanidine hydochloride. Other protocols use urea, and sometimes a reducing reagent is included. Denatured material is renatured by rapid dilution in a buffered saline solution such as PBS. Because it is necessary to reduce the concentration of denaturant to nondenaturing levels, the dilution factor is often as high as 100-fold. After this step, conventional chromatography can be applied to the task of purification. One particular purification step deserves mention. Interleukin-2 chimeras have been purified using receptor-affinity chromatography (2). This step has the advantage that only correctly folded material will bind to immobilized receptor. Bound IL-2 chimeras can then be eluted and further purified by ion exchange and gel filtration steps. Affinity chromatography for IL-2 chimeras was developed by immobilizing the p55 subunit of the IL-2 receptor on silica. Crude IL-2 chimera was passed over the column and later eluted with either potassium thiocyanate or low pH. A drawback with this approach is that the capacity of such columns is not very high. Also, no attention is paid to interactions with the other subunit (p75) of this receptor. This approach may be ideal for ligand interaction with single subunit receptors but becomes problematic when multiple subunits are involved. The p55 column has been used to purify IL-2 and IL-2 chimeras, but could also be used to purify antibody-toxin chimeras directed to the same subunit (such as anti-Tac). Receptor affinity purification has not been developed yet for other ligands, but theoretically would be a very useful way to separate ligand-toxin chimeras that are folded correctly from those that are not.

V. CHARACTERIZATION OF CHIMERIC TOXINS

To be cytotoxic, chimeric toxins must have binding, translocating, and enzymatic activities. If the protein lacks any one of the three, it will not kill the intended target cell or tissue. To some extent each of these activities can be assessed independently.

A. Enzymatic Activity

The enzymatic activity of DT and PE is well known. Both toxins catalyze the ADP-ribosylation of elongation factor 2 (EF-2). The ADP-ribosylating domain of DT is located at its N-terminal end, and for PE it is located at the C-terminal end. Therefore, chimeric molecules exhibiting full enzyme activity can be assumed to have a functional N-terminus for DT and C-terminus for PE. The reasons for measuring ADP-ribosylation activity are severalfold: to check for inactivating mutations, to check for problems with purification such as proteolysis, and to rapidly monitor toxin activity in column fractions. Several excellent reviews describe the details of how ADP-ribosylation activity is measured (7,24).

Under certain circumstances, it may be desirable to eliminate enzymatic activity. The glutamic acid residues at positions 148 of DT and 553 of PE are each key amino acids involved in NAD binding (5,6). By deleting these amino acids it is possible to make full-sized chimeras that can bind and be internalized but cannot kill cells. Such chimeras are often used as control proteins. It would be expected that the enzyme-inactive version of a growth factor–toxin recombinant chimera would still be able to promote cell proliferation. When it has been tested this appears to be so (38).

B. Binding Activity

Together with enzymatic activity, measuring cell-binding activity is relatively straight forward. When generating recombinant chimeras, certain constraints are placed on the binding portion of the hybrid molecule which need to be assessed. Most importantly, binding ligands are constrained by the peptide bond that joins them to the truncated toxin. In addition, chimeras made in *E. coli* will not be glycoslyated, whereas cloning strategies may introduce "extra" amino acids or delete a few wild-type sequences. Other sources of binding abnormalities include the failure to fold novel proteins correctly, the mismatch of disulfide bonds, and residual binding activity by the toxin portion of the chimera. Despite this substantial list of potential problems, it should be noted that many recombinant chimeras bind to cell surface receptors with near normal affinity.

Various assays can be used to characterize binding. Perhaps the simplest of these is to measure the displacement of a trace amount of radio-labeled ligand. Usually, displacement by the chimeric toxin is compared to displacement by the natural ligand. Results are then compared on a molar basis. This approach, while simple in concept, is often rendered more complicated when multiple receptor subunits are involved. For instance, IL-2–PE40 binds to the p55 subunit of IL-2 receptor with close to full activity. However, it binds to the p75 subunit with a 5- to 10-fold lower affinity (34). Of the chimeric toxins examined, most bind 3- to 10-fold less well than the corresponding native ligands (34,38,45). The situation with IL-2–PE40 is worthy of discussion. This material is purified using the p55 receptor affinity column (mentioned above). Clearly, this will select chimeric material that has refolded to allow binding to this subunit. This column does not select for binding to the p75 subunit. It would be of some interest to determine the cytotoxic activity of IL-2–PE40 after it had been passed over both a p55 and p75 affinity column.

C. Translocating Activity

A widely held belief is that translocation to the cell cytoplasm is the rate-limiting step in the action of toxins and toxin chimeras. Therefore, gaining a greater understanding of this process seems essential for the design of future chimeras. Translocation is very difficult to measure in a direct assay. In fact, current efforts are focused on mutational analysis of such toxins as DT, PE, and ricin to discover the sequences which are most important for translocation. Presumably, similar mechanisms for translocation are used by both the native toxins and toxin chimeras, although this remains to be shown.

To date, deletion analysis of PE and PE chimeras has revealed that domain II (residues 253–364) is important for translocation, whereas sequences in domain Ib (365–399) are not. However, domain II is necessary but not sufficient for translocation. Chimeras made by placing binding ligands at the C-terminal end of PE40 were shown to bind normally and have full ADP-ribosylating activity, yet they failed to kill cells. It was noted that blocking the C-terminal of PE or PE40 inhibited cell killing (11). When this was investigated, it was determined that the sequence RDELK at the C-terminal of PE was necessary for translocation to the cell cytoplasm. Only this sequence or a related sequence, KDEL, could mediate translocation.

Recently, it was shown that *Pseudomonas* exotoxin could translocate "foreign" proteins and peptides into the cytosol (15,43). Diphtheria toxin, ricin, anthrax, and other toxins might be expected to possess the same ability, but to date this has not been documented.

VI. IN VITRO CYTOTOXIC ACTIVITY

The ability of a chimeric toxin to kill a tissue culture cell line depends on many factors. A positive result indicates that the chimeric toxin bound a surface receptor, was internalized, and then translocated the enzymatic domain to the cytoplasm. However, when comparing the results with different cell lines, it has been difficult to assess every step. A number of cell lines were assessed both for IL-6 binding and for their susceptibility to IL-6–PE40. There was a good general correlation between the number of binding sites and susceptibility to the chimeric toxin (46) (Table 1).

Chimeric toxins have been made with PE40, with a full-sized (66 kd) mutant form of PE, termed PE^{4E}, and with toxins of intermediate size (11,33,48). We have shown that PE40 has both translocating and ADP-ribosylating activity. However, some conjugates employing larger forms of PE show greater activity even though the binding activity of the toxin has been eliminated by mutagenesis. In the absence of assays for toxin activation, intracellular stability or susceptibility to proteases, it is difficult

Table 1 Cytotoxicity of IL-6–PE40 and IL-6–PE^{4E} Against Cell Lines

Cell line	Type	ID$_{50}$ (ng/ml) IL-6–PE40	IL-6–PE^{4E}
U266	Myeloma	8–15	0.9–1.5
H929	Myeloma	8–12	1.5–3.0
PLC/PRF/5	Hepatoma	5–7	1.5–2.0
HEP3B	Hepatoma	18–30	40–50
HEPG2	Hepatoma	450	70
A431	Epidermoid carcinoma	>625	8
LNCaP	Prostate carcinoma	9	1.7
DU145	Prostate carcinoma	>1000	40

to pinpoint the reasons why one form of the toxin is more active than another. Development of such assays, therefore, seems essential to our further understanding of toxic processes.

One puzzle with PE chimeras focuses on the issue of why IL-2–PE40 can kill murine activated T cells but not their human or monkey counterparts (33). Despite using human recombinant IL-2 to make IL-2–PE40, which retains its binding activity for the human IL-2 receptor, it has not been possible to kill primate T cells activated with either phytohemaglutinin (PHA) or with alloantigens (see below). Explanations for this idiosyncratic result are being sought.

VII. TGFα TOXINS AGAINST THE EGF RECEPTOR

Transforming growth factor α, which is similar to EGF, appears to be involved in the growth and maintenance of many solid tumors, including those of lung, breast, head and neck, prostate, brain, hepatic, bladder, endometrial, renal, and gastrointestinal origin (21,25,30). Therapeutic trials of TGFα–PE40 or related derivatives in nude mice carrying human epidermoid, hepatocellular, prostatic, or colon carcinoma have demonstrated a therapeutic window for antitumor activity (20,40). A derivative of TGFα–PE40 without disulfide bonds in PE40 (TGFα–PE40δcys) is currently being prepared for a clinical trial in patients with bladder cancer (20).

VIII. IL-6 TOXINS

Because it was reported that the IL-6 receptor (IL-6R) is expressed on human myeloma cells and various tumor cell lines (26), we constructed the chimeric toxin IL-6–PE40 (45) in which IL-6 is fused to the amino-terminus of PE40. We found IL-6–PE40 to be cytotoxic not only toward human myeloma cell lines, but also toward hepatocellular carcinoma lines (48). We also made IL-6–PE[4E], which contains the ligand fused to the amino-terminus of PE[4E]. The IL-6–PE[4E] was not only more cytotoxic than IL-6–PE40 toward myeloma and hepatocellular carcinoma cell lines, but also showed cytotoxicity toward human epidermoid and prostate carcinoma lines (46,49).

Because hepatocytes express IL-6R (4), we wished to test whether we could safely administer enough IL-6–PE[4E] to an animal to cause antitumor activity. Since murine IL-6 binds the human IL-6 receptor, we treated nude mice implanted with the human hepatocellular carcinoma line PLC/PRF/5 (47). By administering the chimeric toxin by continuous infusion for 1 week from pumps implanted in the peritoneal cavity, we were able to inhibit growth of tumor in treated animals (Figure 2). This suggested that a therapeutic window for IL-6–PE^{4E} treatment might exist. Since fresh human myeloma cells have fewer IL-6 receptors per cell than continuous lines (26), we decided to test IL-6–PE^{4E} against myeloma samples obtained directly from patients. Of 10 samples, 6 were sensitive to IL-6–PE^{4E} with an ID$_{50}$ of 50 ng/ml or less. Other samples were less sensitive. IL-6–PE^{4E} is being considered for the treatment of multiple myeloma, either to purge malignant cells from the marrow or as a novel in vivo treatment.

A. IL-2 Receptor Toxins

The chimeric toxin IL-2–PE40 is toxic for cells and cell lines that display IL-2 receptors on their surface. This chimeric toxin can also prevent the growth of an IL-2 receptor–bearing tumor in mice (28), has activity against adjuvant-induced arthritis in rats (8), and can inhibit cardiac allograft rejection (31). Because of these results and similar findings with IL-2–DT, toxin-based therapy directed to the IL-2 receptor looks very promising. However, as mentioned above, IL-2–PE40 does not kill activated human lymphocytes.

Alternative strategies are now being considered for the generation of PE-based chimeras that can be targeted to the human IL-2 receptor. One approach has been to place IL-2 at the amino-terminus of PE^{4E}. The IL-2–PE^{4E} was not only more toxic than IL-2–PE40 against several cell lines,

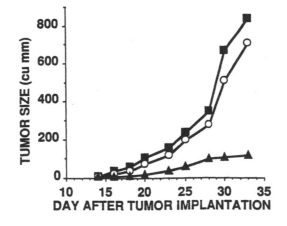

Figure 2 Hepatoma treatment with IL-6 toxin. The human hepatoma cell line PLC/PRF/5 was injected subcutaneously in nude mice. Treatment was by continuous infusion via an ALZA pump implanted in the peritoneal cavity. The mice were treated beginning on day 15 with either IL-6–PE^{4E} 350 μg/kg/day × 7 days IP (▲), or an ADP-ribosylation–deficient IL-6–PE^{4E} mutant (○), or were untreated (■).

but was also active against human activated lymphocytes (33). Another strategy has been to use an antibody to the IL-2 receptor. Anti-Tac binds the p55 component of the IL-2 receptor and has been chemically coupled to PE and PE40 to make active immunotoxins (3,18,27). When the cDNA of this antibody became available, anti-Tac(Fv)PE40 was generated (13). Anti-Tac(Fv)–PE40 was at least 10-fold more cytotoxic toward target cells than either IL-2 toxins or anti-Tac chemical conjugates (Table 2). We have been able to achieve anti-Tac(Fv)–PE40 serum levels exceeding 1500 ng/ml in mice (3) and 400 ng/ml in a monkey without apparent toxicity (Kreitman, Parenteau, FitzGerald, Waldmann, and Pastan, unpublished data). Clearly from a therapeutic standpoint, anti-Tac(Fv)–PE40 appears a promising reagent, perhaps more promising than the corresponding IL-2 toxins.

To determine if anti-Tac(Fv)–PE40 would not only kill cell lines and activated T lymphocytes but also malignant cells directly from patients, we tested the recombinant immunotoxin against peripheral blood mononuclear cells (PBMCs) from patients with adult T-cell leukemia (ATL) (29). The PBMCs of each of six ATL patients (with lymphocytive-Tac positivity $\geq$ 40) were shown to be sensitive to anti-Tac(Fv)–PE40, with ID_{50}s of 1.6–16.0 ng/ml. In contrast, the PBMCs from normal donors, or normal PBMCs from ATL patients without blood involvement, were resistant. IL-2–PE^{4E} was at least 50-fold less cytotoxic than anti-Tac(Fv)–PE40 against the malignant cells. IL-2–PE40 was inactive in similar assays (see Table 2).

Table 2　Activity of Various Chimeric Toxins Against Human Cells

No.	Diagnosis	Anti-Tac(Fv)–PE40	IL-2–PE40	IL-2–PE^{4E}
		ID_{50} (ng/ml)		
1	ATL	4	>1000	250
2	ATL	16	>1000	610
3	ATL	7.5	>1000	1000
4	ATL	1.6		
5	ATL	10		
6	ATL	6		
7	ATL, nonleukemic	>1000	>1000	>1000
8	ATL, nonleukemic	>1000		
9	T-cell lymphoma	170		
10	Childhood ALL	>1000		
11	Normal	>1000	>1000	>1000
12	Normal	>1000	>1000	>1000
13	HUT 102 (ATL cell line)	0.2	17	0.8

IX. ANTIVIRAL ACTIVITY OF CHIMERIC TOXINS

A major clinical focus for the use of chimeric toxins is the treatment of cancer and immunological disorders. However, it may also be possible to use these reagents for treating viral infections. Killing infected cells before they can release new virus should reduce the viral load of an infected individual and help prevent the spread of infection to others. Chimeric toxins have been developed for the treatment of aquired immune deficiency syndrome (AIDS) (12,52,53). CD4 represents a receptor for human immunodeficiency virus (HIV) on lymphoid tissue. Viral gp120 binds CD4 as part of the infectious process. Later, gp120 is expressed on the surface of infected cells. At this point in the life cycle of HIV, soluble CD4 can be used to target reagents to infected cells. CD4–PE40 kills HIV infected cells and, together with azidothymidine (AZT), can "sterilize" cultures of infected cells (1). Clinical trials will begin shortly to test whether the encouraging results seen in tissue culture can be duplicated in patients.

X. CONSIDERATIONS FOR IN VIVO USE OF CHIMERIC TOXINS

Recombinant chimeric toxins have unique and unusual properties. Listed below are several issues that need to be considered with regard to the use of chimeric toxins in therapeutic settings. Their size and shape will undoubtedly be quite different from either of the parent molecules from which they are derived. This will affect half-life and clearance rates which will impinge greatly on toxicity and drug delivery. To some degree, cytokine-toxin conjugates will have both agonist and cytotoxic activity; i.e., one portion of the chimera will "instruct" the cell to grow, whereas the other portion "attempts" to kill it. This may pose problems. The region where the "natural" binding ligand joins the toxin will constitute a unique structure, and may be recognized as a novel antigenic site. Antibodies elicited to such an epitope may react not only with the chimeric toxin, but also with the binding ligand in its native state. This could result in the generation of an autoimmune state.

XI. CONCLUSIONS AND PROSPECTS FOR THE FUTURE

Toxins are potent cytotoxic proteins. They can be targeted to growth factor receptors and antigens on the surface of diseased and disease-causing cells. One way of preparing material for targeting is to fuse the genes for binding ligands (including single-chain antibody molecules, growth factors, and cytokines) with the gene fragments for toxins. These fusions can be expressed in bacteria and produced as novel pharmaceutical reagents. The administration of these agents may kill cancer cells, diminish autoimmune states, or eliminate parasitic infections. Only clinical trials will determine the true worth of this approach and these agents.

ACKNOWLEDGMENT

The authors thank the members of the Laboratory of Molecular Biology for their contributions to this work and our collaborators whose valuable contributions have made this work possible.

REFERENCES

1. Ashorn, P., Moss, B., Weinstein, J. N., et al. Elimination of infectious human immunodeficiency virus from human T-cell cultures by synergistic action of CD4-Pseudomonas exotoxin and reverse transcriptase inhibitors. Proc. Natl. Acad. Sci. U.S.A., *87*(22): 8889–8893, 1990.

2. Bailon, P., Weber, D. V., Gately, M., et al. Purification and partial characterization of an interleukin 2-Pseudomonas exotoxin fusion protein. Biotechnology, *6*: 1326–1329, 1988.

3. Batra, J. K., FitzGerald, D., Gately, M., Chaudhary, V. K., and Pastan, I. Anti-Tac(Fv)-PE40, a single chain antibody Pseudomonas fusion protein directed at interleukin 2 receptor bearing cells. J. Biol. Chem., *265*(25): 15198–15202, 1990.

4. Bauer, J., Lengyel, G., Bauer, T. M., Acs, G., and Gerok, W. Regulation of interleukin 6 receptor expression in human monocytes and hepatocytes. FEBS Letts, *249*: 27–30, 1989.

5. Carroll, S. F., and Collier, R. J. Active site of Pseudomonas aeruginosa exotoxin A. J. Biol. Chem., *262*: 8707–8711, 1987.

6. Carroll, S. F., and Collier, R. J. Amino acid sequence homology between the enzymic domains of diphtheria toxin and Pseudomonas aeruginosa exotoxin A. Mol. Microbiol., *2*: 293–296, 1989.

7. Carroll, S. F., and Collier, R. J. Diphtheria toxin: Quantification and assay. Methods Enzymol., *165*: 218–225, 1988.

8. Case, J. P., Lorberboum, G. H., Lafyatis, R., FitzGerald, D., Wilder, R. L., and Pastan, I. Chimeric cytotoxin IL-2-PE40 delays and mitigates adjuvant-induced arthritis in rats. Proc. Natl. Acad. Sci. U.S.A., *86*(1): 287–291, 1989.

9. Chaudhary, V. K., Batra, J. K., Gallo, M. G., Willingham, M. C., FitzGerald, D. J., and Pastan, I. A rapid method of cloning functional variable-region antibody genes in Escherichia coli as single-chain immunotoxins (published erratum appears in Proc. Natl. Acad. Sci. U.S.A., *87*(8): 3253, 1990). Proc. Natl. Acad. Sci. U.S.A., *87*(3): 1066–1070, 1990.

10. Chaudhary, V. K., FitzGerald, D. J., Adyha, S., and Pastan, I. Activity of a recombinant fusion protein between transforming growth factor type alpha and pseudomonas Toxin. Proc. Natl. Acad. Sci. U.S.A., *84*: 4538–4542, 1987.

11. Chaudhary, V. K., Jinno, Y., Gallo, M. G., FitzGerald, D., and Pastan, I. Mutagenesis of Pseudomonas exotoxin in identification of sequences responsible for the animal toxicity. J. Biol. Chem., 265(27): 16306–16310, 1990.

12. Chaudhary, V. K., Mizukami, T., Fuerst, T. R., et al. Selective killing of HIV-infected cells by recombinant human CD4-Pseudomonas exotoxin hybrid protein, Nature, 335(6188): 369–372, 1988.

13. Chaudhary, V. K., Queen, C., Junghans, R. P., Waldmann, T. A., FitzGerald, D. J., and Pastan, I. A recombinant immunotoxin consisting of two antibody variable domains fused to Pseudomonas exotoxin. Nature, 339: 394–397, 1989.

14. Colombatti, M., Greenfield, L., and Youle, R. J. Cloned fragment of diphtheria toxin linked to T cell-specific antibody identifies regions of B chain active in cell entry. J. Biol. Chem., 261(7): 3030–3035, 1986.

15. Debinski, W., Siegall, C. B., FitzGerald, D., and Pastan, I. Substitution of foreign protein sequences into a chimeric toxin composed of transforming growth factor alpha and Pseudomonas exotoxin. Mol. Cell. Biol., 11(3): 1751–1753, 1991.

16. Edwards, G. M., DeFeo-Jones, D., Tai, J. Y., et al. Epidermal growth factor receptor binding is affected by structural determinants in the toxin domain of transforming growth factor-alpha-Pesudomonas exotoxin fusion protein. Mol. Cell. Biol., 9: 2860–2867, 1989.

17. FitzGerald, D., and Pastan, I. Targeted toxin therapy for the treatment of cancer. J. Natl. Cancer Inst., 81(19): 1455–1463, 1989.

18. FitzGerald, D. J., Waldmann, T. A., Willingham, M. C., and Pastan, I. Pseudomonas exotoxin-anti-TAC. Cell-specific immunotoxin active against cells expressing the human T cell growth factor receptor. J. Clin. Invest., 74(3): 966–971, 1984.

19. Ghetie, V., Till, M. A., Ghetie, M. A., Uhr, J. W., and Vitetta, E. S. Large scale preparation of an immunoconjugate constructed with human recombinant CD4 and deglycosylated ricin A chain. J. Immunol. Methods, 126(1): 135–141, 1990.

20. Heimbrook, D. C., Stirdivant, S. M., Ahern, J. D., et al. Transforming growth factor alpha-Pseudomonas exotoxin fusion protein prolongs survival of nude mice bearing tumor xenografts. Proc. Natl. Acad. Sci. U.S.A., 87(12): 4697–4701, 1990.

21. Hendler, F. J., and Ozanne, B. W. Human squamous cell lung cancers express increased epidermal growth factor receptors. J. of Clinical Invest., 74: 647–651, 1984.

22. Hertler, A. A., and Frankel, A. E. Immunotoxins: A clinical review of their use in the treatment of malignancies. J. Clin. Oncol., 7(12): 1932–1942, 1989.

23. Hwang, J., FitzGerald, D. J., Adhya, S., and Pastan, I. Functional domains of Pseudomonas exotoxin identified by deletion analysis of the gene expressed in E. coli. Cell, 48: 129–136, 1987.

24. Iglewski, B. H., and Sadoff, J. C. Toxin inhibitors of protein synthesis: Production, purification, and assay of Pseudomonas aeruginosa toxin A. Methods Enzymol., 60: 780–793, 1979.

25. Jones, N. R., Rossi, M. L., Gregoriou, M., and Hughes, J. T. Epidermal growth factor receptor expression in 72 meningiomas. Cancer, *66*: 152–155, 1990.

26. Kawano, M., Hirano, T., Matsuda, T., et al. Autocrine generation and requirement of BSF-2/IL-6 for human multiple myelomas. Nature, *332*: 83–85, 1988.

27. Kondo, T., FitzGerald, D., Chaudhary, V. K., Adhya, S., and Pastan, I. Activity of immunotoxins constructed with modified pseudomonas exotoxin A lacking the cell recognition domain. J. Biol. Chem., *263*: 9470–9475, 1988.

28. Kozak, R. W., Lorberboum, G. H., Jones, L., et al. IL-2-PE40 prevents the development of tumors in mice injected with IL-2 receptor expressing EL4 transfectant tumor cells. J. Immunol., *145*(8): 2766–2771, 1990.

29. Kreitman, R. J., Chaudhary, V. K., Waldmann, T., Willingham, M. C., FitzGerald, D. J., and Pastan, I. The recombinant immunotoxin anti-Tac(Fv)-Pseudomonas exotoxin 40 is cytotoxic toward peripheral blood malignant cells from patients with adult T-cell leukemia. Proc. Natl. Acad. Sci. U.S.A., *87*(21): 8291–8295, 1990.

30. Lau, J. L. T., Fowler, J. E. J., and Ghosh, L. Epidermal growth factor in the normal and neoplastic kidney and bladder. J. Urol., *139*: 170–175, 1988.

31. Lorberboum, G. H., Barrett, L. V., Kirkman, R. L., et al. Cardiac allograft survival in mice treated with IL-2-PE40. Proc. Natl. Acad Sci. U.S.A., *86*(3): 1008–1012, 1989.

32. Lorberboum, G. H., FitzGerald, D., Chaudhary, V., Adhya, S., and Pastan, I. Cytotoxic activity of an interleukin 2–Pseudomonas exotoxin chimeric protein produced in Escherichia coli. Proc. Natl. Acad. Sci. U.S.A., *85*(6): 1922–1926, 1988.

33. Lorberboum, G. H., Garsia, R. J., Gately, M., et al. IL2-PE664Glu, a new chimeric protein cytotoxic to human-activated T lymphocytes. J. Biol. Chem., *265*(27): 16311–16317, 1990.

34. Lorberboum, G. H., Kozak, R. W., Waldmann, T. A., Bailon, P., FitzGerald, D. J., and Pastan, I. Interleukin 2 (IL2) PE40 is cytotoxic to cells displaying either the p55 or p70 subunit of the IL2 receptor. J. Biol. Chem., *263*(35): 18650–18656, 1988.

35. Marsh, J. W., Srinivasachar, K., and Neville, D. J. Antibody-toxin conjugation. Cancer Treat. Res., *37*(213): 213–237, 1988.

36. Murphy, J. R., Bishai, W., Borowski, M., Miyanohara, A., Boyd, J., and Nagle, S. Genetic construction, expression, and melanoma-selective cytotoxicity of a diphtheria toxin-related a melanocyte hormone fusion protein. Proc. Natl. Acad. Sci. U.S.A., *83*: 8258–8262, 1986.

37. O'Hare, M., Brown, A. N., Hussain, K., et al. Cytotoxicity of a recombinant ricin-A-chain fusion protein containing a proteolytically-cleavable spacer sequence. FEBS Lett., *273*(1–2): 200–204, 1990.

38. Ogata, M., Chaudhary, V. K., FitzGerald, D. J., and Pastan, I. Cytotoxic activity of a recombinant fusion protein between interleukin 4 and Pseudomonas exotoxin. Proc. Natl. Acad. Sci. U.S.A., *86*(11): 4215–4219, 1989.

39. Olsnes, S., Stenmark, H., McGill, S., Hovig, E., Collier, R. J., and Sandvig, K. Formation of active diphtheria toxin in vitro based on ligated fragments of cloned mutant genes. J. Biol. Chem., *264*(22): 12747–12751, 1989.

40. Pai, L. H., Gallo, M. G., FitzGerald, D. J., and Pastan, I. Anti-tumor activity of a transforming growth factor alpha-Pseudomonas exotoxin fusion protein (TGFα-PE40). Cancer Res., *51*: 2808–2812, 1991.

41. Pastan, I., and FitzGerald, D. Pseudomonas exotoxin: Chimeric toxins. J. Biol. Chem., *264*(26): 15157–15160, 1989.

42. Pastan, I., Willingham, M. C., and FitzGerald, D. J. Immunotoxins. Cell, *47*(5): 641–648, 1986.

43. Prior, T. I., FitzGerald, D. J., and Pastan, I. Barnase toxin: A new chimeric toxin composed of pseudomonas exotoxin A and barnase. Cell, *64*(5): 1017–1023, 1991.

44. Prior, T. I., Helman, L. J., FitzGerald, D. J., and Pastan, I. Cytotoxic activity of a recombinant fusion protein between insulin-like growth factor I and Pseudomonas exotoxin. Cancer Res., *51*(1): 174–180, 1991.

45. Siegall, C. B., Chaudhary, V. K., FitzGerald, D. J., and Pastan, I. Cytotoxic activity of an interleukin 6–Pseudomonas exotoxin fusion protein on human myeloma cells. Proc. Natl. Acad. Sci. U.S.A., *85*(24): 9738–9742, 1988.

46. Siegall, C. B., FitzGerald, D. J., and Pastan, I. Cytotoxicity of IL6-PE40 and derivatives on tumor cells expressing a range of interleukin 6 receptor levels. J. Biol. Chem., *265*(27): 16318–16323, 1990.

47. Siegall, C. B., Kreitman, R. J., FitzGerald, D. J., and Pastan, I. Anti-tumor effects of IL6-Pseudomonas exotoxin chimeric molecules against the human hepatocellular carcinoma PLC/PRF/5 in mice. Cancer Res., *51*: 2831–2836, 1991.

48. Siegall, C. B., Nordan, R. P. FitzGerald, D. J., and Pastan, I. Cell-specific toxicity of a chimeric protein composed of interleukin-6 and Pseudomonas exotoxin (IL6-PE40) on tumor cells. Mol. Cell. Biol., *10*(6): 2443–2447, 1990.

49. Siegall, C. B., Schwab, G., Nordan, R. P., FitzGerald, D. J., and Pastan, I. Expression of the interleukin 6 receptor and interleukin 6 in prostate carcinoma cells. Cancer Res., *50*(24): 7786–7788, 1990.

50. Siegall, C. B., Xu, Y. H., Chaudhary, V. K., Adhya, S., Fitzgerald, D., and Pastan, I. Cytotoxic activities of a fusion protein comprised of TGF alpha and Pseudomonas exotoxin. FASEB J., *3*(14): 2647–2652, 1989.

51. Siegall, C. B., Xu, Y. H., Chaudhary, V. K., Adhya, S., FitzGerald, D., and Pastan, I. Cytotoxic activities of a fusion protein comprised of TGF alpha and Pseudomonas exotoxin. FASEB J., *3*(14): 2647–2652, 1989.

52. Till, M. A., Ghetie, V., May, R. D., et al. Immunoconjugates containing ricin A chain and either human anti-gp41 or CD4 kill H9 cells infected with different isolates of HIV, but do not inhibit normal T or B cell function. J. Acquir. Immune Defic. Syndr., *3*(6): 609–614, 1990.

53. Till, M. A., Zolla, P. S., Gorny, M. K., Patton, J. S., Uhr, J. W., and Vitetta, E. S. Human immunodeficiency virus–infected T cells and monocytes are killed by monoclonal human anti-gp41 antibodies coupled to ricin A chain. Proc. Natl. Acad. Sci. U.S.A., *86*(6): 1987–1991, 1989.

54. Vitetta, E. S., Fulton, R. J., May, R. D., Till, M., and Uhr, J. W. Redesigning nature's poisons to create anti-tumor reagents. Science, *238*: 1098–1104,
1987.
55. Williams, D. P., Parker, K., Bacha, P., et al. Diphtheria toxin receptor binding sustitution with interleukin 2: genetic construction and properties of a diphtheria toxin-related interleukin-2 fusion protein. Protein Eng., *1*: 493–498, 1987.

VII
Conclusions

25
Genetically Engineered Toxins in Perspective

Sjur Olsnes *Institute for Cancer Research at the Norwegian Radium Hospital, Oslo, Norway*

Arthur E. Frankel *Altamonte Springs, Florida*

The discovery of hybridoma monoclonal antibodies and the expression of recombinant cell-selective peptide hormones provided a wealth of protein ligands for targeting a variety of dysfunctional or neoplastic cells in humans. Initial successes in treatment of digitalis toxicity with antidigitalis monoclonal antibody (1) and of kidney transplant rejection with anti–T-cell monoclonal antibody (2) were heartening. Further, anti-idiotype therapy of B-cell lymphoma provided a significant number of clinical responses (3). Treatment of hairy cell leukemia with interferon-alpha has also produced a number of lasting remissions (4). Both recombinant G-CSF and GM-CSF have had a dramatic impact on the duration of neutropenia post–bone marrow transplant and postchemotherapy (5,6). Recombinant erythropoietin, insulin, and factor VIII have had major impacts in anemia (7), diabetes (8), and hemophilia A (9), respectively. However, the impact of recombinant proteins for most autoimmune and malignant diseases has been more modest to date. Partly, this may be attributed to the lack of an effector or killing mechanism for causing cancer cell apoptosis after antibody or hormone binding. While the use of humanized or chimeric mouse-human monoclonal antibodies should improve effector cell recruitment and preliminary results with Campath-1H monoclonal antibody for lymphoma support this hypothesis (10), in cases of poorly vascularized solid tumor metastases or limited human effector cell populations, the enhanced tumor killing may be modest. In an attempt to design a therapeutic molecule which could act independently, once administered, to target and eliminate receptor positive cells, investigators in the last 3 decades have attached a variety of poisons to the antibody or hormone ligand. The clinical efficacy of covalently attached cytotoxic drugs has been poor to date, but radionuclides and toxin conjugates have provided a few clinical successes. ^{131}I-labeled antilymphoma (anti-CD37) mouse monoclonal antibody administered in the setting of autologous marrow transplantation has led to a high rate of maintained complete responses in B-cell lymphoma (11). ^{90}Y-conjugated mouse anti-idiotype monoclonals have also produced a high rate of complete responses without requiring marrow rescue (12). Chemical conjugates of mouse monoclonal antibodies with ricin toxin A chain or blocked whole ricin toxin have shown a significant number of partial responses in patients with graft-versus-host disease (13), rheumatoid arthritis (14), and B-cell lymphoma (15,16). The most sophisticated and difficult form of targeted therapeutics involves the genetic engineering of a single protein molecule so that it will be soluble, recoverable in reasonable yields, amenable to simple purification, possess high-affinity cell-selective binding, and cause cell death subsequent to binding with only a few molecules/cell. The initial exploratory steps to produce such drugs were the subject of this volume.

Limitations in size and time limited the scope of this monograph. Chemical conjugates of toxins with antibodies and hormones were not reviewed. Since the publication of the previous book in this series (17), a significant amount of research has been accomplished with antibody-toxin conjugates. New antibodies, new toxins, and new linkers have been tried. Monoclonal antibodies to laryngeal carcinoma (18), retinal pigment epithelium (19), HIVgp120 (20), and *Trypanosoma cruzi* (21) were conjugated to ricin A chain and CD4 was conjugated to *Pseudomonas* exotoxin (22). In each case, cell-selective cytotoxicity was demonstrated. The plant holotoxin mistletoe lectin I A chain (23), plant hemitoxins trichosanthin (24) and barley toxin (25), and the fungal ribotoxin α-sarcin (26) have been conjugated to monoclonal antibodies. The conjugates each were selectively cytotoxic to receptor positive cells. Lambert and Blattler and colleagues developed a new covalent linker which contains an oligosaccharide with a reactive dichloro-triazine group (27). The linker blocks ricin normal cell binding and provides a thioether linkage to monoclonal antibodies. Thus, new toxin conjugates with both in vitro and, in some cases, clinical activity have been produced in the last 3 years. However, neither an update on these advances nor comparisons with the genetically engineered toxins have been attempted. In the next few years, enough data should be available to permit meaningful comparisons between chemically and genetically engineered toxin conjugated in the clinic.

Another important area not addressed in the book because of space is the use of nonprotein synthesis-inhibiting peptide toxins. Cholera A1 peptide has been linked by disulfide exchange to *Wisteria floriburda* lectin (28). The conjugate activated adenylate cyclase in K562 cells. *Clostridium perfringens* phospholipase C was genetically linked to the Fab–anti-TAC and the chimera killed cells at concentration of 10^{-11} M (29). This recombinant product was secreted in an active form from *E coli*—a marked advantage over anti-Tac(Fv)–PE40. An anti-G_{D7} ganglioside monoclonal antibody conjugated to cobra venom factor killed two logs of human neuroblastoma cells (30). Previous concerns over nonspecificity of membrane-lytic toxin conjugates has not been substantiated in vivo. Glucose oxidase conjugated to monoclonal antibodies also shows selective cytotoxicity, which interestingly was dependent on internalization of the conjugate for killing (31). The use of these alternative toxins is likely to increase both with the use of a greater variety of toxophore domains and their application to more diseases. The translocation domains of toxins offer the possibility of delivering other nonlethal enzymes to the cytosol. The internalization of polypeptides such as cholera and pertussis toxins that act on the α-subunits of G-proteins, invasive adenylate cyclases, tyrosine phosphatases, toxins that modify *rho*-proteins, and toxins that act as growth factor may lead to normalization of dysfunctional tissues (see review in Ref 32).

Finally, toxins can be employed to deliver foreign (nontoxic) peptides to the cytosol for vaccine development (33). The introduced peptides could be of virus or intracellular parasite origin. Once in the cytosol, they would be broken down to smaller peptides, enter the endoplasmic reticulum, and then associate with nascent major histocompatibility (MHC) class I antigens. In this way, the peptides could be presented at the cell surface and induce amplification of the relevant population of $CD8^{+}$ cytotoxic T lymphocytes. If this approach is successful, a nontoxic variant of diphtheria toxin could prove to be a better choice for vaccine development than vaccinia virus, which is currently being tried as a vector with the relevant peptide cDNA cloned into the virus genome.

A final series of studies, omitted here, concern Shiga toxin and Shiga-like toxins. The DNA for these toxins has been cloned and expressed in *Escherichia coli* (34). Mutational analysis has been performed which identified critical active site residues (35). Recently, chimeras of Shiga-like toxin with diphtheria B chain and interleukin-2 have been synthesized and expressed (see Chapter 18). Remarkably, this bacterial toxin has the same tertiary structure and mechanism of action as the plant holotoxin A chains and plant hemitoxins (36). It is not known whether this represents divergent evolution from a precursor to bacteria and plants or convergent evolution. The Shiga toxins are an important source of disease throughout the world, including not only diarrheal disease, but also epidemic cases of hemolytic-uremic syndrome (37). We will next focus on the topics covered by this monograph and attempt to note recurrent observations, highlight missing data, and summarize ongoing studies.

Part I described molecular biology, cell biology, and pharmacology techniques essential to the field. The plethora of approaches at each step can often be confusing and delay progress. Cloning methods include the polymerase chain reaction (PCR) of chromosomal DNA and λ genomic and cDNA libraries screened by oligonucleotides or antibodies. Review of the experience with ricin, abrin, trichosanthin, momorcharin, α-sarcin, restrictocin, saporin-6 (SO6), and luffin suggest that probing cDNA libraries with degenerate large oligonucleotides is most likely to yield the DNA sequence. This reduces the complexity of the DNA being screened and avoids the selection of irrelevant cross-reactive protein sequences. Expression systems for recombinant toxins include in vitro transcription of PCR-amplified genes with subsequent cell-free translation, translation of injected mRNA in *Xenopus* oocytes, and in vivo expression in *Escherichia coli, Saccharomyces cerevisiae, Aspergillus, Spodoptera frugiperda* cells, and COS cells. The first two methods provide a rapid source of protein with sufficient material for selective biochemistry and cell physiology studies. The cell-free approach often provides active protein even when the protein folds improperly and is precipitated or degraded in bacteria; this was used successfully with diphtheria toxin (38). The *Xenopus* system provides eukaryotic posttranslational processing. This approach worked well for ricin B chain (39). *E. coli* expression plasmids utilize the *lac*, λ, *tac*, or T7 polymerase promoters, and usually the pBR322 ori and β-lactamase gene for plasmid maintenance. Both the ompA and Bla signal sequences have been used. Many expression vectors produce fusion proteins to stabilize the recombinant product in vivo and aid in purification. The attached polypeptides range from short six or seven amino acid additions which are recognized by monoclonal antibodies or bind heavy metals, to large proteins—glutathione synthetase, β-galactosidase, and maltose-binding protein. Ribosome-inactivating (RIPs) were well expressed in *E. coli* and the secretion system with the Bla signal peptide for *Mirabilis* protein appeared the best for "wild-type" RIPs. Mutant proteins each have unique properties complicating purification and the fusion expression system with β-galactoside, the FLAG peptide, or glutathione synthetase appeared most useful. Bacterial toxins (diphtheria toxin, DT; and *Pseudomonas* exotoxin, PE) and chimeras expressed in *E. coli* were recovered in highest yield from inclusion bodies after expression from the Studier T7 promoter plasmids. Minimal work was described using the yeast, insect cells, or mammalian cell expression systems. The sensitivity of eukaryotic ribosomes and elongation factor 2 (EF-2) to these toxins complicates such approaches and requires either resistant cells (39) or excellent segregation of toxin to the secretory pathways (40). Toxin purification utilizes denaturation-renaturation from inclusion bodies, size exclusion chromatography, ion exchange or hydrophobic interaction chromatography, or affinity chromatography using antibodies or substrate analogs (Cibacron Blue or NAD^+) or receptors (IL-2 receptor). Wild-type recombinant RIPs and fungal ribotoxins and plant A chains were soluble in bacteria or media and were readily purified by S-Sepharose, size exclusion, and Blue Sepharose chromatography. Mutant or variant plant toxins were more readily purified using fusion proteins (41). Plant lectin polypeptides have not yet been purified in significant quantities, although when prepared, we assume lactosyl-affinity chromatography will permit easy purification. However, mutant plant B chains may require antibody affinity chromatography. To date, denaturation/renaturation from inclusion bodies is the method of choice for bacterial toxins and chimeras. Yields have not been documented, but the large dilutions required and the likelihood of significant mismatching of disulfides has complicated scale-up for clinical trials. While methodologies for measuring protein synthesis inhibition in vitro and vivo and cell-binding affinities were described, no direct measure of membrane translocation were reported. Modification of toxin A moieties by engineering CAAX-boxes to their C-terminal ends may allow identification of those molecules that have reached the cytosol where they become prenylated (42). Other modifications to generate targets for enzymes of known cellular localization may allow tracing of the toxin along the endocytic and vesicular routes. Overall potency and efficacy of chimeras were rarely quantitated. To permit comparisons between chimeras and chemical toxin conjugates, the detailed kinetic analysis and measurement of clonogenic cell kill is recommended (43). The in vivo studies to date have been even more sparse. Efforts to quantitate half-lives in different species, tissue distributions, and methods of clearance and catabolism will be very useful at the time of clinical trials. The careful search for toxicities in vivo will be well merited based on the frequent and serious neurotoxicities and vascular leak syndromes seen with antibody-toxin conjugates in vivo (44). At the time of this monograph, the clinical behavior of genetically engineered toxins is still anecdotal. Nevertheless, future well-planned clinical protocols with adequate pharmacological monitoring will be essential. Emphasis should be placed on quantitating humoral immune responses to the chimeras in addition to pharmacology, toxicity, and clinical response. Development of receptor negative or poorly internalizing resistant tumor cells should be sought. The remainder of the chapter addresses future experiments in the field.

A question which often arises is how safe is cloning of toxins? The initial strict regulations limiting the handling of genes for active toxin retarded the development in the field. With time these regulations have become more liberal and a number of active toxins have now been cloned and expressed. Molecular techniques have, in fact, entered most areas of toxin research. Some of the most active toxins are, however, still considered too dangerous to be cloned in *E. coli*. This is particularly the case with diphtheria toxin, botulinum toxin, and tetanus toxin.

Various approaches have been taken to circumvent the potential dangers in cloning active toxins. In a number of cases, nontoxic mutants have been cloned and used in studies on various aspects of toxin action. In the case of diphtheria toxin, active molecules have been formed from the separately cloned A and B fragments by mixing and dialyzing the translation products (33). Another approach involves ligation of the genes for the separate fragments, amplification of the ligation product by PCR, and transcription of the amplified DNA in vitro with subsequent translation of the mRNA in a cell-free protein synthesizing system (38). Since the specific activity of toxins is often very high, sufficient material to carry out toxicity experiments may be obtained in this way.

Some toxins are still considered too dangerous to clone and others can only be cloned under highly restricted conditions. The chances of producing pathogenic *E. coli* strains by expressing a toxin alone may not be great, as the mechanism of pathogenicity is usually complex and requires several different genes to be expressed. Examples are noninvasive strains of *Shigella dysenteriae* 1, which produce Shiga toxin in good yield, but which appear to be nonpathogenic if they do not carry functional genes coding for invasiveness (45). However, until the mechanisms of pathogenesis of bacteria are known in more detail, it is advisable to be careful, particularly with the highly lethal botulinum and tetanus toxins.

Structure-function studies of toxins have been greatly aided by the availability of x-ray crystallographic structures for ricin and *Pseudomonas* exotoxins. With the eventual solution of diphtheria toxin and cholera toxin three-dimensional structures (46,47), a wealth of information will be available for hypotheses on mechanisms of toxin action at the molecular level. The testing of recombinant toxins with single amino acid residue modifications will support or nullify some of these hypotheses. X-ray diffraction analysis of co-crystals of toxin and substrate analogs would rapidly yield information on the depurination reaction and the ADP-ribosylation reaction. Alternatively, a system to isolate mutant toxin genes producing protein with altered binding to sugars, rRNA, or EF-2, but intact immunological cross-reactivity would permit an unbiased sampling of modified amino acid residues critical in binding but not in maintaining proper protein folding. We are currently employing an expression system producing fusion proteins with the C-terminus of the gene III product of M13 and Fd phages (48–51). Over 10^{14} phage particles can be examined and binding to sugars and nucleotides assessed. The enzymatically active part of toxins has proven a useful tool in studies of development. Defined organs can be inhibited in their development by engineering the A fragment of diphtheria toxin behind selected promoters, such as that of crystallin, and then introducing the constructs as transgenes into mice (52). This principle could have wide applications in the future.

The fascinating homology between plant RIPs and the A chain of plant holotoxins and the ubiquity of the ribosome-inactivating genes throughout the plant kingdom suggests they have an important normal function in plants. They are part of a multigene family in many plants and their products constitute greater than 5% of total cell protein frequently. The physiological role for the proteins remains unknown. They may function to protect the plant from herbivores or microorganisms. No studies have directly addressed this question. One approach suggested by Piatak (see Chapter 6) would be to introduce RIP genes into plants lacking them and test for conferred resistance to pathogens.

The clinical applications of genetically engineered toxins are likely to be diverse. In addition to malignant disease, cell-selective ablation may be useful in a number of clinical settings. Fibrosis from orbital trauma may be reversed with intraocular targeted toxins (53). Multiple sclerosis and acute allergic encephalomyelitis may respond to anti–T-cell toxins (54). Rheumatoid arthritis, lupus, and dermatomyositis may also respond to removal of activated T cells (55). Parasitic and chronic viral infections may respond to toxins targeted to pathogen cell surface products (56). Allergic diseases may be treatable with the Fc fragment of immunoglobulin E fused to toxins (57). Arteriosclerosis may respond to toxins specific for plaque antigens (58). Graft-versus-host disease after bone marrow transplantation may be controllable both with donor marrow purging of T cells and removal of activated graft T cells in the recipient (59). Rejection of organ transplants may be prevented or diminished with anti–T-cell toxins (60). Unmodified plant toxins possess potent activity against the human immunodeficiency virus (HIV) (61). The mechanism for the antiviral properties is unknown. Clinical trials of trichosanthin combined with azidothymidine (AZT) are underway. Plant hemitoxins have been used for centuries in China as an abortifacient and for treatment of trophoblastic disease (62). Again, the drug pharmacology is unknown. Both the antiviral behavior and antitrophoblastic potential of hemitoxins may be due to enhanced endocytosis and permeabilized vesicle membranes permitting entrance and escape of hemitoxins to the cytosol.

In the field of chimeric toxins the use of molecular techniques has opened up a variety of possibilities. The most important advantage is that chimeric toxins can now be made in a well-defined manner without batch variations. Chemical coupling involves, in most cases, linkages between groups (most often lysines) that are present in a number of copies per protein molecule. Therefore, the chimeric product obtained represents a mixture of molecules linked together in different ways. The relative amounts of the different conjugates in the mixture are likely to vary among batches due to minor differences in preparation. It is also likely that the different chimeras in the mixture are handled differently by cells and organisms, and it is difficult to establish which of these are responsible for the observed biological effect. This puts a serious limitation on experimentation to improve the properties of the conjugate.

In contrast to this, chimeric toxins fused at the genetic level are well defined and easily amenable to systematic modifications in order to increase their activity at the cellular level, reduce their side effects, increase their protease resistance and stability in circulation, and so forth. Systematic experimentation along these lines may reveal general rules on how to obtain conjugates with high stability, good ability to penetrate tissues and tumors, and potent cytotoxic activity once the conjugates are bound to the target cells. This is the way that future reseach is likely to proceed.

Another advantage is the possibility to eliminate groups that could interfere with the desired action of the conjugate. An example is ricin A chain, which is normally glycosylated, but which can be produced unglycosylated in bacteria (65). This avoids undesired toxic effects due to uptake by the mannose receptors present on macrophages and related cells.

The intricate mechanism of action of protein toxins has made them interesting tools in cell biology and versatile effector molecules in targeted therapy. In both fields, genetically engineered toxins have opened up a variety of new possibilities and challenging questions.

REFERENCES

1. Smith, T. W., Bullet, V. P., Jr., Haber, E., Pozzard, H., Marcus, F. I., Bremner, W. F., Schulman, I. C., and Phillips, A. Treatment of life-threatening digitalis intoxication with digoxin-specific Fab antibody ligaments: experience in 26 cases. N. Engl. J. M., *307*: 1357–1362, 1982.
2. Hesse, U. I., Wienand, P., Baldamus, C., and Arns, W. Preliminary results of a prospectively randomized trial of ALG vs OKT3 for steroid-resistant rejection after renal transplantation in the early postoperative period. Transplant. Proc., *22*: 2273–2274, 1990.
3. Brown, S. L., Miller, R. A., Horning, S. J., Czerwinski, D., Hart, S. M., McElderry, R., Basham, T., Warnke, R. A., Merigan, T. C., and Levy, R. Treatment of B cell lymphoma with anti-idiotype antibodies alone and in combination with alpha interferon. Blood, *73*: 651–661, 1989.

4. Quesada, J., Hersh, E., Manning, J., Reuben, J., Keating, M., Schnipper, E., Im, L., and Guttenman, J. Treatment of hairy cell leukemia with recombinant α-interferon. Blood, 68: 493–497, 1986.

5. Gabrilove, J., Jakubowski, A., Scher, H., Sternberg, C., Wong, G., Grous, J., Yagoda, A., Fain, K., Moore Mas, Clarkson, B., Oetthen, H., Alton, K., Welte, K., and Souza, L. Granulocyte colony stimulating factor reduces neutropenia and associated morbidity of chemotherapy for transitional cell carcinoma of the urothelium. N. Engl. J. M., 318: 1414–1422, 1988.

6. Fruman, W. L., Fairclough, D. L., Hugh, R. D., Pratt, C. B., Stute, N., Petros, W. P., Evans, W. E., Bowman, L. C., Douglass, E. C., Santana, V. M., Meyer, W. H., and Crist, W. M. Therapeutic effects and pharmacokinetics of recombinant human granulocyte-macrophage colong-stimulating factor in childhood cancer patients receiving myelosuppressive chemotherapy. J. Clin. Oncol., 9: 1022–1028, 1991.

7. Watson, A. J., Gimenez, I. F., Colton, S., Walser, M., and Spivak, A. Treatment of the anemia of chronic renal failure with subcutaneous recombinant human erythropoietin. Am. J. M., 89: 432–435, 1990.

8. Patrick, A. W., Collier, A., Matthews, O. M., Macintyre, C. C., and Clarke, B. F. The importance of the time interval between insulin injection and breakfast in determining postprandial glycaemic control—a comparison between human and porcine insuline. Diabetic Med., 5: 32–35, 1988.

9. Aronson, D. L. The current status of recombinant human factor VIII. Semin. Hematol. 28: 55–56, 1991.

10. Hale, G., Clark, M. R., Marcus, R., Winter, G., Dyer, M., Phillips, J., Riechmann, L., and Waldman, H. Remission induction in non-Hodgkins lymphoma with reshaped human monoclonal antibody CAMPATH-1H. Lancet, 2: 1394–1399, 1988.

11. Press, O. W., Eary, J. F., Badger, C. C., Martin, P. J., Appelbaum, F. R., Levy, R., Miller, R., Brown, S., Nelp, W. B., Krohn, K. A., Fisher, D., DeSantes, K., Porter, B., Kidd, P., Thomas, E. D., and Bernstein, I. D. Treatment of refractory non-Hodgkins lymphoma with radiolabeled MB-1 (anti-CD37) antibody. J. Clin. Oncol., 7(8): 1027–1038, 1989.

12. Miller, R. Personal Communication, 1991.

13. Byers, V., Henslee, P., Kernan, N., Blazar, B., Gingrich, R., Phillips, G., LeMaistre, C., Gilliland, G., Antin, J., Martin, P., Tutsche, P., Trown, P., Ackerman, S., O'Reilly, R., and Scannon, P. Use of an anti–pan T-lymphocyte ricin A chain immunotoxin in steroid-resistant acute graft-versus-host disease. Blood, 75: 1426–1432, 1990.

14. Byers, V., Strand, V., Saria, E., Ma, J., and the XOMA Rheumatoid Arthritis Treatment Group. Patients with rheumatoid arthritis treated with a pan-T lymphocyte immunotoxin: Phase II studies. Proceedings of the Second International Symposium on Immunotoxins, 1990, p. 12.

15. Vitetta, E. S., Stone, M., Amlot, P., Fay, J., May, R., Till, M., Newman, J., Clark, P., Collins, R., Cunningham, D., Ghetie, V., Uhr, J. W., and Thorpe,

P. E. Phase I immunotoxin trial in patients with B cell lymphoma. Cancer Res., *51*: 4052–4058, 1991.

16. Grossbard, M., Freeman, A., Ritz, J., Coral, F., Goldmacher, V., Eliseo, L., Spector, N., Dean, K., Lambert, J., Blattler, W., Taylor, J., and Nadler, L. Serotherapy of B cell neoplasm with anti-B4 blocked ricin: A phase I trial of daily bolus infusions. Blood, in press, 1992.

17. Frankel, A. (ed). Immunotoxins. Boston: Kluwer Academic, 1988.

18. Zenner, H. P. A monoclonal immunotoxin against laryngeal carcinoma cells. Otolaryngol. Pol., *44*: 214–215, 1990.

19. Davis, A. A., Whidby, D. E., Privette, T., Houston, L. L., and Hunt, R. C. Selective inhibition of growing pigment epithelial cells by a receptor-directed immunotoxin. Invest. Ophthalmol. Vis. Sci., *31*: 2514–2519, 1990.

20. Pincus, S., Cole, R., Hersh, E., Lake, D., Mashuo, Y., Durda, P., and McClure, J. In vitro efficacy of anti-HIV immunotoxins targeted by various antibodies to the envelope protein. J. Immunol., *146*: 4315–4324, 1991.

21. Teixeira, A. R., and Santana, J. M. Chagas disease. Immunotoxin inhibition of *Trypanosoma cruzzi* release from infected cells in vitro. Lab. Invest., *63*: 248–252, 1990.

22. Ashorn, P., Moss, B., Weinstien, J. N., Chaudhary, V. K., FitzGerald, D. J., Pastan, I., and Berger, E. A. Elimination of infectious human immuno-deficiency virus from human T-cell cultures by synergistic action of CD4-Pseu-domonas exotoxin and reverse transcriptase inhibitors. Proc. Nat. Acad. Sci. U.S.A., *87*: 8889–8893, 1990.

23. Widelocha, A., Sandvig, K., Walzel, H., Radzikowsky, C., and Olsnes, S. Internalization and action of an immunotoxin containing mistletoe lectin A-chain. Cancer Res., *51*: 916–920, 1991.

24. Wang, Q. C., Ving, W. B., Xie, H., Zhang, Z. C., Veng, Z. H., and Ling, L. Q. Trichosanthin-monoclonal antibody conjugate specifically cytotoxic to human hepatoma cells in vitro. Cancer Res., *51*: 3353–3355, 1991.

25. Ovadia, M., Hager, C. C., and Oeltmann, T. N. An antimelanoma-barley ribosome inactivating protein conjugate is cytotoxic to melanoma cells in vitro. Anticancer Res., *10*: 671–675, 1990.

26. Wawrzynczak, E. J., Henry, R. V., Cumber, A. J., Parnell, G. D., Derbyshire, E. J., and Ulbrich, N. Biochemical, cytotoxic and pharmacokinetic properties of an immunotoxin composed of a mouse monoclonal antibody Fib75 and the ribosome-inactivating protein alpha-sarcin from *Aspergillus giganteus*. Eur. J. Biochem., *196*: 203–209, 1991.

27. Lambert, J. M., Goldmacher, V. S., Collinson, A. R., Nadler, L. M., and Blattler, W. A. An immunotoxin prepared with blocked Ricin: A natural plant toxin adapted for therapeutic use. Cancer Res., *51*: 6236–6242, 1991.

28. van Heyningen, S. A conjugate of the A1 peptide of cholera toxin and the lectin of *Wisteria floribunda* that activates the adenylate cyclase of intact cells. FEBS *164*: 132–134, 1983.

29. Chovnick, A., Schneider, W. P., Yun Tso, J., Queen, C., and Nan Chang,

C. A recombinant, membrane-acting immunotoxin. Cancer Res., *51*: 465–467, 1991.

30. Juhl, H., Petrella, E. C., Cheung, N-K. V., Bredehorst, R., and Vogel, C. W. Complement killing of human neuroblastoma cells: A cytotoxic monoclonal antibody and its f(ab)$_2$-cobra venom factor conjugate are equally cytotoxic. Mol. Immunol., *27*: 957–964, 1990.

31. Muzykantov, V. R., Trubetskaya, O. V., Puchnina, E. A., Sakharow, D. V., and Domogatsky, S. P. Cytotoxicity of glucose oxidase conjugated with antibodies to target cells: Killing efficiency depends on the conjugate internalization. Biochim. Biophys. Acta, *1053*: 27–31, 1990.

32. Olsnes, S. Kozlov, J. V., van Deurs, B., and Sandvig, K. Bacterial protein toxins acting on intracellular targets. Semin. Cell Biol., in press, 1992.

33. Stenmark, H., Moskaug, J. O., Madshus, I. H., Sandvig, K., and Olsnes, S. Peptides fused to the amino-terminal end of diphtheria toxin are translocated to the cytosol. J. Cell Biol., *113*: 1025–1032, 1991.

34. Calderwood, S. B., Auclair, F., Donohue-Rolfe, A., Keusch, G. T., and Mekalanos, J. J. Nucleotide sequence of the Shiga-like toxin genes of *Escherichia coli*. Proc. Natl. Acad. Sci. U.S.A., *84*: 4364–4368, 1987.

35. Hovde, C. J., Calderwood, S. B., Mekalanos, J. J., and Collier, J. R. Evidence that glutamic acid 167 is an active-site residue of Shiga-like toxin I. Proc. Natl. Acad. Sci. U.S.A., *85*: 2568–2572, 1988.

36. Collins, E. J., Robertus, J. D., LoPresti, M., Stone, K. L., Williams, K. R., Wu, P., Hwang, K., and Piatak, M. Primary amino acid sequence of α-trichosanthin and molecular models for abrin A-chain and α-trichosanthin. J. Biol. Chem., *266*: 8665–8669, 1990.

37. Furutani, M., Ito, K., Oku, Y., Takeda, Y., and Igarashi, K. Demonstration of RNA N-glycosidase activity of a Vero toxin (VT2 variant) produced by Escherichia coli 091:H21 from a patient with the hemolytic uremic syndrome. J. Microbiol. Immunol., *34*: 387–392, 1990.

38. Olsnes, S., Stenmark, H., McGill, S., Hovig, E., Collier, R. J., and Sandvig, K. Formation of active diphtheria toxin in vitro based on ligated fragments of cloned mutant genes. J. Biol. Chem., *264*: 12749–12751, 1989.

39. Richardson, P. T., Gilmartin, P., Colman, A., Roberts, L. M., and Lord, J. M. Expression of functional ricin B chain in *Xenopus oocytes*. Biotechnology, *6*: 565–570, 1988.

40. Perentesis, J. P., Genbauffe, F. S., Veldman, S. A., Galeotti, C. L., Livingston, D. M., Bodley, J. W., and Murphy, J. R. Expression of diphtheria toxin fragment A and hormone-toxin fusion proteins in toxin-resistant yeast mutants. Proc. Natl. Acad. Sci. U.S.A., *85*: 8386–8390, 1988.

41. Habuka, N., Akiyama, S., Tauge, H., Miyano, M., Mataumoto, T., and Noma, M. Expression and secretion of *mirabilis* antiviral protein in *Escherichia coli* and its inhibition of in vitro eukaryotic and prokaryotic protein synthesis. J. Biol. Chem., *265*: 10988–10992, 1990.

42. Frankel, A., Welsh, P., Richardson, J., and Robertus, J. D. Role of arginine

180 and glutamic acid 177 of ricin toxin A chain in enzymatic inactivation of ribosomes. Mol. Cell. Biol., *10*: 6257–6263, 1990.

43. Powers, S. Protein prenylation: a modification that sticks. Curr. Biol., *1*: 114–116, 1991.

44. Sung, C., Wilson, D., and Youle, R. J. Comparison of protein synthesis inhibition kinetics and cell killing induced by immunotoxins. J. Biol. Chem., *266*: 14159–14162, 1991.

45. Oeltmann, T. N., and Frankel, A. E. Advances in immunotoxins. FASEB J., *5*: 2334–2337, 1991.

47. Fontaine, A., Arondel, J., and Sansonetti, P. J. Role of Shiga toxin in the pathogenesis of bacillary dysentery, studied by using a tox⁻ mutant of *Shigella dysenteriae* 1. Infect. Immun., *56*: 3099–3109, 1988.

48. Kantardjieff, K., Dijkstra, V., Westbrook, E., Barbieri, J., Carroll, S., Collier, R. J., and Eisenberg, D. Structural studies of diphtheria toxin. *In*: D. Oxender (ed.), Proteins Structure, Folding and Design, Vol. II, pp. 187–200. New York: Alan R. Liss.

49. Maulik, P. R., Reed, R. A., and Shipley, G. G. Crystallization and preliminary x-ray diffraction study of cholera toxin B-subunit. J. Biol. Chem., *263*: 499–501, 1988.

50. Scott, J. K., and Smith, G. P. Searching for peptide ligands with an epitope library. Science, *249*: 386–389, 1990.

51. McCafferty, J., Griffiths, A. D., Winter, G., and Chiswell, D. J. Phage antibodies: Filmentous phage displaying antibody variable domains. Nature, *348*: 552–554, 1990.

52. Bass, S., Greene, R., and Wells, J. Hormone phase: An enrichment method for variant proteins with altered binding properties. Proteins, *8*: 309–311, 1990.

53. Barbas III, C. F., Kang, A. S., Lerner, R. A., and Benkovic, S. J. Assembly of combinatorial antibody libraries on phage surfaces: The gene III site. Proc. Natl. Acad. Sci. U.S.A., *88*: 7978–7982, 1991.

54. Breitman, M. L., Rombola, H., Maxwell, I. H., Klintworth, G. K., and Bernstein, A. Genetic ablation in transgenic mice with an attenuated diphtheria toxin A gene. Mol. Cell. Biol., *10*: 474–479, 1990.

55. Hermsen, V. M., Fulcher, S. F., Spiekerman, A. M., Phinizy, J. L., and Di Tullio, N. W. Long-term inhibition of cellular proliferation by immunotoxins. Arch. Ophthalmol., *108*: 1009–1011, 1990.

56. Rose, J. W., Lorberboum-Galski, H., FitzGerald, D., McCarron, R., Hill, K. E., Townsend, J. J., and Pastan, I. Chimeric cytotoxin IL2-PE40 inhibits relapsing experimental allergic encephalomyelitis. J. Neuroimmunol., *32*: 209–217, 1991.

57. Lorberboum-Galski, H., Lafyatis, R., Case, J. P., FitzGerald, D., Wilder, R. L., and Pastan, I. Administration of IL-2-PE40 via osmotic pumps prevents adjuvant induced arthritis in rats. Improved therapeutic index of IL-2-PE40 administered by continuous infusion. Int. J. Immunopharmacol., *13*: 305–315, 1991.

58. Barnett, B. B., Burns, N. J., Park, K. J., Dawson, M. I., Kende, M., and Sidwell, R. W. Antiviral immunotoxins: Antibody-mediated delivery of gelonin inhibits *Pichinde* virus replication in vitro. Antiviral Res., *15*: 125–138, 1991.

59. Slater, J. E., Boltansky, H., and Kaliner, M. IgE immunotoxins. Effect of an IgE-ricin a chain conjugate on rat skin histamine content. J. Immunol., *140*: 807–811, 1988.

60. Sato, R., Komine, Y., Imanaaaka, T., and Takano, T. Monoclonal antibody EMR1a/212D recognizing site of deposition of extracellular lipid in atherosclerosis. J. Biol. Chem., *34*: 21232–21236, 1990.

61. Vallera, D. A., Carroll, S. F., Snover, D. C., Carlson, G. J., and Blazar, R. R. Toxicity and efficacy of anti–T-cell ricin toxin A chain immunotoxins in a murine model of established graft-versus-host disease induced across the major histocompatibility barrier. Blood, *77*: 182–194, 1991.

62. Herbort, C. P., de Smet, M. D., Roberge, F. G., Nussenblatt, R. B., FitzGerald, D., Lorberboum-Galski, H., and Pastan, I. Treatment of corneal allograft rejection with the cytotoxin Il-2-PE40. Transplantation, *52*: 470–474, 1991.

63. McGrath, M. S., Hwang, K. M., Caldwell, S. E., Gaston, I., Luk, K.-C., Wu, P., Ng. V. L., Crowe, S., Daniels, J., Marsh, J., Deinhart, T., Lekas, P. V., Vannari, J. C., Yeung, H. W., and Lifson, J. D. GLAZZ3: An inhibitor of human immunodeficiency virus replication in acutely and chronically infected cells of lymphocyte and mononuclear phagocyte lineage. Proc. Nat. Acad. Sci. U.S.A., *86*: 2844–2848, 1989.

64. Lu, P., and Jin, Y. C. Trichosanthin in the treatment of hydatidiform mole. Clinical analysis of 52 cases. Chin. Med. J. (Peking), *103*: 183–185, 1990.

65. O'Hare, M., Roberts, L. M., Thorpe, P. E., Watson, G. J., Prior, B., and Lord, J. M. (1987) Expression of ricin A chain Escherichia coli. FEBS leH., *216*: 73–78.

Appendix
Primary Amino Acid Sequences of Toxins

Paul Sehnke *Florida Hospital Cancer and Leukemia Research Center, Altamonte Springs, Florida*

Alexander Tonevitsky *All-Union Research Institute of Genetics and Selection of Industrial Microorganisms, Moscow, Russia*

The explosion of the toxins field and its applications in the last few years has created an enormous wealth of molecular biological data presented in various scientific publications. For the benefit of the reader, we present several of the available primary amino acid sequences of toxins, including references. Many of the nucleotide sequences are also published in these papers; however, due to spatial constraints, they are not presented here.

Abrin A chain

```
EDRPIKFSTE GATSQSYKQF IEALRERLRG GLIHDIPVLP DPTTLQERNR

YITVELSNSD TESIEVGIDV TNAYVVAYRA GTQSYFLRDA PSSASDTLFY

GTDQHSLPFY GTYGDLERWA HQSRQQIPLF LQALTHGISF FRSGGNDNEE

KARTLIVIIQ MVAEAARFRY ISNRVRVSIQ TGTAFQPDAA MISLENNWDN

LRGVQESVQD TFPNQVTLTN IRNEPVIVDS LSHPTVAVLA LMLFVCNPPN
```

Funatsu et al. (1988) Agric. Biol. Chem. **52**, 1095

Barley ribosome–inactivating protein

```
MAAKMAKNVD KPLFTATFNV QASSADYATF IAGIRNKLRN PAHFSHNRPV

LPPVEPNVPP SRWFHVVLKA SPTSAGLTLA IRADNIYLEG FKSSDGTWWE

LTPGLIPGAT YVGFGGTYRD LLGDTDKLTN VALGRQQLAD AVTALHGRTK

ADKPSGPKQQ QAREAVTTLL LMVNEATRFQ TVSGFVAGLL HPKAVEKKSG

KIGNEMKAQV NGWQDLSAAL LKTDVKPPPG KSPAKFAPIE KMGVRTAVQA

ANTLGILLFV EVPGGLTVAK ALELFHASGGK
```

Leah et al. (1991) J. Biol. Chem. **266**, 1564

Dianthin 30

signal peptide
←─────────────────────→
```
mkiylvaaia wilfqssswt tdaATAYTLN LANPSASQYS SFLDQIRNNV

RDTSLIYGGT DVAVIGAPST TDKFLRLNFQ GPRGTVSLGL RRENLYVVAY

LAMDNANVNR AYYFKNQITS AELTALFPEV VVANQKQLEY GEDYQAIEKN

AKITTGDQSR KELGLGINLL ITMIDGVNKK VRVVKDEARF LLIAIQMTAE

AARFRYIQNL VTKNFPNKFD SENKVQFQVS WSKISTAIFG DCKNGVFNKD

YDFGFGKVRQ AKDLQMGLLK YLGRPKSSSI EANSTDDTAD VL
```

Legname et al. (1991) Biochim. Biophys. Acta **1090**, 119

Diphtheria toxin A fragment

```
GADDVVDSSK SFVMENFSSY HGTKPGYVDS IQKGIQKPKS GTQGNYDDDW

KGFYSTDNKY DAAGYSVDNE NPLSGKAGGV VKVTYPGLTK VLALKVDNAD

TIKKDLGLSL TEPLMEQVGT EEFIKRFGDG ASRVVLSLPF ADGSSSVDYI

NNWDQAKALS VELEINFDTR GKRGNDAMYD YMAQACAGNR VRR
```

Greenfield et al. (1983) Proc. Natl. Acad. Sci. U.S.A. **80**, 6855

Luffin

```
DVRFSLSGSS STSYSKFIGD LRKALPSNGT VYNLTILLSS ASGASRYTLM

TLSNYDGKAI TVAVDVSQLY IMGYLVNSTS YFFNESDAKL ASQYVFKGST

IVTLPYSGNY EKLQTAAGKI REKIPLGFPA LDSALTTIFH YDSTAAAAAF

LVILQTTAEA SRFKYIEGQI IERISKNQVP SLATISLENS LWSALSKQIQ

LAQTNNGTFK TPVVITDDKG QRVEITNVTS KVVTKNIQLL LNYKQVA
```

Islam et al. (1990) Agric . Biol. Chem. **54**, 1343

Maize ribosome-inactivating protein

```
MAEITLEPSD LMAQTNKRIV PKFTEIFPVE DANYPYSAFI ASVRKDVIKH

CTDHKGIFQP VLPPEKKVPE LWFYTELKTR TSSITLAIRM DNLYLVGFRT

PGGVWWEFGK DGDTHLLGDN PRWLGFGGRY QDLIGNKGLE TVTMGRAEMT

RAVNDLAKKK KMATLEEEEV KMQMQMPEAA DLAAAAAADP QADTKSKLVK

LVVMVCEGLR FNTVSRTVDA GFNSQHGVTL TVTQGKQVQK WDRISKAAFE

WADHPTAVIP DMQKLGIKDK NEAARIVALV KNQTTAAAAT AASADNDDEE

A
```

Walsh et al. (1991) J. Biol. Chem. 266, 23423

Mirabilis toxin

```
MAPTLEGTIA SLDLNNPTTY LSFITNIRTK VADKTEQCTI QKISKTFTQR

YSYIDLIVSS TQKITLAIDM ADLYVLGYSD IANNKGRAFF FKDVTEAVAN

NFFPGATGTN RIKLTFTGSY GDLEKNGGLR KDNPLGIFRL ENSIVNIYGK

AGDVKKQAKF FLLAIQMVSE AARFKYISDK IPSEKYEEVT VDEYMTALEN

NWAKLSTAVY NSKPSTTTAT KCQLATSPVT ISPWIFKTVE EIKLVMGLLK

SS                                    Habuka et al. (1989) J. Biol Chem. 264, 6629
```

α-Momorcharin

```
          signal peptide
    ◄─────────────────────────────►
msrfsvlsfl ilaiflggsi vkgDVSFRLS GADPRSYGMF IKDLRNALPF

REKVYNIPLL LPSVSGAGRY LLMHLFNYDG KTITVAVDVT NVYIMGYLAD

TTSYFFNEPA AELASQYVFR DARRKITLPY SGNYERLQIA AGKPREKIPI

GLPALDSAIS TLLHYDSTAA AGALLVLIQT TAEAARFKYI EQQIQERAYR

DEVPSLATIS LENSWSGLSK QIQLAQGNNG IFRTPIVLVD NKGNRVQITN

VTSKVVTSNI QLLLNTRNIA EGDNGDVSTT HGFSSY
```

```
                              Ho et al. (1991) Biochim. Biophys. Acta 1088, 311
```

Pseudomonas exotoxin domain III

```
SGDALLERNY PTGAEFLGDG GDVSFSTRGT QNWTVERLLQ AHRQLEERGY

VFVGYHGTFL EAAQSIVFGG VRARSQDLDA IWRGFYIAGD PALAYGYAQD

QEPDARGRIR NGALLRVYVP RSSLPGPYRT SLTLAAPEAA GEVERLIGHP

LPLRLDAITG PEEEGGRLET ILGWPLAERT VVIPSAIPTD PRNVGGDLDP

SSIPDKEQAI SALPDYASQP GKPPREDLK
```

```
                              Gray et al. (1984) Proc. Natl. Acad. Sci. USA 81, 2646
```

Ricin

signal peptide

myavatwlcf gstsgwsftl ednnIFPKQY PIINFTTAGA TVQSYTNFIR

AVRGRLTTGA DVRHEIPVLP NRVGLPINQR FILVELSNHA ELSVTLALDV

TNAYVVGYRA GNSAYFFHPD NQEDAEAITH LFTDVQNRYT FAFGGNYDRL

EQLAGNLREN IELGNGPLEE AISALYYYST GGTQLPTLAR SFIICIQMIS

EAARFQYIEG EMRTRIRYNR RSAPDPSVIT LENSWGRLST AIQESNQGAF

ASPIQLQRRN GSKFSVYDVS ILIPIIALMV YRCAPPPSSQ FSLLIRPVVP

NFNADVCMDP EPIVRIVGRN GLCVDVRDGR FHNGNAIQLW PCKSNTDANQ

LWTLKRDNTI RSNGKCLTTY GYSPGVYVMI YDCNTAATDA TRWQIWDNGT

IINPRSSLVL AATSGNSGTT LTVQTNIYAV SQGWLPTNNT QPFVTTIVGL

YGLCLQANSG QVWIEDCSSE KAEQQWALYA DGSIRPQQNR DNCLTSDSNI

RETVVKILSC GPASSGQRWM FKNDGTILNL YSGLVLDVRA SDPSLKQIIL

YPLHGDPNQI WLPLF

Lamb et al. (1985) Eur. J. Biochem. **148**, 265

Saporin-6

VTSITLDLVN PTAGQYSSFV DKIRNNVKDP NLKYGGTDIV IGPPSKEKFL

RINFQSSRGT VSLGLKRDNL YVVAYLAMDN TNVNRAYYFR SEITSAESTA

LFPEATTANQ KALEYTEDYQ SIEKNAQITQ GDQSRKDLGL GIDLLSTSME

AVNKKARVVK DEARFLLIAI QMTAEAARFR YIQNLVIKNF PNKFNSQNKV

IQFEVNWKKI STAIYGDAKN GVFNKDYDFG FGKVRQVKDL QMGLLMYLGK

PKSSNEAN

Benatti et al. (1989) Eur. J. Biochem. **183**, 465

Shiga toxin A chain

```
KEFTLDFSTA KTYVDSLNVI RSAIGTPLQT ISSGGTSLLM IDSGTGDNLF

AVDVRGIDPE EGRFNNLRLI VERNNLYVTG FVNRTNNVFY RFADFSHVTF

PGTTAVTLSG DSSYTTLQRV AGISRTGMQI NRHSLTTSYL DLMSHSGTSL

TQSVARAMLR FVTVTAEALR FRQIQRGFRT TLDDLSGRSY VMTAEDVDLT

LNWGRLSSVL PDYHGQDSVR VGRISFGSIN AILGSVALIL NCHHHASRVA

RMASDEFPSM CPADGRVRGI THNKILWDSS TLGAILMRRT ISS
```

Kozlov et al. (1988) Gene 67, 213

Shiga-like toxin A1 fragment

```
KEFTLDFSTA KTYVDSLNVI RSAIGTPLQT ISSGGTSLLM IDSGSGDNLF

AVDVRGIDPE EGRFNNLRLI VERNNLYVTG FVNRTNNVFY RFADFSHVTF

PGTTAVTLSG DSSYTTLQRV AGISRTGMQI NRHSLTTSYL DLMSHSGTSL

TQSVARAMLR FVTVTAEALR FRQIQRGFRT TLDDLSGRSY VMTAEDVDLT

LNWGRLSSVL PDYHGQDSVR VGRISFGSIN AILGSVALIL NCHHHASR
```

De Grandis et al. (1989) J. Bact. 169, 4313

α-Trichosanthin

```
DVSFRLSGAT SSSYGVFISN LRKALPNERK LYDIPLLRSS LPGSQRYALI

HLTNYADETI SVAIDVTNVY IMGYRAGDTS YFFNEASATE AAKYVFKDAM

RKVTLPYSGN YERLQTAAGK IRENIPLGLP ALDSAITTLF YYNANSAASA

LMVLIQSTSE AARYKFIEQQ IGKRVDKTFL PSLAIISLEN SWSALSKQIQ

IASTNNGQFE SPVVLINAQN QRVTITNVDA GVVTSNIALL LNRNNMA
```

Chow et al. (1990) J. Biol. Chem. 265 8670

Index

485